BK 384. T2673
TELECOMMUNICATIONS : AN INTERDISCIPLINARY TEX
C1984 66.00 FV
3000 710117 30011
St. Louis Community College

D0082862

WITHDRAWN

St. Louis Community
College

Telecommunications: An Interdisciplinary Text

Telecommunications: An Interdisciplinary Text

Edited by Leonard Lewin

ARTECH HOUSE, INC.
610 Washington Street
Dedham, MA 02026

International Standard Book Number: 0-89006-140-8
Library of Congress Catalog Card Number: 84-070225

CONTENTS

BIOGRAPHIES OF CONTRIBUTORS

G. Gail Crotts Arnall, was a communications policy analyst in the Office of Plans and Policy at the Federal Communications Commission. She obtained her B.A. from Baylor University. She received her M.S. in journalism in 1969 and her Ph.D. in communications in 1974 from the University of Illinois, Urbana-Champaign. Her thesis was on the public information function of the FCC. She joined the agency in 1974, and helped to establish the Consumer Assistance Office in 1976. In 1977 she joined the Office of Plans and Policy. In 1980 she joined the staff at National Public Radio and recently has become a private consultant.

Floyd K. Becker is currently a lecturer in the Department of Electrical and Computer Engineering at the University of Colorado at Boulder, and teaches the telecommunications course in Telephone Systems. Before he joined the faculty at the University, he had 30 years of experience with Mountain Bell and Bell Laboratories, where he supervised and developed the Data Communications Laboratory and later became the Department Head of the Customer Switching Laboratory. He has 32 U.S. patents and several publications to his name.

George A. Codding, Jr. is a professor of political science at the University of Colorado. He attended the University of Washington where he earned the degrees of Bachelor of Arts and Master of Arts in 1943 and 1948, respectively; and the Graduate Institute of International Studies of the University of Geneva, Switzerland, where he earned the degree Docteur ès Sciences Politiques in 1952. The next year he joined the faculty of the University of Pennsylvania, where he taught until he took up his present position in 1961. Codding was one of the founders of the University of Colorado's Master of Science in Telecommunications Program. He has worked in the Secretariat of the International Telecommunication Union and acted as a consultant to one of its Secretary Generals. His books include *The International Telecommunication Union: An Experiment in International Cooperation; The Universal Postal Union: Coordinator of the International Mails;* and *Broadcasting Without Barriers.*

Warren L. Flock is a professor of electrical engineering at the University of Colorado. He received a B.S. in electrical engineering from the University of Washington, and from 1942 to 1945 he was a staff member of the Radiation Laboratory of M.I.T. He received an M.S. in E.E. from the University of California, Berkeley and a Ph.D. in Engineering from the University of California, Los Angeles in 1960. From 1960 to 1964, he was with the Geophysical Institute, University of Alaska, and he then accepted his present position. His interests include telecommunications, remote sensing, radar systems, and solar energy.

Harvey M. Gates is technical vice president, Telecommunications, of the BDM Corporation, Boulder, Colorado; adjoint associate professor to the Graduate Telecommunications Program at the University of Colorado and a registered professional engineer. Dr. Gates has been associated with the BDM Corporation for the past fourteen years with a three year career break with the U.S. Department of Commerce, National Telecommunications and Information Administration, Institute for Telecommunications Sciences, Boulder, Colorado from 1979 to 1982. He is a member of the United States, CCITT Study Group 18, Integrated System Digital Network (ISDN), Joint Working Party; American Defense Preparedness Association; Armed Forces Communications and Electronics Associates; American Association for the Advancement of Science; American Institute for Aeronautics and Astronautics; the Institute of Electrical and Electronic Engineers; Sigma-Xi, and Eta Kappa Nu. He has extensive background in DoD C^3 systems and non-DoD communication programs.

Hussain A. Haddad has been involved in the research and management of communications systems for several years. His background also includes consulting in microwave and mobile communications systems. He holds a number of patents and is the author of many papers. He obtained M.S.E.E. and Ph.D. degrees from the University of Colorado at Boulder where he also teaches courses in satellite communications systems.

Dale N. Hatfield was a program specialist with the National Telecommunications and Information Administration, U.S. Department of Commerce. He received a B.S. in electrical engineering from Case Institute of Technology in 1960 and an M.S. in Industrial Management from Purdue University in 1961. He has held telecommunications policy research positions with various government agencies and was formerly the Chief of the Office of Plans and Policy, Federal Communications Commission. He is currently a Telecommunications Consultant.

John E. Hershey is a telecommunications principal staff member of the BDM Corporation in Boulder, Colorado. Dr. Hershey received his Ph.D. from Oklahoma State University in 1981 and holds an engineering degree from George Washington University; a M.S. in electrical engineering from the University of Arizona; a B.S. in electrical engineering from MIT; and a B.S. in Physics from MIT. He is a part-time faculty member of the Graduate Telecommunications Program and the Electrical Engineering Department of the University of Colorado, Boulder. From 1968 to early 1983, Dr. Hershey had been affiliated with various departments and agencies of the United States Government. Dr. Hershey is a specialist in information sciences as applied to defense and nondefense systems.

Harold E. Hill was professor and chairman, Department of Communication, University of Colorado-Boulder. He has B.S. and M.S. degrees from the University of Illinois-Urbana, and his experience includes: several years (at all levels from announcer to program director) in both commercial and public broadcasting; service as a communication officer (terminated as Major) during World War II; executive vice president of the National Association of Educational Broadcasters in Washington, D.C. for 13 years; has written some 50 articles and made over 100 presentations on communication, particularly media; served as consultant for many broadcasting projects, e.g., helped design the educational television systems in American Samoa; 18 years teaching experience at the University of Illinois and Colorado; recently received the Distinguished Service Award from the National Association for Educational Communicaton and Technology. He is currently a professor in the C.U. School of Journalism.

Leonard Lewin is a professor of Electrical Engineering at the University of Colorado at Boulder, where he is coordinator of the M.S. Interdisciplinary Program in Telecommunications. He was formerly head of the Microwave Department, and later assistant manager of the Telecommunications Research Division of Standard Telecommunication Laboratories in the United Kingdom. Earlier, he had worked on radar at the British Admiralty during World War II. His research has focused on antennas, waveguides and propagation, and he was awarded the 1963 International Microwave Prize and also the IEEE W. G. Baker award for a research paper on waveguides. He joined the Boulder faculty in 1966 and was awarded an honorary D.Sc. the following year. Lewin is interested in many aspects of the educational process and in the development of meaningful teaching techniques at all grade levels.

S. W. Maley is professor of Electrical Engineering at the University of Colorado where he has been teaching telecommunications since 1962. He also conducts research in waveguiding systems and in electromagnetic propagation. He organized and taught courses for the Bell Telephone System Regional Communication Engineering School at the University of Colorado between 1962 and 1970.

Lawrence M. Mead received his B.A. in political science from Amherst College in 1966 and his Ph.D. in government from Harvard University in 1973. He served as a policy analyst in the Department of Health, Education, and Welfare and as a speechwriter for Secretary Kissinger at the Department of State. In 1975, he joined a new research group at the Urban Institute studying implementaton problems in federal social programs. There he participated in a study comparing the decision processes of the FCC and the Environmental Protection Agency for the National Science Foundation. He has published several studies on implementation, bureaucratic reform, and health policy. He recently became deputy director of research for the Republican National Committee, and is currently with the Department of Politics at New York University.

William J. Pomper holds a B.S. degree in Electrical Engineering from the University of Pittsburgh. He has over 19 years experience in communication systems engineering with both military and civilian agencies of the U.S. Government. His experience covers most areas of telecommunications ranging from voice communications to advanced satellite communications, network design, and includes secure communications applications.

Robert J. Williams is professor of Mechanical and Industrial Engineering at the University of Colorado where he teaches courses in engineering economy and engineering management. In addition to his involvement with the M.S. Interdisciplinary Program in Telecommunications since its inception, he has taught engineering economy courses to a variety of special interest groups including the Regional Communications School at CU. He has served as senior industrial engineer with the Boeing Company and for two years as a full-time consultant to the Government of India on industrial productivity. He continues his interest in Third World and Development issues. He recently served on a team of five international consultants who were invited by the province of Hunan, China to plan for the establishment of the Hunan Management College for factory managers. Professor Williams is a registered professional engineer.

Wesley J. Yordon is a professor of economics at the University of Colorado where he teaches courses in microeconomic theory and industrial organization. He attended Wesleyan University from 1949-51, served in the U.S. Army, then received his B.A. from the University Colorado in 1956. His Ph.D. is from Harvard in 1960, and he has been a Fulbright lecturer in Argentina and Mexico. His research interests have ranged from the mechanics of inflation to the problems of cost analysis in the regulated industries. He is currently Chairman of the Economics Department at C.U.

PREFACE

This material is offered by way of a sequel to the earlier *Telecommunications: An Interdisciplinary Survey*. A few of the chapters in the previous book have been dropped, the remainder rewritten and brought up-to-date, and several important new ones have been added on data transmission, local area networks, telephone systems, and satellites. Although one reason for publishing this material has been to put on record a current presentation of material taught in the M.S. Interdisciplinary Telecommunications Program at the University of Colorado at Boulder, the contents should be of much wider interest than simply to students at the University. Telecommunications managers, system operators, and telecommunications professionals in many diverse areas should find this material invaluable. It is, of course, not possible in a single volume to cover in an adequate way the entire and rapidly expanding field of telecommunications; neither is it possible to teach it in a 12 to 18 month university program. What has been attempted is to provide a balanced survey with representative material selected to provide a wide and reasonable coverage, described by experts in the field.

Leonard Lewin
December 1983

INTRODUCTION

Telecommunications and Academia

Leonard Lewin
University of Colorado

As I have remarked in the previous edition, it would be a great mistake to see telecommunications too narrowly as only the technical aspect of the design and operation of the communication network. Features concerned with government regulation at the international, national, and local level, the economics, the management, the legal and social impact; these are but a few of the important factors that determine the shape and growth of the industry. Until about 1970 the training of telecommunications professionals was largely the responsibility of in-house programs in the telephone companies, which were the main source for industrial recruits. It is only in the last ten to fifteen years that telecommunications, as an academic subject, has been taken seriously in American universities. Although some of the purely engineering aspects, such as transmission, spectrum, signal processing, noise, etc., have long been part of standard electrical engineering courses, the last decade has seen the emergence of *interdisciplinary* telecommunications programs at quite a few U.S. universities. It is well recognized that the shape of the telecommunications industry is molded not only by technological developments but also by regulatory, political, financial, social, and other forces. To understand this impact, and also some of the more recent changes, particularly the interaction of computers and the telecommunication network, a much broader curriculum is essential. This has necessitated the development of interdisciplinary programs, both at the undergraduate and at the graduate levels.

It is not that easy to construct such programs since much of the expertise needed may be absent from the regular campus faculty. Very few universities, for example, have a resident expert on international telecommunications regulation, or on the workings of the F.C.C. Some flexibility in internal university funding is also clearly necessary.

The first such comprehensive program was implemented at the Master's level at the University of Colorado at Boulder in 1971. This was followed in 1974 by an interdepartmental program at Southern Methodist University, and which was later transformed into a fully interdisciplinary structure reporting to the dean of engineering. Evening classes, also at the Master's level, were introduced at George Washington University and Golden Gate University about this time, whilst an undergraduate program, with support from the International Communications Association (ICA), was set up at Texas A and M. The ICA also supported some of the other programs, and has been instrumental in current developments to encourage further such programs at both the graduate and undergraduate levels. A fairly recently instituted program at New York University is concerned particularly with teleconferencing and information display. From this brief summary it can be seen that quite a variety of programs have arisen, all to fulfill the need of professionals schooled in the broad spectrum of telecommunications-related skills. The differences in the programs reflect the demographic features of the locale, the availability of expertise, local employment needs, and the teaching philosophy of each school and faculty.

All the programs are in a state of rapid evolution. The industry having changed enormously in the last ten years, the need for curriculum up-date and extension is probably greater here than almost anywhere else. Coming now at a time of nationwide financial retrenchment this has created real difficulties in terms of flexibility, staffing and, in particular, the institution of whole new programs. The expansionist opportunities of the sixties are no longer with us, and the ability, with limited resources, to respond appropriately to rapid change will determine, in the long run, which programs will survive and flourish, and which will not. There has, in fact, already been one major casualty, and with new programs arising, and with competition for students and funds, only time will show which programs have the most useful format and support.

As a personal assessment, I think that I cannot emphasize too strongly the distinction that can — and should — be made between *education* and *vocational training*; between *learning* and *mere schooling*. Failure to do this, I feel, is at the root of much of the malaise currently afflicting "education" in this country, and abroad, too. It shows up in student disaffection, in curriculum fragmentation, in "Mickey-mouse" courses, in poor discipline, and so on. In times of economic stress, when the emphasis shifts in the direction of shorter-term survival capabilities, student demand is likely to move towards immediately marketable skills at the expense of longer-term and more profound individual development. Hence the demand for exposure to course material on, say, the latest PBX equipment, even though the detailed material will almost surely be outdated in a few years. Of course, learning does not take place in a vacuum, and currently available equipment may be as good a subject matter as any other on which the development of understanding can proceed. *How* it is done is the point at issue; it is very much a matter of emphasis and attitude, among other things. In the long run we need people who are both knowledgeable *and* flexible and creative in their thinking and action. The development of the latter qualities as part of a life-long learning process is essential both to the health of the individual and of society. One of the great advantages of an interdisciplinary program is that it gives an opportunity for a holistic approach to otherwise disparate subject matter, and prevents the student from specializing too narrowly. I have very little patience with a student who, say, tries to avoid a course on computer programming on the grounds that he or she may never need to use it. Such an excuse is likely to be only a rationalization for a fear of trying something new and formidable. On the contrary, succeeding in such a course would "stretch" the student's mind, a permanent acquisition that spills over into almost everything else encountered. To achieve such desiderata a well-orchestrated and dynamic interaction between students, teachers, curriculum, and industrial input is essential. The form this takes from moment to moment depends on both the perceptivity of the individuals concerned and the nature of external pressures which determine, among other things, the nature of the student body itself. It is sometimes very hard to see exactly what is happening when one is in the middle of things; to distinguish between longer-term trends and short-term transients. One thing, I think is quite clear, however; there will continue to be a demand during at least the present decade, for a variety of telecommunications-related programs in the universities, as well as short, intensive courses, seminars and presentations in other formats through which detailed, current information can be channeled. As already mentioned, the ICA, among several other voluntary bodies, has been instrumental in helping to get these programs going, assisting with funding, publicity, and other forms of encouragement. It sponsored a recent thesis study at Berkeley,* aimed at a comprehensive review of all the published university-related academic and training programs in telecommunications. This compilation of courses is a most valuable summary of the programs available at this time, and should prove a useful

*Jane M. Clemmensen, "Telecommunications Education in the United States: An Informational Guide and Assessment" University of California, Berkeley, February, 1983.

reference work both for students, employers, teaching professionals, and others engaged in the many ramifications of the telecommunications industry. It can be strongly recommended to anyone wishing to investigate the various university-related programs. However, its publication date is February, 1983, and much of the survey itself dates back to 1980. Its contents, valuable though they are, must gradually cease to remain current. With technology in the information-processing field developing as it is, one may suspect that in a few years a continuously up-dated version should become available through one of the several accessible national data bases. I think that when this happens we will finally have become of age; at which time we may better be able to judge whether the fear of a division of the populace into information haves and have-nots will be realized. For my part, I am still unclear on what will be the relationship of the electronic to the printed page; and with the imminent and widespread advent of personal computers we may soon be able to come to closer grips with this emerging problem.

chapter 1

INTERNATIONAL CONSTRAINTS ON THE USE OF TELECOMMUNICATIONS: THE ROLE OF THE INTERNATIONAL TELECOMMUNICATION UNION

George A. Codding, Jr.
Department of Political Science
University of Colorado
Boulder, Colorado

1.0 INTRODUCTION

The ultimate use of technological developments depends upon a number of variables, other than the purely technical, including economic, social, and political considerations. This is nowhere more true than in the field of telecommunications. From the introduction of the electromagnetic telegraph in the mid-19th century to the present, telecommunications have provided the lifeblood of commerce, government, and the military. The importance of telecommunications to these critical areas of human endeavor was such that in a majority of countries telecommunications were made a government monopoly from the beginning, and in all others they were placed under strong government control. When telecommunications crossed national frontiers, as they did almost from the beginning because of the needs of international commerce and international politics, a mechanism was needed to ensure that it would be used for the maximum benefit of all nations.

The mechanism was the International Telecommunication Union (ITU) which has its headquarters in Geneva, Switzerland. Not only was the ITU the first true international organization among the hundreds now in existence, but over the years it pioneered a number of cooperative procedures which were adopted by the others when they came into being, including the League of Nations and the United Nations itself. Because of these procedures, and the importance of telecommunications, the ITU was able to survive many international crises, including two world wars — a record few other international organizations have achieved.

To this day the ITU is serving the needs of nations in the ever-changing, ever-expanding world of telecommunications. Basically, the ITU carries out four major tasks: 1) the establishment of binding regulations dealing with certain uses of telecommunications; 2) the creation of non-binding standards; 3) the dissemination of information concerning new developments and uses of telecommunications; and 4) aid to developing nations. This study will investigate all four of these functions, but first there will be a look at the manner in which the ITU has adapted over the years to the continuing telecommunications revolution and a description of its structure, which has had a major influence on international cooperative action.

Before proceeding, it should be noted that the ITU lays claim to all the world of telecommunications, including almost any method whatsoever for the transmission of information from one point to another— telegraph, telephone, radio, and television as well as the transmission of computer data and the like. It also encompasses all techniques of transmission: wire, radio, microwave, and satellites. About the only sphere of the transmission of information in which the ITU does not have primary interest is postal communication; but if certain trends continue, including the ever-rising cost of letter mail, it might become occupied with that too.

1.1 HISTORICAL BACKGROUND

How the ITU became involved in the various aspects of international telecommunications is interesting. Unlike the concerns of most other UN Specialized Agencies, the subject matter under the competence of the ITU evolved over a long period of time. This section will trace this evolution, concentrating on the international needs that gave rise to each new technological innovation, and the manner in which the ITU responded.

1.1.1 The Telegraph

The International Telegraph Union, the direct parent of the International Telecommunication Union, came into being in order to provide governments of Western Europe with a vehicle that would enable them to exploit the international posibilities of the newly developed electromagnetic telegraph. By the second half of the 19th century, the electromagnetic telegraph was beginning to meet the domestic needs of European governments as a rapid and dependable system of communication. It also had obvious advantages for international communication from the point of view of both government and commerce; but it was prevented from exercising its true international potential by a number of factors, including differences in equipment and procedure, and a complicated and confusing rate structure.

An international conference to address the problem was convened in Paris in 1865 by the French Imperial Government and attended by representatives of twenty interested European countries. This conference had two important products, the first was the International Telegraph Convention of Paris (1865) and its appended Telegraph Regulations. In these documents the European nations agreed to interconnect their networks in such a manner as to ensure uninterrupted traffic between their major cities and to use compatible equipment. A single table of tariffs was drawn up, and the French franc was chosen as the monetary unit for the payment of international accounts. This new public international telegraph network to be created by the Convention and Regulations was dedicated "to the right of everybody to correspond by means of the international telegraphs."[1]

The second major product of the Paris conference was the International Telegraph Union. While this new international organization, which became the International Telecommunication Union in 1934, was originally composed almost exclusively of European states, it quickly began to take on a worldwide character as additional states in the world innaugurated telegraph systems and connected them either directly or by cable to those already in existence. The two outstanding holdouts were the United States and Canada. For many years these countries did not sign the Telegraph Convention, nor did they accept the provisions concerning telegraph under the International Telecommunication Union until after World War II, and then only reluctantly. They did, however, send observers to conferences and meetings as did American private operating companies. The United States defended its refusal to accept the telegraph regulations, despite heavy pressure to do otherwise, on the basis of the rather obvious argument that the telegraphs in the United States were operated by private enterprise. Observers have hinted that the U.S. position may actually have been motivated by pressure from the telegraph companies, which did not want any outside interference in their rate structures.

The structure of the International Telegraph Union was quite simple. Prior to World War I, it consisted of periodic conferences and a small secretariat, called the Berne Bureau, located in Berne, Switzerland. The functions of the conferences were to provide an arena where telegraph experts could meet periodically to discuss telegraph matters and to revise the rules and regulations concerning the international telegraph. The task of the secretariat, which was created at a second international telegraph conference in Vienna in 1868, was primarily to aid in the arrangements for conferences and to act as a central office for the dissemination of information about telegraph matters to the members between conferences. As an example of this latter function, the Vienna Telegraph Conference of 1868 directed the secretariat to gather information relating to the international telegraph, publish regularly a table of telegraph rates, collect general statistics on the international telegraph, and publish a journal on telegraph matters. The expenses of the secretariat were borne by the member governments.

The only major change in the structure of the International Telegraph Union occurred in 1925 when it was decided to create the International Telegraph Consultative Committee and the International Telephone Consultative Committee as sub-organs to study various problems affecting the telegraph and telephone between conferences. The International Telegraph Consultative Committee was patterned on the Telephone

Committee, which the International Telegraph Union inherited when it took on the regulation of telephones, a subject which will be treated in the next section. In support of his successful proposal to create the Committee, the German delegate at the Paris telegraph conference of 1925 pointed out that there were still many telegraph systems in use, each of which differed from the others in a number of details. Further, there were many questions of operation and procedure that had to be resolved before the international telegraph could be used to its fullest potential. In 1938, in a hotly contested move, it was decided to add rate questions to the competence of the International Telegraph Consultative Committee.

With this basic organization, the International Telegraph Union, and later the International Telecommunication Union, was able to carry out satisfactorily its function regarding the international service, including the setting of appropriate tariffs, and the study of ways and means to make the international telegraph service as rapid and dependable as possible.

1.1.2 The Telephone

The telephone did not come into being until the late 1870's, and its expansion was slow for three reasons. First, in Europe at least, its exploitation was taken over by the already existing postal and telegraph administrations, and was treated as an adjunct to the two older types of communication rather than an important new means of communication, except perhaps in cities. Second, for many years there was no method to connect the telephones of different continents, thus the expansion of telephone took on a national and a regional character and not an international one. Third, the international telephone was confronted almost everywhere by the language barrier. For all of these reasons there was less need for international cooperation in telephone matters in its earlier stages than there had been for the telegraph.

The International Telegraph Union reluctantly accepted the telephone as a subject deserving of its attention. The first proposal for regulation came from the German government at the Telegraph Union's administrative conference in Berlin in 1885. The public had shown its appreciation of telephonic connections between large cities within several countries and was requesting that it be extended to foreign countries. International regulation was therefore in order. Opponents of the German proposal, including the delegations of Austria, Italy, and Portugal, argued that it was premature since the service was so new. It was suggested that international connections should be left to bilateral agreement until they became more widespread. The compromise solution was to add a short five-paragraph chapter of a general nature on telephone to the Telegraph Regulations, which left almost complete freedom to administrations, except that it fixed the time unit for charges at five minutes and limited the length of a total call at ten minutes if there were other requests for the use of the line.

It was not until the Administrative Telegraph Conference of London (1903) that it was agreed there might be enough experience in international telephonic communications to justify the drafting of a set of regulations. Rather than start from base zero, it was decided to take advantage of the similarities of the two types of wire communication and draft regulations only for those aspects of the telephone service which were different from the telegraph service. These new regulations, consisting of fifteen articles, became the new Chapter 17 of the Telegraph Regulations. One of the articles made all provisions of the Telegraph Regulations which were not inconsistent with Chapter 17 also applicable to the telephone.

The telephone was finally given its own separate set of regulations at the 1932 Madrid Plenipotentiary Telegraph Conference. As mentioned earlier, the Madrid conference was the one where the decision was made to join forces with the International Radiotelegraph Union of which the United States was a member. The United States made it known that its telephone system was run by private enterprise and not by the government. Consequently, the U.S. Government could not be a signatory to an agreement that would result in its regulation. The U.S. was supported in this stand by Canada and a number of Latin American countries. As a result, the delegates of the Madrid Conference made the new Telephone Regulations apply only to the "European regime." As pointed out in the ITU's centennial publication, the Madrid delegates made it clear the "European regime" meant all of the world except for the Americas.[2]

The United States finally agreed to be bound by the Telegraph Regulations and the Telephone Regulations in 1973.[3]

1.1.3 Radio

The second major addition to the subject matter of the International Telecommunication Union, and the one which has had the most revolutionary impact on its operation, was radio. Radio has two major characteristics which made it a likely candidate for international regulation from the beginning: first, if two stations transmit on the same or close to the same frequency, they may cause harmful interference; and second, radio waves do not respect national frontiers.

The first few conferences on radio (radiotelegraph in those days) were not necessarily called for this reason, however. Radio had come into practical being close to the end of the 19th century and was used first commercially for communication with and between ships at sea because it worked well for such communications and it filled an important need. The first company to exploit radio commercially, the Marconi Wireless Telegraph Company, in an attempt to preserve its lead and perhaps even to create for itself a monopolistic position in the field, ordered its operators not to communicate with any other station which did not use the Marconi apparatus. Although there were other motives, the first international radio conferences were called to end this practice and thus enable Marconi's competitors to carve out a place for themselves in this developing, lucrative market. Safety of life at sea was the major excuse used, because a number of embarrassing, if not dangerous incidents, had occurred when Marconi operators had refused to communicate with others.

The use of radio for other services which soon occured resulted in the first major task to be performed by the ITU for radio: the allocation of bands of frequencies to services. The first allocations were for coast stations and ship stations. These were followed in fairly rapid order by allocations for weather reports, time signals, radio beacon stations, broadcasting, aeronautical stations, and on through the myriad of radio services that have been designed for the benefit of mankind including, most recently, those services using satellites.

Two additional developments concerning radio are important to us at this point, namely, the creation of the International Radio Consultative Committee (CCIR) and the International Frequency Registration Board (IFRB). The Radio Consultative Committee had been created by the 1927 Conference of the International Radiotelegraph Union and became a part of the ITU when the International Radiotelegraph Union was merged with the International Telegraph Union. In contrast to what happened with the other two committees, the Radio Committee had a difficult time being approved. The opposition to the Committee at the 1927 conference was especially strong from the United States, France, and the United Kingdom. The delegate from the United States argued that because radio was evolving so rapidly, the Committee might hold back its progress by establishing too rigid principles. The French were afraid that private companies might attempt to obtain the official approval of the committee and use it for commercial gain. The British, who had originally submitted a proposal to establish a technical committee, changed their minds and registered their disapproval on the basis that since no changes could be made to the regulations between conferences such a committee would be useless.

The German and Italian delegations argued that the establishment of such a committee was needed precisely because radio was advancing at such a rapid state and because the committee could relieve the delegates of many of the exhaustive technical studies that they were forced to undertake during conference time. The proponents barely carried the day by a vote of twenty-six to six, with eleven abstentions and thirteen delegates absent. The new Committee was to study "technical and related questions" submitted to it by "participating administrations and private enterprises."[4]

Despite their initial opposition to the creation of the International Radio Consultative Committee, the United States, France, and the United Kingdom became active participants in its work and the CCIR rapidly gained the genuine respect of all of the members of the Union.

Prior to World War II, with the limited use of radio that existed then, governments rarely had difficulty in finding places for individual radio stations in the allocations that had been made for various radio services at ITU conferences. The frequency spectrum is finite, however, and little by little certain bands of frequencies tended to fill up and interference become more prevalent, especially in the high frequency bands which for a long time were the only bands which could be used for long distance radio-communication.

At the Atlantic City Plenipotentiary Conference in 1947, it was decided to eliminate this problem once and for all by taking two actions: the creation of a master list of frequency assignments for all nations engineered to their present and future needs, and the creation of the International Frequency Registration Board which would have the power to approve or disapprove any additions or changes to that list. In the words of Harold K. Jacobson, the Americans who were responsible for the creation of the IFRB saw it "as something of a cross between the Federal Communications Commission and the International Court of Justice."[5]

As we shall see later, the IFRB never became equivalent to either the FCC or the International Court of Justice, but nevertheless carries out an increasingly valuable function for the members of the ITU.

1.2 STRUCTURE

The structure that has developed over the years to carry out the functions of the ITU may be a little complex when compared to other international organizations, but they are far from complex when compared to any national post and telegraph administration. In essence, the structure of the ITU has five major components: 1) conferences made up of delegates from member states; 2) the Administrative Council, which is in effect a conference of delegates but one of restricted membership; 3) the two international consultative committees, which are similar to conferences except they do not draft treaties but only make recommendations to member states; 4) the International Frequency Registration Board; and 5) a secretariat. This section will treat each of these elements of the ITU's structure and, in addition, the extremely important item of financing.

1.2.1 Conferences

There are three types of ITU conferences: plenipotentiary, worldwide administrative conferences, and regional administrative conferences. The plenipotentiary conference is the supreme organ of the ITU and has the exclusive power to amend the basic treaty, the Telecommunication Convention. Plenipotentiary conferences meet in principle every five years at a place decided upon by the preceeding plenipotentiary conference, and are made up of delegations from all member states wishing to be represented. Specific powers include establishing the budget of the Union and approving its accounts, electing the Administrative Council, electing the IFRB, electing the Secretary-General and the Deputy Secretary-General and the Directors of the CCIR and CCITT. Delegates to plenipotentiary conferences take their duties seriously, and there are always hundreds of proposals at each conference to amend the Convention. The last plenipotentiary was held in Nairobi in 1983.

Administrative conferences, which are convened to consider specific telecommunications matters, are of two types, world and regional. The worldwide administrative conference can take up any telecommunications question of a worldwide character including the partial or complete revision of the Radio, Telegraph, and Telephone Regulations. The specific agenda is determined by the Administrative Council with concurrence of a majority of the members of the Union and must include any question that a plenipotentiary conference wants placed on its agenda. World administrative radio conferences also review the activities of the International Frequency Registration Board and give it instructions concerning its work. A world administrative conference can be convened by the decision of a plenipotentiary conference, at the request of at least one quarter of the members of the Union, or on a proposal by the Administrative Council. Examples of world administrative conferences include the World Administrative Radio Conference held in Geneva in 1979, and the World Administrative Radio Conference for the Mobile Services held in Geneva in 1983.

The agenda of a regional administrative conference must deal only with specific telecommunications questions of a regional nature. Regional administrative conferences may be convened by a decision of a plenipotentiary conference, on the recommendation of a previous world or regional administrative conference, at the request of one quarter of the members belonging to the region concerned or on a proposal of the Administrative Council.[6]

1.2.2 The Administrative Council

With the failure to make the IFRB the organ intended by its originators, the creation of the Administrative Council was probably the most important adition ever made to the structure of the ITU. Prior to the creation of the Council, the day-to-day work of the Union was carried out by the Secretary-General under the guidelines laid down by the periodic plenipotentiary conferences. The Swiss government audited the Union's accounts without charge. There were a number of reasons behind the successful 1947 proposal to create an Administrative Council to carry on the administrative work of the ITU between meetings of the plenipotentiary conference. The most emphasized reason was the need for continuity, the need to make decisions between such conferences in the light of the rapid advances in radio communications and the ensuing problems. Another was to have a body which could help the ITU coordinate its work with that of the United Nations and the other international organizations that were being created immediately after World War II. Along these lines, all the new international organizations which were to become the Specialized Agencies of the United Nations were being supplied with an administrative council, or its equivalent, made up of representatives of states. Finally, although it was not expressed openly, there was a desire to eliminate the control of Switzerland over the secretariat.

In effect, the Administrative Council of the ITU is the agent of the plenipotentiary conference in the relatively long intervals between meetings of that body. The 1982 convention articulates three major areas of concern for the Council: ensuring the efficient coordination of the work of the Union, taking steps to facilitate the implementation of the Convention, Regulations and decisions of the various ITU conferences by member countries, as well as determining the ITU's policy of technical assistance. While the Convention goes into considerable detail about these duties, the work of the Council in the past can be summarized in three major categories: external relations, coordination of the work of the permanent organs of the Union, and administration. As regards external relations, the Council is the agency which makes formal contact with the United Nations, the Specialized Agencies, and other inter-governmental agencies, and can conclude provisional agreements with them. Coordination includes the review of the annual reports on the activities of the permanent organs of the Union, temporarily filling vacancies among elected officials of the permanent organs, and arranging for the convening of plenipotentiary and administrative conferences. Administration is the most time consuming occupation of the Council. It is responsible for drawing up regulations for administrative and financial activities of the Union; supervising the functions of the various organs, including the Secretariat, the Consultative Committees and the IFRB; reviewing and approving the annual budget of the Union in order to ensure "the strictest possible economy;" arranging for the annual audit of the accounts of the Union; and submitting a report on its activities and those of the Union to the plenioptentiary conference for its consideration. Personnel problems probably account for the majority of the council's efforts. The Council is responsible for the staff regulations and rules. It sets job classifications, and it supervises the application of the UN system of salaries and allowances, the UN Joint Staff Pension Fund, and the Union Staff Superannuation and Benevolent Funds. The Administrative Council may also resolve any question not covered by the Convention which cannot await the next conference for settlement, upon agreement of a majority of the members of the Union. Also development assistance activities are taking an increasingly important share of the Administrative Council's time.

The Administrative Council meets every spring in Geneva for a three-to-four-week session and is composed of forty-one members (up from the original eighteen of 1947) elected "with due regard to the need for equitable distribution of the seats on the Council among all the regions of the world."[7] Each member of the Council has one vote, the Council adopts its own rules of procedure, and the Secretary-General acts as its secretary. The Secretary-General, his deputy, the Chairman and Vice Chairman of the IFRB, and the Directors of the Consultative Committees all have the right to participate in the deliberations of the Council, without voting privleges, but the Council has the right to hold meetings confined to its own members. Additional meetings may be held if so decided by the Council or by its Chairman at a majority's request.

1.2.3 The International Consultative Committees

The next element in the structure of the ITU is the International Consultative Committees. The International Radio Consultative Committee

(CCIR) and the International Telegraph and Telephone Consultative Committee (CCITT), are close to being international organizations in their own right. The supreme organ in each is the Plenary Assembly which normally meets three years and which is made up of delegates from all interested administrations and any recognized private operating agency approved by a member of the Union. The Plenary Assembly chooses questions that it wishes to study and creates study groups to deal with them. Questions can also be referred to a Consultative Committee by an ITU conference, by the IFRB, by the plenary assembly of another Consultative Committee, or by any twenty of its members. Both CCI's have their own specialized secretariat, and the CCITT has its own telephone laboratory. Plenary assemblies may submit proposals to administrative conferences and may study and offer advice to member countries on their telecommunication problems.

There are some differences in the mandates of the two Consultative Committees as well as in their method of operation. According to the Telecommunication Convention the duties of the CCIR are "to study technical and operating questions relating specifically to radiocommunication without limits of frequency range and to issue recommendations on them."[8] The duties of the CCITT are similar regarding telegraph and telephone except that in addition to technical and operating questions it can also study "tariff questions." A study group which receives a question to investigate normally prepares a "study program" which is sent out for comments. The comments and the preliminary response are then discussed in a meeting of the whole group and a report, and possibly a recommendation, is drafted which is sent to all participating parties and to the next Plenary Assembly. The Plenary Assembly can accept, reject or modify the draft recommendations. Only the Plenary Assembly has the right to make recommendations to administrations.

In recognition of the fact that the distinctions between telegraph, telephone, and radio are fading as a result of new techniques and new technologies, the two Consultative Committees are increasingly forming joint committees and working groups. There is, for example, a Joint Study Group on Circuit Noise and Availability administrated by the CCITT and a Joint Study Group on Television and Sound Broadcasting and a Joint Study Group on Vocabulary administrated by the CCIR.

Further in order to keep a larger perspective of the evolution of telecommunications services, the Montreux Plenipotentiary Conference of 1965 asked the two CCI's to establish a Joint World Plan Committee and regional planning committees as necessary to "develop a General Plan for the international telecommunication network to help in planning international telecommunication services."[9] In pursuance of this request, the CCI's created a World Plan Committee and regional planning committees for Africa, Latin America, Asia and Oceania, and Europe and the Mediterranean Basin.

1.2.4 The International Frequency Registration Board

As mentioned earlier, for a number of reasons the IFRB never achieved the stature that was envisioned for it by the 1947 Atlantic City Plenipotentiary Conference. The most important single reason was the failure of administrations in a series of conferences, called shortly after Atlantic City, to create an overall international frequency list engineered to the needs of all member countries. Without such a list, reflecting the current usage and future needs of all the members on the ITU, the IFRB was unable to perform all of its function as arbitrator of the frequency uses of all the nations of the world. Because of the change of function of the IFRB and other reasons including financial ones its size was reduced from eleven to five by the 1965 Montreux Plenipotentiary Conference and might have been abolished had it not been also performing a function essential to the developing countries. (The specific function that the IFRB performs for the members of the ITU will be outlined in the next section.)

As regards structure, the IFRB is made up of five individuals" thoroughly qualified by technical training in the field of radio. . .(who) posses practical experience in the assignment and utlization of frequencies" elected by the Plenipotentiary Conference from candidates sponsored by member countries" in such a way as to ensure equitable distribution amongst the regions of the world." In addition, each member is required to be familiar with" geographic, economic and demographic conditions within a particular area of the world."[11]

The members of the IFRB are considered as "custodians of a public trust" and not as representatives of "their respective countries, or of a region." To make certain that the members of the IFRB maintain objectivity, the country of which a member is a citizen "must respect the international character of the Board and of the duties of its members and shall refrain from any attempt to influence any of them in the exercise of their duties."[11]

The IFRB elects its own Chairman and Vice-Chairman, who serve in that capacity for one year, and it has its own specialized secretariat.

1.2.5 The Secretariat and the Coordination Committee

The foundation on which every international organization rests — and the ITU is no exception — is the secretariat. It is the secretariat that gives an international organization its permanence and provides the support activities that make its existence possible. The Telecommunication Convention provides for a General Secretariat to be directed by a Secretary-General, who is assisted by a Deputy Secretary-General. Both are elected by the Plenipotentiary Conference. The Secretary-General is the agent of the Administrative Council because he is responsible to it "for all the administrative and financial aspects of the Union's activities."[12] Thus, while the Plenipotentiary conference is the supreme organ of the Union and sets long-term goals, and the Administrative Council meets once a year to make certain that all goes well, the Secretary-General is responsible for the day-to-day operation of the Union. In particular the Secretary General appoints the staff of the General Secretariat, organizes its work, and publishes numerous documents, reports and studies, the responsibility for which he has accumulated over the years. The Secretary-General is also the legal representative of the Union and is the official welcomer of individuals, both offical and unofficial, who call at the headquarters of the Union in Geneva.

The General Secretariat of the ITU is divided into six departments: Personnel and Social Protection, Finance, Conferences and Common Services, Computer, External Relations, and Technical Cooperation. In numbers, the General Secretariat in 1981 was made up of 355 individuals with permanent contracts, 136 with fixed term contracts, and almost a thousand additional employees on short-term contracts. Officials above the General Service category, for which the principle of geographical distribution applies, represented some 50 different countries. Among those on the regular budget, Switzerland had the highest representation, followed by France, the United Kingdom and the United States.[13]

Up to this point, the secretariat of the ITU is quite similar to that in most other comparable international organizations. But here the similarity ends, and a situation develops which is unique and has given rise to much criticism. The Telcommunication Convention provides that the IFRB, the CCIR and the CCITT shall have a "specialized secretariat." The specialized secretariats are not as large as that of the General Secretariat, but impressive nevertheless. The IFRB in 1981 had a total of 100 employees (excluding the five elected members), the CCITT had 46 employees, and the CCIR employed 32.

These secretariats are under the direct control of their respective Directors, or the Board in the case of the IFRB, as far as their duties are concerned, but under the administrative control of the Secretary-General. To further complicate the situation, the Board and the Directors of the CCITT are given the power to choose whom they wish regarding the technical and administrative members of their secretariats, but this must be done "within the framework of the budget as approved by the Plenipotentiary Conference or the Administrative Council." This is a general budget, applicable to all the organs of the Union including the General Secretariat. Further, the actual appointment of these members of the specialized secretariats is done by the Secretary-General, albeit "in agreement with the Director." But: "The final decision for appointment or dismissal rests with the Secretary-General."(14)

It was obvious even in 1947 that conflict could arise if members of the secretariat were under the authority of more than one head, and it did.(15) In an attempt to straighten out the lines of authority, at least a little, the Geneva Plenipotentiary Conference of 1959 added the sentence mentioned above which gave the Secretary-General the final decision for appointment or dismissal. Although it would have been welcomed by some, the 1965 Montreux Plenipotentiary Conference would go no further in strengthening the hand of the Secretary-General, but did attempt to provide a formal procedure for consultation between the heads of the various organs on matters of mutual interest, especially personnel matters. Article II of the Montreux Convention established an official Coordination Committee to advise the Secretary-General on "administrative, financial and technical cooperation matters affecting more than one permanent organ and on external relations and public information."(16) The committee is now made up of five individuals: the Secretary-General as Chairman, the Deputy Secretary-General, the Director of the CCIR, the Director of the CCITT, and the Chairman and Vice Chairman of the International Frequency Registration Board. The committee makes its decisions by majority vote, although it is requested to reach conclusions unanimously whenever possible. If this cannot be done, and the matter is important but not urgent, it must be referred to the next session of the Administrative Council. However, if the Secretary-General judges the question to be of an urgent nature he is authorized to make a decision in "the absence of the support of the majority of the Committee." In such cases the Secretary-General must report promptly, in writing, to the members of the Administrative Council "setting forth his reasons for such action together with any other written views submitted by the other members of the Committee."(17)

1.2.6 Finances

The ITU has a primitive method of financing that remains basically what it was when introduced in 1868. The great bulk of the funds necessary to the ITU's operation come from contributions of member governments. The rest comes from such activites as participation in the United Nations Development Program and the sale of documents. The amount that each country contributes to the budget depends upon the contributory class that it selects, ranging from a minimum of 1/8 unit to a maximum of 40 units. The 1/8 unit class category is reserved for the least developed countries as determined by the United Nations and the I.TU's Administrative Council. The members are free to choose any other class of contribution, although a resolution of the Montreux Plenipotentiary Conference suggested that members choose a category "most in keeping with their economic resources".[18] The total of the expenses of the Union not covered by other sources, such as the sale of documents, is divided by the total number of contributory units to obtain the amount of a single contributory unit. Each member then multiplies this amount by the number of units that it has chosen. For instance, in 1981 there was a total regular budget of 65,728,000 Swiss francs not covered by other sources. This amount was divided by the number 427½, which represented the total number of contributory units subscribed to by all the members of the ITU in that year. The result was 135,700 Swiss francs, the amount of one contributory unit. Thus, a country such as the Kingdom of Losotho, which subscribed to a ½ unit contriutory class, paid 67,850 Swiss francs as its share of the 1981 expenses, approximately, $34,000 at the exchange rate of the time. In the same year the United States subscribed to the 30-unit class, thus paying 4,071,000 Swiss francs or about two million dollars to the 1981 regular budget. The U.S.S.R. was the largest single contributor, paying for 34 units, of which 30 was for itself, 3 for the Ukrainian S.S.R., and one for the Byelorussian S.S.R.[19]

Until 1973 the ITU Convention did not provide for any sanctions against a state that failed to pay its contributions to the expenses of the Union. As a result there was always a fairly large amount of money owed to the I.T.U. In 1971, for instance, thirty member countries and eight private companies were delinquent for a total of 9,848,574.00 Swiss francs. The largest single amount was the 2,242,543.80 Swiss francs owed by Bolivia. The Torremolinos Conference added a provision to the Convention which provided that members will lose the right to vote in ITU conferences and meetings, and in "consultations carried out by correspondence" if "the amount of arrears equals or exceeds the amount of contribution due from it for the preceding two years."[20] In 1981, although the number of delinquents had risen to thirty-six countries and seventeen other agencies, the total amount owed to the ITU was only 6,678,806.80 Swiss francs. The largest amounts in 1981 were the 755,303.40 Swiss francs owed by Zaire and the 681,951.75 Swiss francs owed by Chad.[21]

1.3 THE PRODUCT

All international organizations perform a number of tasks important to the nation states that make up their membership, but few perform more than the ITU. Indeed, it would be difficult to envision international telecommunications without it. As stated in Article 4 of the ITU Convention, the overall purpose of the ITU is:

> a) to maintain and extend international cooperation between all members of the Union for the improvement and rational use of telecommunications of all kinds, as well as to promote and to offer technical assistance to developing countries in the field of telecommunications;
> b) to promote the development of technical facilities and their most efficient operation with a view to improving the efficiency of telecommunications services, increasing their usefulness and making them, as far as possible, generally available to the public;
> c) to harmonize the actions of nations in the attainment of those ends."[22]

The more specific tasks that the ITU performs in order to achieve those purposes can be summarized under the five rubrics: Regulation, Research and Standards, Information, the International Frequency Register, and Development Assistance.

1.3.1 Regulation

The rules for the use of telecommunications that have been agreed upon by the members of the ITU over the years are contained in five main documents: the International Telecommunication Convention, the Telegraph Regulations, the Telephone Regulations, and the Radio Regulations. In signing and ratifying the Convention, the state members of the ITU agree in principle to be bound by the various Regulations.[23] In point of fact, however, many governments including that of the United States treat the Regulations as separate treaties and carry out the ratification process that is provided for in their respective constitutions.[24] In addition, administrations can and do refuse to be bound by certain of the Regulations. While most states consider themselves bound by the Regulations, to be absolutely certain it is necessary to consult the reservations made by the delegations when signing the Convention and the various Regulations and any which have been attached to the instruments of ratification.[25]

The Convention

The International Telecommunication Convention is the ITU's basic document, in effect its constitution. The Convention gives the ITU its legal existence, sets forth its purposes, establishes its structure, defines its membership[26] and fixes its relations with the United Nations.[27] It also sets forth the basic regulations concerning telecommunications in general and radio in particular.

Since World War II, the Convention has been revised at intervals ranging from five to eight years. One of the major tasks of recent plenipotentiary conferences has been to attempt to reduce the number of transitory details in the Convention to eliminate the necessity for so many expensive plenipotentiary conferences.

The main emphasis of the general rules on telecommunications in the Convention is on maintaining a truly universal telecommunication network. Members of the ITU recognize the right "of the public to correspond by means of the international service of public correspondence" and pledge to establish facilities adequate to meet demands, to use the best technologies, to maintain their parts of the system "in proper operating condition," and to use the best operational methods and procedures. Provisions are made for the settlements of accounts between states, and the gold franc and the monetary unit of the International Monetary Fund are established as the monetary units in the composition tion of tariffs and the settlement of accounts. Secrecy of communication is pledged and message dealing with "safety of life at sea, on land, in the air, or in outer space, as well as epidemiological telecommunications of exceptional urgency of the World Health Organization" are given absolute priority. (Government messages are given second priority.)[28]

While the emphasis is on maintaining a smoothly functioning universal system, members of the ITU reserve the right to make it less than perfect if they so desire. States retain their right to stop private telecommunications "which may appear dangerous to the security of the state or contrary to their laws, to public order or to decency." In the case of private telegrams, the state stopping messages must immediately notify the office of origin except "when such notification may appear dangerous to the security of the State."[29] As if this was not enough, the Convention permits a state to suspend its portion of the international telecommunication service for an indefinite period of time provided only that "it immediately notifies such action to each of the other Members through the medium of the Secretary-General."[30] Finally, member governments refuse any responsibility to the users of the international telecommunication network, especially concerning claims for damages.

The section of the Convention dealing with radio is shorter and more specific. Member administrations agree to "limit the number of frequencies and the spectrum space used to the minimum essential to provide in a satisfactory manner the necessary services." They also agree to bear in mind "that radio frequencies and the geostationary satellite orbit are limited natural resources" thus they must be used as efficiently and economically as possible "so that countries or groups of countries may have equitable access to both, taking into account the special needs of the developing countries and the geographical situation of particular countries."[31]

The Radio Regulations portion of the Convention also requires the users of mobile radio services to exchange radiocommunications without distinction as to the radio system used by them, and to operate all radio stations in such a manner as to avoid harmful interference to the radio services of stations in other countries. Radio stations of all countries must give absolute priority to distress messages, and the governments of members agree to take all necessary steps to prevent the transmission of false or deceptive distress, urgency, safety, or identification signals. Even radio used by the military is not completely exempt from regulation. Although member states "retain their entire freedom with regard to military radio installations of their army, naval, and air forces," these installations must "so far as possible" observe all provisions concerning distress messages, measures to prevent harmful interference, and provisions of the Radio Regulations concerning the type of emissions and frequencies to be used according to the nature of the service performed by such stations. Moreover, if military stations take part in the service of public correspondence they must comply with all of the regulations dealing with the public service.[32]

The Radio Regulations

Because of the propagation characteristics of radio and the finite nature of the radio frequency spectrum, a great deal of coordination of activities of nations and their stations is necessary to ensure that the best interests of all are served adequately. This concern is reflected in the breadth and detail of this interesting document.[33]

The most important single section of the Radio Regulations is Chapter II which sets forth the conditions for the use of the radio spectrum and allocates sections of it to some two dozen different radio services throughout the world. Some allocations are on a regional basis and others on a universal basis depending mainly on the propagation characteristics of the frequencies involved. Unless nations have made a specific reservation concerning a specific service, which many have done, they are obliged to make certain that all of the radio stations operating under their authority use frequency assignments in accordance with the Table of Frequency Allocations. This chapter is followed by one dealing with notification and registration procedures and another detailing measures to be used to prevent harmful interference between radio stations.

A large portion of the Radio Regulations concerns the manner in which public radio services are to be operated, with special attention to the mobile service. Chapter VI, for instance, includes rules concerning operators' certificates, the class and minimum numbers of operators for ships and aircraft stations, and the working hours of stations in the maritime and aeronautical mobile services. Chapter VII deals with various aspects of working conditions in the mobile services including the frequencies to be used for various purposes and standard procedures. A final section deals with the procedures for handling distress, alarm, urgency and safety signals, and traffic. It is in this section that one finds the international recognition of the familiar distress signals SOS and MAYDAY.

The Radio Regulations also contain a chapter dealing with radiotelegrams and radiotelephone calls, including such items as priority of messages, routing, and accounting.

The Telegraph and Telephone Regulations

When contrasted with the Radio Regulations, the Telegraph and Telephone Regulations are meager indeed. The 1973 Telegraph Regulations, for instance, are only fifty-eight pages in length including some twenty pages devoted to the names of the delegates who signed it. The 1973 Telephone Regulations are only 20 pages in all.

The small size of the Telegraph and Telephone Regulations is the result of the monumental decision of the 1973 World Administrative Telegraph and Telephone Conference to replace the great bulk of the regulations concerning wire communications with the non-binding recommendations of the CCITT.

For many years preceeding the 1973 conference there had been a growing feeling among ITU member administrations that the two Regulations were being filled with useless and overly restrictive details, and that the process for revising the Regulations was too cumbersome and slow to keep up with the rapidly changing technology and the new services being offered to consumers.

Based on the work of the CCITT and its study groups, the 1973 World Administrative Telegraph and Telephone Conference carried out a wholesale substitution of CCITT recommendations for most of the provisions of the two Regulations. The essence of this revolutionary action is contained in section one of Article 1 of the 1972 Telegraph Regulation (the Telephone Regulations are similar):

> "(1) The Telegraph Regulations lay down the general principles to be observed in the international telegraph service."

"(2) In implementing the principles of the Regulations, Administrations (or recognized private operating agencies) should comply with the C.C.I.T.T. Recommendations, including any Instructions forming part of those Recommendations, on any matters not covered by the Regulations."[34]

The remainder of the regulations contains provisions dealing with the types of services offered and certain operational methods, including the keeping of archives. Both regulations also have a section concerning rates, including accountability and the collection of charges, and Annexex dealing with the payment of accounts.

1.3.2 Research and Standards

The second major product of the ITU is the output of the two Consultative Committees. Few if any international organizations can boast of a more dedicated and hard-working research complex than that represented by the CCIR and the CCITT. Their study groups are composed of some of the finest technical experts in the governments of the developed countries, supplemented by experts from industry, and the reports and recommendations that are produced are universally held in high respect.

The scope of the activities of the Consultative Committees can be seen from the following lists of their Study Groups:

CCIR Study Groups:

1. Spectrum utilization and monitoring
2. Space research and radioastronomy services
3. Fixed service at frequencies below about 30 MHz
4. Fixed-satellite services
5. Propagation in non-ionized media
6. Propagation in ionized media
7. Standard frequencies and time-signals
8. Mobile radiodetermination and amateur services
9. Fixed service using radio-relay systems
10. Broadcasting service (sound)
11. Broadcasting service (television)

CCITT Study Groups:

1. Definition and operational aspects of telegraph and telematic services (Facsimile, Telex, Videotext, etc.)
2. Telephone operation and quality of service

3. General tariff principles
4. Transmission maintenance of international lines, circuits and chains of circuits; maintenance of automatic and semi-automatic networks
5. Protection against dangers and disturbances of electromagnetic origin
6. Protection and specification of cable sheaths and poles
7. Data communication networks for telematic services (Facsimile, Telex, Videotext, etc.)
8. Terminal equipment for telematic services (Facsimile, Telex, Videotext, etc.)
9. Telegraph networks and terminal equipment;
10. Telephone switching and signalling
12. Telephone transmission performance and local telephone networks
15. Transmission systems
16. Telephone circuits
17. Data communication over the telephone network
18. Digital networks

The amount of work that the CCI's produce is also formidable. The Fifteenth Plenary Assembly of the CCIR, for instance, produced sixteen volumes of research and recommendations totalling more than 6,000 pages. The largest was the 881-page volume from Study Group Eight, Mobile Services and one of the shortest was the 107-page Volume produced by CMV Vocabulary Study Group.[35]

The 5,533 pages produced by the Seventh Plenary Assembly of the CCITT in 1980 compares favorably with the work of the CCIR. In this twenty-eight volume work the largest amount, eight volumes for a total of 1,488 pages was produced by Study Group Eleven (Telephone Switching and Signalling).[36]

The results of the work of the Consultative Committees have only the status of recommendations and are not legally binding on the member states as are the Radio and Telegraph and Telephone Regulations. Nevertheless, because of the high quality of the CCI's work, their recommendations are taken very seriously by the member states. The reason for the willing acceptance of CCITT recommendations by the member administrations was explained by the Director of Danish Telecommunications at the opening ceremony of the 1973 World Administrative Telegraph and Telephone Conference when he said: "The recommendations have been formulated by the best specialists from countries all over the world and they represent good advice. Of course, the Administrations are free to accept and use these recommendations but are not compelled to do so. However, we have seen that the recommendations meet a wide acceptance in all quarters simply because they represent the best possible advice in the field of telecommunications."[37]

Because of the dynamic nature of the subject matter with which it deals, for many years the work of the CCIR tended to overshadow that of its sister committee. The CCIR's recommendations in the areas of propagation and spectrum utilization are a case in point. With the development of new technologies, and the action of the 1973 World Administrative Telegraph and Telephone Conference noted above, however, the CCITT has taken on a new visibility. One participant in the work of the CCITT for example gives the Committee especially high marks for its recent efforts to achieve common standards which permit international direct telephone dialing despite "seemingly forbidding obstacles presented by a multitude of national configurations and different signaling systems" and "the strong interests of national supplier industries competing for a share of the world market."[38] One of the newer problems that the CCI is currently investigating is transnational data communication. In order to ensure the effective flow of such communication it is necessary to achieve some sort of system compatibility and common operating methods and, on the commercial level, some common rules concerning payment for services. The 6th CCITT Plenary Assembly gave primary responsibility in the field of data transmission to Study Groups VII and XVII and made provisions for liaison and coordination between these two groups and any other Study Groups where data transmission is involved.

1.3.3 Information

Making technical information available to its membership has always been one of the more important functions of international organizations. This exchange occurs in a number of ways: between delegates to conferences and meetings, between officials of national administrations and individuals in the secretariat, and through the publications of secretariats. The ITU is no exception. The ITU hosts a proportionately high number of conferences and meetings, expecially in the field of radio, and the headquarters in Geneva is constantly receiving the visits of officials from telegraph, telephone, and radio administrations as well as officials from private operating companies.

The ITU is especially well known for the quality and the quantity of its publications. The distribution of these publications by the secretariat to the members of the Union, in the words of the Deputy Secretary-General, "guarantees the continued flow of international communications on a worldwide basis."[39]

The secretariat of the ITU receives its authorization to publish from the Convention, the various service regulations, the Administrative Council, Plenipotentiary Conferences, and Administrative Conferences. The Convention, for instance, requires that the secretariat publish the records of all conferences and meetings, all principal international and regional telecommunication agreements communicated to it by the parties thereto, "data, both national and international, regarding telecommunication throughout the world," technical and administrative information especially useful to the developing countries, information regarding technical methods for the most efficient operation of telecommunication services, and a "journal of general information and documentation concerning telecommunication."[40] The latter is, of course, the prestigious *Telecommunication Journal* which appears monthly in three languages. For the IFRB, the secretariat publishes the weighty *International Frequency List* which contains the particulars of the frequency requirements recorded in the Master International Frequency Register and the weekly, airmail IFRB circulars. The secretariat also publishes IFRB technical standards, "as well as any other data concerning the assignment and utilization of frequencies and geostationery satellite orbit positions prepared by the Board in the performance of its duties."[41]

The following is a list of the service documents that the ITU secretariat is directed to publish and keep up to date by the Radio Regulations:

1. The International Frequency List
2. List of Fixed Stations Operating International Circuits
3. List of Coast Stations
4. List of Ship Stations
5. List of Radiodetermination and Special Service Stations
6. Alphabetical List of Call Signs Assigned from the International Series to Stations Included in Lists 1, 2, 3, 4, 5, and 8.
7. List of International Monitoring Stations
8. List of Stations in the Space Radiocommunication Services and in the Radio Astronomy Service
9. Map of Coast Stations which are open to Public Correspondence or which Participate in the Port Operations Service
10. Chart in Colours Showing Frequency Allocations

11. Manual for Use by the Maritime Mobile and Maritime Mobile-Satellite Service

The Telephone Regulations direct the secretariat to publish: 1)a Yearbook of Common Carrier Telecommunication Statistics; 2) Codes and Abbreviations for the Use of the International Telecommunication Services; 3) a List of International Telephone Routes; 4) a List of Definitions of Essential Telecommunication Terms; and 5) Telecommunication Statistics.

The Telegraph Regulations require the publication of the following:

1. Yearbook of Common Carrier Telecommunication Statistics
2. Transferred Account Booklet
3. International Credit Card for Telegraph Services
4. Codes and Abbreviations for the Use of the International Telecommunication Services
5. List of Destination Indicators for the Telegram Retransmission System and of Telex Network Identification Codes
6. List of Telegraph Offices Open for International Service
7. List of Cables Forming the World Submarine Network
8. List of Point-to-Point Radio Telegraph Channels
9. List of Definitions of Essential Telecommunication Terms
10. Telecommunication Statistics
11. Routing Table for Offices Connected to the Gentex Service
12. Transferred Account Table
13. Table of International Telex Relations and Traffic
14. Table of Service Restrictions
15. Tables of Telegraph Rates

The secretariat also publishes the voluminous reports of the CCIR and the CCITT mentioned above. As a result, the ITU is one of the larger publishers among the international technical organizations. In 1981, for instance, the ITU secretariat had a publications budget of 10.2 million Swiss francs and it turned out some fifty-three publications.[42]

1.3.4 The Master International Frequency Register

The International Frequency Board's unique task deserves a separate and distinct place in our inventory of the ITU's services. This task, although far from that envisioned by the delegates to the Atlantic City Plenipotentiary Conference in 1947 who created it, will continue to make an important contribution as long as the radio frequency spectrum remains crowded by radio stations and services. In view of the fact that the radio spectrum is finite and the number of radio stations still continues to increase, this will be a long time indeed.

In essence the IFRB maintains a univeral register of the frequencies used by radio stations in all countries of the world. Article 12 of the Radio Regulations provides that administrations shall notify the IFRB of any frequency assignment to a fixed, land, broadcasting, radionavigation land, radiolocation land or a standard frequency and time signal station, or a ground-based station in the meterological aids service: 1) if the frequency is capable of harmful interference to a station of another administration, 2) if the frequency is to be used for international radiocommunication or 3) if it is desired to obtain international recognition of the use of the frequency.[43] The Board examines each notification for conformity which the Convention and the Table of Frequency Allocations, and for the possibility that its use will cause interference to the stations of another administration. If the finding is favorable, the Board enters the particular frequency assignment in the Master Register with the date of receipt.[44]

If the notification should not conform with the regulations, or might cause interference with a prior assignment, the Board returns it to the administration, along with findings and any suggestions it may have to offer to solve the problem. If the administration in question should resubmit the notice with adequate modifications, the assignment is entered by the Board in the Master Register with the date of receipt of the original notification. If the notice is resubmitted in its original form, or with modifications that are still inadequate, the administration in question may nevertheless insist that the frequency by entered in the Register. This the Board will do upon proof that the assignment has been used for at least sixty days without any complaints of harmful interference, and the data recorded will be the original date of notification with certain qualifications. The Board is also required to survey the Register and, with the agreement of the notifying administration, cancel any entry that is not being used.

Through the auspices of the IFRB, the Master International Frequency Register, and the weekly circulars, the telecommunication administrations of the world are provided with a central register of the frequencies in use, a means of obtaining international recognition of national frequency usage, and the expert advice of a highly qualified group of radio frequency specialists.

The task of registering frequency assignments is more complicated than one might expect. In 1981, for example, the IFRB reported the receipt of 147,041 frequency assignments notices. A "full technical examination" was made on 44,829 of them in bands below 28 MHz.[(45)]

The IFRB carries out a similar function for frequency assignments to radio astronomy and space radiocommunication stations and processess and coordinates the seasonal high frequency broadcasting schedules of member administrations. In recent years the IFRB has also become involved in aiding administrations in their efforts to create space communication systems. This includes advanced notification of planned satellite networks to the IFRB, and coordination of frequency assignments for such networks with the help of the IFRB.

1.3.5 Development Assistance

The fifth area in which the ITU provides a distinct service for its members, and the newest, is development assistance. Before the beginnings of the United Nations technical assistance program, the ITU had traditionally emphasized activities directed mainly towards the maintenance of a rapid and efficient international telecommunications network and was not overly concerned with what went on within national systems as long as their major cities were a part of the international system. The ITU in effect was dominated by the nations which were the large users of telecommunications, and it functioned to meet their requirements.

The new states which achieved independence after World War II and proceeded to join the United Nations and the ITU had different needs. Many of the new states had little in the way of internal communications, and the only foreign links they had were with cities of the former colonial power.

The first actions of the ITU to help meet these new needs for member states came about through outside pressures. In the late 1940's the United Nations became cognizant of the fact that the newly independent states had numerous needs that they could not meet with their own resources, and hence started the UN technical assistance program. After a certain amount of hesitation, the ITU joined the program in 1951 and was allotted a junior partner's share of the available funds. Over the years, the ITU's share of UN-sponsored assistance, now the UNDP, has continued to increase. In 1981 the ITU was responsible for a total of

$33,302,488 in UNDP funds, $15,773,870 to Africa, $4,280,115 to the Americas, $9,581,615 to Asia and the Pacific, $10,242,031 to Europe and the Middle East, and $415,438 was spent on interregional projects. With the addition of $6,900,581 from other sources, the ITU was responsible for arranging for the sending of 727 experts on missions to developing countries, the dispatch of 726 students from developing countries abroad, and the purchase of $12,011,143 in equipment for various field projects.[46]

As more and more developing countries became members of the ITU, more and more emphasis has been placed on development assistance. The 1959 Geneva Plenipotentiary Conference, for instance, formally acknowledged the ITU's commitment to development assistance by adding a paragraph to Article 4 of the ITU Convention, "Purposes of the Union," to the effect that the ITU shall "foster the creation, development, and improvement of telecommunication equipment and networks in new or developing countries by every means at its disposal, especially its participation in the appropriate programmes of the United Nations."[47] The Montreux Conference of 1965 saw so many proposals by the developing countries to get the ITU more fully involved in development assistance, such as the organizing of seminars for the telecommuniation technicians of developing countries, that the U.S. delegate to that conference was obliged to comment: "It may be stated that if there was one dominant note at the Conference it was the oft-stated belief that the ITU had as one of its basic obligations the requirement to assist the new and developing nations with their individual telecommunication problems."[48] The same trend was also more than evident at the Malaga-Torremolinos Plenipotentiary Conference in 1973.

One proposal that upset the United States govenment advocated the establishment of a special ITU development fund supported by contributions of member states to provide assistance for the developing countries to improve their domestic telecommunication networks. While the proposal was defeated at Montreux, it was passed at Malaga-Torremolinos over the strong objections of the United States delegate who declared that if it passed the United States would not contribute to it. The United States still has not contributed to it, and it remains very small.

The trend was continued with enthusiasm at the Nairobi Plenipotentiary in 1982. Eighteen resolutions were passed at that conference dealing with development assistance, including directives to the permanent organs of the ITU to spend more of their time helping the developing nations and upgrading the development assistance services of the ITU. Of special interest were decisions to finance a number of such activities from the regular ITU budget and to create an Independent International Commission for Worldwide Telecommunication Development to explore additional ways that the ITU can help developing countries improve their domestic telecommunication networks.[49]

With the developing nations in a majority in the ITU, and in view of the fact that each state has one vote in ITU conferences and meetings, there is very little possibility that development assistance will not continue to remain one of the ITU's major functions.

1.4 CONCLUSIONS

The International Telecommunication Union has been heavily involved in telecommunications since the development of the electromagnetic telegraph a little over a century ago. This involvement has consisted primarily of establishing binding rules and regulations necessary for the functioning of the international telecommunication system, doing research and setting standards to permit the integration of new technologies into the system, disseminating information necessary for the smooth functioning of the system, and more recently providing assistance to the developing countries to create workable domestic telecommunication systems. All of these functions meet a genuine need of the international community.

The machinery has been fairly simple: delegates from member countries meet to revise the rules and regulations as the need has arisen and study committees composed of experts from member countries carry out the necessary research and set standards. The secretariat provides the infrastructure necessary for holding conferences and meetings and for dissemination of information; the IFRB helps administrations to select and use radio frequencies without causing harmful interference with the stations of other countries. And the cost is relatively small for the services rendered.

All in all, despite some imperfections, the International Telecommunication Union has performed a vital task for the nations of the world and gives all indications of continuing to respond to the changing needs of nations well into the future.

TABLE 1.1
ITU Organizational Structure*

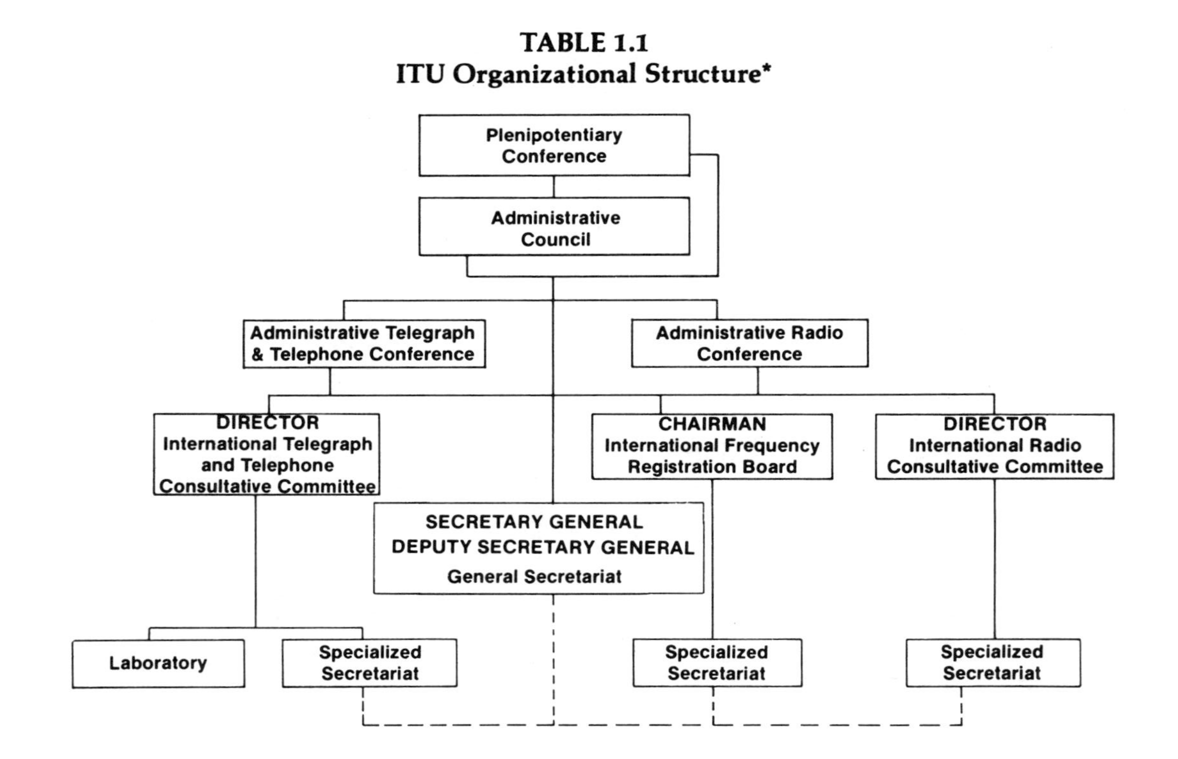

*The ITU Convention designates the CCIR, CCITT, IFRB, and the General Secretariat as the "permanent organs" of the ITU. The ITU also has a Coordination Committee, not shown in this diagram, made up of the Secretary General, Deputy Secretary General, Director of the CCIR, Director of the CCITT, and Chairman and Vice Chairman of the IFRB whose major task it is to "assist and advise" the Secretary General. See ITU, *International Telecommunication Union*, Nairobi, 1982, Arts. 5 & 12.

TABLE 1.2
The General Secretariat of the ITU*

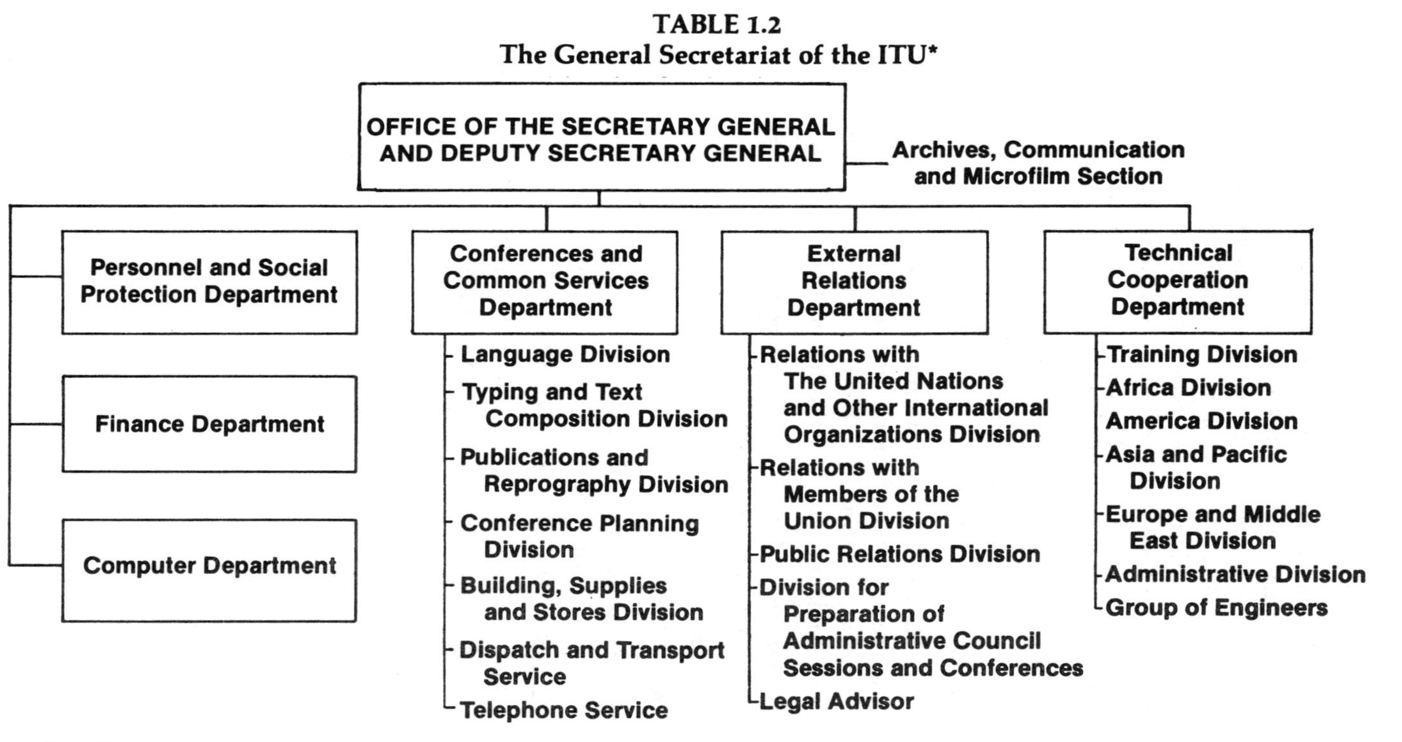

*as of January, 1983.

REFERENCES

1. The early history of the International Telecommunication Union, unless otherwise referenced, is from the author's *The International Telecommunication Union: An Experiment in International Cooperation* (Leiden: E.J. Brill, 1952). This book was republished in 1972 by Arno Press of New York. For more recent information concerning the ITU and what it does, consult George A. Codding, Jr. and Anthony Rutkowski:, *The International Telecommunication Union in a Changing World,* Dedham, MA, Artech House, 1982.
2. ITU, *From Semaphore to Satellite* (Geneva: I.T.U., 1965), p. 110.
3. See Codding and Rutkowski, pp. 36 & 53
4. Codding, *op. cit.,* p. 122.
5. Jacobson, Harold K., "The International Telecommunication Union: ITU's Structure and Functions," in *Global Communications in the Space Age: Toward a New ITU* (New York: The John and Mary R. Markle Foundation and Twentieth Century Fund, 1972), p. 49.
6. ITU, *International Telecommunications Convention, Nairobi, 1982,* (Geneva, 1982), Arts. 5, 6, 7, 53, and 54.
7. *Ibid.,* Art. 8, para. 1.
8. *Ibid.,* Art. 11, para. 1.
9. ITU, General Secretariat, *International Telecommunication Convention Montreux, 1965,* (Geneva, 1965), Art. 14, para. 5.
10. *ITU Convention, Nairobi, 1982,* Art. 57.
11. *Ibid.,* Arts. 10 and 57. Members of the IFRB are forbidden to request or receive instructions relating to the exercise of their duties from any government, or "any public or private organization or person."
12. *ITU Convention, Nairobi, 1982,* Art. 9, para. 1.
13. ITU, *Report on the Activities of the International Telecommunication Union in 1981* (Geneva, 1982), pp. 116-119.
14. *ITU Convention, Nairobi, 1982,* Art. 74.
15. See Codding, *op. cit.,* pp. 433-435.
16. *ITU Convention, Montreux, 1965,* Art. 11, para. 1.
17. *ITU Convention, Nairobi, 1982,* Arts. 12 and 59.

18. *ITU Convention, Montreux, 1965*, Resolutions, Recommendations and Opinions, No. 15.

19. *Report*, 1981, pp. 122-123.

20. *ITU Convention, Malaga-Torremolinos, 1973*, Art. 2, para. 2 and Art. 15, para. 7.

21. *Report*, 1981, pp. 195-196.

22. *ITU Convention, Nairobi, 1982*, Art. 4.

23. *Ibid.*, Art. 42.

24. See, for instance, the notification in the Report on the Activities of the ITU in 1974 concerning the "deposit of the instrument of ratification of the Radio Regulations" on the part of the United States of America. ITU, *Report on the Activities of the International Telecommunication Union in 1974*, (Geneva), 1974, p. 112. See also David M. Leive, *International Telecommunications and International Law: The Regulation of the Radio Spectrum*. (Dobbs Ferry, N.Y.: Oceana Publications, Inc., 1970).

25. States can and do attach reservations to particular parts of the Regulations at the time of negotiation or refuse to be bound by the whole Regulation that they helped draft, but the latter is not often the case.

26. Any member of the United Nations may become a member of the ITU by formally acceding to the Convention. A non-UN member must apply for membership and, if its application receives a two-third affirmative vote, may accede to the Convention. In point of fact the membership provisions no longer have any great importance since almost every independent state is already a member. Membership in the ITU as of 31 December 1982, stood at 157.

27. In essence of two organizations pledge their cooperation in matters of mutual concern, agree to exchange documents and other information, and give each other a limited right to participate in conferences and meetings. See *ITU Convention, Nairobi, 1982*, Art. 39 and Annex 3.

28. *Ibid.*, Arts. 18, 22, 23, 25, 26, 28, 29, and 30.

29. *Ibid.*, Art. 19.

30. *Ibid.*, Art. 20.

31. *Ibid.*, Art. 33.

32. *Ibid.*, Art. 38.

33. The latest Radio Regulations are those drafted in Geneva in 1979. In the past it was the custom of the ITU to re-work the entire Regulations at one conference but, as mentioned earlier, in more recent times the evolution of the technology has been such that it has been necessary to convene conferences between times to make necessary changes in the regulations concerning particular services. Consequently, the 1979 Regulations are supplemented and brought up to date by the final acts of the various WARCS that have taken place since 1979 such as the World Administrative Radio Conference for Mobile Services, Geneva, 1983.

34. See ITU, *Final Acts of the World Administrative Telegraph and Telephone Conference Geneva, 1973*, (Geneva), 1974. The signatures of the delegates to the conference alone account for 32 pages of the total text.

35. See CCIR, *XVth Plenary Assembly Geneva, 1982*, Geneva, 1982, Vol. VIII and XIII.

36. See CCITT, *Yellow Book - Seventh Plenary Assembly Geneva, 1980* Geneva, 1981, Vol. VI.

37. See *Telecommunication Journal*, Vol. 40, No IX (September, 1973), p. 567.

38. Gerd D. Wallenstein, "Collaboration without Coercion: The I.T.U. as a Model for Worldwide Agreement-Making," in his *International Telecommunication Agreements* (Dobbs Ferry, N.Y.: Oceana Publications, Inc., 1977), Vol. 1, p. 59.

39. From a speech by Richard E. Butler, Deputy Secretary-General of the ITU, entitled "World Telecommunication Development and the Role of the International Telecommunication Union" delivered in Brussels, February 9, 1978, to the International Conference on Trans-National Data Regulation.

40. See *ITU Convention, Nairobi, 1982*, Art. 56.

41. *Ibid.*

42. Report, 1981, pp. 163 and 164-168.

43. *Radio Regulations*, 1982, Vol. I, Art. 12, para. 1.

44. For the definitive work on the IFRB consult David M. Leive, *International Telecommunications and International Law.*

45. *Report,* 1981, p. 26.
46. *Ibid.*, pp. 97-99.
47. ITU, General Secretariat, *International Telecommunication Convention, Geneva, 1959,* (Geneva), 1960, Art. 4, para. 2, sec. d.
48. U.S. Department of State, Office of Telecommunications, *Report of the United States Delegation to the Plenipotentiary Conference of the International Telecommunication Union, Montreux, Switzerland, September 14 to November 12, 1965* (TD Serial No. 973), Washington, D.C., December 15, 1965 (mimeo.), p. 10.
49. ITU, *Convention, Nairobi, 1982,* Resolutions, Nos. 16-33.

chapter 2

DECISION MAKING AT THE FCC

Gail Crotts Arnall
Telecommunications Consultant
Washington, D.C.
and
Lawrence M. Mead
Department of Politics
New York City University
New York, NY

2.0 INTRODUCTION

We usually do not think of the FCC as an institution. We think of it as a group of seven commissioners establishing policy or handing down rulings *ex cathedra*. The literature on the FCC is mostly about its regulations, not about the agency that produces them. Discussions of the FCC as such will usually mention its legal mandates and describe the basic organization of its bureaus rather than analyze how it really functions. Speaking generally, "... the literature on the regulatory commissions is replete with formalistic, legalistic, and purely descriptive accounts of how such agencies are structured, what their legal powers and authority are, and what they have done or not done."[1]

The regulatory literature usually assumes that regulations were decided for the reasons stated in the official decisions — or at least by some *intellectual process* (see Cox, 1970, for a typical description of FCC policy during the 1960's in these terms).[2] In fact, the administrative and political processes surrounding decision-making matter just as much. Newspaper accounts of major decisions — and common sense — tell us this. Nevertheless, the *process* of regulation has been studied very little. Economists have academic theories of how commissions should regulate and political theories of why they do otherwise. Yet we still know very little of why the regulatory commissions do or do not live up to the norms critics propose for them. The answers lie somewhere in the goals and behavior of the commissions as organizations.[3]

Our purpose here is to begin to fill this void. We mean to show that the Federal Communications Commission can be understood not only as a *rational actor*, making policy for intellectual reasons (the model mentioned above), but also as an *organization* and as a *political actor* which tend to make policy for reasons of their own. Each of the three models is developed using elements of social, organizational, and political theory, and is meant to suggest hypotheses for concrete research. We will illustrate the models with descriptions of the FCC and its major decisions in order to show that they can be applied usefully to the agency. Our purpose, however, is more to suggest than to verify them.

We are heavily indebted to Graham Allison, who first applied the three models together in an effort to explain a governmental decision.[4] We have made some adjustment to his approach, which are explained below. Allison's work opened up powerful new avenues for research on foreign policy decision-making. In a more modest way, we hope to do the same for the study of regulatory decisions.

Each of the following three sections describes one of the three models, shows how it applies to the FCC, and uses it to analyze several major FCC decisions. Some of the decisions are discussed more than once, since the different models illuminate different aspects of the same case. The fifth section uses the models to analyze and assess major proposals for the reform of the FCC. The conclusion argues for the importance of institutional development in the reform of regulatory policy.

2.1 MODEL 1: THE RATIONAL ACTOR

As mentioned above, most writing about the FCC assumes intellectual reasons for the decisions made by the FCC. Such accounts treat the FCC as if it were an individual, or a cohesive group of five individuals, making policy for stated reasons. This "rational actor" or "single actor" model assumes that the FCC, like "economic man," acts rationally to maximize values of importance to itself. To understand its actions, according to this model, we need only to imagine the calculations which led the FCC to its conclusions. In the literal sense, we understand FCC action by *rationalizing* it.[5]

We use the rational actor model here in a more normative form than Allison does. His treatment assumes that political actors pursue self-interested objectives; the focus is on the calculations of interest leading to action. Here we focus on the goals the actors are presumed to maximize. We take as goals those norms to which the FCC is commonly held by the professional groups within the Commission and by external authorities and critics. In our use, Model 1 says that these norms explain all of what the FCC *should* do and much of what it *does* do. The reasoning is that policy is decided as if by individuals; that these actors favor actions which may be publicly justified; and that the reasons for the actions derive from the central values pursued by the Commission.

What values does the Commission serve? The professional groups and external authorities involved with the FCC tend to assert different values. Technically-minded people, such as engineers and economists, generally say the FCC should obey its *legal mandate* to enforce the public interest on the industry and should promote technical and economic *efficiency*. Lawyers and judges say it should follow *due process* in its rule making, and that its decisions should satisfy norms of *jurisprudence* (that is, they should mold past precedents and current cases into coherent regulatory law). Public interest advocates say it should follow the *popular will* and achieve the *public good* in ways not confined to the dictates of either economic calculation or law. These characterizations are of course only approximate.

These three normative positions all imply that FCC actions will have authority to the extent they adhere to the appropriate values. The three positions correspond quite closely to the three forms of political authority discussed by social theorist Max Weber. The following subsections take up the three positions, explain each in terms of Weber's theory, and describe the influence each appears to have had on the FCC's structure and decisions. Each offers a different version of the rational actor view of the Commission.

2.1.1 Rational Authority

The ethic for the FCC which states that it should execute its legal mandates and achieve efficiency seems the most self-evident of the three. These premises are so deeply embedded in our culture that they seem obvious.

They reflect what Weber called the "rational" or "legal" form of authority. In modern government, according to Weber, citizens typically give allegiance to impersonal laws. There is a "government of laws rather than men." Where administrative discretion is necessary, governance is by expert functionaries chosen on the basis of merit, not by political favorites or by others chosen by ascriptive criteria. Allegiance is to an impersonal order composed of laws and bureaucracies ruling with the authority of expertise.[6]

Weber presents rational authority as an "ideal type." No government and no commission fully lives up to it. The ideal has political weight, however, because of its unquestioned legitimacy.

Rational authority has been especially important for the independent regulatory commissions. These agencies were set up by Congressional mandate in order to solve difficult allocation problems by technical means. Much of the rationale for establishing independent regulatory agencies was the assumption that experts acting with delegated authority could regulate business in a less partisan way than politicians would by themselves.

For an FCC official, the norms of rational authority command (1) adherence to the FCC's legal mandate, insofar as it dictates specific decisions and (2) expert calculation, insofar as the official is left to make decisions himself. Calculation means to make policy in the light of consequences. It means choosing the option that is most advantageous in technical or economical terms.

The formal structure of the FCC reflects the norms of rational authority. The Chairman and the four other commissioners are named and confirmed in ways set out by law. They are nominated by the President and confirmed by the Senate. They serve seven-year terms, staggered so no more than one commissioner is replaced in any year. The Chairman is appointed from among the commissioners by the President and serves at his pleasure. A Chairman removed by the President, however, would remain a commissioner until the end of his term unless he chose to resign. The commissioners rule by delegated authority and may be overruled by statute. They make decisions in their own name but with the advice of an expert staff of some 1,900 persons organized in bureaus and offices. (The bureaucratic structure is discussed fully in Model 2 below.)

The norms of rational authority, as Model 1 predicts, have had great influence on the FCC's evolution. The legal basis of the FCC's authority is simply assumed by most students of the agency. The history of communications regulation can be understood as a search for the right kind of agency resting on the right kind of legal mandate. The Commission's authority is based on legislation, and deficiencies in its authority can also be attributed to legislation.

The FCC was formed by an amalgamation of earlier agencies and authorities which had proven insufficient for the job. The Commission was created by the Communications Act of 1934 through a merger of the Federal Radio Commission set up in 1927 and those aspects of the Interstate Commerce Commission dealing with telephone and telegraph.[7] An important later enactment is the Satellite Communications Act of 1962, which defines the FCC's authority over satellite communications.[8] Other legislation to clarify, strengthen, or limit the Commission's authority has repeatedly been urged by its friends and critics. All make the assumption — consensual in modern politics — that statute law is the strongest authority an agency can have.

Legislation directly governs some specific FCC decisions. For example, the Communications Act specifies that the FCC must enforce technical requirements for radiotelephone equipment ships, must approve the extension of lines in the common carrier service, and must make an annual report to Congress.[9] Other things that it cannot do, such as censor broadcasting, are defined by the Communications Act or the Constitution.[10] Sometimes the Commission acts or refuses to act because of a threat of legislation in Congress. When the Commission cites criteria like these in justifying its actions or inactions, it is clearly acting in the manner supposed by Model 1.

The FCC and its decisions have also been molded by the other aspect of rational authority — expert calculation. The Communications Act in fact leaves the agency much discretion; the Commission seeks to exercise it on a basis of expertise. It has its own staff of experts in the various communications fields — broadcasting, common carrier, cable television, land mobile radio, etc., — and it solicits advice from outside experts and interests before it makes major decisions. Its decisions are usually justified in technical and economic terms, even when bureaucratic or political influences have been important (see below). That is, the decision claims to lead to the most favorable technical or economic consequences.

The expert side of the FCC is most clearly seen in its more technical decisions, notably those involving spectrum management. Since the radio spectrum has limited range of usable frequencies, only a limited number of frequencies may be allocated to the various radio services in an area without interference between them. As the demands for spectrum space grow and shift, the Commission must decide anew how to divide up a scarce resource. The assumption that it could do this scientifically was a major rationale given for the FCC at its founding.

Several examples will show that the rationalist ethic clearly influences spectrum decisions, whatever other motives are involved. In 1945 the Commission shifted FM radio from the 42-50 megahertz (MHz) band to 88-108 MHz on grounds that interference from skywaves would be less at the higher frequencies. While the move inflicted considerable economic cost on established FM stations, the Commission viewed the long-term technical benefits as overriding.[11]

In 1952, after seven years of deliberation, the Commission decided to allocate 70 UHF channels for the use of television in addition to the existing 12 VHF channels. The options of a television system using VHF only or UHF only were rejected for a host of complex reasons. While economic interests favoring VHF or UHF were involved, the issue was heavily technical and was fought out in technical terms. In addition, the decision to use both VHF and UHF was based on the prediction that the UHF stations could develop despite VHF competition. These technical and economic projections committed the Commission to several later actions to foster the development of UHF stations.[12]

In decisions of 1970 and 1974-75, the Commission withdrew the 806-960 megahertz (MHz) band from the use of UHF-TV channels and reallocated it for land mobile radio use. The initial reason was simple — the pressure of increasing demand for land mobile radio services. However, the decision followed a lengthy inquiry by the Commission, a careful consideration of options, and a call for technical studies into the demand for land mobile in the future.[13]

More recently, the FCC has authorized cellular mobile radio telephone service as a result of technical innovations achieved by AT&T. Prompted by severe crowding of existing mobile telephone service (MTS) channels, AT&T developed over a ten-year period a new network radio design which uses low power transmitters to serve small areas ("cells") within a geographical region. The FCC's adoption and licensing of this new technology has been justified mainly on the technical grounds established by AT&T development research and the pressures of market demand.[14]

The FCC often listens to outside expert opinion about the best way to achieve effective and efficient policy. In 1971, the Commission authorized the licensing of private companies to compete with AT&T in the specialized common carrier business (that is, wire and microwave communication for computer signals and other specialized uses).[15] The decision, which rested on a series of precedents, broke with the tradition of regulating AT&T as a public utility. Instead, AT&T (or at least its specialized services) would be held accountable to the public interest by increased market competition. The decision reflected, in part, a growing consensus among economists that competition might be the best way to discipline regulated industries, given their propensity to capture the regulating commissions and use regulation to protect themselves against innovation.[16]

These decisions show some of the plausibility of Model 1. Only a total cynic would deny that the rationalist values of the FCC importantly influence its decisions. Model 1 argues that the Commission acts as an individual would in searching for actions which may be publicly justified. Commissioners know that to act in the name of rational authority confers legitimacy second to none.

2.1.2 Traditional Authority

The second set of norms specified by a Model 1 analysis of the FCC are those of lawyers: decisions should follow due process and be well-argued in the light of precedent.

These values are akin to those Weber associates with the "traditional" form of political authority. If rational authority is impersonal and expert, traditional authority (which usually precedes it historically) is conservative and personal. A traditional ruler, such as a feudal chieftain, does not *make* laws; he discovers them in the past. He rules in the light of precedents. And to the extent his own judgment is necessary, he gives it as a judge rather than an expert. He does not calculate; he construes past law to fit present circumstances. He does not make general policy; he issues rulings in particular cases. While the norms of rational authority are adherence to law and expertise, those of the traditional ruler are tradition and discretion.[17]

At the FCC, descretion is necessary because of the very vagueness of the legislative mandate. The Communications Act imposes very few specific policies or actions on the Commission. The best known injunction in the Act is that the FCC must regulate communications according to "public interest, convenience, and necessity." This frees the Commission to set or change policy on the basis of its own reasoning, as in the specialized common carrier case just mentioned. It also weakens the Commission against the pressure of the regulated interests. Commissioners can rarely say their hands are tied by higher legal authority. They *have* to respond to demands for change, in part, because they have the authority to do so. This, plus changing membership on the Commission, means that stable policies are difficult to sustain. Inevitably, policy comes to mean a sequence of case decisions responsive to precedent and discretion, but without any overall rationale.

Traditional-legal norms appear to have influenced the FCC's structure and procedures at least as much as rational ones. Indeed, at first glance the FCC appears to be a court rather than a modern bureaucracy. Alongside the staff bureaus, which advise the Commission on the basis of expertise, are the administrative law judges, who advise using adversary proceedings in the legal manner. The Commission is required to order a hearing when it confronts "a substantial and material issue of fact," for example in the renewal of a broadcasting license or in a rate making case.[18] The decision of the administrative law judge may be appealed to a Review Board and then to the Commission itself. Commission decisions, in turn, may be appealed to the U.S. Court of Appeals usually in the District of Columbia and to the Supreme Court.

The Commission must use court-like procedures to judge individual cases, but it often uses them to make general rules as well, even though this is not required. On major issues, the commissioners themselves may hold oral arguments and listen to lawyers argue for the various sides. In the past, the Commission has held public meetings at which anyone could present grievances or proposals for change — a merger of the governing and judicial functions closely akin to a medieval court. In March 1983, the Commission held oral hearings on a pending rulemaking to eliminate its financial interest and syndication rules, even before the Reply Comments were due in the proceeding. The proposed rule would allow the TV networks to acquire financial interests in programs produced by others and participate in syndicating such programs. Although the timing was very unusual, this action demonstrates how useful hearings are to the Commission in generating a record on which to make a judgment.[19]

The rational and traditional modes of decision making are to some extent in tension. When comparative hearings are not involved, the FCC's bureau staff may advise the commissioners about policy *ex parte* (off the record) and play the dominant role that rational authority assigns to it. When there are hearings, however, the bureau may only participate and advise the Commission on the record, as if it were a private party. In the specialized common carrier case mentioned above, the Commission could have chosen either route. If it had ordered comparative hearings, pitting AT&T against the specialized common carrier companies, the common carrier bureau, which favored competition, would have been only one party among many. Since it chose the rulemaking procedure, there was no adjudicatory proceeding. The Commission acted in response to written submissions and oral presentation, both formal and informal. Hence, the common carrier bureau was able to summarize the viewpoints in its own terms, remain in control, and persuade the Commission to adopt a major policy change.[20] *Ex parte* contacts have historically been considered helpful to the Commission in rulemaking proceedings, but have recently been challenged as inappropriate. In 1977 the Appeals Court suggested that rules against *ex parte* contacts might have to be applied even in rulemaking proceedings.[21]

A commitment to adversarial methods is one reason the FCC's expert staff capability is not greater than it is. The Commission has lacked the expertise to analyze most highly technical issues on its own. It has relied instead on data and analysis submitted by the affected parties or by outside experts. One instance is regulation of AT&T's common carrier rates.[22] Another is the design of rules for granting certificates to cable systems.[23] According to some critics, in both cases heavy reliance on the industry for information may have made policy less effective than otherwise. Until the 1970's, there were virtually no economists at the FCC.[24] In 1972, Chairman Burch appointed a special assistant for planning and policy, and in 1973 established an Office of Plans and Policy, to try to fill this void. Even OPP at first was headed by an attorney or engineer. The appointment of an economist as chief, in the spring of 1978, signaled a change of focus.

The Commission's traditional style of decision making is best seen in the relatively non-technical context of broadcast license renewals. Stations supposedly hold their licenses as a public trust and must renew them every five or seven years. On the one hand there is a history, amounting almost to law, that stations are renewed regardless of how closely they have fulfilled their promises made at the previous renewal about public service broadcasting, attentiveness to community needs, and other obligations.[25] Most renewals are granted routinely by staff using delegated authority. On the other hand, renewal can be capricious. When community groups or rival stations file a petition to deny a renewal application, the commissioners must personally rule on the petition, sometimes after a hearing to determine fact. In such decisions, the Commission rules with discretion, as a judge would. The cases may result in unexpected decisions that represent departures in policy.

In 1961, in the KORD, Inc. decision, the FCC announced that stations would henceforth have to take their public obligations seriously or have their renewals denied.[26] In 1969, the Commission denied the application of WHDH-TV in Boston in favor of another applicant on grounds that this would diversify ownership of communications facilities in the area.[27] In 1971, the Commission rejected a petition to deny the renewal of WMAL-TV, Washington, D.C., on grounds of employment discrimination, but propounded a "zone of reasonableness" doctrine that has been evolving ever since. This doctrine requires that the composition of the work force at a broadcast station must reflect, within specified "zones", the compositon of the population in the market being served.[28] All of these decisions sent shock waves through the industry, but none has altered the practice of granting most renewals routinely.

Another force for discretion is the power of the courts to review FCC decisions. Occasionally the Appeals Court has ordered the FCC to hold a hearing on an application when community groups have lodged a petition to deny renewal.[29] The courts will sometimes use FCC cases to launch major policy departures of their own. For example, when the Appeals Court reviewed the WLBT-TV, Jackson, Mississippi, decision in 1966, it formally extended legal standing to contest renewals to community groups with no economic interest in the outcome other than that of a consumer. This decision opened the doors to many public interest challenges to renewals and, in a broader sense, fundamentally widened the range of interests represented in FCC decisions.[30]

The essence of the renewal process is that the FCC has never reduced it to rules, as it would if rational norms were operative. The Commission has never been able to quantify the standards that stations must meet for renewal nor established the types of cases it will personally review.[31] An attempt to do so after the WHDH case, under Congressional pressure, was overturned by the courts.[32] Rules for review proposed internally by the Broadcast Bureau were accepted by the commissioners in 1973, then ignored.[33] Only after considerable public pressure were internal guidelines used by the staff in reviewing EEO compliance made public.[34] The Commission wants to preserve its discretion to respond to external pressure or personal whim, and this is sanctioned by traditional norms of authority.

The prevalence of traditional norms at the FCC is not surprising. About half the commissioners and much of the staff have routinely been lawyers, a background attuned to precedent, adversary proceedings, and judicial discretion as a basis for policy making.[35] If, as Model 1 argues, agency action will be driven toward actions which individual commissioners can justify in terms of their dominant values, traditional norms will color much of what the FCC does.

2.1.3 Charismatic Authority

The third set of values useful for a Model 1 analysis are those of the public interest advocates. They assert that the FCC should follow the popular will in its decisions and should serve the public in ways not confined by legal precedent or economic calculation.

We would call these norms populist. They appeal in a general way to what Weber called charismatic authority. The charismatic person commands allegiance through the display of remarkable personal gifts rooted in a sense of higher calling. Weber himself speaks of charisma in connection with religious prophets. We use the term here in a looser sense to describe any political leader who commands authority more on the basis of personal qualities and appeal to "the people" than on the basis of specific actions or policies.[36]

The commissioners are driven toward a populist style by the collegial nature of the FCC. No one of them can make decisions without the support of the others. Each has to build a personal constituency inside and outside the FCC if he or she wants to lead. This is also true of the chairman, even though that office controls the agenda and manages the routine administration of the Commission. The chairman's power, like that of the President within the larger political system, is mainly the power of persuasion. In large part, such authority is personal, and potentially charismatic.

Commissioners easily see their job as political. Most have been associated with politics before going to the FCC. For some, to appeal over the head of industry to the people is second nature. To do this gives them personal identity in an otherwise collective leadership. Politically, charisma has the advantage that it centers authority in the individual, not in bureaucratic or judicial structure.

The populist appeal is also endorsed by the larger political system. Democratic political theory supports the idea that the meeting of the leader and the led yields a more basic form of authority than bureaucratic or judical institutions. If commissioners seek actions they can justify publicly, as Model 1 predicts, democratic values will tend to drive them toward populist behavior.

Populist norms had increasing influence over FCC action beginning in the 1960's. Since Newton Minow's chairmanship, the FCC has been under growing pressure from public interest lawyers, community groups, and popular forces generally to be more and more responsive. Individual chairmen and commissioners have tried to respond by making direct personal appeals to the public.

The clearest example is Nicholas Johnson, a highly controversial activist lawyer, who served on the Commission from 1966 to 1973. Johnson was viewed by the industry as a subversive and a zealot. He made himself spokesman for the public interest advocates and minority and community groups who were seeking to make broadcasting and FCC regulation more accountable to the public. He actively encouraged these elements to take the public service obligations of broadcasters seriously and to challenge the license renewals of unresponsive stations.[37] While still a commissioner, he wrote a book on how to mobilize popular opinion against the broadcasting establishment.[38] After he left the Commission, he became head of the National Citizens Committee for Broadcasting, one of the major public interest groups involved in FCC proceedings.

Johnson was typically charismatic in his isolation from established institutions. He sat on the Commission only in order to reform it. He did not enjoy or seek great influence among the FCC staff, and he refused to engage in political compromise with the industry. His influence among the other commissioners and outside groups was real, but it rested almost entirely on his personal ability, integrity, and eloquence — that is, on charismatic qualities.

Richard Wiley, Chairman in 1974-77, was also charismatic, but he used his ability to master the institutions rather than attack them. His unusual energy and control of detail gave him, perhaps, more power over the Commission staff than any other Chairman. Because of his political ability, he usually had the support of the other commissioners and even the regulated industries. For instance, he encouraged the networks and the NAB Television Code to institute the Family Viewing Hour, as a way of limiting violence on television, without ever having to make it mandatory.

Some Chairmen have made explicitly populist appeals to the people as a way of balancing the dominant power of the communications industry. Newton Minow, chairman 1961-63, dealt with what he called the "hostile environment" of his position by articulating the resentments of people against broadcasting. His speech characterizing most television programming as a "vast wasteland," voiced the disillusionment of many with the commercial banality of television and has provided a slogan for critics ever since.[39] Wiley instituted a series of regional open meetings throughout the country in 1974-76, to hear the grievances of people firsthand.

Populist or charismatic authority appeals to the feelings of ordinary people, not the support of established interests. Despite their own influence over the Commission, boadcasters often perceive the FCC as a radical juggernaut which reifies every popular resentment toward broadcasting into some coercive policy.[40] Indeed, during the 1970's, many of the Commission's more important decisions affecting broadcasting sprung directly from perceived popular distaste for excessive sex, violence, and commercialism on television. The 1974 Policy Statement on Children's Television expressed the Commission's intention to limit commercialism in children's programming.[41] In its 1975 Pacific Foundation WBAI or "seven dirty words" decision, the FCC sought to restrict the use of obscene language in broadcasting.[42] And in 1976, Chairman Wiley promoted the Family Viewing Hour with the networks in order to exclude gratuitious violence from the early portion of prime time.[43]

However, charismatic authority also tends to be unmeasured. When leader and followers connect in a direct, populist way, the one tends to promise and the other to demand everything they feel. The agenda often goes beyond what is allowed by other forms of authority — specifically, by law or tradition.Conflicts can arise. Popular resentments toward broadcasters tend to demand of the FCC a power to censor programming which, in fact, the Communications Act and the First Amendment do not allow.[44] Hence, for example, the Children's Television policy seems to commit the Commission to programming policies that it could never enforce. Initially, the WBAI decision was struck down in appellate court as an act of censorship; the Supreme Court, however, overturned the lower court's decision.[45]

Other times, popular decisions offend not law but reason. That is, they are not expert in the terms of rational authority. They lead to unfortunate consequences. This happens because charismatic authority resists restraint by the norms of economic rationality.[46] The prime time access rule, 1970, limited network programming to three hours nightly in an attempt to make more evening time available to community-oriented local programs.[47] In fact, such programming was not forthcoming, and the dominance of the networks over the schedule probably increased.[48]

Although Chairman Burch dissented from this ruling, he pushed through another with a similar rationale. The cable television rules of 1972 were supposed to foster cable development. Cable, which promised

to diversify control over broadcasting, had become a symbol for those seeking to break the power of the networks over television. However, cable did not develop as rapidly as expected. The reasons, besides the 1970's economic recession, may include the explicit public service obligations which were initially laid on cable in order to win the support of commissioners Johnson and H. Rex Lee.[49]

The appointment of Mark Fowler as FCC Chairman in 1981 ushered in a new thrust of political action based upon a strong commitment to marketplace ideology. Fowler was named by President Reagan, who had been elected on a platform attacking excessive regulation of the economy. The result has been a reversal of policies established only a few years before. As late as 1980, for example, the Justice Department confirmed the financial interest and syndication rules adopted by the FCC in 1970 by having the three broadcast networks sign consent decrees containing the essence of these rules. Then, in 1982, the Fowler Commission, supported by the Justice Department, issued a *Notice of Proposed Rule Making* calling into question the need for these rules.[50]

Chairsmatic values are in some ways the most — and in other ways the least —usable norms for a Model 1 analysis of the FCC. No others trace political action so clearly to individual actors rather than to institutions. On the other hand, no others make action so unpredictable. The populist leader must follow the popular whim of the moment. Because he has little connection to structure, his actions do not follow clear patterns, as do those of the expert civil servant or judge. Hence, it is difficult to state hypotheses about what action will occur.

2.1.4 Summary

To sum up, the single actor model has obvious explanatory power for the FCC. It is true that Commission decisions proceed importantly from the mind of the individual commissioners. Policy is clearly affected by the individuals who happen to sit on the Commission at a given time.[51] It is also true that commissioners approach their decisions in part as intellectual and moral problems. Options which can be rationalized in terms of the dominant values of the Commission do have an advantage in the decision process.

A Model 1 view would trace the complexity of FCC behavior, not as much to the organizational and political influences considered below, as to the complexity of the Commission's norms. The values of economists and engineers, lawyers and judges, and populist leaders and followers, are all available and, to some degree, in tension. Any or all the norms may play the rationalizing role that is central to the Model 1 explanation of official behavior. It seems that different norms tend to come to the fore for different kinds of decisions. Rational norms are dominant for heavily technical decisions (spectrum management, common carrier); traditional-legal ones for broadcast license renewal; and charismatic ones for broadcasting policy decisions.

Methodologically, Model 1 has the advantage that the analyst's imagination can do much of his research for him. Since action is attributed to individual officials rationalizing their behavior, the student of the FCC need only imagine how the effort to maximize relevant values would have led to the action in question. There is no need to pick apart the "black box" of the FCC to understand its intricacies; one can intuit its workings from the outside.

However, this advantage implies a danger. With Model 1, the analyst seeks understanding by projecting his own reasonings on the phenomenon he wants to explain. What assurance can he have that the commissioners really acted for the apparent reasons, even if they claim these reasons publicly? There may be a plausible *parallel* between the apparent reasoning and the agency action, but how can we know that the connection is *causal* ?[(52)]

Therefore, there is no substitute for on-the-spot research inside the agency to see how it actually functions — the kind of research we noted in our introduction to be lacking. The moment we look at actual decisions close up, it is clear that individual officials rationalizing their actions do not dictate policy by themselves. Bureaucratic and political factors not discussed in Model 1 come to the fore. These influences are emphasized, respectively, in our second and third analytic models.

2.2 MODEL 2: ORGANIZATIONAL PROCESS

A second approach to understanding the FCC is to look at it as an organization. Bureaucracy came into the picture often under Model 1, but not as an independent influence. Rather, official action was attributed to the reasoning of individual officials, and organization was merely the place where they worked. In this second approach, FCC decisions are attributed to the influence of organization *as such*. This model argues that organization has inherent strengths and weaknesses which affect decisions quite apart from the choices of individual actors. Often, this model contends, policy is not *chosen* at all. It is simply an *output* of a bureaucratic process with a dynamic of its own.

According to theories of public administration, bureaucracy should have no influence of its own. Civil servants are supposed to carry out laws and policies with perfect obedience and efficiency. Experience clearly tells us they do not. Organizational theory is an attempt to formulate why not. Major concepts drawn from organizational theory generate Model 2's hypotheses about the causes of FCC behavior.

2.2.1 The Formal View of Bureaucracy

In the formal view, bureaucracy behaves according to the norms of rational authority already considered above. In Weber's theory of bureaucracy, an organization of officials, just like any one individual official, is accountable to two standards: (1) obedience to policy embodied in laws and the decisions of higher officials, and (2) expertise in serving the public interest where discretion is allowed. Bureaucrats follow rules laid down for them by higher authority, and they are trained, recruited, and promoted on the basis of merit. As Weber puts it, bureaucracy means "A continuous organization of official functions bound by rules" and "the exercise of control on the basis of knowledge".[53]

Weber adds a third norm for bureaucracy peculiar to it as an organization which is very significant. It is that the officials should have no independence of higher authority. In Weber's terms, they must not be able to "appropriate" their positions. Officials must not have a right to their jobs or anything they work with. Just as a modern economy depends, in Weber's view, on the total expropriation of the worker by the capitalist, so modern government depends on the expropriation of office holders by the state. Political modernization means the end of rule by semi-autonomous potentates, as under feudalism. Civil servants then become totally dependent on their superiors. The superiors can hire, fire, promote them, and give them orders at pleasure, subject only to rationalist norms that law and efficiency be served.[54]

Because of its obedience to rational norms and its inability to resist from within, Weber's bureaucracy has no identity of its own. It is the shadow of the Model 1 policy maker acting on rational values. It has the faceless and efficient character of a machine:

> "Precision, speed, unambiguity, . . .continuity, discretion, unity, strict subordination, reduction of friction and of material and personal costs — these are raised to the optimum point in the strictly bureaucratic administration. . ."[55]

The formal theory might be called the "organization chart view of bureaucracy." If reality followed the theory, the FCC would function exactly the way it appears on the organization chart (see Figure 2.1). On paper, the commissioners sit over the entire FCC. All are equal in voting on decisions, but the Chairman has the main administrative authority over the staff. As of 1982, the bureaucracy consists of three bureaus covering the main areas of Commission responsibility (Mass Media, Common Carrier, and Private Radio), plus the Field Operations Bureau and seven offices with cross-cutting functions (General Counsel, Plans and Policy, Public Affairs, Review Board, Administrative Law Judges, Science and Technology, and Managing Director). The distinction between the bureaus and offices is close to that between "line" and "staff."

In theory, communication lines among the units should be fairly simple, much as they appear on the chart. The bureaus and offices are functionally differentiated. The main overlaps should occur between "line" and "staff" offices dealing with the same issue. Most communication should be, not among units, but between them and the commissioners whom they advise on decisions and who in turn tell them what to do.

In theory, communication between the commissioners and staff is mostly about rules and the making of rules. Regulations are essential to formal bureaucracy. While FCC rules are meant, of course, to govern the industry, the FCC also needs them to govern itself. Through regulations, superiors tell subordinates how to make decisions. The rules tell officials what information they are to consider and what standards to apply in making judgments.[56] Guidance is essential when staff have to make thousands of routine decisions each year. For example, over 100,000 complaints are received annually about radio or television interference alone. Formalism based on rules is the only way to avoid arbitrary or inconsistent decisions.[57] When there are no clear rules, administration is not properly bureaucratic. Decisions are governed instead by usage and discretion — the norms of traditional authority.

The FCC has unusual latitude to make rules because of its quasi-legislative mandate under the Communications Act. The Commission's rules occupied 734 pages in 1956 and 1,347 pages in 1976, a growth of 84% in 20 years.

The Commission also has rules for the making of rules. The Administrative Procedure and Judicial Review statues prescribe in detail how to go about rulemaking.[58] The Commission issues a Notice of Inquiry (NOI) to seek expert testimony from affected interests on a policy issue. It issues a Notice of Proposed Rule Making (NPRM) when it seeks comment on a proposed new regulation. After reviewing the responses, the Commission may, in the case of an NOI, issue an NPRM, setting forth its response to comments, its own view of the policy issues, and proposed rules. In the case of an NPRM, the Commission will issue a Report and Order which sets out the final rules to be instituted. Parties opposing the

Figure 2.1: FCC Organization Chart (August 1983)

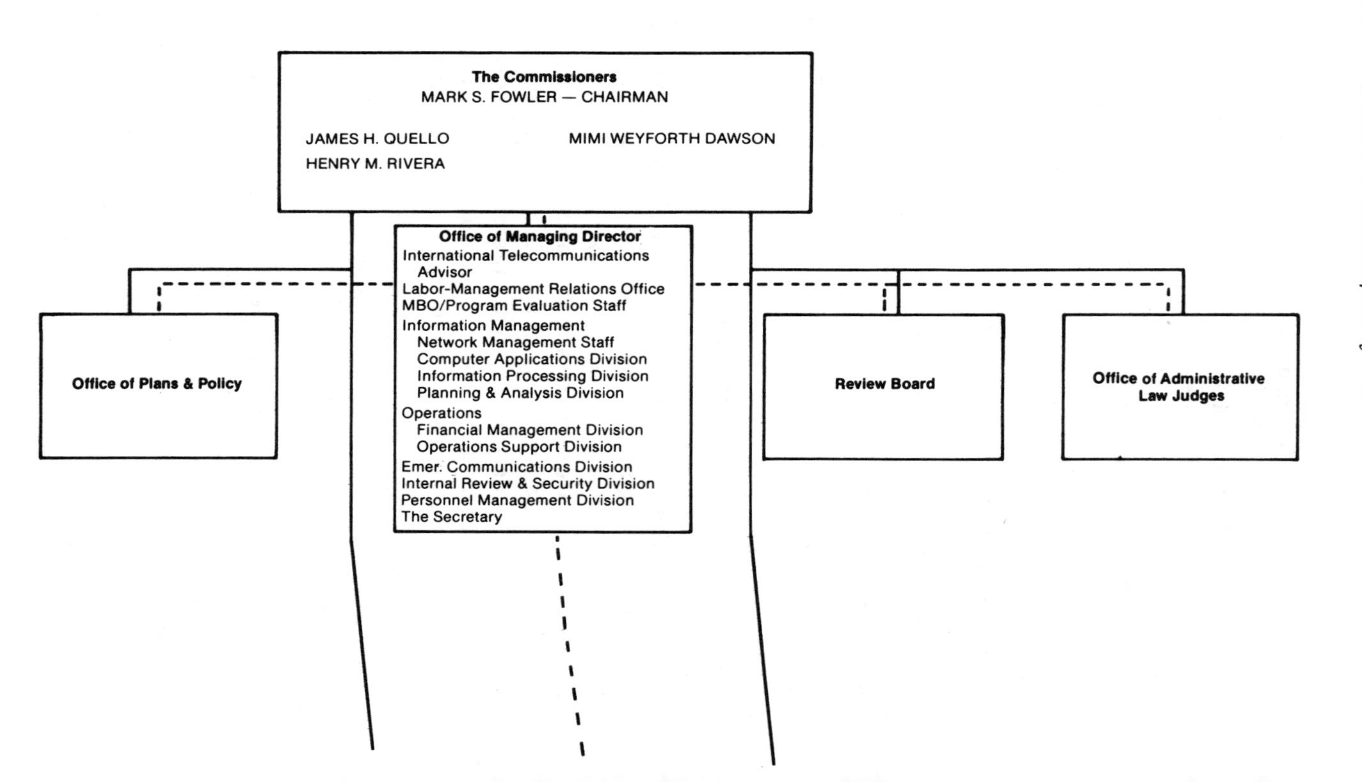

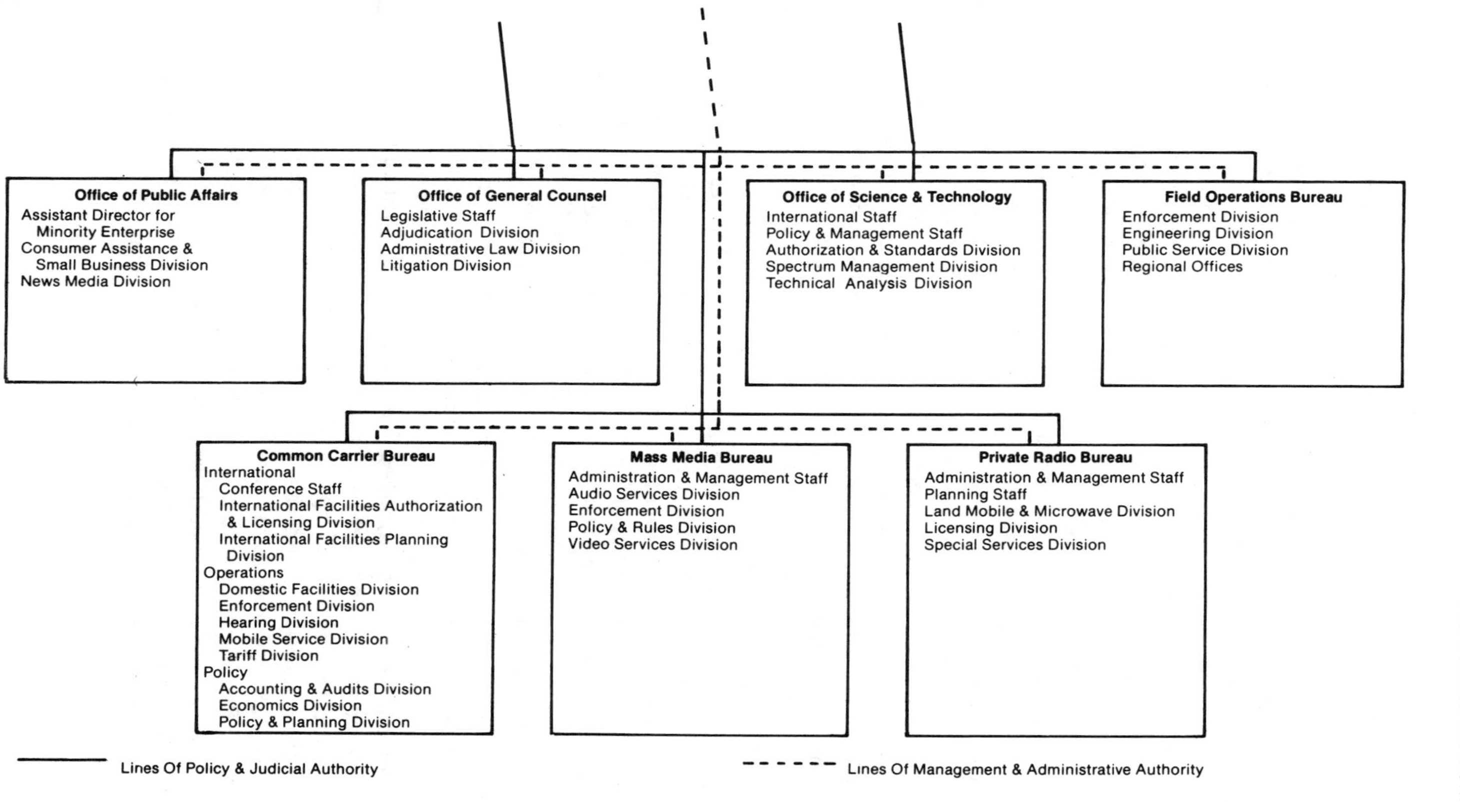
Office of Public Affairs
Assistant Director for
Minority Enterprise
Consumer Assistance &
Small Business Division
News Media Division
Office of General Counsel
Legislative Staff
Adjudication Division
Administrative Law Division
Litigation Division
Office of Science & Technology
International Staff
Policy & Management Staff
Authorization & Standards Division
Spectrum Management Division
Technical Analysis Division
Field Operations Bureau
Enforcement Division
Engineering Division
Public Service Division
Regional Offices
Common Carrier Bureau
International
Conference Staff
International Facilities Authorization
& Licensing Division
International Facilities Planning
Division
Operations
Domestic Facilities Division
Enforcement Division
Hearing Division
Mobile Service Division
Tariff Division
Policy
Accounting & Audits Division
Economics Division
Policy & Planning Division
Mass Media Bureau
Administration & Management Staff
Audio Services Division
Enforcement Division
Policy & Rules Division
Video Services Division
Private Radio Bureau
Administration & Management Staff
Planning Staff
Land Mobile & Microwave Division
Licensing Division
Special Services Division
Lines Of Policy & Judicial Authority
Lines Of Management & Administrative Authority

decison may petition the Commission to reconsider on grounds that it lacked necessary information or that new information has come to light.

In 1974, for example, the FCC faced an overload on the existing 23 Citizens Band radio channels because of the exploding popularity of CB. The Commission issued a Notice of Proposed Rule Making to increase the number of CB channels to 70, tighten technical standards for CB equipment, and simplify CB operating procedures. After comments, the Commission issued a First Report and Order adopting many of the technical and operating standards but postponing the more sensitive issue of new channels. Then in March, 1976, it issued a combined Notice of Inquiry and Further Notice of Proposed Rule Making in which additional technical issues were raised and a proposal for 40 rather than 70 channels was made. After comments were evaluated, a Second Report and Order was issued adopting these proposals. Petitions to reconsider the decision were filed by the Association of Maximum Service Telecasters, Inc., and the American Broadcasting Companies. Although the Commission did not alter its position on the technical standards or the number of new channels, it did amend its Second Report and Order to specify the date beyond which the manufacture and sale of 23-channel CB sets would be prohibited.[59]

While these procedures emerge from a judicial tradition, as we saw earlier, they also provide needed structure for a bureaucratic staff. They tell the large number of people involved how to go about making policy. The Notice of Inquiry and Notice of Proposed Rulemaking serve to focus the rulemaking process for the FCC as well as outside parties. The staff prepares all Commission documents at the direction of the Commission and summarizes the comments received from outside parties.

Other procedures, purely internal to the agency, govern consultation among bureaus on issues of common concern. One unit, for instance, will be given the "lead" in preparing a decision to be recommended to the Commission, but the document, called "an agenda item," will be routed to the others for concurrence before going to the commissioners. In the CB case, the former Safety and Special Radio Services Bureau initiated and guided the rulemaking, but the Office of the Chief Engineer was consulted on technical standards, the Field Operations Bureau on enforcement, the Broadcast Bureau about possible television interference, the Office of Plans and Policy because of the long-range planning implications, and the Office of the Executive Director because of the added personnel and funds necessary to implement the decision.

Rules dealing with substance and procedure work to coordinate officials by limiting the options they may consider in decisions. The rules exclude some options as improper or impolitic and steer action toward others. In the CB case, standard routines for rulemaking caused policy to be made one way and not another, and the new regulations in turn set norms for subsequent regulation of CB radio. Rules define the common patterns of behavior in terms of which officials are organized. In this sense, the FCC as an organization is literally defined by its rules[(60)]

The formal norms of bureaucracy clearly influence the FCC's structure and functioning. A Model 2 analysis of the FCC, which attributes outcomes to organizational process, must clearly pay attention to them. Yet the formal view of bureaucracy, like the ideal of rational authority, is a norm never fully realized. Organizations may be driven toward it by efforts to manage more efficiently, but they never fully attain it. As every citizen knows, the Post Office is simply *not* as orderly, obedient, and efficient as theories of public administration say it could and should be.

Nor is the FCC. The economics and reform literature on the Commission express great impatience with it. Why *is* it so reluctant to do its job? Why is it so suspicious of new technology in communications? Why is it so heavily influenced by the industry? The Commission was set up to solve regulatory problems on a rational basis. Why so often does it fail to do so?[(61)]

2.2.2 Bounded Rationality

Much of modern organizational theory is an effort to explain why formal norms of bureaucracy are never fully achieved. The dominant school, that of Herbert Simon, suggests that the formal view of bureaucracy is utopian. Officials could not live up to it even if they wanted to. The formal view tends to assume that perfect rationality is possible. The norm of perfect adherence to law and policy would be feasible, however, only if superiors could clearly communicate policy downward and verify compliance, and if subordinates could clearly apprehend it. The norm of efficiency or expertise would be feasible only if officials could communicate the necessary information to each other and craft optimal solutions to every problem. In fact, they cannot meet either norm because of fundamental limits to their ability to absorb information. Human reason cannot integrate an unlimited number of facts or variables. Its reach extends to a certain horizon and stops. It can only proceed further through rules, habits, or other conventions that simplify reality. In Simon's term, it is "bounded."[(62)]

Control Problems

Cognitive limits place severe curbs on the ability of bureaucratic superiors to control what goes on in their organization. They are few, and their staff are many. It is mentally impossible for them to direct or monitor what staff members do in detail. The staff, for their part, are absorbed in the organization. They are immersed in the countless interactions with other staff and offices required to solve daily problems. Communication lines are not primarily vertical, as the organization chart shows, but heavily horizontal among bureaucratic units; and this is truest for non-routine issues.(63) For the staff, directions from the Commission are merely one input among many. Even with the best will in the world, it is hard for them to *hear* what the commissioners want, just as it is hard for the commissioners to speak clearly to them.

The control problem gets worse the further down in the organization one goes. The commissioners can usually get the cooperation of the bureau chiefs and other senior staff who directly advise them, since many of these are political appointees directly accountable to them, and all tend to be highly dedicated. More junior officials, however, must be supervised and monitored through intervening bureaucratic layers. At each level orders are subject to distortion and evasion. There is "authority leakage" as orders come down the line. Leaders often have the sense that their directives disappear into an abyss, without any effect on the organization. Lower-level officials, who chiefly implement Commission decisions, may spend very little of their time actually doing what the Commission wants. Bureaucratic sub-units may drift apart and follow policies of their own.(64)

As an example, some notices of rulemakings take the staff months to prepare — even when the commissioners have expressed great interest in them. For example, Chairman Wiley publicly announced in 1976 that the Commission planned to produce a CB handbook to explain CB rules in layman's language to literally millions of new CB-ers. But at the time, the staff had already decided to rewrite the CB rules into plain English, eliminating the need for an explanatory handbook. Although periodically the chairman would inquire about the handbook, the lower echelon staff never took it as a serious assignment. In 1978, the Commission adopted the plain English version of the CB rules; it never issued a handbook.(65)

For Model 2, the control problem suggests the hypothesis that some FCC actions may not have been chosen by the Commission at all. Rather, they may result from decisions by individual employees or bureaus whom the Commission does not fully control. Some decisions by staff may become known to the commissioners and hence controllable only when they are elevated to public view by the press or Congress. The leadership of the FCC, like other agencies, spends a lot of time and energy just reacting to actions by their own personnel.

The publicity surrounding a "bootlegged" copy of the Roberts report on children's television shows how press coverage of actions (in this case recommendations) can compel Commission response. Roberts was the head of the task force on children's television which Chairman Burch set up in 1971. Less than two weeks after Richard Wiley became chairman in 1973, Jack Anderson ran a column about Roberts' recommendations to curtail commercials on children's programs, among other proposals. Cole and Oettinger point out that the "explosive" details in the report were in fact known to members of Congress and had been discussed in the trade press; it was the forceful publicity given to the report that demanded a public response. The column appeared in the *Washington Post* the morning that the commissioners appeared before the Senate Subcommittee on Communications for their annual oversight hearings, compelling them to answer questions about it.[66]

Such examples are not unusual, although most often the publicity occurs in the trade press. In fact, internal memos between bureaus, and from bureaus to commissioners, are often written with the full expectation that they will be highlighted in the trade press.

Information Overload

Related to the control problem is information overload. Because the commissioners are overburdened, they cannot give full attention to every decision they face. The weekly Commission agenda may include up to fifty or more separate items. A single, day-long Commission meeting may take up 25 or more of these — many of them highly technical — for which commissioners have had less than a week to prepare.[67] The press of business means that FCC staff have to compete with each other to get the time and attention of the Commission. Sometimes they cannot break through the log jam with urgent information. Hence, the Commission may be unadvised about the full implications of a decision or the fact that resources are unavailable to implement it. The basic problem is that the FCC, like other agencies, is hierarchical, and coordination depends on the action of a handful of leaders at the top.[68]

Because of the amount of information to be processed, commissioners are heavily dependent upon the staff. They cannot attend to all the information coming up from below. What items staff members decide to show them heavily influences what they do. According to Cole and

Oettinger, the staff usually give the Commission only one recommendation on each issue, rather than offering a range of options. When faced with an "option paper," commissioners will inevitably ask the staff to recommend one course of action. They simply have no time to do otherwise. Obviously, this gives the staff great power over decisions.[69]

The bureau chiefs are the most powerful gatekeepers of information. It may be that junior staff have knowledge or opinions that differ from those of the chief, but a bureau chief can keep them from getting through to the Commission. For example, the head of what was then called the Safety and Special Radio Services Bureau wanted to discontinue annual equipment measurements for land mobile transmitters. His staff wrote him memos opposing the idea — in part because parties who filed comments in the case had opposed it. The bureau chief, however, held to his view and finally produced a document arguing for discontinuation. With the staff sitting quietly in the meeting, he presented his view to the commissioners and they accepted it.[70] When, shortly after, Chairman Wiley appeared before the Association of Public-Safety Communications Officers (APCO), which had opposed the change, he was greatly surprised by the industry's opposition.

From the information overload problem Model 2 derives the hypothesis that FCC actions may result, not from premeditation, but from the problems or facts which are selectively reported to the commissioners from the staff below.

The more salient an issue is politically, the more likely the commissioners are to hear about it from outside sources, escape control by staff, define their own options, and make the decision themselves. The cable televison rules of 1972, for example, were hammered out of the commissioners among themselves because of heavy pressures from the affected industries, Congress, and the Office of Telecommunications Policy.[71] On the other hand, the more technical and obscure the issue, the less likely the Commission is to act on its own and the more likely to defer to staff. In 1971, the FCC issued new rules authorizing competition with AT&T in the specialized common carrier field almost entirely because the Commission's Common Carrier Bureau recommended this course. The decision was a major policy departure but attracted little public attention at the time, and, therefore, little personal attention from the commissioners.[72]

Satisficing

Another consequence of bounded rationality is that officials do not seek perfectly optimal solutions to every problem, as the rationalist ideal prescribes. To do this would require processing too much information too often. Instead, they "satisfice;" that is, they seek solutions which are not perfect but known to be satisfactory, either because they can easily be changed later or because they build on the agency's established ways of doing things.[73]

The FCC often satisfices by making short-term rather than long-term decisions. The tentative decision can be revised in the light of experience. To make a final decison would require calculating all of the consequences in advance. The incremental approach allows the consideration of consequences to be spread over time, not concentrated in advance of the decision.[74]

In 1974, the Commission issued a policy statement on children's television based on a voluntary agreement with the broadcasters to restrict commercials in children's programming, but postponed promulgating the policy as rules. In 1970 the Commission reallocated the 806-960 MHz band to land mobile radio but postponed the issue of what to do with existing television translators operating in this band. (Translators are used to transmit television signals beyond the primary service areas of a station.) Eventually the 700 or so translators already in the band were "grandfathered" — allowed to remain there — while all new translators were assigned to channels below 806 MHz.[75]

In 1976, the FCC voted to permit "closed captioning" of television programs for the benefit of the hearing-impaired. ("Closed captioning" is a technique for transmitting caption signals, to be decoded by translators in special TV sets purchased by the hearing-impaired.) There was considerable pressure from the deaf community to require captioning, but the Commission did not go beyond permitting it.[76] Chairman Wiley told representatives from various organizations for the deaf that the Commission could not go further, and the obligation was now on them to lobby broadcasters to provide captioning. Obviously, the decision could be revised if the organizations convinced the commissioners it was insufficient.

A second way to satisfice is to base decisions on past policies which have proved workable. A fresh solution is not calculated for every new problem. Rather, old solutions are adapted incrementally. One reason is that the organization already has heavy investments in established ways of doing things. To change policy for every new issue would involve not only information costs but the learning and psychological costs of shifting the organization to new procedures.[77]

Bureaucratic inertia may explain why some FCC functions seem to go on as they always have regardless of policy changes by the Commission. The Broadcast Bureau, now the Mass Media Bureau, has consistently followed a policy of renewing virtually all broadcast licenses, despite occasional Commission decisions appearing to hold stations more accountable for meeting their public service obligations.[79] The Common Carrier Bureau still depends upon the telephone companies for much of the data and analysis necessary to set common carrier rates, even though advancing technology and the advent of competition in specialized common carrier create a greater need for analytic capability within the FCC.[79] These bureaus simply have established ways of doing things that are unlikely to change quickly.

The commissioners are each appointed for seven years and serve an average of three. They confront an organization where standard operating procedures are already established. Bureaucratic inertia favors the power of the career officials, who are permanent, over the commissioners, who come and go. Lee Loevinger, a former FCC commissioner, has written:

> The middle management of career executives know that the announced policies will change as top political appointees change, but that the routine work of the bureaucracy will go on regardless... So middle management makes minimum concessions to policy pronouncements by top management, secure in the knowledge that before any very significant change in policy can be fully implemented it will be changed as a result of a political change in top management.[80]

One career executive, particularly secure through his expertise and tenure, simply refused to change an internal interpretation of the definition of broadcasting. He retorted to an angered Commissioner Wells, "I was here before you arrived and I'll be here when you are gone." Commissioner Wells, in fact, left the Commission a few months later.

At the heart of a Model 2 analysis of the FCC is the idea that organizational routine seriously constrains the options policy makers have available for decisions. The bureaucracy is not infinitely flexible. It cannot do whatever leaders want, as the formal view tends to assume . The leaders can only choose among, or incrementally adjust, the routines the organization already knows. The computer is already programmed, so to speak, and the decision makers can only choose among the programs.[81]

As a result, policy change and bureaucratic change often have to go hand in hand. The bureaucracy tends to resist innovation. If the policy maker means to change policy, he has to fight bureaucratic battles. Sometimes, the agency must be reorganized so that the bureaucratic mold is broken and fresh thinking can occur. If new options are to be available for decisions, new units must be created to represent them in the bureaucratic policy process.

On occasion, the FCC has had to create task forces to explore new policy areas because the established structures were inhospitable. For example, in 1966, under pressure from cable television interests, the Commission established a Cable Television Task Force to look into ways to expedite the development of cable. Previously, responsibility for cable had been lodged in the Broadcast Bureau, which looked upon it as a competitor of broadcasting and tended to stifle its growth. Not by coincidence, when Chairman Burch saw the need for new cable rules in 1970, the task force was elevated to bureau status and played the major staff role leading to the 1972 cable regulations. Correspondingly, the Broadcast Bureau at the time played a lesser role than before.[82] In 1971, the Broadcast Bureau was hostile to the idea of an inquiry into children's television. Therefore a task force on the issue was lodged officially in Broadcast but actually attached directly to Chairman Burch's office; it provided the staff work for the 1974 policy statement on children's television[83] In 1978, Chairman Ferris established a Network Inquiry Special Staff, apart from the Broadcast Bureau, to examine whether the major commercial television networks had engaged in practices that were anticompetitive. Ironically, the Fowler Commission is using much of the work of this Special Staff to justify elimination of the financial interest and syndication rules.[84]

2.2.3 Social Psychology

The control and information problems of bureaucracy and its satisficing behavior all result ultimately from the limited rationality of officials. But this analysis still assumes that breaucrats are at least trying to satisfy the formal norms for bureaucracy. The more radical critique, stemming from social psychology, suggests that officials are not trying to be obedient or efficient at all. They are serving social purposes of their own.

This reasoning extends the idea of bureau routines mentioned above. Organizations develop not only standard operating procedures but distinctive "subcultures." These attitudes not only justify standard procedures but come to constitute an entire world view for staff members. In order to operate efficiently, bureaus need an internal consensus on goals. But consensus often becomes an end in itself.[85] The ethos deepens the more the bureau is isolated from a need to change and the more it ages.[86] Eventually, the culture can condition officials to the point where they are positively incapable of dealing rationally with new challenges.[87]

Political appointees who come into an agency at the top discover that they cannot change the bureaucratic subculture, only use or avoid it.[88] The FCC commissioners sidestepped the entrenched attitude of the Broadcast Bureau by setting up the Cable and Children's Television Task Forces mentioned above. For more incremental policy changes, they have tried to win over the relevant bureau and encourage it to implement the new policy with the same enthusiasm as the old.

For example, the Commission has tried to persuade the Field Operations Bureau to bolster its public service activities in the field, due to the rapid increase in numbers of citizens band radio operators (from 1 million in 1974 to over 13 million in the spring of 1978) and the national prominence of consumerism. The engineers, who comprise most of the bureau, would much prefer to continue routine tasks of enforcement and licensing. However, recent chairmen have produced gradual change by allocating the bureau more funds for public outreach and publicly praising bureau performance.

Commissioners have also tried to change the attitude of the Broadcast (now Mass Media) Bureau toward petitions to deny broadcast license renewals. Cole and Oettinger report that the staff will not generally investigate renewal applications that look weak unless a petition to deny prompts them to. They also usually advise the Commission to deny or set aside such petitions. Commissioners Johnson and Hooks advocated a more critical attitude by staff, with limited success. Hooks argued that "... [The Commission] should not stand behind a procedural barrier on the apparent side of a licensee and let the matter ride simply because the complainant is without the ... resources or legal acumen to mount a perfect attack." The renewal staff simply does not perceive itself as an investigative arm of the Commission. It does not have the instincts or habits of prosecutors.[89]

The more basic point is that organizations cannot avoid having to adjust to their members as people. Individuals seek not only to perform well on the job but to find generalized meaning for their lives. They need and want the emotional security afforded by the bureau ideology. They seek to round out their jobs, however narrowly the organization may define them, so that their lives are personally satisfying.[90] They want to use the authority of their positions to participate in the decision-making process. The formal structures of bureaucracy must continually adjust to the social needs and processes of its members. The adjustment may lead to new problems, new structures, then new adjustments, in a continual interplay. In principle, organizations simply *cannot* be as rigid and static as the formal norm supposes.[91]

In theory, staff members are supposed to be subservient to the organization. In practice, the agency must negotiate implicitly for their support.

Employees have their own reasons for seeking to work in the organization and cooperate with others. These objectives must be met, or they will be satisfied in ways damaging to the agency mission.[92] Employee concerns have to do with prestige as well as pay. Amenities such as the color of offices, the size of desks, or whether the floor is carpeted can become symbols of esteem that affect people's performance. Such needs are probably more pressing today than in the past because people have become more individualized and may be less well socialized into organizations than previously.[93]

Under Model 1, it was argued that individuals may exercise *authority* at the FCC through charismatic personal qualities. Under Model 3 below, we will argue that they may exercise *power* if they are major stakeholders in the game of breaucratic politics. Here, the argument is that individuals influence the organization at all levels in a much more mundane sense. Tolstoy argued in his novels that history is determined, not by the leaders known to history, but by the countless actions of myriad ordinary people just living out their lives. The same could be said of policy at the FCC. The actions of the Commission are affected significantly by who happens to occupy a given office at a given time — and even how they feel on a given day. Important functions are often shifted between offices, every bureaucrat knows, not for reasons of high policy (Model 1) or bureaucratic organization (Model 2), but simply because of the special competence — or incompetence — of individual staff members.

Of course, personal influences are random by definition. They can only be told about through journalism or the gossip of insiders. They take no single direction; they could never be described in theory, which must appeal to general causes. For the same reason, Model 2 says they can never be entirely controlled. To a degree, people are simply *disorderly*, and no amount of organization can make them otherwise. The final limits of organizational rationality lie in "... limitations on the power of individuals and groups to influence the actions of other individuals and groups."[94]

2.2.4 Explicit Constraints

Social psychology says that employees do not, as even bounded rationality assumes, seek to satisfy the formal norms of bureaucracy. But at least, only their nature as human and social animals prevents them from doing so. A yet more pessimistic critique would point to the many ways the very structure of the bureaucracy aids and abets disinterest in doing the job.

In Weber, again, bureaucracy requires that officials have no rights to their jobs; they must be totally dependent on their superiors. American bureaucracy fails to meet this standard in several respects. The political culture has always been hostile to hierarchical authority. In several ways, the dominant pluralism has denied bureaucratic managers effective authority over their staffs.

Before the advent of civil service with the Pendleton Act of 1883, American public administration at all levels of government was staffed though patronage. Public jobs were viewed as "spoils" to be distributed to the followers of the victorious party. This accorded with populist attitudes which were suspicious of governance by experts and favored rule by amateurs.[95] These attitudes live on in such politicized hiring practices as "political" appointments at the top of bureaus, the politicized nature of some civil service hiring criteria (preferences for veterans and handicapped), and the informal preferences for women and minorities legitimized under affirmative action programs.

Patronage compromised one of Weber's norms for bureaucrats — expertise — but sometimes served the other — obedience to higher authority. Officials obeyed the politicians who could fire them. When civil service systems arose after the Pendleton Act, expertise was enhanced to some extent, at the expense of obedience.

Civil service was supposed to eliminate patronage personnel practices and instead base hiring and promotions on "merit." In practice, personnel decisions have come to be based on a combination of seniority, test results, and experience ratings that have limited connection to on-the-job performance. The system has taken away most of the supervisors' discretion to hire and fire for their own reasons, and with it most of their power to motivate their staff to perform. While the Federal civil service is more meritocratic and less rigid than most state and local systems,[96] it does represent one of the major problems facing FCC managers. These constraints were only marginally eased by the Civil Service Reform Act of 1978.

This constraint has increasingly been compounded by another — public service unionism. Unions have organized much of local and state government bureaucracy and are making progress in Federal agencies. They tend to support civil service rules strongly just because they tend to shield employees from supervisory authority. Though unions rarely question merit selection and promotion in principle, they want these norms construed in ways that strongly favor seniority and restrict the discretion of managers. Increasingly, pay and work conditions for government employees are set wholly or partly through collective bargaining.[97]

Civil service is a reality at the FCC, and unionism came in 1978. An Employee Representatives Board had long existed to bring employees' grievances to the attention of management. A union attempt to organize the Commission failed in 1971. In June, 1978, employees voted for the National Treasury Employees Union as their bargaining agent for non-supervisory personnel. Before, personnel grievances had already taken collective form. In 1977, the broadcast license examiners caused a work slowdown because they were not paid as much as other examiners doing, in their view, equivalent work. The move to unionize the Commission arose from widespread opposition to a shift to later working hours decreed by Chairman Ferris in November, 1977.

All of these constraints have the effect of denying bureaucratic superiors control over their staff. They give employees exactly the entitlement to their jobs that they should not have in formal bureaucratic theory. The result is that in the FCC, as in all government agencies, there are "deadwood" employees who do little and whom supervisors must simply work around. And the Carter reforms, which gave somewhat greater authority to managers, were strongly opposed by public service unions. This verified the view of Weber, who well knew his model was only an ideal type, that ". . . historical reality involves a continuous . . . conflict between chiefs and their administrative staffs for appropriation and expropriation in relation to one another."[98]

However, a more important explicit constraint on FCC organization may be a more subtle one — the influence of professional groups. Of the FCC's 1914 staff in September, 1983, 245 (or 13 percent) were lawyers and 432 (23 percent) were engineers or electronic technicians. Professional people bring to their work qualities of intellectual and economic independence that may be contrary to bureaucratic organization. Their ways of thinking are molded to professional, not organizational, norms. Because their skills are transferable, they can often leave bureaucratic settings they find uncongenial. They tend to be a force for innovation in organization, not for order or stability[99]

At the FCC, lawyers tend to argue for policy options that can be defended in an adversary setting in court, not those that might be effective or efficient in terms of practical consequences. In 1977, for instance, the staff recommended that an unauthorized broadcast antenna be ordered demolished and then rebuilt after a construction permit had been issued, in order to avoid an undesirable legal precedent. The four commissioners who were *not* lawyers carried the vote, 4-3, to let the antenna stand. The news reports read, "common sense won."[100]

Lawyers also insist on obeying the letter of the law. Changes in international law in the 1970's required that every radio operator permit contain a photograph of the holder. Engineers in the Field Operations Bureau recommended that photos be required only for licenses likely to be used abroad — where international law would apply. The lawyers on the Commission could not bring themselves to support what would have been a minor breach of a treaty. They eventually yielded to staff arguments that to require photos on all licenses would inconvenience millions of operators and require funding the fifty FCC field offices to emboss the licenses. The new policy was adopted by consent, rather than in a public meeting, to minimize the publicity.[101]

Engineers, for their part, can be equally rigid on technical issues. Their professional ethic tells them to argue for the solution that is technically the most satisfactory, regardless of the economic, legal, or political consequences. Strong arguments from the FCC engineers have sometimes helped push through important technical changes in the face of strong resistance. One example is the decision in 1945 to shift FM radio to a higher frequency in the spectrum. Another is the preparations for the 1979 World Administrative Radio Conference. WARC meets every 20 years to settle spectrum management issues on which nations must cooperate, such as which frequencies to use for international aircraft and shipping. FCC engineers prepared position papers outlining optimal solutions to the technical problems, blind to the fact that some Third World nations might resist these proposals merely because they came from the United States. For engineers, the idea that contingency positions should be prepared in deference to political realities was difficult to accept.

Of course, lawyers and engineers serve values — traditional and rational ones, respectively — which we found above to be deep-seated at the

FCC. In this sense, these groups are very much part of the organization. The point here is that they are not subject to close *control* as bureaucratic theory says they should be. The ethos of each group is part of the subculture of the FCC as an organization, not something the leadership has chosen on grounds of policy. A Model 2 analysis of FCC action assumes that it will be colored by these beliefs, whatever the commissioners choose.

In summary, the FCC bureaucracy is altogether less hierarchical, efficient, responsive, and cohesive than the formal theory suggests. Not only the inherent limits of organization but explicit constraints on bureaucratic authority limit the ability of the commissioners to obtain what they want from the staff. The FCC, like the rest of the Federal bureaucracy, is substantially a law unto itself.[102]

2.3 MODEL 3: POLITICAL PROCESS

According to Model 1, FCC actions can be understood as if they proceeded from officials acting as individuals. According to Model 2, bureaucracy, which in theory should be the invisible servant of the policy maker, in fact has influence of its own. Our third perspective explains FCC action by looking to political rivalry and compromise among the FCC's leaders within the Commission and its various political masters without. If Model 1 says action results from *individual* choice and Model 2 from *organization*, Model 3 says it results from *collective* choice among the *leaders* of organizations.

The last section used Weber's formal requirement that bureaucratic officials have no independence of their superiors as a touchstone. Organizational influences on decisions all result from breaching this condition in some way. The analogous requirement for Model 3 is that organizations should receive unambiguous political direction. Weber usually assumes that bureaucracy has only a single master. The only political, as opposed to career, official in a bureau is the minister who heads it, and he is the conduit for laws and orders emanating from the larger political system.[103] From him comes the clear-cut political will that the bureau needs in order to satisfy the norm of obedience and, also, the norm of efficiency, since optimal means cannot be calculated until specific goals have been set.

2.3.1 A Plural Executive

In public administration theory, unity of command is commonly achieved by placing a single executive at the head of an organization. The FCC had a seven-person plural executive until 1982, when Congress reduced the number to five.[104] Collective rule is the first of several ways politics enters to make a unified political will unavailable to the agency. Weber says that collective leadership serves the political values of representativeness and extended deliberation before decision. But it compromises the executive's authority over staff and the precision and consistency of decisions.[105]

That assessment is valid for the FCC. A political process, as much as an intellectual or organizational one, must occur before the Commission can make policy. Important FCC decisions tend to be preceded by vigorous horse-trading among the commissioners. A classic example is the cable television rules of 1972. Chairman Burch wanted new regulations that would allow the cable industry to expand, but he could get the support of commissioners H. Rex Lee and Nicholas Johnson only by requiring that the cable operators provide access for educational programming and the general public.[106]

The whole trend of FCC decisions changed around 1969-70, when Republican appointments tipped the Commission majority in a conservative direction. The earlier liberal Democratic majority, symbolized if not led by Nicholas Johnson, had initiated populist, anti-industry actions such as the prime time access rule and rules limiting ownership of newspapers and broadcast stations in the same market by the same company. Though the new chairman, Dean Burch, carried through some of these changes, the Republican majority generally intervened less in broadcasting.[107]

Shifting politics helps explain why the Commission in 1969 granted WHDH-TV to a competing applicant in order to diversify media ownership in Boston — a landmark liberal decision — and then, under Congressional pressure in 1970, issued a policy statement on license renewals very favorable to established stations. (The Court of Appeals later disallowed the policy.)[108]

Another shift occurred in 1977 with the appointment of Charles Ferris as Chairman. Democrats recovered their majority on the Commission.

With the appointment of senior-level staff by a liberal Chairman, the agenda of critical issues before the Commission changed. Concerns about minority ownership, childrens' television, and network monopoly were highlighted. Fowler's appointment as Chairman in 1981 did not bring about as immediate a change, largely because of the diversity of political views on the Commission. Throughout Fowler's first two years, he has had to cajole and compromise because of the lack of a solid political majority among the commissioners.

When decisions are political compromises, they do not result from choice in the Model 1 manner. The commissioners all participate in the political process, but no one of them usually chooses the outcome. The result often cannot be rationalized by the values of any one of the players.

2.3.2 Bureaucratic Politics

Even if the FCC were headed by a single executive, however, outcomes would be swayed by bureaucratic politics at lower levels. We saw above that policy can be influenced by the established mindset of bureaus within the FCC. Model 3 adds to this the expectation that units will actively compete for power in a political process. In their drive for security and autonomy, the units will compete more vigorously for "turf" than is rational for the organization as a whole.[109] In the early 1970's, the competition of the Broadcast Bureau and the Cable Television Task Force (later bureau) for control of cable regulation was a struggle of this kind.

For Model 3, the key figures in official action are the political executives atop the bureaucratic units. Bureaucratic politics is preeminently bargaining among people in high offices. Bureaucratic leaders tend to act as politicians seeking to build a constituency for themselves and their organizations. They also tend to be, for Model 2 reasons, captives of the flow of information given them by their staffs. A steady stream of issues and decisions presented by the bureau tends to orient them to its viewpoint and wean them away from purely personal action. Both as politicians and bureaucratic spokesmen they tend to become advocates of their organizations — and, frequently, of the segments of their industry they are charged to regulate."Where you stand depends on where you sit." Further, the need to bargain forces players to stake out their views more strongly and with greater confidence than they would if they could make policy without competition. Each can assume responsibility only for *one* viewpoint. The outcome may be in the general interest, but it will be a compromise no one of them has argued for[110]

Several of the FCC's major decisions show the importance of effective or ineffective leadership in the bureaucratic infighting. Between 1971 and 1974, there was a struggle between the Broadcast Bureau and the Children's Television Task Force over whether the Commission should issue rules to restrict commericals on children's programs and encourage programming designed for children. The task force leaders, who wanted new policies, strengthened their position by effective advocacy within the Commission and leaks of their proposals to the press outside. The Broadcast Bureau, which opposed change, was driven on the defensive despite its greater size and influence. The Bureau was able to persuade the Commission to issue a policy statement rather than rules, but this concession came primarily because of changes in the Commission's membership.[111]

The 800 MHz spectrum management decision pitted the Safety and Special Radio Services Bureau, which favored reallocation of UHF-TV frequencies to land mobile use, against the Broadcast Bureau, which opposed it. Land mobile demands were sufficiently strong to persuade the Commission finally to reallocate the space, but it did so on condition that a spectrum management task force be established to ensure that the additional space was used efficiently. The Broadcast Bureau wanted the space used well to prevent future land mobile demands, and hence supported the task force. However, neither Broadcast nor Safety and Special controlled it. The Chief Engineer, with a well-timed proposal, got control of it and used it to add some 70 positions to his office.

As the program grew, a second dispute pitted the task force, which wanted to use its own regional offices to assign mobile frequencies, against the Field Operations Bureau, which wanted its field offices used, and the Executive Director, who was concerned about the increasing budgetary demands and wanted the required computer services centralized in Washington. After extended controversy, the Executive Director, with support of the Office of Plans and Policy, won over the Chairman and Commission, causing the task force to be absorbed in the Safety and Special Bureau and most of the work of its Chicago regional office moved back to Washington.

In 1976, the issue arose whether there was a more efficient way to license CB operators. Established procedure was for the FCC to issue licenses to CB operators, just as for other kinds of land mobile radio. The alternatives were to allow self-licensing, under which new operators would simply mail in applications to be filed by the FCC, or to abolish licenses altogether. Safety and Special supported the existing procedure, because the necessary personnel added resources to the bureau. The other offices and bureaus favored the other options, but their failure to agree clearly on any one of them weakened their position. Also, the commissioners tend naturally to give extra weight to the opinion of the lead bureau on an issue. For these reasons, Safety and Special was able to dominate the arguement before the commissioners, and the decision went to its favor.

Interestingly, the Commission gave the lead on the issue initially to an inter-bureau committee headed by the Deputy Executive Director, exactly to prevent Safety and Special from dominating the decision. This was the first time a task force structure had been used to evaluate an issue relating to only one bureau. The committee was supposed to settle the issue and save the Commission from having to make more than the formal decision. But bureaucratic habit was too strong. The committee presented options, rather than a single recommendation, for fear of preempting the decision of the Commission.[112] And the committee members were allowed to argue their separate cases before the commissioners, just as if there had been no committee at all. Hence, the committee generated research on the issue but had hardly any effect on the outcome. This shows how organizational routines can mold the environment for bureaucratic politics.

The staff offices of the FCC tend to be closely associated with the Chairman, and sometimes are used by him to counter the weight of the service bureaus. Dean Burch established the Office of Plans and Policy in 1973 to make sure he would get planning advice independent of the bureaus (and sometime ahead of the other commissioners). The General Counsel's office did the main staff work on the cable rules for Burch before the cable task force was promoted to bureau status. The General Counsel also supported Burch against the Broadcast Bureau on the children's television issue by arguing that the Commission had full authority to regulate on the question.[113] This pattern resembles that of larger Federal departments, in which the Secretary commonly has staff units comprising an "Office of the Secretary" to help him control the operating bureaus.

So far, all our models have looked inside the FCC to explain its behavior. The justifications available to decision-makers, organizational structure, or the internal contest of bureaucratic players.[114]

2.3.3 External Politics

In addition, Model 3 looks to political forces impinging on the FCC from the outside. This is where Model 3 departs most clearly from the other two approaches.

A major influence on the FCC is, of course,the regulated industries, most prominently broadcasting and common carriers. The major "players" include the networks and the National Association of Broadcasters (NAB). The major common carrier has been the American Telephone and Telegraph Co. (AT&T). In the future because of the AT&T divestiture, the seven regional Bell operating companies will likely become major players as well. Industry power is based on a political version of the control of information which we saw, in Model 2, gives bureaucrats power over the commissioners. In theory, the Commission is accountable to Congress and the public, and the industries are defenseless before it. In fact, influence goes to whoever gives the FCC most attention. Congress and the public, whose interests are diffuse, give the Commission only sporadic attention and hence have less power over it than it appears. The industries, whose interests are concentrated on what the FCC does, give it a great deal of attention and gain, in return, disproportionate influence.[115] The constant drumfire of their lobbyists, lawyers, and publications tends to overwhelm other inputs to the commissioners and attune them to an industry viewpoint.[116]

Industry influence depends on constant, informal contact with the Commission. An important medium is the Washington communications lawyers who represent businesses before the FCC. They practice "law by telephone," in which they both inquire about policy and seek to influence it. The power of the law firms is enhanced by the fact that many commissioners and staff go to work for them or the regulated industries when they leave the FCC.[117] Of the 33 commissioners who left the FCC between 1945 and 1970, 21 went to work for the communications industry or communications law firms.[118]

As we saw under Model 1, rules against *ex parte* contact with the commissioners are designed to restrain unofficial influence on the FCC. However, until 1978 they applied only to adjudication cases. That year the Commission initiated a rulemaking to establish the extent to which the rules should be applied to rulemaking proceedings and adopted an Interim Policy that was finally adopted as part of the rules in 1980. In one important instance — the 1975 rules for pay TV — the Appeals Court overturned a decision because of the extensive off-the record lobbying that was allowed.[119]

Industry influence, as well as technical considerations, help explain the Commission's decision of 1952 to force UHF television to develop against the competition of VHF stations. Heavy lobbying by RCA, owner of the NBC network, precluded the option of an all-UHF system even in some localities.[120] To a lesser extent, a desire to protect AM radio interests may have influenced the FCC's 1945 decision to shift FM radio to a higher spectrum band, a move which made existing FM equipment obsolete and delayed the development of FM stations.[121] On the whole, FCC policy may be said to have delayed the development of cable and pay television as competitors of broadcasting.

These cases all suggest that industry influence is a reason why, as critics have pointed out, the FCC is suspicious of new technology. Industry interests naturally defend the technology they have. The FCC tends to protect the existing economic investment of companies. It hesitates to put at risk a system spoken for by economic interests, in favor of a new system spoken for only by enthusiasts and academics.[122] There seems to be a notion of economic standing parallel to that of legal standing. Ideas are taken seriously only when an industry has given them economic reality "on the ground." Cable television has been able to get even partial attention and support from the Commission only because it has achieved some economic presence as an industry. By definition, new technology is not yet developed in industry.

A second political influence of growing importance has been consumer groups interested in making broadcasting more responsive to the general public. These groups came into prominence following the landmark 1966 decision of the Court of Appeals in the WLBT case which granted community representatives legal standing in license renewal cases. Consumer groups were inspired by the oratory of Nicholas Johnson when he was on the Commission. They have been represented by, and are barely distinguishable from, public interest law firms, such as the Citizens Communications Center (CCC). They include the National Citizen's Committee for Broadcasting (NCCB), now the Telecommunications Research and Action Center (TRAC), the NAACP, Action for Children's Television (ACT), the National Black Media Coalition and the United Church of Christ (the plaintiff in the WLBT case), among others.

These groups challenge license renewals on the grounds that the licensee has not met standards of community responsiveness, public service, or affirmative action. They also contend for general broadcast policies favorable to the public interest as they see it. A notable example has been ACT's campaign since 1968 to restrict commercials and improve programming for children's television. ACT parlayed an attractive populist appeal, strong advocacy, and effective publicity into considerable pressure on the FCC. Counterpressure by the NAB caused the Commission to confine itself to a policy statement stating standards for commercials voluntarily agreed to by the broadcasters. ACT appealed unsuccessfully to the Court of Appeals to require the Commission to set binding rules.[123]

Like the industry, the public interest groups seek close contact with the FCC. A bit of a "revolving door" has already grown up between them and the Commission, parallel to the one between the communications law firms and companies and the Commission. At the end of his tenure in 1973, Nicholas Johnson left the Commission to become the head of NCCB. The former head of the Citizen's Communications Center became administrative assistant to Charles Ferris when he became Chairman in November, 1977.

Rather than exercise influence within the FCC, these groups aim publicity at the public. This indicates a basically different political style. As we saw above, the industry is most able to protect highly concentrated economic interests if it is given a quiet atmosphere in which it can dominate the flow of information reaching the commissioners. The public interest groups, on the other hand, seek to protect the diffuse interests of the general public. To do this, they must break open the closed world of industry influence and force the commissioners to pay attention to a wider constituency. They seek to use as weapons the politicians and the press — the two political forces speaking for the undifferentiated public.

They seek, like industry lobbies, to find out about Commission deliberations and try to influence them, but they use these resources to mobilize outside political forces. They work to give the general public the resource it cannot have by itself — a degree of day-by-day attention to the Commission rivaling that of industry. Hence, the position of the groups is paradoxical. They seek success as lobbyists only in order to undercut lobby politics. Presumably, if they could ever mobilize the public permanently, their function would disappear.[124]

The advent of the public groups has no doubt been salutary for the FCC. They tend to make real what might otherwise be mainly formal — the Commission's accountability to the general public. They give organizational presence to the general interest and hence tend to prevent capture of the agency by the industry.[125]

However, a Model 3 analysis allows no heroes. Good policy cannot be chosen independently of the political process. No one actor, however well-intentioned, can cause FCC policy to serve the general interest. The groups can have influence only by engaging in a political struggle with the industry. To influence the outcome they must sometimes state their views at an extreme that would not serve the public interest if it prevailed. Some public interest advocates sound as if they would destroy the industry if they could. In this sense, political competition corrupts even players with disinterested motives. The outcome may in fact serve the public interest, but despite — perhaps *because* of — the fact that none of the competitors have argued for it.

2.3.4 Multiple Masters

The pressure of interest groups is only one dimension of external politics. A second is competition among the official institutions who have authority over an agency. Here is where the departure from Weber's presumption of a unified political will is most blatant.

The FCC has three official masters — Congress, the Executive, and the courts. Traditionally Congress has been the most powerful. Although Congress has delegated near-total authority to the FCC to deal with communications, it may always revoke it or overrule the Commission by statute. It also has the power to confirm commissioners' appointments, determine the agency's budget, and investigate its policies. The mere threat, as well as the use, of these powers has given Congress pervasive influence over much that the Commission does. Congressional oversight is concentrated in the Communications Subcommittees of the Commerce Committees of both House and Senate, but extends also to the Appropriations Subcommittees that deal with the FCC and to individual Congressmen or staff.[126]

Traditionally, Congressional influence has been exercised on behalf of broadcasters and against FCC attempts to regulate in a populist spirit. One reason is that Congressmen are heavily dependent on local broadcast stations for the media coverage they need to be reelected.[127] The Senate Commmunications Subcommittee, long headed by Rhode Island Senator John Pastore, often took this stance, for instance by influencing

Dean Burch to revise the proposed cable television rules in the interests of broadcasters and broadcast copyright owners.[128] In the late 1970's, the House Communications Subcommittee, under San Diego Congressman Lionel Van Deerlin, pursued a much more populist line, reflecting the increasing influence of public interest lobbies and attitudes. In June 1978, the House Subcommittee produced a bill which, if adopted, would have revamped significantly the authority and practices of the FCC, along with the Public Broadcasting Corporation and the National Telecommunications and Information Administration.[129] Since assuming Chairmanship of the House Subcommittee in 1981, Colorado Congressman Timothy Wirth has continued that stance, but in the face of considerable opposition from other members of the subcommittee and the House. At this writing, a deregulation bill has passed the Senate Committee on Communications, and is now before the House Subcommittee, which would significantly reduce the obligations of broadcasters to provide public interest programming as favored by the public interest lobbies.[130]

Some of the FCC's more adventurous decisions have been taken under threat of Congressional action and have led to negotiations with Congress. In 1961, the Commission proposed to expedite the development of UHF Television by "deintermixing" UHF and VHF stations in eight local markets. Six markets would have been given over wholly to UHF, reviving an option rejected for the whole country in the UHF decision of 1952. Alarmed, VHF broadcasters mobilized Congressional support to stop the action. However, FCC Chairman Minow obtained in return passage of legislation to require that all new television sets be capable of receiving both VHF and UHF channels — another reform long desired by UHF interests.[131] In 1963-64, broadcasters used Congressioinal support, plus a weak majority on the Commission itself, to stop an FCC move to limit permissible advertising time on the air.[132] In 1969, the Commission gave the license of WHDH in Boston to a competing applicant in order to diversify media ownership — its first-ever revocation on overtly populist grounds, since the station was innocent of obvious malfeasance. As mentioned above, Congressional pressure, combined with a changed political balance on the Commission, forced the Commission to issue guidelines for comparative hearings involving regular license renewals which were more favorable to established licensees.[133]

Common carrier regulation, which has been too technical to attract much political interest, has received increasing Congressional attention because of FCC efforts to introduce competition into the field. The full import of the 1971 decision allowing specialized common carriers to compete with AT&T was not at first apparent to anyone outside the

FCC's Common Carrier Bureau, which sponsored it. However, the decision survived routine court challenges. The Commission followed it up by denying petitions to reconsider, questioning the rate submissions AT&T proposed in response to competition, and favoring competition in domestic satellite and land mobile communications.[134] AT&T and its supporters became alarmed and appealed to Congress. A Consumer Communications Reform Bill, known as the Bell Bill, was introduced in 1976 which essentially would have rolled back the FCC competition policy and restored AT&T's position as a public utility monopoly. The bill was reintroduced in 1977 and 1978, but was never enacted.[135] Since then AT&T has divested its operating companies because of a consent decree ending the Justice Department's antitrust suit against the company.

In 1982, Congress changed the make-up of the Commission due to a purely political squabble. The White House, with the support of Chairman Fowler, nominated Stephen Sharp to replace Commissioner Abbot Washburn, whose term had expired. Senator Stevens from Alaska was adament that his candidate, Marvin Weatherly, be appointed. The White House nominee was confirmed, but Senator Stevens secured passage of a bill that reduced the Commission from seven to five members, beginning in June 1983. Commissioner Sharp served less than a year. Thus, in a matter of weeks, Congress accomplished with seemingly little awareness what various study commissions and task forces had debated for years.[136]

The Commission, on occasion, seeks Congressional permission for policy departures. Recently it asked Congress to allow the use of lotteries to grant licenses in cases where the applicants' capacity to serve the public interest seemed indistinguishable. Lotteries were to be used to award licenses for lower power television stations. Congress concurred in 1981, but included restrictions which the FCC has since sought to relax.[137]

The Executive's powers over the FCC are less pervasive than those of Congress but increasingly important. Their basis is the President's authority to nominate the commissioners, to name the FCC Chairman (who serves at his pleasure), and to make communications policy through channels which are competitive with the FCC. Also, the President's staff reviews senior staff appointments at the FCC, and the Office of Management and Budget (OMB) reviews its personnel ceilings, budget requests to the House, and legislative proposals.[138]

Presidents have sometimes used task fores to preempt FCC policy making. President Johnson's Task Force on Communications Policy, 1967-68, held up FCC action on domestic communications satellites. Presidents have also used task forces to evaluate the performance of the FCC and other regulatory commissions. The Landis Committee reported to President-elect Kennedy, and the Ash Council reported to President Nixon.

President Nixon's creation of the Office of Telecommunications Policy (OTP) in 1970 gave the Executive a point of leverage on FCC affairs. OTP presided over the final negotiation of the Consensus Agreement that resolved the dispute over the Commission's 1971-72 cable television rules.[139] A rivalry eventually grew up between Chairman Burch and Clay Whitehead, the head of OTP. In 1978, the Carter Administration moved OTP to the Commerce Department to become the National Telecommunications and Information Administration (NTIA), headed by Henry Geller, former General Counsel of the FCC and advisor to Dean Burch during the cable controversy. Removing NTIA from the office of the President significantly reduced its power in the policy process.

A reason for executive influence is that communications policy has become too complicated for the FCC to make alone. The issues extend far beyond spectrum management, broadcast licensing, and common carrier rate regulation — the core Commission activities. New technology raises a host of legal and economic issues that bring many other agencies into the political game. OTP, now NTIA, has appeared as a party to FCC proceedings. The Justice Department's anti-trust division put pressure on the Commission to extend its rules against multiple ownership of media facilities to cover newspapers and to stop the merger of ABC and ITT.[140] The minute the number of agencies multiplies, the FCC tends to lose its preeminent position in communications policy. Power gravitates to the President or White House staff who have the authority to coordinate all the executive players.

The most influential external institutions for the FCC may now be the Courts of Appeals, particularly the Appeals Court for the District of Columbia. While the courts traditionally have reviewed Commission decisions for adherence to due process and the FCC mandate, during the 1970's they began increasingly to decide the merits of cases as well.

To a great extent, legal appeals are an extension of the interest group politics reviewed earlier. The FCC itself is a quasi-judicial body, as we saw under Model 1. It uses adversary hearings for adjudication cases and solicits comments of an adversary nature before making rules. Hence, to appeal a decision usually results in an adversary proceeding in court not very different from the original proceeding before the FCC. Often, losing parties who appeal are merely trying to win on legal grounds the case they have first lost on substance before the Commission. Legal action is the continuation of politics by other means.

When the courts rule on due process they ask whether the FCC has adequately considered all the interests and viewpoints relevant to a decision, and specifically whether it has complied with the Administrative Procedure Act. Precedents govern the range of interests entitled to "standing" (representation) in cases. The importance of the Appeals Court's 1966 decision in the WLBT case was that it gave community groups standing even though they lacked a clear economic interest in broadcasting. The specialized common carrier decision of 1971 had to survive a legal challenge alleging that the FCC should not have used rule making procedure, which requires only written comments, but adjudication, which would have required lengthy comparative hearings among AT&T and the applicant companies.[141] In 1977, the courts overturned the restrictions on violence in early prime time (the "family viewing hour") which Chairman Wiley had negotiated with the broadcasters, on grounds that the agreement was tantamount to rule making and the Commission had not consulted other interests.[142] In 1977, ACT unsuccessfully challenged the FCC's 1974 Policy Statement to force it to issue binding rules on children's television, rather than just a policy statement.[143]

The courts also hear challenges to the Commission's legal authority to regulate. Such questions arise because the Communications Act is vague and does not clearly give the Commission authority over the new types of communications that have arisen since 1934. The precedents, some of which go back before the Communications Act, have become in effect part of the FCC's legal mandate. One of the most vexing questions has been whether the Commission had the authority to regulate cable television. Since cable is not mentioned in the Communications Act, the courts have held in decisions since 1968 that the FCC may regulate cable only for purposes "ancillary" to the protection and development of broadcasting, for which the Commission's responsibility is clear. But how wide a range of cable regulation this doctrine covers is still unclear.[144]

Federal courts increasingly decide questions of substance as well. Their traditional criteria of due process and statutoriness have become so discretionary that they cannot be applied without, in effect, judging the merits of the case. Once legal standing is extended to groups with no immediate economic interest in a case, private suits for redress of grievance can be turned into wide-ranging, quasi-legislative inquiries. When community interests are defined this broadly, how to balance their interests depends more on judgment than law. Another reason judicial decision is necessary is simply that Congress so often delegates authority to agencies in very broad terms, as with the FCC, leaving it unclear what they can and cannot do.[145] These changes have given the Federal courts an increasing role in regulatory policy, at the expense of both Congress and the agencies.[146]

An uncertain mandate, among other factors, has allowed the Appeals Court virtually to take over cable television regulation since 1972. Judges have affirmed the FCC's authority to regulate cable but have decided many of the details for themselves. In 1972, the Court of Appeals approved the Commission's mandatory cable origination rules as "reasonable ancillary" because they increase the number of outlets for community expression and augment the public's choice of programs and type of service.[147] In 1977, the court overturned rules preventing cable operators "siphoning" broadcast signals out of the air for cable transmission, on grounds that siphoning had created no real economic threat to broadcasters.[148] In 1978, the court ruled that the imposition of minimum public access and channel capacity standards on cable systems was improper.[149] Also in 1978, the court held that the FCC has the authority to preempt state and local price regulation of special pay cable programming, and that the way the Commission has done this is adequate and effective.[150]

2.3.5 The Pluralist Problem

Model 3 traces FCC outcomes to political bargaining in the absence of a clear political will. The lack of clear-cut goals seems initially due to the bargaining itself, among the commissioners, FCC bureaus, outside interests, and other institutions. The players see to it by their competition that the policy is set by a political process, not by any one interest acting alone. However, the politicking is due to a larger cause — the inability of the overall political system to agree closely on goals. Regulatory commissions like the FCC reflect the problem at an extreme. If the FCC had a more specific legislative mandate, as some other regulatory bodies do (such as the Environmental Protection Agency), its leaders could base policy on clear obligations to law, in the manner of Weber's rational authority, and pay less attention to political pressures. But the FCC arose precisely from Congress' failure to agree. In fear that politics would infect communications regulation, Congress delegated the area wholesale to the FCC.

The uncertain mandate has merely shifted the political pressures from Congress to the FCC, not excluded them. The Commission is subject to political pressures precisely because players know it has the discretion to yield to them. If an agency operates under no authoritative commands, the effort to influence is worthwhile. Congress itself, as we saw above, seeks to influence the FCC. The Commission's main problems, a vague mandate and undue Congressional pressure, go hand in hand.[151] Only closer collective agreement about goals for communications could eliminate them.

The lack of agreement on policy also means that organizational arrangements are politicized. Traditionally, Congressional committees and interest groups have shaped the structure of Federal agencies to fit their own needs, not the norms of public administration theory. Programs or units are reorganized or their departmental location changed in order to change the interests they serve.[152] The FCC has periodically faced such threats. One or another bill to reorganize the Commission has been pending in Congress in most years since its founding.

The Commission has so far escaped major rearrangement. Its main structural problem is something more subtle — lack of a clear relationship to the executive or legislative branches. Agencies with a clear mission usually receive it from the Executive or Congress, and are organizationally linked accordingly. Thus, OMB is clearly accountable to the President and the Government Accounting Office (GAO) to Congress. The major departments are accountable to the President through their political appointees and to Congress through their legislation and budgets. The political authorities accept some responsibility for protecting these entities from contrary political pressures. The agencies are viewed as *agents* of government.

The FCC, however, is a kind of subordinate government. It is supposed to be "independent." It has no clear relationship to either the Executive or Congress, or to the courts. Rather, it has some of the nature of all three. It has quasi-legislative functions, but is closely associated with the Executive and behaves like a court.[153] In this position, as we have seen, the FCC gets very little protection. All three branches of government can influence the FCC, but none takes responsibility for shielding it from the others. All can sway it for short-run reasons, but none helps it develop a coherent long-term policy for communications. The FCC has to do this alone.

The FCC's lack of a clear political or organizational "place" opens the door to pluralist pressures. The agency is not part of government in such a way that it knows clearly who to follow and what to do. Rather, it is one of a number of organizations and interests interacting to determine communications policy without any clear hierarchy. The environment is uncomfortable for the FCC, but endorsed by the larger political system.

Pluralism is a deeply American approach to achieving social order, maybe more American than either representative democracy or hierarchical, bureaucratic authority.[154]

To the extent pluralism is a reality, the Model 3 analysis of the FCC becomes plausible, and the viewpoints of Model 1 and 2 implausible. Leaders cannot rationalize actions according to their own values, as Model 1 imagines, if they must continually adjust to the demands of others. Expert calculation of the kind stressed by rationalist values can play a role only insofar as competing interests use it as a weapon to influence decisions.[155] Nor will decision result primarily from bureaucratic dynamics, as Model 2 says, if the major determinants are political pressures from outside the organization. Political executives are not, it turns out, very interested in organization at all. They spend most of their time reacting to external pressures.[156]

In short, the policy process cannot be rationalized and cannot be organized coherently as long as political bargaining is dominant. Leaders can only adjust; they lack the security to calculate or to manage.[157] Bargaining cannot even be understood in as theoretical a way as the dynamics specified by the other two models. The nature of pluralism is to provide, and to reveal, very little structure of its own.

2.4 THE REFORM OF THE FCC

The foregoing analysis can help us understand the proposal usually made to reform the FCC. We will find that it is usually based on one or another of the analytic models reviewed above. The models underlie reformers' views of what is wrong with the FCC and how it could be made right. Each model explains what a particular class of reforms can accomplish —and what it cannot.

2.4.1 The Critique of the FCC

The reformers agree closely about the shortcomings of the FCC, aside from details. In theory, the Commission is supposed to enforce the public interest on the communications industries by requiring responsive and efficient service. Regulation, rather than the market, has to do this, because the industries must for technical reasons (e.g., the limited radio spectrum, the need for an integrated national phone system) be oligopolies subject to only limited competitive pressures. The goal is not simply that the industries should thrive in market terms; it is that they should be responsive to the public and new technology in ways the market would not, by itself, require.

Critics have claimed that, in practice, the FCC has been much more concerned to protect established industrial economic interests than to enforce either responsiveness or efficiency.[158] The Commission has allowed broadcasters to cater to a mass commercial market rather than serve local communities, and it has to some extent shielded them from competition from alternative technologies (e.g., UHF and cable television, FM radio). It has allowed AT&T to serve some common carrier needs exclusively, excluding until recently competitors who might have served them better. These outcomes indicate poor performance by the FCC even if the industries are profitable and, by international standards, technologically advanced.[159] The Commission's public goals have not been fully achieved.

2.4.2 Will and Expertise

Some critics, following Model 1 assumptions, argue that the problem can be traced to the minds of the commissioners and, by extension, all FCC personnel. The Commission does not sincerely will the public interest, or it does not know how to achieve it. If the FCC were more principled and more expert, it could achieve its goals.

We saw above that the single actor viewpoint speaks of maximizing at least three different sets of values. Economist critics of the FCC have been those arguing most forcefully for rational values. They see better FCC performance primarily as a matter of intelligence. They want to improve the quality of policy analysis, planning, and evaluation done at the Commission. All these terms mean to make policy in the light of foreseeable economic consequences. Since economists are those most skilled at economic calculation, they should exercise increased authority at the FCC. One suggestion is that economists sit on the Commission itself, but none ever has.[160] The Commission has gone part way in instituting the Office of Plans and Policy in 1973, which all by itself vastly increased the agency's economic expertise.

Lawyers, the advocates of common-law or traditional values, seek to reform the FCC by improving its internal processes and judicial review. Lawyers were very active in the 1930's and 1940's when a leading reform issue was how to reconcile admisistrative law with judicial notions of due process. The Administrative Procedure Act of 1946 codified the legal view of administrative due process for the FCC and most other Federal agencies.[161] Lawyers have argued, further, that active judicial review of FCC actions is an effective way to hold the Commission accountable and force it to perform. This view is shared by some economists.[162] It was endorsed in the Ash Council reorganization proposals, which specified a special Administrative Court to take over review from the Court of Appeals.[163]

Public interest critics, for their part, assume populist values. Their diagnosis of the FCC problem is that the commissioners are coopted by the regulated industries and detached from due allegiance to the general public. "Better men," more able and more principled, should be appointed. Their separation from the industry should be enforced by giving them longer terms than those currently in force and forbidding them from going to work for the industry for an extended period after they leave the Commission.[164] Also, representatives of the general public or consumers could be added to the Commission.[165] The rewrite of the Communications Act proposed in 1978 reflected this view to the extent that a commissioner's term would have been changed from seven years with the possibility of reappointment to ten years without reappointment, and the present one-year prohibition on representing industry interests after leaving the Commission would have been extended to include supervisory personnel as well as commissioners.

All these proposals are defensible in themselves. But they ignore the reality that the commissioners, however able and principled, must carry out their decisions through an *organizational and political process.* This was the criticism that Model 2 and 3 made of Model 1. As Weber puts it, leaders cannot exercise authority by themselves; they need an administrative staff.[166] Just as much, they need to be politicians.

The economists' critique tends to neglect the importance of organization. They reject, for instance, the Ash Council's contention that the organization of the regulatory commissions is a problem separable from the substance of regulation.[167] This implies that if policy is the product of good intentions and expertise, its implementation and success can be taken for granted. We have seen how important organization and politics actually are for the FCC. Such factors lie outside the normal range of economic reasoning.[168]

Lawyers' suggestions for improved due process and review tend to ignore the adverse institutional consequences. If regulation could be reduced to court-like deliberations about policy, legal approaches could be nothing but salutary. But deliberation has costs in time and effort, and rules once made must be administered. Judicial procedures for rule making and adjudication emphasize precise representation of affected interests over speed and flexibility. They also give the regulated industries a dominant role in proceedings, as against inputs which are tough to represent, such as economic calculations and the interests of the unorganized public. That is, they emphasize traditional over rational or populist values. Critics of the regulatory commissioners commonly find that they have been "over-judicialized" — tied up in knots by court-like procedures. The Administrative Procedure Act was a triumph for the legal values in policy making, but costly to others. Its structures may have to be relaxed before a more balanced approach can emerge.[169]

Public interest reform proposals sometimes seek to disrupt the institutional apparatus they would need to be effective. Reformers of the Nicholas Johnson or Ralph Nader type tend to campaign *against* the institutions rather than seeking to use them for more enlightened purposes. Populist leaders may be effective as critics and prophets; however, to bring about change they must come to terms with organization and politics. They have to compromise with existing institutions, and as charismatics, they resist this.[170]

2.4.3 Managerial Reform

A second group of reformers, far from ignoring organization, makes it central. The perspective is managerial; the FCC's major weakness is that the bureaucracy is not controlled by the political executives. Even an advocate like Nicholas Johnson argues, after experience on the Commission, that the commissioners are at the mercy of their agenda and staff.[171]

The answers are to be found in public administration orthodoxy. The formal principles of bureaucracy must become fact as well as norm. The commissioners must communicate their will to staff more clearly through more specific delegations of authority, and they must devise a management information system to see that orders are carried out.[172] Better communications channels up and down the hierarchy, in other words, will overcome the control and information blockages that work to separate subordinates from their managers.

More than this, some reformers have questioned the basic principles of the Commission form. They say that the plural executive inherently prevents decisive management and should be replaced with a single executive atop each agency. Further, the commissioners should be made accountable to the President like other agencies, since their independence (and, in practice, considerable subservience to Congress) prevents the President from developing and implementing consistent policy.

A series of official studies of the regulatory commissions has proposed to (1) strengthen Presidential authority over the commissions; (2) absorb them into the executive departments, perhaps with continuing independence for adjudication functions; (3) strengthen the managerial role of the Chairman and Executive Director within each commission; and (4) strengthen policy planning and evaluation capability. The Landis Report of 1961 argued forcefully for the first change, recommending special offices within the White House to coordinate regulatory policy.[173] The Brownlow Report, 1937, and the Ash Council Report, 1971, argued aggressively for the second, although Ash specifically left the FCC independent because of the political sensitivity of communications regulation.[174] The Hoover Commission studies of 1949 and 1950 gave detailed attention to the commissions' internal management.[175]

All these recommendations appear to have had some influence on the FCC. The President's role in communications policy was strengthened by the creation of the Office of Telecommunications Policy. OTP, now the National Telecommunications and Information Administration, which also constituted a new administrative authority (under the Executive), separate from adjudication (left independent in the FCC). The Chairman's managerial role has been strengthened, and the founding of the Office of Plans and Policy gives him greater planning capability. Some argue that the reduction of the Commission from seven to five will increase the managerial ability of the commissioners.

The second Hoover Commission, 1950, helped bring about a shift of much of the FCC from professional to functional organization. Offices of Accounting, Engineering, Law, etc., ceded to functional bureaus for broadcasting, common carrier, etc., leaving only the Office of Chief Engineer and a relatively small Office of General Counsel organized on a professional basis.[176] This weakened the grip of the professional castes on the organization and strengthened the hand of higher management.

But we must ask how much has changed. The criticisms of the FCC are almost unchanged from 30 years ago. The commissioners still lack the authority over their organization that managerial norms say they should have. The FCC remains slow-moving and set in its ways.[177] The same proposals for reorganization are made repeatedly. The reason may be that the suggested changes have not gone far enough. Reform, especially, has barely touched the explicit limits on supervisors' managerial authority, notably the civil service.

Two explanations are more fundamental. One is that managerial reforms are self-limiting. Most of the bureaucratic problems discussed under Model 2 are inherent in organization itself. Efforts to overcome them through reorganization encounter the same problems. Efforts to clarify communications up and down the hierarchy run into the same distortions of orders and messages they are designed to overcome.[178] Strengthening the formal position of the leadership can accomplish only so much when the real problem is control of the lower levels. Ultimately, the inherent limitations of coordination through formal rules cannot be overcome through the imposition of more formal rules. Bureaucracy is not a panacea; there are hard limits to what it can accomplish.[(179)]

The other explanation is politics. As we saw under Model 3, the main problem of bureaucratic managers may not be management at all. It is, rather, to assemble a consensus for their policies among diverse players inside and outside the organization who are *not* clearly subject to their authority. Outcomes apparently due to ineffective management may really be due to a balance of political forces which executives and administrative reformers cannot change.

2.4.4 Political Change

A third method of reform seeks to improve the FCC by changing its political environment. Unlike the other two, this approach has not had a coherent school behind it. The authors propose it as a construct to organize random suggestions here and there in the FCC's history and literature. Rather than reorganize the FCC from within, the political tactic aims to expose it to pressure to improve from without.

One variant is to mobilize the public interest groups which already play an increasing role in Commission policy making. The theory behind the idea has been developed by disillusioned reformers of social programs who have given up trying to change bureaucracy from within. If pressure from advocate groups and experts from the ouside can induce agencies to change, the argument goes, there will be less need to tinker with the "black box" of agency organization.[(180)]

Applied to the FCC, organized public pressure seeks to counter the dominant influence of the regulated industries. As discussed above, public interest goups seek to break the industries' grip on the commissioners' attention and information, and hence make politically real to them what is otherwise merely formal — their accountability to the general public.[(181)] One way to expand the groups' influence on adjudication cases would be to reimburse the expenses of public participants or groups taking part. In June 1978, the Commission issued a Notice of Inquiry into this possibility.[(182)] A way to expand their role in rule making would be to appoint their representatives to task forces or advisory bodies exploring specific issues. For example, in the summer of 1977, a special notice about the World Administrative Radio Conference proceedings was mailed to public interest groups to acquaint them with the issues under consideration and solicit their participation.

A second political tactic is to mobilize the institutions that influence the FCC within the Federal establishment, and perhaps create new ones. A standard tactic of bureaucratic politics is to create competitors for the agency one wants to pressure or weaken. To a degree, President Nixon did this when he created the Office of Telecommunications Policy. Congress did it when, in 1962, it divided control communications satellites between the FCC and the Communications Satellite Corporation. An Administration which seriously wished to reform the FCC might be well advised not to try to persuade the Commission or reorganize it, but to orchestrate pressure on it from the other executive branch players in communications policy. To some extent the 1978 proposed rewrite of the Communications Act uses this tactic by curbing the FCC's authority relative to the Corporation for Public Broadcasting and the National Telecommunications and Information Administration.

The major hope of this approach is probably the courts. We saw above that Federal judges have taken control of some aspects of communications policy away from the FCC. Some critics view this not as a threat to the agency, but as a salutary check likely to lead to improved rule making. Out of judicial case law may come the coherent policies which the FCC seems unable to formulate by itself.[183]

The limitation of political reform is that pluralism tends to allow only incremental change. To add or strengthen one group or institution in the pluralist interaction cannot produce large results when the other players remain unchanged. Hence, political solutions may be more symbolic than real.[184] It is implausible to suggest that public interest groups or the courts can transform the FCC when these forces have had only limited influence on the Commission to date.

A more basic objection is that pluralism is not subject to direction. The nature of it is to deny any one interest control over the outcome. The proposed act reflects inevitable compromise between the various interest. No reformer will be able to control the FCC political process for his own ends, however enlightened. How could he know that the public interests groups, or rival agencies, or the courts would serve the purposes supposed for them in the reform plan? The reformer cannot shape the process as if he stood apart; he can only join in it as one player among many.

2.5 INSTITUTIONAL DEVELOPMENT

Our review of FCC reform proposals has been sobering. The gains of reforms have been limited in practice. The theoretical analysis seems to indicate why. The proposals grounded in one model seem to leave out necessary perspectives drawn from the other two. The organizational and political proposals have limitations specified in their own models. Prospects for change seem *inherently* circumscribed.

The problem, however, is really that reform ideas tend to be piecemeal and short-term. They propose gimmicks which will improve the FCC's performance if applied right now. All will be well if the Chairman is an economist, has more authority to manage his staff, or can call on public interest groups to balance the industry. Such suggestions must have limited effect, since each leaves unchanged the other institutional features of the FCC.

But suppose all institutional limitations were eased at once. Suppose there were, not marginal changes, but better appointments to the Commission, better public administration, and better politics. Then changes in one dimension would not be limited by rigidity in another. If the commissioners were more capable and dedicated, they would make better rules.[(185)] If Congress gave the FCC a more specific legislative mandate, and a more responsive civil service system, the FCC would perform better as an organization.[(186)] If the public cared enough for regulation so that politicians routinely made an issue of it, regular representative democracy would balance the influence of the industry without a need for public interest groups.[(187)]

Changes on this scale would transform the nature and performance of the FCC. Such changes have seemed to be beyond the reach of deliberate reform. Even to suggest them seems utopian. And yet, from a more *historical* perspective, they have occurred. What deliberate change has not achieved has been realized partially by a process of development over time.

As a group, the commissioners probably are more able and dedicated, their staff more expert, their political environment less dominated by the industry, than they were 10, 20, or 30 years ago. Certainly, the Commission has regulated in a more activist and probably more effective manner since about 1960 than it did before. The same could probably be said of American government and politics in general. Many institutions really do perform better today than they did 50 or 100 years ago.

Why do we usually dismiss improvement of this evolutionary kind? In part because it occurs too slowly to meet the short-run personal and political needs of reformers. Moreover because it seems due more to the rising education, professionalism, and affluence of the country as a whole than to specific reform actions. Far from causing us to think better of the FCC, or any agency, such change merely elevates the standards by which we criticize.

A more basic reason is that the theories underlying our analytic models have difficulty perceiving and valuing institutional development. The theories tend to assume "economic man" or self-interested behavior on the part of people. Institutional improvement, on the contrary, requires an increase in public morality. Improvement, for instance, requires that public interest groups and the general public take an interest in regulatory issues in which they have no immediate economic interest — something inexplicable to an economic analysis.[(188)] To perform better as citizens or officials in *any* collective endeavor, people commonly need non-economic, even altruistic inducements.[(189)]

A yet more basic reason is that Americans do not view public institutions in terms of development at all. The Federal government is perceived to have been created all at once with the writing of the Constituion in 1787. It has not evolved; it was "founded" once and for all. Even the numerous social institutions created at the New Deal and since, such as Social Security, are perceived to have been legislated, not developed over time. Static political beliefs have provided consensus in a nation without many other bases of agreement. In a society where all other structures perpetually change and modernize, stability is provided by political institutions that are perceived to change hardly at all.[(190)]

The regulatory commissions have been viewed in especially static terms. Their original purpose was to take an area of policy "out of politics" and entrust it to the hands of "experts." Legitimized by powerful rationalist values, the FCC and the other commissioners have not been allowed human failings. Their deficiencies have been recognized, but viewed more as scandals to be rooted out than as mandates for long-term change. Reformers have shown limited patience, and limited determination, working with the Commission's problems over time. There is a tendency to decry evident abuses and propose urget reforms, but not for long enough or in terms practical enough to have effect.

Reformers succumb to impatience when dealing with the FCC as an *institution*. Economists flirt with the idea of abolishing much communications regulation and trying to control the industry through market competition instead.[(191)] Public interest advocates would like to subject broadcasting to direct "community" control. The assumption is that these steps would end the need for a public agency with all its problems of competence and control. Consumers, or voters, would tell the industry what to do directly through the economic or political marketplace, without the need for the FCC as an intermediary. Control would be accomplished effortlessly through the "invisible hand," ending the need to wrestle with the public hand of government.

These proposals *are* utopian. The FCC and other public agencies are necessary for practical reasons. They *must* be wrestled with. Americans too readily turn away from low-performing organizations in search of alternatives. They need to stick with the institutions until reforms are pushed through.[192] What the FCC needs more than any single change is sustained commitment from its constituency.

REFERENCES

1. Krasnow, Erwin G. and Lawrence D. Longley, *The Politics of Broadcast Regulation* (New York: St. Martin's Press, 1973), p. 73.
2. Cox, Kenneth A., "The Federal Communications Commission," *Boston College Industrial and Commercial Law Review*, Vol. 11, No. 4, May 1970, pp. 595-688.
3. Capron, William M. (ed.), *Technological Change in Regulated Industries*, Studies in the Regulation of Economic Activity (Washington, D.C.: The Brookings Institution, 1971), pp. 200, 223.
4. Allison, Graham T., *Essence of Decision: Explaining the Cuban Missile Crisis* (Boston: Little, Brown and Company, 1971).
5. *Ibid.* Chapter 1.
6. Weber, Max, *From Max Weber: Essays in Sociology*, ed. and trans. by H. H. Gerth and C. Wright Mills (New York: Oxford University Press, 1958), Chapters 4 and 8, and *The Theory of Social and Economic Organization*, trans. by A. M. Henderson and Talcott Parsons, ed. by Talcott Parsons (New York: Free Press, 1964), pp. 329-41.
7. *Communications Act of 1934*, 48 Stat. 10064 (1934), Title 47, U.S.C.; *Federal Radio Act of 1927*, 44 Stat. 1162 (1927); *Interstate Commerce Act of 1887*, 24 Stat. L 379 (1887).
8. *Communications Satellite Act of 1962*, 72 Stat. 419, 47 U.S.C. 701-744.
9. *Communications Act*, Sections 356, 214(a), and 4(k).
10. *Communications Act*, Sections 326.
11. *FCC Report on Allocations from 44 to 108 Megacycles*, Docket No. 6651, June 27, 1945, reprinted in *Broadcasting*, July 2, 1945; see Krasnow and Longley, pp. 85-95.
12. *Sixth Report and Order*, Docket Nos. 8736, 8975, 9175 and 8976, 41 FCC 148 (1952).
13. *First Report and Order and Second Notice of Inquiry*, Docket No. 18262, 35 FR 8644, June 4, 1970; *Second Report and Order*, Docket No. 18262, 46 FCC 2d 752 (1974); *Memorandum Opinion and Order*, Docket No. 18262, 51 FCC 2d 945 (1975).

14. *Report and Order,* CC Docket No. 79-318, Fed. Reg. 27655 (May 21, 1981); *Memorandum Opinion and Order on Reconsideration,* Fed. Reg. (March 1982); *Memorandum Opinion and Order on Further Reconsideration,* Fed. Reg. 32537 (July 28, 1982).
15. *Specialized Common Carriers, First Report and Order,* Docket No. 18920, 29 FCC 2d 870 (1971).
16. Strassburg, Bernard, "Case Study: FCC's Specialized Common Carrier (SCC) Decision" in *Analysis of the Regulatory Process: A Comparative Study of the Decision Making Process in the Federal Communications Commission and the Environmental Protection Agency,* Report prepared for the National Science Foundation under Grant No. APR 75-16718, unpublished (Washington, D. C.: The Urban Institute, November 30, 1977). pp. III 29-40.
17. Weber (1964), pp. 341-58.
18. *Communications Act,* Section 309 (e).
19. *Tentative Decision and Request for Further Comments,* BC Docket No. 82-345, FCC 83-377 (August 12, 1983). See also *Notice of Proposed Rule Making,* BC Docket No. 82-345, 47 Fed. Reg. 32959 (1982).
20. Strassburg, pp. III 3-5.
21. *Home Box Office, Inc. v. FCC,* No. 76-1280 (D.C. Cir. Mar. 25, 1977), *cert. denied,* 46 U.S.L. W. 3216 (U.S. Oct. 3, 1977) Dkt. Nos. 76-1841 and 1842; see *Policies and Procedures Regarding Ex Parte Communications During Informal Rulemaking Proceedings,* Order, Notice of Inquiry and Interim Policy, Gen. Docket 78-167, June 9, 1978, *Ex Parte Communications in Rulemaking Proceedings,* Report and Order, Gen. Docket No. 78-167, 78 FCC 2d 1384 (1980); *Memorandum Opinion and Order,* 78-167 (Released May 20, 1983).
22. Cox, pp. 671-74, and Strassburg, pp. III 29-34.
23. Geller, Henry, "Case Study: 1972 Cable TV Rules," in *Analysis of the Regulatory Process: A Comparative Study of the Decision Making Process in the Federal Communications Commission and the Environmental Protection Agency,* Report prepared for the National Science Foundation Under Grant No. APR 75-16718, unpublished (Washington, D.C.: The Urban Institute, November 30, 1977), p. IV 11-17.
24. *Ibid.,* pp. IV 13, 14.
25. Cole, Barry, and Mal Oettinger, *Relucant Regulators: The FCC and the Broadcast Audience* (Reading, Mass.: Addison-Wesley Publishing Co., 1978), Part III, pp. 131-241.

26. *KORD, INC.*, 31 FCC 85 (1961); see also Geller, Henry, *A Modest Proposal to Reform the Federal Communications Commission* (Washington, D.C.: The Rand Corp., 1974), pp. 15.
27. *WHDH, Inc.*, 16 FCC 2d 1 (1969).
28. *The Evening Star Broadcasting Co.*, 27 FCC 2d 316 (1971); *Chuck Stone et al. v. FCC*, 466 F.2d 316 (1972); see also Cole and Oettinger, pp. 222.
29. *Black Broadcasting Coalition of Richmond v. FCC*, 556 F. 2d 59 (1977); see also Cole and Oettinger, pp. 224-25.
30. *Office of Communicatiions of the United Church of Christ, et al. v. Federal Communications Commission*, 359 F. 2 (1966); see also Krasnow and Longley, pp. 36-41.
31. Cole and Oettinger, Part III.
32. *Policy Statement on Comparative Broadcast Hearings Involving Regular Renewal Applications*, 22 FCC 2d 424 (1970); *Citizens Communications Center v. FCC*, 447 F. 2d 1201 (1971); see also Krasnow and Longley, pp. 118-24.
33. Cole and Oettinger, pp. 138-42.
34. "FCC Directs Staff to Use Revised EEO Processing Criteria." News Release, Mimeo. No. 79238 (March 10, 1977). In 1980, the "zone of reasonableness" was made more stringent. See "EEO Processing Guidelines Changed for Broadcast Renewal Applicants," *Public Notice*, FCC 80-61 (February 13, 1980).
35. See Cole and Oettinger, p. 7; Krasnow and Longley, pp. 26-8; and Noll, Roger G., *Reforming Regulation: An Evaluation of the Ash Council Proposals*. Studies in the Regulation of Economic Activity (Washington D.C.: The Brookings Institution, 1971), p. 43.
36. Weber (1964), pp. 358-63, 386-92.
37. Brown, Les, *Television: The Business Behind the Box* (New York: Harcourt Brace Jovanovich, Inc., 1971), pp. 167-69, 255-57.
38. Johnson, Nicholas, *How to Talk Back to Your Television Set* (Boston: Little, Brown and Company, 1967).
39. Krasnow and Longley, p. 31; see also Minow, Newton M., *Equal Time: The Private Broadcaster and the Public Interest* (New York: Atheneum, 1964).
40. Brown, pp. 160-67.
41. *Children's Television Programs, Report and Policy Statement*, Docket No. 19142, 50 FCC 2d 1 (1974); *Action For Children's Television v. FCC*, 564 F. 2d 458 (1977).

42. *Citizen's Complaint Against Pacifica Foundation Station WBAI (FM), Memorandum Opinion and Order,* 56 FCC 2d 94 (1975).
43. *Writers Guild of America, West, Inc. et al. v. FCC.* The court of appeals vacated the district court judgment with instructions that the case be referred back to the FCC for resolution in administrative proceedings. See *Writers Guild of America, West, Inc. v. American Broadcasting Cos., Inc.,* 609 F. 2d 353, 358 (9th Cir. 1979) *cert. denied,* 449 U.S. 824 (1980); also *Order,* FCC 83-414 (released September 23, 1983).
44. Cole and Oettiner, pp. 113-15.
45. *Pacifica Foundation et al. v. FCC and U.S.A.,* 556 F. 2d 9 (1977); see also *FCC v. Pacifica Foundation, et al.,* No. 77-528 (S.C., July 3, 1978).
46. Weber (1958), p. 247.
47. *Prime Time Access Rule, Report and Order.* 23 FCC 2d 395 (1970); *Second Report and Order (PART III),* 50 FCC 2d 829 (1974).
48. Brown, pp. 357-60.
49. *Cable Television Report and Order,* 36 FCC 2d 141 (1972); see Geller (1977), pp. IV 5, 18, and *passim.*
50. *Notice of Proposed Rulemaking,* BC Docket No. 82-345, 47 Fed. Reg. 32959 (1982); *Tentative Decision and Request for Further Comments, BC Docket No. 82-345, FCC 83-377 (Released August 12, 1983).*
51. *Krasnow and Longley, pp. 29-31.*
52. *Allison, pp. 18-9, 22, 24, 247 and 251.*
53. Rourke, Francis E., *Bureaucracy, Politics, and Public Policy* (Boston: Little, Brown and Company, 1969), pp. 1-8, 39-61; Weber (1964), pp. 330-31, 333-34, 339: Weber (1958), pp. 90-1, 196, 198.
54. Weber (1958), pp. 82, 197, 221-24; Weber (1964), pp. 331-32, 334, 341-58.
55. Weber (1958), pp. 214.
56. Simon, Herbert A., *Administrative Behavior: A Study of Decision-Making Processes in Administrative Origanization,* Second Edition (New York: The Free Press, 1957), pp. 4-16.
57. Weber (1964), pp. 340.
58. *Administravtive Procedure and Judicial Review,* 80 Stat. 381-388, 392-393; 5 U.S.C. 551-559, 701-706; and 81 Stat. 54-56; 5 U.S.C. 552.

59. *Revision of Operating Rules for Class D Stations in the Citizens Radio Service,* Docket No. 21020; Notice of Proposed Rulemaking, 47 FCC 2d 1022 (1974); First Report and Order, 54 FCC 2d 841 (1975); Notice Inquiry and Further Notice of Proposed Rulemaking, 58 FCC 2d 928 (1976); Second Report and Order, 60 FCC 2d 762 (1976); and Memorandum Opinion and Order, 62 FCC 2d 646 (1976).
60. Weber (1958), pp. 198, 215-16; Weber (1964), p. 330; Simon, pp. 103-8, 222-26, 240-41.
61. Noll, Chapter 3; President's Advisory Council on Executive Organization (Ash Council), *A New Regulatory Framework: Report on Selected Independent Regulatory Agencies* (Washington, D.C.: Government Printing Office 1971), pp. 13-27, 115-19; James H. Landis, *Report On Regulatory Agencies to the President-Elect* (Washington, D.C.: U.S. Government Printing Office, 1960), pp. 53-4, and *passim.*
62. March, James G., and Herbert A. Simon, *Organizations* (New York: John Wiley & Sons, Inc., 1958), pp. 137-50; Simon, Chapter 5.
63. Leavit, Harold J., William R. Dill, and Henry B. Eyring, *The Organizational World* (New York: Harcourt Brace Jovanovich, Inc., 1973), Chapter 3.
64. Simon, Chapter 7; Tullock, Gordon, *The Politics of Bureaucracy* (Washington, D.C.: Public Affairs Press, 1965), Chapters 6, 11-20, 25; Downs, Anthony, *Inside Bureaucracy* (Boston: Little, Brown and Company, 1967), Chapter 11.
65. *First Report and Order,* Docket No. 21318, 67 FCC 2d (1978).
66. Cole and Oettinger, pp. 272-73.
67. Johnson, Nicholas and John Jay Dystel, "A Day in the Life: The Federal Communications Commission," *Yale Law Journal,* Vol. 82, No. 8., July 1973.
68. Simon, Chapter 8; Downs, Chapter 10.
69. Cole and Oettinger, pp. 11-12, 215-17; Krasnow and Longley, pp. 24-26.
70. *Report and Order,* 60 FCC 2d 591 (1976).
71. Geller (1977), pp. IV 17-8, 22, 25.
72. Strassburg, pp. III 36, 39 and *passim.*
73. Simon, Chapters 4-5; March and Simon, Chapters 6-7; Downs, Chapter 14-6.

74. Lindblom, Charles E., *The Intelligence of Democracy: Decision Making Through Mutual Adjustment* (New York: The Free Press, 1965), and "The Science of 'Muddling Through'," *Public Administrations Review*, Vol. 19, No. 2, Spring 1959, pp. 79-88.

75. First Report and Order and Second Notice of Inquiry, Docket No. 18262, 35 FR 8644, June 4, 1970; see Docket No. 18861 and FCC *Rules and Regulations*, Section 2.106, footnote NG63.

76. *Report and Order*, Docket No. 20693, 63 FCC 2d 378 (1976).

77. Simon, Chapter 5; March and Simon, pp. 141-50; Downs Chapter 16.

78. Cole and Oettinger, Part III.

79. Strassburg, pp. III 29; Cox, pp. 671-74, 677.

80. Loevinger, Lee, "The Sociolo 94 of Bureaucracy," Speech to IEEE International Conference on Communications, Philadelphia, Pa June 13, 1968.

81. Allison, Chapter 3

82. Geller (1977), pp. IV 10-11, 17.

83. Cole and Oettinger, pp. 261-65.

84. Network Inquiry Special Staff, *An Analysis of the Network-Affiliate Relationship in Television, Preliminary Report, (October 1979).*

85. Simon, Chapter 10; Downs, Chapters 18-19.

86. Downs, Chapters 2, 13.

87. Merton, Robert K., "Bureaucratic Structure and Personality," *Social Theory and Social Structure*, Enlarged Edition (New York: The Free Press, 1968), pp. 249-60.

88. Seidman, Harold, *Politics, Position, and Power: The Dynamics of Federal Organization*, Second Edition (New York: Oxford University Press, 1975), Chapter 5.

89. Cole and Oettinger, pp. 215-16.

90. Downs, pp. 61-3, 69-70.

91. Blau, Peter M., *The Dynamics of Bureaucracy: A Study of Interpersonal Relations in Two Government Agencies*, Revised Edition (Chicago: University of Chicago Press, 1963), and *On the Nature of Organizations* (New York: John Wiley & Sons, 1974), Chapter 2; Peter M. Blau and W. Richard Scott, *Formal Organizations: A Comparative Approach* (San Francisco: Chandler Publishing Co., 1962).

92. March and Simon, Chapters 3-4.

93. Leavitt, pp. 30-44, 128-47, 167-8; Berkley, George E., *The Administrative Revolution: Notes on the Passing of Organization Man* (Englewood Cliffs, N.J.: Prentice-Hall, Inc. 1969).

94. Tullock, pp. 160.

95. Weber (1958), pp. 88, 108-10, 200-2, 242.

96. Savas, E.S., and Sigmund G. Ginsburg, "The Civil Service: A Meritless System?" *The Public Interest,* No. 32, Summer 1973, pp. 7-85.

97. Stanley, David T., "What are Unions Doing to Merit Systems?" *Civil Service Journal,* vol. 12, No. 13, January-March 1972, pp. 10-14.

98. Weber (1964), pp. 384.

99. Merton, Robert K., "Role of the Intellectual In Public Bureaucracy." *Social Theory and Social Structure,* Enlarged Edition (New York: The Free Press, 1968), pp. 261-78; Downs, pp. 203; Berkely, pp. 66-90.

100. "FCC Issues Post-Facto Permit to FM," *Broadcasting,* May 30, 1977, pp. 39-40.

101. "Radiotelegraph Operator License Ceritificates to Bear Photographs," *Public Notice,* FCC 77-600, Mimeo No. 83072, August 31, 1977.

102. Rourke, *passim.*

103. Weber (1958), pp. 90-1; Weber (1964), pp. 333-35.

104. U.S. Congress, *Omnibus Budget Reconciliation Act,* 96 Stat. 763, PL 97-253, 97th Cong. 2d sess. (September 8, 1982).

105. Weber (1968), pp. 392-404.

106. Geller (1977), pp. IV 18-19.

107. Brown, pp. 253-54, 259-60.

108. Krasnow and Longley, pp. 112-120; *Citizens Communications Center v. FCC,* 447 F. 2d 1201 (1971).

109. Downs, Chapter 17.

110. Allison, Chapter 5.

111. Cole and Oettinger, p. 264.

112. *Program Review of the Citizens Radio Service,* FCC Staff Report, December 6, 1976.

113. Cole and Oettinger, p. 264.
114. Allison, p. 257.
115. Bernstein, Marvin H., *Regulating Business by Independent Commission* (Princeton, N.J.: Princeton University Press, 1955), p. 219; Noll, pp. 39-45; Olson, Mancur, *The Logic of Collective Action: Public Goods and the Theory of Groups* (Cambridge, Mass: Harvard University Press, 1971).
116. Cole and Oettinger, Chapter 3; Krasnow and Longley, pp. 31-5.
117. Cole and Oettinger, pp. 8-10, 30-34.
118. Noll, Roger G., Merton J. Peck and John J. McGowan, *Economic Aspects of Television Regulation.* Studies in the Regulation of Economic Activity (Washington, D.C.: The Brookings Institution, 1973), p. 123.
119. Cole and Oettinger, pp. 44-8; see *Policies and Procedures Regarding Ex Parte Communications During Informal Rulemaking Proceedings,* Order, Notice of Inquiry and Interim Policy, Gen. Docket No. 78-167, June 9, 1978; *Ex Parte Communications in Rulemaking Proceedings,* Report and Order, Gen. Docket No. 78-167, 78 FCC 2d 1384 (1980); *Memorandum Opinion and Order,* 78-167 (Released May 20, 1983).
120. U.S., Congress, House, Committee on Interstate and Foreign Commerce, Subcommittee on Oversight and Investigations, *Federal Regulations and Regulatory Reform,* Subcommittee Print, 94th Cong. 2d sess., 1976, pp. 251-53.
121. Krasnow and Longley, Chapter 5.
122. Noll (1971), pp. 23-7.
123. Cole and Oettinger, Part IV; *Action For Children's Television v. FCC,* 564 F. 2d 458 (1977).
124. Crotts, G. Gail, "The Public Information Function of the Federal Communications Commission," PH.D. dissertation, University of Illinois at Urbana-Champaign, 1974, pp. 253-58.
125. Bernstein, pp. 156, 285.
126. Krasnow and Longley, Chapter 3.
127. *Ibid.,* pp. 55-6.
128. Geller (1977), pp. IV 19-21.
129. See U.S., Congress, House, Committee on Interstate and Foreign Commerce, Subcommittee on Communications, *Option Papers,* Subcommittee Print, 95th Cong., 1st sess., May 1977; *The Communications Act of 1978,* H.R. 13015, 95th Cong., 2d sess., June 7, 1978.

130. U.S. Senate, *Bill to Amend Communications Act of 1934*, S. 55, 98th Cong. 1st sess. (January 1983); U.S. House, HR 32382 and H.R. 2370, 98th Cong., 1st sess. (March 24, 1983).
131. Krasnow and Longley, Chapter 6; *All Channel Television Receiver Act*, P.L. 87-529, 87th Cong., July 10, 1962.
132. Krasnow and Longley, Chapter 7.
133. *Ibid.*, Chapter 8.
134. Strassburg, pp. III 24-30.
135. *Consumer Communications Reform Act of 1976*, S. 530, 95th Cong., 1st sess. 1976.
136. U.S. Congress, *Budget Reconcilliation Act*, 96 Stat. 763, PL 97-253, 97th Cong., 2d sess. (September 8, 1982).
137. U.S. Congress, *Ominibus Budget Reconcilliation Act of 1981*, PL 97-35, 97th Cong., 1st sess. PL97-259, 97th Cong. 2d sess.
138. Krasnow and Longley, pp. 44-7.
139. *Ibid.*, pp. 48-50.
140. *Ibid.*, pp. 47-8; see *FCC v. National Citizens Committee For Broadcasting et al.*, No. 76-1471 (S.C. June 12, 1978).
141. Strassburg, pp. III 25-6.
142. *Writers Guild of America, West., Inc. et al. v. FCC et al.*, 423 F. Supp. 1064 (1976); The appellate court sent this decision back to the FCC for review. See ff 43.
143. *Action For Children's Television v. FCC*, 564 F. 2d. 458 (1977).
144. *United States v. Southwestern Cable Co.*, 392 U.S. 157, 178 (1968); see also *Midwest Video Corp. v. FCC*, No. 76-1496 (8th Cir. Feb. 27, 1978).
145. Lowi, Theodore J., *The End of Liberalism: Ideology, Policy, and the Crisis of Public Authority* (New York: W. W. Norton & Co., 1969).
146. Chayes, Abram, "The Role of the Judge in Public Law Litigation," *Harvard Law Review*, Vol. 89, No. 7, May 1976, pp. 1281-1316; Stewart, Richard B., "The Reformation of American Administrative Law," *Harvard Law Review*, Vol. 88, No. 8, June 1975, pp. 1669-1813.
147. *United States v. Midwest Video Corp.*, 406 U.S. 649, 667-69 (1972).
148. *Home Box Office, Inc. v. FCC*, No. 76-1280 (D.C. Cir. March 25, 1977), *cert. denied*, 46 U.S.L.W. 3216 (U.S. Oct. 3, 1977) (Dkt. Nos. 76-1841 and 1842).

149. *Midwest Video Corp. v. FCC,* No. 76-1496 (8th Cir. Feb. 27, 1978); FCC has filed an appeal.

150. *Brookhaven Cable et al. v. Robert F. Kelly et al,* Docket Nos. 77-6156, 77-6157 (2nd Cir. March 29, 1978).

151. Noll (1971), pp. 34-8, 101-02; Krasnow and Longley pp. 16-17, 57.

152. Seidman, Chapters 1 and 10.

153. President's Advisory Council on Executive Organization (Ash Council), p. 13.

154. Madison, James, *The Federalist Papers #10;* Dahl, Robert A., *Democracy in the United States: Promise and Performance,* Second Edition (Chicago, Rand McNally & Co., 1972); Dahl, Robert A., and Charles E. Lindblom, *Politics, Economics and Welfare* (New York: Harper & Row, 1963); Leavitt, Chapter 17.

155. Parks, Rolla Edward (ed.), *The Role of Analysis in Regulatory Decision Making: The Case of Cable Television* (Lexington, Mass.: D.C. Heath and Company, Lexington Books, 1973).

156. Kaufman, Herbert, *Administrative Feedback: Monitoring Subordinates' Behavior* (Washington, D.C.: The Brookings Institution, 1973), Chapter 9; Hargrove, Erwin C., *The Missing Link: The Study of the Implementation of Social Policy,* (Washington D.C.: The Urban Institute, July 1975), pp. 110-17.

157. Rourke, pp. 11-37, 63-86.

158. U.S., Congress, House, Committee on Interstate and Foreign Commerce, Subcommittee on Oversight and Investigations, *Federal Regulation and Regulatory Reform,* Subcommittee Print, 94th Cong. 2d sess., October 1976, pp. 245-77.

159. Noll (1971), Chaper 3.

160. Noll (1971), pp. 93-4, 108; Rourke, pp. 122-28.

161. *Administrative Procedure Act,* P.L. 404, 79th Cong., 2d Sess., June 11, 1946, 60 Stat. 237, 5 U.S.C. 1001-1011.

162. Noll, pp. 103-5.

163. President's Advisory Council on Executive Organization (Ash Council), pp. 21-2, 53-5.

164. Geller (1974), pp. 48-9; *The Communications Act of 1978,* Sec. 236 (a)(1).

165. Noll (1971), pp. 94-5.

166. Webber (1964), pp. 324.
167. Noll (1971), pp. 2, 4-5, 13-14, 97-8.
168. Leibenstein, Harvey, "Allocative Efficiency vs. 'X-Efficiency', *American Economic Review*, Vol. 56, No. 3, June 1966, pp. 392-415.
169. Noll (1971), pp. 1-2, 79-80; President's Advisory Council on Executive Organization (Ash Council), pp. 20-2, 36-40, 48-52.
170. Weber (1958), Chapter 11.
171. Johnson and Dystel, *passim.*
172. *Ibid.*, pp. 1633-34.
173. Landis, *passim.*
174. The President's Committee on Administrative Management (Brownlow Committee), *Administrative Management in the Government of the United States* (Washington, D.C.: U.S. Government Printing Office, January 1937), Senate Document No. 8, 75th Cong. 1st sess.; President's Advisory Council on Executive Organization (Ash Council).
175. Commission on Organization of the Executive Branch of the Government (Hoover Commission), *Task Force Report on Regulatory Commissions*, Appendix N (Washington, D.C.: U.S. Government Printing Office, January 1949); Pritchett, C. Herman, "The Regulatory Commissions Revisited," *American Political Science Review*, Vol. 43, October 1949, pp. 978-89.
176. *Federal Communications Commission Annual Report to Congress* (Washington, D.C.: Government Printing Office, 1950).
177. Minow, pp. 290-306.
178. Downs, pp. 118-27; Tullock, pp. 217-20.
179. Wilson, James Q., "The Bureaucracy Problem," *The Public Interest*, No. 6, Winter 1967, pp. 3-9.
180. Levine, Robert A., *Public Planning: Failure and Redirection* (New York: Basic Books, Inc., 1972), pp. 82-102, 174-75; Murphy, Jerome T., *State Education Agencies and Discretionary Funds: Grease the Squeaky Wheel* (Lexington, Mass.: D.C. Heath and Co., Lexington Books, 1974), pp. 149-56; Kaufman, pp. 74-9.
181. Johnson, *passim.*
182. *Reimbursement of Expenses For Participation in FCC Proceedings*, Notice of Inquiry, Gen. Docket No. 78-205, FCC No. 78-478 (June 30, 1978);

Notice of Proposed Rulemaking, FCC 79-852 (Released January 8, 1980).

183. Noll (1971), pp. 103-5; Johnson and Dystel, p. 16334.

184. Lowi, Theodore J., *The Politics of Disorder* (New York: Basic Books, 1971), Chatper 2.

185. Geller (1974), *passim.*

186. Lowi (1969), *passim.*

187. Lowi (1971), *passim.*

188. Noll (1971), *passim.*

189. Olson *passim.*

190. Hartz, Louis *The Founding of New Societies* (New York: Harcourt, Brace & World, 1964), Chapters 1-4.

191. Noll (1971), *passim.*

192. Hirschman, Albert O., *Exit, Voice, and Loyalty: Responses to Decline in Firms, Organizations, and States* (Cambridge, Mass.: Harvard University Press, 1970).

chapter 3

FCC REGULATION OF LAND MOBILE RADIO — A CASE HISTORY

Dale N. Hatfield
Dale N. Hatfield Associates
Boulder, Colorado

3.0 INTRODUCTION

Radio is the only practical way of communicating with vehicles that move about on the land, on the sea, in the air, or through outer space. Communication with ships at sea was one of the earliest applications of radio, and lives were saved through its use when the art was still in its infancy. There are always alternatives for communicating between two fixed points on land (e.g., by wirelines or underseas cable). Thus, land mobile, aeronautical mobile, maritime mobile, and, more recently, space, communications have always been regarded as being among the highest priority uses of the radio spectrum. This chapter will concentrate solely on *land* mobile radio in the United States and its regulation by the Federal Communications Commission (FCC). Land mobile radio includes communication through portable radio units as well as through mobile units permanently installed in vehicles. Only business and personal type communications will be considered, not communications for entertainment or as a hobby.

FCC regulation of land mobile radio is a particularly interesting case history for a number of reasons. First, it has involved a series of extremely difficult and far-reaching decisions in a fundamental area of FCC responsibility — the allocation of the scarce radio spectrum among competing users. Second, it provides important examples of the profound influence of FCC actions on communications industry structure and the types, quality, quantity, and costs of services ultimately provided to the public. Third, and finally, the delays in the introduction of new technologies and the uncertainties produced by FCC involvement in land mobile radio are often cited as reasons for proposing basic changes in the present regulatory scheme.

3.1 DEFINITIONS

There are currently three basic types of land mobile radio services: one-way signaling (paging), dispatch, and mobile telephone.

The first, one-way signaling or paging, uses a radio signal to merely alert or instruct the user to do something. The user (an office machine repairman or doctor, for example) carries a small receiver which is automatically actuated when a message is directed to it. There are three types of paging systems: tone-only, tone-voice, and digital display. In the tone-only system, the receiver emits a tone and the user takes some predetermined action such as calling his or her office or telephone answering service. In the tone-voice system, the tone is followed by a short voice message which allows alternative instructions; e.g., call a specific number. Tone-only is cheaper than tone-voice because less "air-time" is used and the equipment can be somewhat less complex. In the digital display system, a short numeric message (such as the telephone number to be called) is delivered to the receiver and displayed on a small readout device. Recent systems of this type are capable of storing and displaying much longer messages consisting of alphanumeric characters.

Dispatch communications allow two-way communications between a base station and mobile units or between mobile units without access to the regular, switched public message telephone system. Dispatch communications are used to coordinate and control a fleet of mobiles, e.g., police cars, taxis, or cement trucks. The mobile user can normally talk only to the dispatcher or another mobile; they cannot dial a telephone in the regular landline network. Messages on a dispatch system are typically short, usually one minute or less.

Mobile telephone services allow the mobile user to receive or place regular calls through the exchange (local) and message toll (long distance) facilities of the regular landline telephone system. The service can be exactly equivalent to regular landline service except that the telephone can be mounted in a vehicle or perhaps carried in a briefcase. Messages on a mobile telephone system are typically of longer duration.

These three types of services can be provided on a private or common carrier basis and there are also shared systems. In a private system, the user — a taxicab company for example — is independently licensed and owns (or leases) the base station and mobile equipment for his own exclusive use. In a cooperatively shared system, several users may band together and share the use and cost of a system. In a common carrier system, the service is provided for hire by the common carrier who owns the base station equipment and other facilities. The subscriber can own his own mobile equipment or lease it from the carrier.

3.2 EARLY TECHNOLOGICAL DEVELOPMENTS

In a number of articles,[1, 2, 3, 4] the history of the technological development of land mobile radio has been traced, and hence it will only be summarized here. One of the earliest uses of land mobile radio was by the Detroit Police Department who began experimenting with a one-way (i.e., base-to-car) system in early 1921. By the late 1920's they had achieved widely publicized success, and by 1930 there were police radio stations in 29 cities. In 1932, the first license for a mobile transmitter in a vehicle was issued. This permitted two-way communications. Growth from this point was rapid, and by the onset of World War II several other services besides police radio were in existence. These were private dispatch systems as described above; that is, communications occurred only between the dispatcher and the associated vehicles, and no interconnection with the conventional telephone network was provided.

In the late 1930's and early 1940's, the first Very High Frequency (VHF) band came into use and Frequency Modulation (FM) was introduced. These specific advances and the extensive research and development to support military requirements for mobile communications during the war years brought about considerable improvements in performance and greatly stimulated further growth of land mobile radio.

3.3 REGULATORY AND FURTHER TECHNOLOGICAL DEVELOPMENTS

Problems of channel congestion caused by the rapid growth of land mobile communications plagued the industry and the regulators from the beginning. In the early 1930's, the Federal Radio Commission, the predecessor of the FCC, allocated frequencies just above the broadcast band for police use, and the FCC allocated the VHF channels, referred to above, soon after it was formed in 1934. Because of the rapid growth, the FCC formally recognized several categories of mobile radio services in the general frequency allocation proceeding (Docket 6651) of 1945. Late in this proceeding, the Bell System requested the allocation of exclusive channels for common carrier mobile telephone service. The FCC then initiated a rulemaking proceeding to address the entire question of frequency allocations for land mobile radio. This was entitled the General Mobile Radio Proceeding (Docket 8658) and the Commission reached its decision in 1949. During the period from 1945 to 1949, the Bell System and the independent telephone companies started plans to provide mobile telephone service throughout the nation. The first urban mobile telephone system was established in 1946 in St. Louis, Missouri, on an experimental basis.

The growth in private specialized uses was dramatic during this period, and, in the decision in Docket 8658, the Commission recognized a number of specific subcategories of service including Police Radio, Land Transportation, Automobile Emegency, and several others. At the same time, the Commission allocated channels for mobile telephone service. Significantly, it provided separate sets of frequencies for the landline (wireline) common carriers (WCCs) and miscellaneous (radio) common carriers (RCCs), thus consciously providing for competition in the provision of public mobile telephone service.

The next major regulatory decision affecting land mobile radio came in 1958 under FCC Docket 11991. This proceeding established the business radio service, a category that yielded a substantially greater number of eligible licenses. This service rapidly outgrew the others in terms of numbers of licensed transmitters in service. Another major milestone for the RCCs occurred in 1961 when they were able to negotiate an interconnection agreement with the Bell System. Up until that time they were unable to effectively compete with the WCCs because they could not interconnect their base station facilities with the public message landline network. This meant that an operator had to relay each message between a mobile and a wireline telephone. In the Bell mobile telephone system of the era, each call was handled manually by an operator, but once connected, the mobile could carry out essentially a normal telephone conversation with any landline telephone. Dial mobile operation was introduced much earlier than this, but did not come into wide use until the advent of the Improved Mobile Telephone Systems (IMTS) in the mid-1960's.

Paging was another major growth area. This type of one-way service to vehicles was offered by the Bell System on an experimental basis as early as 1946. In 1960, small portable receivers were developed and extension of the service to these convenient types of units added greatly to the appeal. In Docket 16776, the FCC reallocated channels for private paging systems in the Industrial Radio Service. The RCCs in particular exploited the common carrier paging market and, recognizing the greatly increased demand, the FCC in 1968 (Docket 16778) allocated two new channels to the RCCs and two new channels to the WCCs for exclusive use in paging systems.

By the mid-1960's, the continuing rapid growth of land mobile again produced congestion on the channels. This crowding was especially severe in large urban areas. The problem was addressed in reports by a number of groups including the President's Task Force on Communications Policy,[5] the Advisory Committee for the Land Mobile Radio Services (ACLMRS),[6] and the Joint Technical Advisory Committee.[7] In 1967, the Commission established a Land Mobile Relief Committee to study various options. As a result of these studies, the Commission took two steps in the late 1960's to alleviate the situation. On one hand, they established a Spectrum Management Task Force to design a system to effectively manage the existing land mobile frequency resources (Docket 19150).[8] On the other hand, the Commission moved to allocate additional frequencies for land mobile use in Dockets 18261 and 18262.

In Docket 18261, the Commission proposed in 1968 to provide for almost immediate relief by reallocating to the land mobile services the lower seven UHF television channels (i.e., channels 14-20) in the largest urban areas. These seven channels, which lie in the frequency range of 470 to 512 MHz, are immediately adjacent to an existing land mobile radio allocation (450 to 470 MHz). This proximity to an existing allocation greatly facilitated equipment availability.

Broadcasters strongly opposed the proposal, and sought to thwart it entirely or to sharply limit its scope. This confrontation between broadcasters and land mobile radio interests has been a continual one and persists to this day. It has been compounded by the rapid growth of land mobile radio from the early police systems serving a few hundred mobiles on a one-way basis to a market where several million mobiles are being served. Faced with this growth, and acknowledging both the importance of the land mobile service and that further technological improvements could do little to improve the utilization of existing allocations, the Commission finally voted to provide short-term relief by geographically reallocating two of the lower seven UHF television channels in each of ten major urban areas (Boston, Chicago, Cleveland, Detroit, Los Angeles, New York, Philadelphia, Pittsburgh, San Francisco, and Washington, D.C.). This decision was reached in June of 1970.

During the time it proposed the limited geographic reallocation in Docket 18261, the Commission also proposed in another proceeding to reallocate the uppermost fourteen UHF television channels (i.e., channels 70-83) to the land mobile service. Because of the importance of this proceeding (Docket 18262) to the future of land mobile radio, it will be described in more detail in the following section. Before turning to that description, however, it may be useful to briefly review the frequency allocations for land mobile radio as they stood prior to the reallocations of additional spectrum in Dockets 18261 and 18262.

Private land mobile radio systems are regulated under Part 90 of the FCC Rules and Regulations. There are four broad categories of private land mobile radio services defined in the rules: Public Safety, Special Emergency, Industrial, and Land Transportation. These categories are, in turn, divided into subcategories as shown in Table 3.1.

Table 3.1

Private Land Mobile Radio Services

Public Safety Radio Services

Fire
Forestry Conservation
Highway Maintenance
Local Governemtn
Police
Special Emergency
State Guard

Special Emergency Radio Services

Industrial Radio Services

Business
Forest Products

Industrial Radiolocation
Manufacturers
Motion Picture
Petroleum
Power
Relay Press
Special Industrial
Telephone Maintenance

Land Transportation Radio Services

Automobile Emergency
Motor Carrier
Railroad
Taxicab

Before the reallocation of Dockets 18261 and 18262, the Public Safety, Industrial, and Land Transportation categories of private land mobile services had access to radio channels in the 30-50 MHz, 150 MHz and 450 MHz regions of the spectrum. These regions are referred to in the land mobile community as the low band, high band, and UHF band, respectively. The actual amount of spectrum allocated to the private land mobile services in these three bands is summarized in Table 3.2.[9]

Table 3.2

Pre-Docket 18261/18262 Spectrum Allocations to Private LMR Services

	Spectrum in Megahertz (MHz)			
Band	Public Safety	Industrial	Transportation	Total
Low Band	7.160	5.880	1.420	14.460
High Band	4.110	2.295	2.610	9.015
UHF Band	3.375	10.475	1.750	15.600
	14.645	18.650	5.780	39.075

In addition to the approximately 39 MHz of spectrum allocated to private systems, another 3 MHz of spectrum spread among the same bands was (and is) allocated to the WCCs and RCCs for the provisions of radio telephone and radio paging service. The spectrum allocated to this service, which is formally known as the Domestic Public Land Mobile Radio Service, is divided evenly between the two types of common carriers.

In the early 1970's, not long after the initiation of Docket 18262, Motorola estimated the total land mobile radio market as shown in the following table:[10]

Table 3.3

Relative Shares of the LMR Market
(Based on Figures from the FCC Annual Report for 1971)

	Mobiles in Service	Percent
Private Dispatch	2,537,996	94.2
Common Carrier —		
Bell System	66,372	2.2
Non-Bell WCC	11,130	.4
RCCs	84,025	3.1
Total	2,695,523	100.0

Growth of the land mobile radio market is estimated to be over 10 percent per year except for radio paging which is somewhat higher. Today, it is estimated that there are over eight million mobile units in service.

3.4 DOCKET 18262: REALLOCATION OF SPECTRUM SPACE NEAR 900 MHz FOR LAND MOBILE SERVICES

3.4.1 Background

As noted in section 3.3, Docket 18262 stemmed from a number of studies of land mobile radio conducted in the mid-1960's. The most direct impetus, however, came from the release on March 20, 1968, of the previously referenced report of the Land Mobile Frequency Relief Committee. The report consisted of three FCC staff studies, the last of which addressed the problems and costs associated with reallocating all or a part of the 806-960 MHz band to the land mobile radio services. This staff study was the genesis of Docket 18262. The remainder of this section is devoted to (1) a description of the issues that were raised as a result of the subsequent reallocation, and (2) a review, in chronological order, of the Commission's actions in resolving these issues.

3.4.2 Major Issues

Land Mobile Radio Versus Competing Uses

The threshold question in Docket 18262 was whether or not spectrum in the vicinity of 900 MHz should be reallocated to land mobile radio. The radio spectrum is a scarce natural resource; unlike most other natural resources such as oil or coal, it is not depleted by use. However, if two or more users attempt to use the same frequency at the same time and in the same vicinity, harmful interference can result. Since the radio spectrum is limited in extent, and since various portions in certain applications have more valuable characteristics than others, society must apportion spectrum among competing users in some coordinated fashion to avoid chaotic interference. Under the Communications Act of 1934, the FCC had the responsibility for allocating spectrum for all non-federal government users. This responsibility is described in general in Chapter 2 of this book.

In Docket 18262, the competing service was predominantly television broadcasting, since the proposal eventually called for a direct reallocation of UHF television channels 70 through 83, a total of 84 MHz of spectrum. Thus the Commission was faced with the fundamental issue of whether the public interest would be better served by reallocating the spectrum to land mobile radio or preserving it for UHF television.

The competition between land mobile radio and broadcast interests over frequency space in the UHF portion of the spectrum began before the initiation of Docket 18262. As early as 1949 AT&T proposed a high capacity, common carrier mobile radio system for operation in the 470-500 MHz band. In a July 11, 1951 decision in Docket 8976, the Commission allocated the spectrum to television broadcasting instead. In deciding the needs of television were paramount in the 470-500 MHz band, the Commission stated that: "in arriving at this conclusion we are forced to resolve a conflict between two socially valuable services for the precious spectrum space involved. We find that needs of each of the two services are compelling."[(11)]

The conflict appeared again in Docket 11997, a proceeding initiated by the Commission in April of 1957. This proceeding dealt with the broad issues of non-federal government spectrum allocations in the entire 25 to 890 MHz band, and AT&T once again indicated the need for a substantial amount of spectrum for an efficient, high-capacity domestic public land mobile service. Although the Commission acknowledged the importance of land mobile service, in 1964 it denied the request and reaffirmed its stance that all 70 UHF television channels (i.e., channels 14 to 83) were necessary to insure a competitive and effective system of television broadcasting.

As an alternative to encroaching on the UHF channels, the Commission sought technical solutions short of outright reallocation. These proved insufficient, and the Commission was caught between a burgeoning demand for land mobile service on the one hand and a disappointing growth of UHF television broadcasting on the other.

The disappointing growth in UHF television was originally attributed to a number of factors including (a) the Commission's decision to allow both VHF and UHF stations in the same markets, (b) the lack in many households of television sets with UHF reception capability, (c) significant technical limitations of UHF relative to VHF, and (d) the frequent lack of network affiliation or other access to attractive programming by UHF stations. The Commission tried a policy of "selective deintermixture" and the All Channel Receiver Act was passed. The latter required that all television sets manufactured after 1964 be equipped to receive both VHF and UHF channels. Even with these and other policies favoring UHF development, growth continued to be slow and a large percentage of UHF channels remained unused. This lack of use, coupled with the large numbers of UHF channels that were unused because of television receiver "taboos," presented an inviting target for the advocates of additional spectrum for land mobile radio. At the time Docket 18262 was initiated, UHF channels 70-83 were used almost exclusively for unattended translator stations* rather than by regular broadcast stations.

Groups associated with land mobile radio interests pointed to this apparent underutilization and advocated reallocations. Broadcasting groups, including educational broadcasters, argued that the congestion on existing land mobile channels was the result of poor frequency management techniques rather than lack of spectrum, and that reallocations could be avoided by improvements in these methods. They argued that the Commission's policies toward UHF broadcasting were beginning to bear fruit and that eventually all 70 channels would be needed to provide a sufficient number of outlets in urban areas and fulfill educational broadcasting needs.

Additional spectrum in the vicinity of 900 MHz for land mobile use was also contemplated through reducing the amount of spectrum allocated to the Industrial, Scientific, and Medical (ISM) Service. This band near 900 MHz was used primarily for microwave ovens. Reducing the size of the ISM band increases the costs of the microwave ovens. The Commission also proposed reducing the amount of spectrum allocated to broadcasters for studio-to-transmitter links (STLs) in the vicinity of 900 MHz.

In summary, then the threshold question in Docket 18262 was dominated by the UHF television broadcasting versus the land mobile spectrum allocation issue. Other issues centered around the STL and ISM band reallocations.

---

*Translators are used to extend the coverage area of a broadcast station. A translator station receives a signal from a distant television station and retransmits it on another channel. When the translator is located at an advantageous place (on a hilltop, for example), viewers who cannot pick up the station directly can receive it via the translator.

Types and Sizes of Suballocations

Following the resolution of the threshold reallocation issue just described, the next major issue centered around how the new allocation should be subdivided among the various land mobile services: domestic public (common carrier), public safety, industrial, and land transportation. The Commission simplified the debate by lumping the latter three categories under the general heading of private systems. Essentially, the competition for spectrum within the new allocation was between the domestic public land mobile radio service offered by the WCCs, such as AT&T, and the private land mobile systems. In addition, the RCCs also sought spectrum for public offerings. The resulting intra-LMR service rivalry was in many ways as intense as the inter-service rivalry between UHF television broadcasting and land mobile radio. This rivalry was further intensified when the Commission provided for the licensing of "specialized mobile radio (SMR) systems," a concept which will be described in a later section. The Bell System argued that their high-capacity mobile telecommunications systems (HCMTS) required substantial amounts of spectrum to achieve economic and spectral efficiency and to accomodate future growth. Proponents of private systems argued that the common carrier growth forecasts were overly optimistic, that growth had been and would continue to be in private conventional and shared dispatch systems, and that Bell's HCMTS was not necessarily the most efficient way of handling dispatch communications.

The Commission changed the proposed suballocations from time to time during the course of the proceeding. These changes will also be described in a later section.

Technology Choices

In addition to the private versus common carrier service dichotomy and the disputes over the amount of spectrum to be allocated to them, the Commission was also faced with a choice among competing technologies for providing LMR services at 900 MHz. Three such classes of systems emerged during the proceeding: conventional, trunked, and cellular.

A private, single-channel system used for dispatch service is perhaps the simplest conventional system to visualize. It consists of a base station transmitter-receiver combination with minor accessories, an antenna tower and antenna, and the individual mobile transceivers. The base station equipment is frequently located on the premises of the business or public agency, but it may be located remotely at an advantageous location for good coverage. The channel may be shared on an informal basis with a number of similar systems in the same service.

Increased geographic coverage, especially for mobile-to-mobile or portable-to-portable communication, can be accomplished through the use of a repeater located on a very high tower, building, or mountain top. A repeater receives a mobile (or base station) signal on one frequency and retransmits it on a second frequency. Because of its advantageous location, the repeater can normally receive and transmit over a large area. Thus, two low-powered mobiles or portables which are only a few miles apart may not be able to communicate directly with each other, but, because they are both within line-of-sight of the repeater, they are able to communicate through it. It also has a cost advantage if several licensees share a repeater because for equivalent coverage each licensee would otherwise need his or her own expensive base station on a high building or mountain top. With a community repeater of this type, the individual licensees can use simple, low power transmitting equipment at their base stations. Frequently, the repeater itself is operated by an equipment manufacturer or other entrepreneur who receives payments from the licensees.

Conventional systems can provide access to more than one channel. They can be distinguished from multi-channel trunked systems described below because the access is provided manually rather than under computer control.

A conventional common carrier mobile telephone system is similar in principle to the non-repeater system described above. Added base station equipment would be required, principally in terms of added control equipment and facilities for interconnecting with the landline network. The mobile units cost significantly more due to added control, dialing, and duty cycle requirements. The second class of system considered by the Commission was the multi-channel trunked system. Both Motorola and General Electric proposed such systems for use in the new band at 900 MHz.

To fully appreciate the advantages of these systems, it is instructive to consider one of the technical objections to conventional single-channel systems: low spectral efficiency. This inefficiency can be visualized by considering a collection of channels each occupied by one or more conventional single-channel systems in the same or different service. Spectral inefficiency can result when some frequencies are heavily used and others are only lightly used in a particular area. This can occur, for example, in a major urban area where forestry frequencies are lightly used while taxicab frequencies are over-crowded. In a long-term sense, this can be relieved to a certain extent by managing frequencies on a regional rather than a national basis. Even then, during a given hour there may be certain channels that are being heavily used and others which are only being lightly used or perhaps not being used at all. It is obvious that pooling a group of channels together and giving the users access to all channels will give them better service, or, for the same quality of service, it will enable a greater volume of traffic to be handled. This is known as trunking, and the resulting improvements can be easily predicted.

In the systems proposed for the new band, access to a large number of channels (say 5 to 20) is controlled by a computer which gives the user a channel for the duration of his message, or, if all channels are busy, places him in a queue stored in the machine. Such systems are essentially precluded from existing bands because of assignment regulations and lack of sufficient channels to fully exploit trunking. There is extensive use of trunking in the common carrier supplied mobile telephone service, but only a few channels are available.

The cellular system is the third class of system considered by the Commission. It was proposed by AT&T and, subsequently, by others. The availability of a wide block of spectrum and the propagation characteristics at 900 MHz made cellular systems ideal for the new band. The cellular system gets its name from the concept of dividing a geographic area into a series of hexagonal cells. The hexagonal cell was chosen because its shape roughly approximates the circular coverage of a base station, and because when they are fitted together they completely cover the area. Base station transmitters and receivers are placed in each cell and connected by wirelines to the central switching computer and from there the into regular telephone network. The base station transceivers provide coverage for the particular cell, but naturally a signal strong enough to provide good service at the edge of the cell will also enter the immediately adjacent cells. This means that the frequencies used in one cell cannot be reused in these adjacent cells, but they can be used again in more distant cells. AT&T proposed a seven-cell repeating pattern.

The key to the cellular concept is that in an initial system covering a market area the cells can be quite large, and as demand develops these large cells can be subdivided into smaller cells. When the cells are large, a particular channel might only be used a few times in the area; while in the final system, it might be used ten or more times. Thus, the system becomes more and more spectrally efficient; i.e., more and more users are accomodated in the same amount of spectrum. In the 900 MHz band, there is a sufficient number of channels available to give each cell an adequate number to achieve good trunking efficiency. Thus, the overall spectral efficiency of the cellular system comes from both trunking and extensive geographic reuse of channels. In a conventional system, reuse of a channel may be precluded over an entire metropolitan area and is spectrally inefficient in that sense.

A major technical problem in the cellular systems is locating a particular mobile and transferring a call from cell to cell as the mobile moves about. For example, when a call is placed to a mobile from the landline network, the mobile must be located in any one of perhaps 50 to 100 cells in the area. Once the conversation is initiated, the mobile may move from one cell to another, and this requires the system to be capable of tracking the mobile. The smallest cell proposed by AT&T has a radius on the order of one mile, hence during a normal length telephone conversation at freeway speeds, the mobile may transit several cells.

Industry Structure

It is clear that the Commission's decisions regarding the suballocations of the spectrum in the 900 MHz band among the different services and among different types of service providers will have a significant impact on the resulting industry structure. In the extreme, for example, the Commission might have excluded either common carrier or private systems entirely. In fact, at one point in the proceeding, the Commission concluded that only wireline common carriers should be licensed to operate high-capacity cellular systems. Similarly, its decisions regarding the number of SMR systems to be allowed in an area and the ease of entry of such systems would have a tremendous effect on the degree of competition among SMR systems and between SMR and private and common carrier systems.

In addition to these rather obvious decisions, the Commission was also faced with other issues concerning industry structure. Some parties in the proceeding feared that the wireline common carriers might use revenues from their monopoly telephone services to cross-subsidize their mobile services. They argued that such cross-subsidies could give wireline carriers an unfair advantage and could destroy desirable competition in the supply of dispatch services and equipment. To partially combat this potential problem, various parties urged the Commission to require the wireline common carriers to set up a separate corporate subsidiary to offer mobile services. The Commission was also urged not to allow the wireline common carriers to manufacture mobile (as opposed to base station) transmitting and receiving equipment. Again the fear was that the wireline common carriers might destroy competition in the provision and maintenance of LMR equipment.

The fear was also expressed that mobile equipment manufacturers (e.g., Motorola, GE, and RCA) might dominate the market for trunked type

SMR system offerings, and the Commission was urged to limit the number of such systems each manufacturer could own.

3.4.3 Chronological Review of Commission Actions

Docket 18262 was formally initiated by the Commission on July 17, 1968, with the adoption of a Notice of Inquiry and Notice of Proposed Rulemaking (NOI/NPRM).[12] The 1968 NOI/NPRM was addressed to the entire spectrum range from 806 to 960 MHz, but the Commission proposed changes in only the 806-947 MHz portion of the band. Based primarily on the work of its Land Mobile Frequency Relief Committee, the Commission proposed to reallocate a total of 115 MHz of spectrum — 40 MHz for private systems and 75 MHz for common carrier systems. This 115 MHz of spectrum was to be provided by reallocating 84 MHz from television broadcasting (UHF channels 70-83), 26 MHz by a reduction in the size of an ISM band, and 5 MHz by a 50 percent reduction in the spectrum allocated for broadcast studio to transmitter links.

The actual proposal called for the reallocation of UHF television channels 70-83 for use in the 25 largest urban areas on a co-equal basis with broadcast translators. In other words, land mobile use of the spectrum would have been confined to designated large cities.

The next Comission action in the proceeding came on May 20, 1970, when it adopted its First Report and Order and Second Notice of Inquiry.[13] In the action, the Commission firmed up its proposed allocations and made an adjustment in the placement of the ISM band. It concluded that the LMR-translator sharing was not practical; it proposed, instead, to make the allocation exclusively for the former. It proposed to accomodate displaced translators on UHF channel 69 and below. The resulting allocations looked like the following:

806-881 MHz	Common Carrier LMR
881-902 MHz	Private LMR
902-928 MHz	ISM
928-947 MHz	Private LMR

The Commission also concluded that development of the common carrier band should be limited to wireline common carriers, and it said that the needs of the radio common carriers would be accommodated in the 470-512 MHz band being treated in Docket 18261.

Most significantly, the Commission called on AT&T and others to undertake a "comprehensive study of market potentials, optimum system configurations, and equipment design looking toward the development and implementation of an effective high capacity common carrier service in the band 806-881 MHz."[14]

In the Second Memorandum Opinion and Order[15] adopted on July 28, 1971, the Commission responded to various petitions for reconsideration of parts of its May 20, 1970 order. In the relatively brief action, it deleted the restriction limiting the development of the 806-881 MHz band to wireline telephone companies. In addition, it clarified the earlier order by stating that it also welcomed studies and proposals for spectrally efficient, high-capacity systems by private land mobile interests.

The next major Commission action was the Second Report and Order [16] adopted on May 1, 1974. In the interim between the July 1971 order and this action, the Commission received the results of a number of studies by parties responding to their earlier request. These included extensive technical and marketing studies for common carrier cellular systems submitted in December of 1971 by AT&T and Motorola. The latter also filed information on a computer controlled trunked system for private systems and, in April 1973, it submitted a further technical report on a portable telephone system. The AT&T cellular system proposal called for the use of 64 MHz of spectrum for the offering of domestic public land mobile service and for 11 MHz of spectrum for a public air-to-ground service.

The second Report and Order of May 1, 1974, was the Commission's major decision in Docket 18262. It rejected AT&T's proposal to use 11 MHz of spectrum for public air-to-ground service. Furthermore, it found that the remaining 64 MHz of spectrum for common carrier cellular systems was excessive, and it reduced it to 40 MHz. It also reinstated the wireline common carrier restriction of the development of cellular systems in the 40 MHz suballocation. At the same time, the Commission reduced from 40 to 30 MHz the amount of spectrum available to conventional and trunked systems operated on a private or shared basis. In doing so, the Commission departed from its past practice of suballocating the available spectrum to the individual land mobile radio services. Instead it merely divided the spectrum by employed technology (i.e., trunked and conventional) and allowed all the eligible groups to have access to the channels on essentially a first-come, first-served basis. In an important aspect of the decision, and one that generated substantial controversy, the Commission created the new class of service providers which eventually became known as Specialized Mobile Radio (SMR) systems. It also made the same spectrum available to the new SMR class of service providers. Thus dispatch users eligible for licensing under Parts 89, 91, and 93 (now Part 90) of the Comission's rules (i.e., the public safety, industrial, and land transportation services) would be able to receive dispatch service on a private basis, on a not-for-profit, cost-shared basis with other eligible users, or from an SMR service provider

— *a commercial firm offering common-user service on a for-profit basis.* In the regulatory scheme developed by the Commission, the SMR systems were to operate on a competitive, open entry basis with a minimum amount of regulation. The Commission found that the SMR systems would not be common carriers for regulatory purposes; and to prevent individual states from taking regulatory action which might thwart its rules and policies, the Commission asserted federal primacy in the area.

Because of the uncertainty regarding the market demand for the various types of services, and in recognition of the possibility of future technological developments, the Commission set aside 45 MHz to be held in reserve.

With regard to other industry structure questions, the Commission decided to require the wireline common carriers to establish separate subsidiaries to offer cellular mobile radio service. It decided against restricting the wireline carriers from offering dispatch service, but it did preclude them from offering a "fleet-call" capability. It also precluded the wireline common carriers from manufacturing, providing, or maintaining mobile equipment. To prevent equipment manufacturers from dominating the SMR service market, the Commission decided to limit them to owning one SMR system per market and a total of five nationwide.

On March 19, 1975, the Commission adopted a Memorandum Opinion and Order(17) dealing with certain petitions for reconsideration of portions of the decision. The Commission generally denied the petitions, except it removed the restriction that only wireline common carriers would be allowed to develop cellular systems. It also discussed at some length the rationale for the creation of the SMR license category.

The Commission terminated Docket 18262 with a final Memorandum Opinion and Order(18) dated July 16, 1975. In the brief order, the Commission disposed of further petitions for reconsideration and clarified its position on the licensing of repeater systems. It said that repeaters could be shared at 900 MHz under existing "multiple licensing" practices used in other bands.

3.5 FURTHER DEVELOPMENTS

In Docket 18262, the Commission nearly quadrupled the amount of spectrum available to the land mobile radio services, and in many respects, it established the basic framework for the evolution of the industry until the end of the century. However, the controversy over spectrum allocations and industry structure for the land mobile radio services did not end with the closing of Docket 18262. Major disputes have continued in such areas as (a) the adequacy of the reallocated spectrum to meet the future requirements of private land mobile telecommunications, (b) the further allocation of the 45 MHz that the Commission set aside as a reserve, (c) the release of private land mobile channels within the original 30 MHz of spectrum allocated to private systems, (d) the legality and specific rules and regulations governing SMR systems, and (e) the respective roles of the RCCs and WCCs in the provision of cellular services and the methods for resolving mutually exclusive applications. These disputes, many of which have not yet been fully resolved, are discussed in the following paragraphs.

3.5.1 Future Land Mobile Radio Spectrum Requirements

With the reallocation of a total of 115 MHz of spectrum in Docket 18262, the Commission clearly felt that it had solved the land mobile radio channel congestion problem. Nevertheless, the controversy did not die down, especially as it related to the spectrum needs of the private land mobile radio services. Because cellular systems can accomodate additional demand within their existing spectrum allocations by going to smaller cells and more frequency reuse, and, as will be described later, because the Commission has at least tentatively designated 20 MHz of the 45 MHz reserve for use by such systems if they encounter unexpected growth, the non-private or common carrier mobile radio providers have appeared relatively satisfied with the outcome of Docket 18262 and subsequent proceedings. They have focused primarily on protecting their potential access to the reserve frequencies rather than requesting additional spectrum.

Private land mobile radio interests, on the other hand, resumed their demand for additional allocations immediately following the landmark decision in Docket 18262. For example, in the preparations for the 1979 World Administrative Radio Conference, the Private Land Mobile Advisory committee submitted a report projecting further rapid growth in the private land mobile services. The private land mobile interests urged the government to obtain international authorization to use UHF television channels for the land mobile services in the United States. The conference authorized such use on a co-equal, primary sharing basis within the United States. It only allows the United States to reallocate additional UHF television spectrum for land mobile use if it chooses to do so.

In January of 1982, the FCC responded to the continuing concerns of the private land mobile radio community by initiating a proceeding to gather information needed to develop a strategy for meeting the future spectrum requirements of private systems.[19] The Commission received comments on the inquiry and on an interim report that had been prepared by the Planning Staff of its Private Radio Bureau. The Private Radio Bureau then released a final report by the Planning Staff entitled *Future Private Land Mobile Telecommunications Requirements.* The comments filed in the inquiry paralleled many of the arguments made a decade earlier in Docket 18262. Private land mobile radio interests emphasized the importance of mobile radio, provided estimates of future growth, and generally argued for retaining the remaining reserves for their use. They also argued for further regional or national reallocations of UHF television spectrum to the land mobile services. Broadcast interests, on the other hand, argued that future spectrum demands of private land mobile radio systems could be met from existing reserves and by the increased use of advanced technologies such as amplitude-compandored single sideband modulation (ACSB), trunking, digital transmission, and cellular type systems. They argued that reallocation of UHF television spectrum would increase the opportunities for growth of the medium and, in particular, restrict opportunities for increased programming and ownership diversity especially among minority groups.

It is not clear what actions the Commission will take in response to this inquiry. It is clear that the Commission has not finally resolved the fundamental issue of competing demands of UHF television and land mobile radio for access to this desirable portion of the spectrum.

3.5.2 Allocation of the 45 MHz Reserve

In Docket 18262, the Commission allocated 40 MHz of the available spectrum to cellular systems and 30 MHz to private systems including SMR systems. This left 45 MHz to be held in reserve. Of this 45 MHz, the Commission established a 20 MHz block that could be used to provide additional spectrum for cellular systems if growth in those systems justified it. However, this spectrum was treated as part of the total reserve.

As noted in section 3.3, paging has been one of the fastest growing parts of the land mobile radio industry. In General Docket No. 80-183, the Commission responded to this growth and released 3 MHz of the reserves for use by paging systems.[20] In particular, it set aside 1 MHz for common carrier systems, 1 MHz for private systems (including private carrier systems which operate like SMR systems to provide service on a third-party basis), and 1 MHz for advanced technology paging systems. These new allocations were made in 1982. In General Docket No. 79-18, the Commission released another 1 MHz of spectrum for multiple address systems. Such systems are used by the electric power industry, for example, to control the distribution of energy by opening and closing switches remotely. Thus, of the 45 MHz originally set aside as a reserve in Docket 18262, 4 MHz has been released for paging and multiple address systems.

The Commission is also faced with numerous additional requests for use of this spectrum. General Electric and others have strongly advocated the allocation of 8 MHz of the reserve for the creation of a Personal Radio Communications Service (PRCS). The PRCS would be a limited range system which could be interconnected with the telephone network at the user's base station. This Citizens Band-like service would be intended to provide very low cost personal and business communications to the general public. Generally speaking, its range and quality would be less than that offered on cellular, SMR, or conventional private systems. In March of 1983, the FCC issued a Notice of Proposed Rulemaking to deal with the PRCS.[21] In General Docket No. 83-30, the Commission has proposed to allocate 4 MHz of the reserve for the creation of air-to-ground telephone service,[22] and in General Docket No. 82-243, the Commission is investigating the possibility of allocating 6 MHz of the reserve for Government and non-Government fixed service operations.[23] In addition, the Commission has received proposals from the National Aeronautics and Space Administration (NASA) and private companies to provide portable and mobile communications services to remote areas via communications satellites. Although these recent proposals have not reached the rulemaking stage, they would, if implemented, use substantial portions of the remaining reserves.

All of these proposals, coupled with the demands for additional spectrum for cellular and private land mobile radio systems, illustrates the difficulty and importance of the Commission's spectrum allocation responsibilities.

3.5.3 Release of Private Land Mobile Channels

In its decision in Docket 18262, the FCC allocated a total of 30 MHz of spectrum for use by private systems. Since the Commission adopted a bandwidth of 25 kHz for private systems, the new allocation provided for a total of 1,200 assignable frequencies. However, the frequencies were assigned in pairs to allow the use of repeaters, to separate low power mobile transmitters from high power base station transmitters, and to facilitate full duplex operation where needed. Thus, the 30 MHz of reallocated spectrum yielded 600 new private radio channel-pairs.

The Commission did not immediately make all 600 channels available for use. Instead, it allocated 100 channels for conventional systems, 200 for trunked systems (including SMR systems), and it held 300 channels in temporary reserve. As explained earlier, the 300 allocated channels were made available to all eligible users and were suballocated by type of technology or system employed, not by service category of the user as had been the tradition. As the Commission later explained, it "expected the 'systems' approach would provide a more effective means of assuring that spectrum was used efficiently, since under this format actual applicant demand was to determine the use of the spectrum."[24] In other words, the Commission sought to reduce the inherent inefficiencies of a rigid block allocation approach that often left certain frequencies underutilized in some localities. The Commission, at the same time, established loading standards and mileage separation rules to help assure the efficient use of the newly reallocated spectrum.

By 1978, the original 100 channels set aside for conventional use had all been assigned in major metropolitan areas, and in August of that year, the Commission released 50 additional channels for conventional use. Various adjustments in the systems approach were made in the next few years, but pressure for release of the remaining 250 channels continued. The Commission also faced pressure from various groups, particularly local government units. These groups argued that they would not be able to secure frequencies needed in the long term because the available channels were all being assigned to business users and SMR system operators who could plan and implement their systems very rapidly. They said that their limited budgets and long lead times for budget approval precluded them from acting quickly. The Commission responded to these "slow growth" users and others by reinstating a limited block allocation approach when it released the remaining 250 channels in August of 1982.[25] It allocated 70 channels to the Public Safety and Special Emergency Radio Services, 50 channels to the Industrial and Land Transportation Radio Services, 50 channels to the Business Radio Service, and 80 channels for SMR systems. The Commission recognized the disadvantages of returning to a block allocation approach and sought to minimize those disadvantages by encouraging sharing among the categories and by stressing that the blocks will eventually be dissolved.

3.5.4 SMR Systems*

As described in section 3.4.3, the Commission in Docket 18262 created a new class of service providers called Specialized Mobile Radio (SMR) systems. The Commission had two major, interrelated reasons for creating the SMR system concept. First, it wanted to encourage the development and use of innovative techniques for more efficient utilization of

*Major portions of this subsection are drawn from reference (26).

the radio spectrum resource. Specifically, it wanted to foster the introduction of trunked systems. It felt that if an entrepreneur were given exclusive use of a block of channels in an area, he or she would have an incentive to make more efficient use of the spectrum in order to increase profits. Second, because it recognized that these more specialized, spectrally efficient systems would require a larger initial investment, the Commission wanted to ensure that smaller private radio users would have access to the new technologies at a reasonable cost. It felt that competition (rather than detailed common carrier-type regulation) would stimulate innovation and also ensure that the smaller user who might not be able to afford the investment in a large sophisticated system would still be able to receive specialized services.

At that time, the Commission thought that private radio systems (including SMR systems) should be used primarily for dispatch-type communications rather than for mobile telephone-type service. Indeed, the Commission severely restricted the ability of SMR system operators to interconnect their systems with the regular switched telephone network. These restrictions effectively precluded SMR system operators from competing with cellular systems and existing RCC systems in the provision of mobile telephone services. Similarly, the restrictions against cellular system operators providing "fleet dispatch" services limited their ability to compete with conventional private and SMR systems.

The Commission not only established the SMR classification in Docket 18262, but it also provided a large enough block of channels to make possible multiple-entry and the competition it sought. At the same time the Commission decided that SMR systems would not be common carriers and, as noted above, pre-empted state regulation of their operations. These latter steps were strongly opposed by the state regulatory commissions represented by the National Association of Regulatory Utility Commissioners (NARUC) and by the National Association of Radiotelephone Systems (NARS),* the trade group representing the RCC industry. Both of these groups (NARUC and NARS) recognized that the provision of unregulated mobile radio services by SMR operators could have a significant impact on the state-regulated common carrier mobile telephone services offered by the RCCs. NARUC was concerned about this impact and the erosion of state regulatory power.

NARUC and NARS challenged the Commission in the federal courts, arguing that SMR systems were really common carriers and should be treated as such. Basically, the court asked whether SMR systems would offer their service to the public indiscriminately. First, the court found that the FCC was not going to require SMR systems to do so. Second, the

*NARS is now known as the Telocator Network of America.

court asked whether SMR systems would voluntarily provide service on a nondiscriminatory basis. Since there were no operating SMR systems at that time, the court could only speculate. Because of the specialized nature of the services to be provided, the court reasoned that the SMR systems would "negotiate and select future clients on a highly individualized basis" rather than offer the services on more or less standard terms and conditions to the public at large.[27] It stated that "a carrier will not be a common carrier where its practice is to make individualalized decisions, in particular cases, whether and on what terms to deal."[28] Thus, in effect, the court upheld the Commission's creation of SMR systems, but it did so in a way that inhibited how SMR systems could do business. In fact, the court stated that if the SMR systems began to act like common carriers, the whole matter could be revisited.

Thus in Docket 18262, the Commission was successful in creating a new class of service provider and the basic concept has since been extended to other services, e.g., paging. This new class of service provider is now referred to more generally as a private carrier. The SMR/private carrier concept took an added significance in March of 1982, when the FCC voted to greatly liberalize the rules and policies governing the interconnection of SMR systems with the public switched network.[29] That decision, in what was formally known as Docket 20846, paved the way for SMR operators to provide not only dispatch-type services, but also mobile telephone-type services as well. This action further blurred the distinction between private and common carrier systems and between dispatch and mobile telephone offerings. Also in 1982, Congress passed legislation that addressed mobile radio communications and, in particular, the distinction between private and common carrier mobile radio services. Both Telocator and NARUC had requested reconsideration of the Commission's decision in Docket 20846 and, in responding to those petitions, the Commission interpreted this legislation, known as "The Communications Amendment Act of 1982," to allow SMR operators to provide service on a non-common basis and to permit the relatively unfettered interconnection of SMR systems with the telephone network.

3.5.5 Cellular Radio Systems

As noted above, the Commission in Docket 18262 allocated a total of 40 MHz of spectrum for cellular systems. In the decision the Commission ruled that only one system using the entire 40 MHz would be licensed in each market because it felt multiple systems would be technically complex and costly. Although the Commission established certain rules to deter the WCCs from using their control over landline facilities to extend their monopoly power into this new industry, major criticism of the decision came from individual RCCs and NARS, their trade organization. NARS was critical "of the FCC's historic allocation policies favoring private radio services"[30] and the lack of their own spectrum allocation at 900 MHz. NARS appealed the portions of the decision dealing with cellular systems to the courts, but, in a decision that dealt with other appeals as well, the courts upheld the Commission's decision.[31] In affirming the decision, however, the court expressed concern about AT&T's marketpower in the mobile radio marketplace even with the protective steps adopted by the Commission.

In its final order in Docket 18262, the Commission decided only to license developmental cellular systems and to leave detailed policies and technical regulations to a further proceeding. This proved to be a very significant step because the Commission later used the further proceeding to again make major changes in its policies. Subsequent to its order in Docket 18262, the Commission authorized two developmental cellular systems, one in the Chicago metropolitan area and one in the Washington, D.C.-Baltimore area. The former was granted to Illinois Bell Telephone Company (a WCC), and the latter was granted to American Radio Telephone Service (an RCC group). Both groups constructed the developmental systems and made regular reports to the FCC on the results.

In January of 1980, the Commission issued a Notice of Inquiry and Notice of Rulemaking in CC Docket No. 79-319.[32] In this new proceeding dealing specifically with cellular radio, the FCC sought comments on a wide range of topics, including some basic policy issues that had been originally addressed in Docket 18262. Among the most important of these issues were the questions of whether more than one system should be licensed per market, what total amount of spectrum should be allocated to cellular, and how competing applications should be treated. After extensive comments from many interested parties, the Commission, in April of 1981, reached its decision in CC Docket No. 79-319.[33] In the decision, the Commission adopted a detailed regulatory framework for the operation and licensing of cellular systems. In doing so, however, it once again made major changes in policy. It divided the original 40 MHz frequency allocation for cellular into two 20 MHz blocks to allow two competing systems in each market. This was a significant departure from its earlier "one system per market" strategy. In essence, it responded to the RCCs' arguments that they should be provided with a block of spectrum of their own and to the court's and other groups' expressed concern about Bell System dominance of the new industry.

Thus, the Commission set aside one block of spectrum for the WCCs and an equal 20 MHz block for the RCCs. This did not entirely end the controversy, however, because it was argued that since there was often only one and at most a handful of WCCs in a given market, they would be able to very quickly reach a voluntary agreement on providing the service in each area. The RCCs, on the other hand, would apply in large numbers in each area and the Commission would be forced into very lengthy hearings to decide which applicant was the most highly qualified to provide the service on the single block of channels available to them. Those who opposed this approach argued that the Commission should allow WCCs and RCCs alike to apply for both blocks of frequencies or, failing that, to preclude the WCCs from entering a market before the RCCs. The Commission generally rejected these arguments and stated that the dual allocation approach "constitutes the best way of accomodating two separate aspects of the public interest standard: the antitrust component...and the need for rapid implementation of cellular service."[34]

Having reaffirmed the new dual allocation approach, the FCC invited applications from both WCCs and non-WCCs for their respective frequency blocks in the 90 largest metropolitan areas. In order to make the process more manageable, the FCC divided the 90 markets into three groups of 30 each. The FCC received nearly 200 applications for the top 30 markets in June of 1982, nearly 400 applications for markets 31-60 in November of the same year, and in March 1983, over 550 applications for markets 61-90. The Commission was literally inundated with crates of applications as interest in the new technology grew.

The first regular offering of cellular mobile telephone service on a commercial basis began in Chicago in October of 1983. The service was offered by Ameritech Mobile Communications, a subsidiary of Ameritech, the regional holding company that evolved from the Bell System operating companies in the mid-West subsequent to their divestiture from AT&T. This first offering of cellular services by a wireline carrier was rapidly followed by a non-wireline offering in the Washington, D.C.-Baltimore metropolitan area where the competing applications were able to reach an agreement to proceed, thus eliminating the need for the FCC to choose from among them. By all accounts, both systems are adding subscribers at a rapid pace, and the public is finally beginning to enjoy the benefits of this advanced technology. Nevertheless, the Commission is faced with the monumental problem of choosing from among many applicants for the non-wireline block in most cities, and while this was being written, they were in the process of deciding whether to use a lottery to award licenses in markets 31-90 and in the smaller markets as well.

3.6 FINAL OBSERVATIONS

To summarize, it is obvious that the FCC has had a dramatic impact on the development of the land mobile radio industry. One of the most obvious impacts is the long delay in completing the proceedings dealing with cellular radio technology. There was a period of about fifteen years from the original Notice of Inquiry in Docket 18262 until the first system went into regular commercial operation in 1983. The period is even longer when it is measured from the time of AT&T's early proposal for a high-capacity mobile radio system. The costs of such delays are difficult to measure, but, with the current pace of change in the electronics industry, delays of this magnitude could easily equal several technological generations.

While much of this delay can be attributed to the regulatory process, it is only fair to recognize that the issues being grappled with are very difficult and the various interest groups are strong and vocal.

Another obvious impact is the direct effect the Commission's actions have on industry structure and performance. For example, in Docket 18262 the Commission created a new class of service provider, the SMR system operator, and then adopted and refined the rules under which these entrepreneurs could do business. Another example is radio paging. By allocating a substantial amount of spectrum for this technology, the Commission stimulated competition to the extent that some states have virtually eliminated the regulation of common carrier paging services.

Finally, after reviewing the last few decades of Commission regulation of land mobile radio, one is struck by the fact that many of the same basic issues arise again and again. As noted earlier, the threshold issue of competition between television and land mobile for access to UHF spectrum has not been resolved even after fifteen years. There is still considerable debate about the respective roles of WCCs and RCCs in the provision of mobile services and over the division of regulatory jurisdiction between the FCC and state regulatory commissions. Nevertheless, the industry appears poised for vigorous growth as cellular systems are installed, SMR systems reach maturity, and increasing use is made of advanced technologies by private systems.

REFERENCES

1. The Origin and Development of Land Mobile Radio," Appendix A of the *Reply Comments of Motorola, Inc.* in FCC Docket 18262, dated July 20, 1972.
2. Noble, Daniel D., "The History of Land Mobile Radio Communications," *Proceedings of the IRE*, May 1962.
3. Tally, David, "A Prognosis of Mobile Telephone Communications," *IRE Transactions on Vehicular Communications*, August 1962.
4. Kargman, Heidi, "Land Mobile Communications: The Historical Roots," Contained in a Report by Raymond Bowers, *et al., Communications for a Mobile Society*, Cornell University, 1977.
5. *Final Report*, President's Task Force on Communications Policy, U.S. Government Printing Office, Washington, D.C., 1968.
6. *Report of the Advisory Committee for the Land Mobile Radio Services*, U.S. Government Printing Office, Washington, D.C., 1968.
7. Joint Technical Advisory Committee, *Spectrum Engineering —The Key to Progress*, The Institute of Electrical and Electronics Engineers, Inc., New York, 1968.
8. King, Diane, "Chronology of the National Spectrum Management Program," Federal Communications Commission, Office of the Chief Engineer, Report No. SMTF 76-01, August 1976.
9. Agy, Vaughn L., "A Review of Land Mobile Radio," Technical Memorandum 75-200, Office of Telecommunications, U.S. Department of Commerce, 1975.
10. *Reply Comments of Motorola, Inc.*, in FCC Docket 18262, July 20, 1972.
11. *Fourth Report of Commission and Order*, Docket 8976, July 11, 1951, Quoted in *Notice of Inquiry and Notice of Proposed Rulemaking*, Docket 18262, July 17, 1968 (14 FCC 2d 311).
12. *Notice of Inquiry and Notice of Proposed Rulemaking*, Docket 18262, July 17, 1968 (14 FCC 2d 311).
13. *First Report and Order and Second Notice of Inquiry*, May 20, 1970 (19 RR2d 1663).
14. *Ibid.*, p. 1677.
15. *Second Memorandum Opinion and Order*, Docket 18262, July 28, 1971 (31 FCC 2d 50).

16. *Second Report and Order,* Docket 18262, May 1, 1974 (46 FCC 2d 75).

17. *Memorandum Opinion and Order,* Docket 18262, March 19, 1975 (51 FCC 2d 945).

18. *Memorandum Opinion and Order,* Docket 18262, July 16, 1975 (55 FCC 2d 771).

19. *Notice of Inquiry,* PR Docket 82-10, FCC 82-8, January 26, 1982.

20. *Second Report and Order,* GEN Docket No. 80-183, May 4, 1983 (91 FCC 2d 1214).

21. *Notice of Proposed Rulemaking,* GEN Docket No. 83-26, March 23, 1983 (48 FR 12094).

22. *Notice of Proposed Rulemaking,* GEN Docket No. 83-30, March 23, 1983 (48 FR 12253).

23. *Further Notice of Proposed Rulemaking,* GEN Docket No. 82-243, March 23, 1983 (48 FR 12267).

24. *Second Report and Order,* PR Docket No. 79-191, 79-107, and 81-703, August 16, 1982 (90 FCC 2d 1281).

25. *Ibid.*

26. Hatfield, D. N., *SMR News,* FutureComm Publications, Vol. 1, No. 2, July 1983.

27. *NARUC v. FCC,* 525 F. 2nd 630 (D.C. Circuit), cert. denied 425 U.S. 992 (1976).

28. *Ibid.*

29. *Second Report and Order,* Docket 20846, August 16, 1982 (89 FCC 2d 741).

30. Excerpts of Testimony by George Perrin and Kenneth Hardman, For Independent Mobile Radio — Preferred Station Classification, *Telocator,* December 1977.

31. *NARUC v. FCC, op. cit.*

32. *Notice of Inquiry and Notice of Proposed Rulemaking,* CC Docket No. 79-319, January 18, 1980 (78 FCC 2d 984).

33. *Report and Order,* CC Docket No. 79-319, May 4, 1981 (86 FCC 2d 469).

34. *Memorandum Opinion and Order on Reconsideration,* CC Docket No. 79-318, March 3, 1982 (89 FCC 2d 58).

chapter 4

BROADCAST CONTROL AND REGULATION

Harold E. Hill
School of Journalism
University of Colorado
Boulder, Colorado

4.0 INTRODUCTION

The regulation of broadcasting has developed, and continiues to do so, by a process of interaction between the broadcast industry and society. Broadcasting has emerged as an institution in American life. It both reflects and molds the mores of society, yet each has a definite effect upon the other, shaping and developing attitudes, opinions, and values. Therefore, it is important to examine and understand the various pressures and influences that determine what we see and hear via the broadcast media. It is the purpose of this chapter to provide a preliminary overview of many of these "controls" in the hope that the reader will be encouraged, even challenged, to seek more detailed information.

4.1 DISTINCTION BETWEEN GOVERNMENTAL AND NON-GOVERNMENTAL CONTROLS

It is necessary that we understand the distinction between governmental and non-governmental controls. The first are a result of legislative or regulatory action, often supported by judicial decision. The latter are those imposed by the various elements of the society: broadcast owners, economics, audiences, pressure groups, sociological and psychological factors, technology, and numerous other factors. Contrary to what one might believe, governmental controls are rather insignificant when compared to all of the others. Interestingly enough, some government controls are the result of changing social interests and needs and thus reflect, in a sense, what may have been formerly considered non-governmental controls or influence.

During the discussion of controls in this chapter, there will be opinions expressed and criticisms raised. But no attempt will be made to "dissect" the pros and cons of the various forces involved, nor of the influence they have on broadcasting content. The primary purpose is to provide some basic information to enable the reader to make judgmental decisions on his own.

4.2 GOVERNMENTAL CONTROLS

The Federal Government has exercised varying degrees of control over broadcasting since the Wireless Ship Act of 1910, which required larger ships to have wireless radio for safety reasons. This was followed in 1912 by the Radio Act which simply stated that no one could operate a radio station unless a license was obtained from the Secretary of Commerce and Labor. However, no standards were provided for the establishment or operation of any station, nor were guidelines provided to assist the Secretary of Commerce and Labor in making a decision. In fact, there was little need for any decisions because there was no legislative provision for denying a license.

Naturally, the proliferation of stations was so rapid under these "non-rules" that the airwaves soon became intolerably crowded and interference was rampant. Subsequently, then-Secretary of Commerce and Labor, Herbert Hoover began to make specific frequency assignments. At the same time (in the early 1920s), Hoover was attempting to convince Congress that regulating legislation was needed in the rapidly expanding field of radio broadcasting.

Finally, with some 800 stations on the air and no control over frequency use, Congress passed the Radio Act of 1927 which established the Federal Radio Commission to license broadcast stations and generally oversee the industry. Seven years later when Congress determined that it was necessary to have coordinated Government regulation of all facets of electronic communication, the old Radio Act of 1927 became, practically word-for-word, the broadcast section of the Communications Act of 1934. Therefore, although there have been numerous amendments to the act of 1934, none of them has been substantive on program control, although through regulation there have been certain program guidelines or standards established. The broadcast industry has been operating under federal legislation passed more than 50 years ago. Later in the chapter, there will be a discussion of more recent efforts by Congress to alter, or even eliminate, many of the regulations which have become outdated with time.

4.2.1 Extent and Desirability

The Communications Act of 1934, as was true of the Radio Act of 1927, deals rather cursorily with program control while dealing extensively with the technical aspects of broadcasting. Congress gave the Federal Communications Commission (FCC), which was established by the act, broad authority to license stations in "the public interest, convenience, and necessity." Such general language has given the FCC, as well as the broadcasters, a considerable amount of leeway, and decisions have generally tended to follow the mores of society at the time, although lagging by a period of months or even years.

The only specific restrictions in the law deal with political broadcasts, lotteries, and obscenity. In brief, the law stipulates that if a station provides time for a political candidate, equal opportunity must be provided for all other qualified candidates for the same office. No station is required to provide time for candidates in the first place. "Equal opportunity" applies to time and day, and simply means that the station cannot put a friend on the air at 7:00 in the evening and an opponent on at 6:30 Sunday morning. The law does not attempt to define "qualified candidate" because state statutes provide the information regarding what conditions must be met for a group to be a "legally qualified political party." Finally, this section of the law applies only to candidates, not spokespersons.

Congress subsequently amended the original act to exempt interviews, bona fide news events, newscasts, and certain types of documentaries from the equal time provision. However, there are still some problems which must be faced. For example, there is an obvious advantage for an incumbent who can benefit from official conferences. This is especially true as far as the Presidency is concerned. In order to provide a modicum of equality, the networks, after covering a major Presidential address or press conference, will often make time for a rebuttal available to leading spokespersons for the opposition party.

In early 1981, the FCC denied Metromedia's application for review of the Broadcast Bureau's August 20, 1980, ruling that *Donahue* is not a bona fide news interview program that is exempt from equal opportunities provisions. Metromedia claimed that each *Donahue* program was produced in an attempt to probe newsworthy issues, not to advance the views of political candidates. The Bureau ruled that while specific episodes dealt with news-type programming, a majority of the *Donahue* programs did not focus on news events of the day.

A related problem stems from the fact that we no longer have just two parties with candidates for office, even for the Presidency, and it is virtually impossible, therefore, to broadcast such important events as Presidential debates. To date, we have only limited exceptions: in 1960, Congress enacted special legislation to exempt the Kennedy-Nixon debates from the provisions of the law, and, in 1976 and 1980, the debates were staged as bona fide news events in a maneuver designed to circumvent the true intent of the law.

In addition to the provisions of the law related to programming, the FCC has the authority to promulgate regulation designed to facilitate their task of regulation. One of the most far-reaching has been the Fairness Doctrine. Quite briefly, the Doctrine evolved from the FCC's "Mayflower Case" in 1941, when the Commission ruled that stations could not editorialize. Naturally, this created considerable concern, and, after extensive hearings, the FCC reversed itself in 1948 and ruled that stations could editorialize as long as they maintained a fair overall balance in their presentation of views. Subsequently, over the intervening years, the FCC determined that stations could present either editorials or statements by outsiders on matters of public concern, provided time was made available for spokespersons for opposing views. This was followed by a directive indicating that making time available for opposing views was not sufficient, but that stations must seek out such views. This directive was later amended to require that stations may not avoid the whole matter by simply not presenting either side, but that stations must seek out and air varying points of view about issues of concern to their communities. Also, the Commission established rather elaborate guidelines to assist the stations in determining more specifically what the local issues were that concerned the various, complex segments of their communities' populations.

However, in 1981, the Commission began the deregulation of radio, and in 1983, started consideration of the deregulation of television in much the same manner. Radio deregulation covered four principle subject areas. (84 FCC 968)

1. *Non-entertainment programming.* Guidelines were eliminated that specified the amount of time stations must devote to non-entertainment programming. The Commission retained a general obligation for stations to offer programming that is responsive to the public issues and, because radio stations tend to cater to specific audiences, the issues addressed need to concern that specialized audience rather than the entire community. All-in-all, this is quite a departure from the earlier specific restrictions of the Fairness Doctrine.

2. *Ascertainment.* The Commission eliminated the requirement of following prescribed methods of determining community issues. Deregulation provided that new applicants must file programming proposals and that stations up for renewal submit lists of issues facing the community. These issues were to be determined "by any means reasonably calculated to appraise them of the issues." (94 FCC 971)

3. *Commercial Guidelines.* The old limit of 18 minutes of commercial advertising per hour was eliminated, leaving it to the marketplace to determine the appropriate level of commercialization. The Commission said the listening audience would act as a watchdog, and that elimination of the regulation would encourage experimentation in commercial advertising.

4. *Program Logs.* Log requirements were eliminated to reduce the paperwork burden on the stations. Stations are required to maintain an annual listing of five to ten issues covered in non-entertainment programming and examples of programming broadcast concerning those issues.

Although the trade press indicated in the summer of 1983 that deregulation of television would closely follow radio deregulation and that specific regulations in each of the foregoing four areas would be eliminated and replaced with general guidelines, no such steps have been taken.

As far as lotteries are concerned, the Communication Act of 1934 explicitly forbids broadcasting information about them. The courts have held that three elements must be present to constitute a lottery: prize, consideration, and chance. It is because all three of these elements are not present that stations are able to broadcast the various contests and game shows they do. In addition, when state lotteries became popular and were instituted by several states, the courts ruled that it was legal for stations to carry commercials for such lotteries.

Obscenity, also forbidden by the law, is a more complicated matter. This is one of the examples of broadcast regulation tending to relax as the mores of the country become more lenient regarding obscenity, sex, etc. A number of years ago, stations would not have broadcast programs that contained even such words as "damn" and "hell." Now these words are relatively mild compared to some of the language we hear. During the summer of 1978, the United States Supreme Court upheld the right of the FCC to reprimand WBAI-FM, a public radio station in New York City, for airing comedian George Carlin's "Seven Dirty Words" monologue. This resulted in an immediate response from spokespersons of the broadcast industry, who claimed this was tantamount to allowing the FCC to censor programming material, which is forbidden by the Communication Act of 1934.

In reply to this reaction, then-FCC Chairman Charles Ferris said that the Commission did not consider the Supreme Court ruling a carte blanche permission to rigidly control broadcast programming, but that the Commission would continue to handle obscenity on a case-by-case basis. And, in fact, the FCC soon thereafter declined to censure public television station KQED, San Francisco, for an alleged obscenity infringement.

Thus, we see that the extent of FCC control over programming content is generally limited to the specific areas indicated in the foregoing, with the exception of that rather nebulous area, "the public interest, convenience and necessity." (The matter of "public interest" will be explored furthur in the matter of KTTL, Dodge City, Kansas, a little later in this section.) However, it is questionable whether it is desirable to have any control over program content (this does not necessarily apply to commercial content which will be discussed in section 4.2.3). Since commercial broadcasting, which constitutes by far the largest portion of broadcasting in the United States, depends for survival upon the selling of time to sponsors, and those sponsors will buy time only in those stations which appeal to a fairly sizeable audience, isn't it reasonable to assume that stations will program material which will appeal to at least a large enough segment of the audience to attract sponsors? Just as there are customers for XXX-rated motion pictures and "girlie" magazines, there will always be a segment of the audience that would like to have similar materials available on radio and television. If, in a given community, there is that sufficient interest in that type of programming to attract sponsors to make the programming economically viable, why should it not be broadcast? For those who claim there is danger to children, the only response is, "There is always the 'off button'."

The principal question is, does the government have the right, either legally or ethically, to dictate what will be available to the broadcast audience? There may be just as many people opposed to loud rock and roll as there are opposed to "dirty" broadcast material. Does that mean that we should ban rock and roll music from the airwaves? Although these may appear to be frivolous questions on the surface, they actually go right to the very heart of the concerns of the extent of the government's control over broadcast program content. Furthermore, as will be discussed later in this chapter, other forms of influence and control are perhaps much more insidious and potentially harmful.

Nevertheless, some critics of broadcasting content may find it difficult to understand why programming like that of KTTL, Dodge City, Kansas, should not be banned by someone. For some time KTTL has been broadcasting material that might be labeled racist, bigoted, and inflammatory. As Ted Koppel, host of ABC's *Nightline,* said on the May 18, 1983, broadcast, "If a station broadcasts blatantly anti-Semitic, anti-black, and anti-Catholic material, should that be protected by the First Amendment, or is it grounds for having that station owner's license revoked? That is not a theoretical problem; it's a reflection of what's happening on a little FM radio station in Dodge City, Kansas.... Some of it, as you can imagine is rather nasty." This statement was followed by a portion of a broadcast from KTTL with a Wisconsin minister, Jim Wickstrom: "'Keep thyself pure for I am pure.' Does that sound like you should have your daughters marry a bunch of boat people? Bunch of niggers? A bunch of spiritless creature creations of the sixth day?" This excerpt is from a syndicated program featuring Wickstrom, and this program and others like it broadcast by KTTL link quotes from the Scriptures with inflammatory comments. Wickstrom has said, "You can tell our own race, the Anglo-Saxons, these mysteries and truths. You can tell they're God's chosen people. And they deny it. They don't want it. They want to give it to Satan's kids, called the Jews. And then they tell you we need more money to send more people to Africa because the niggers are getting hungrier. And they're requesting fatter ones. But not too fat 'cause they can't get in the pot [laughs]."[(1)]

The station is owned by 48-year old Nellie Babbs and her husband. The station has been on the air only two years, and all of the programming is recorded, generally syndicated, except for live weather reports. According to *Nightline* correspondent, Gary Shapiro (KAKE, Wichita, Kansas), "Babbs, a supporter of the radical tax protest group, the Posse Comitatus, says she is simply exercising her first Amendment right to free speech. She says her critics have never discussed the situation with her face-to-face. In any event, there have been attempts to insure that KTTL's license will not be renewed, or will actually be revoked. Kansas Attorney General Bob Stephan has joined with the National Black Media Coalition, which has filed a petition to deny the station renewal of its license. Also, Community Service Broadcasting has filed a competing claim for the KTTL-FM frequency assignment with the FCC. As spokesperson Dodge City resident, Naomi Gunderson, told *Nightline,* 'There's certainly a large group in Dodge City who've been upset about these controversial broadcasts for quite some time. And it became apparent that the most effective means of getting these broadcasts off the air when Mrs. Babbs refused to discontinue them was to file a competing application for her license with the FCC.' "(2)

The dichotomy between minimal government control of program content, and the abuses of that right, as may be exemplified by the KTTL-FM case, presents a strong argument for thoughtful discussion of this problem by many interested parties. Does the right to express inflammatory positions extend just to the "soap box" or to all media? There appears to be little doubt that under the present regulations (and perhaps legal precedence) the FCC has the power to intervene in instances like the KTTL-FM case. As former FCC Chairman Newton Minow told *Nightline,* "In this case, I don't see it as an issue of free speech. The law is very clear: no one has a constitutional right to a broadcast license when there is more than one group competing for a license. It's up to the FCC to decide who will best serve the community in the public interest. And, clearly, in this case the other group ought to have a chance to prove that it will better serve the community than the current licensee." In response to a further question by Koppel as to whether the FCC would have the right to revoke the license if there were no competing application, Minow replied, "In my opinion, yes. Because clearly, these programs that we're talking about were not voluntarily put on for nothing. They were sold, and in some cases, as I understand it, the source of the funds was a violation of the law." In response, Mrs. Babbs indicated that the money was "in the form of contributions from various people within the area, and there were contributions made from out of state. . . . I really couldn't say who they were from because it was brought in, in the form of cash, on various occasions from various meetings that were held within the American agricultural — within the agricultural segment." Thus, we must balance our desire for free speech, meaning no program content control, with what might be deemed a threat to "the American way of life." But control of content of any media might be too large a price to pay. (It is likely the final disposition of the KTTL-FM case by the FCC won't occur until mid-1984 at the earliest. An FCC attorney indicated that the Commission considers this to be a very important case.)

4.2.2 The Federal Communications Commission

Although the FCC has been discussed to some extent in the previous section, it is important to understand more about this body which has so much to say about the broadcasting industry in this country. The Commissioners are appointed by the President, and they must be approved by the Senate. Their terms are for seven years and not more than four may be from the same political party. This commendable effort by Congress to keep politics out of the FCC has been only partially successful, because generally the Commissioners of the same party as the President tend to echo his philosophy. That the Commission does not have sufficient staff or budget to perform its functions properly is a bigger problem. The FCC, in 1983, had a staff of approximately 1,800 and a budget of approximately $68,000,000 with which to regulate some 10,000 broadcast stations, plus consider applications for new stations, respond to challenges by would-be licensees or by the general public who are not satisfied with the job being done by the present licensee, and defend many of their actions in court.

In addition to the problems above, because the Commission is small, much of the work must be done by either subcommittees of one or more Commissioners, by Hearing Examiners, who then make recommendations to the Commission, or by senior staff members. Therefore, although it is fair to say that the Commissioners are generally honest men and women attempting to do their job well, the mere enormity of their task plus the paucity of resources makes it extremely difficult. The result is often decisions which might not be in the best interests of the public, and actions which arouse the ire of the broadcasting industry. If, for example, the FCC needed to concern itself only with the electronic aspects of control, it might be able to function much more efficiently, and the public might be better served in the long run.

Considerations such as these are part of the reason the Congress has been considering, for several years, not only modification of the present Communication Act, but also the adoption of a completely new and rewritten telecommunications act. Actually, this latter step is not so much favored by the broadcast industry as is the former, primarily due to the anticipated delay in completing the process on a new bill, which will likely take years. The broadcasters appear to prefer action that will modify the present law in order to alleviate the extent of control. Of course, as previously indicated, steps in this direction have been taken by the FCC, such as major deregulation of radio and steps toward deregulating television. Certainly, if either new legislation or revised FCC regulations result in less control of program content, we will have a better opportunity to evaluate the pros and cons of governmental content control.

4.2.3 The Federal Trade Commission

Earlier, a distinction was made between the control of program content and the control of commercial content. The Federal Trade Commission (FTC) is charged by law with the responsibility for, among other things, making sure that the consumer is protected from misleading and false advertising.

It is only in recent years that the FTC has concerned itself to any great extent with broadcast advertising. The change may be due to the general toughening of the FTC regarding all advertising, labeling, etc., or it may be the result of more blatant claims on the part of the sponsors. At any rate, during the past several years, the FTC has investigated numerous advertising campaigns which it deemed not in the best interests of the public.

As a result, several consent decrees (wherein the advertiser does not admit any wrongdoing but agrees to cease and desist from further tactics of the same nature) have been entered into between the advertiser and the FTC in which the advertiser agrees not only to stop what is being done but also to admit on the air that misstatements had been made in earlier commercials. In some cases, a certain percentage of the advertising budget for the forthcoming year has been pledged to such corrective actions.

In addition, the FTC issued a decision a few years ago encouraging comparative advertising, while admonishing advertisers to honestly compare products and not to attempt to deceive the public. Subsequently, the manufacturers of Velamints sued the makers of Tic Tacs because of broadcast commercials which said that Tic Tacs have only 1 1/2 calories compared to 9 in Velamints. The suit was filed because the commercial failed to note that the caloric content reflected the differences in size of the mints. The mints contain the same number of calories by weight. The judge denied the injunction request because he said that viewers could see that the commercial compared the products on a mint-to-mint basis. The case was eventually settled out of court.[3]

Although there is little direct evidence of the effect of such FTC action on the consumer, it would appear logical to assume that at least a portion of the viewing and listening audience would be interested in the truth and would possibly alter their buying habits as a result. If this result does occur, it seems like adequate reason for continuing, perhaps even strengthening, the government's role in regulating commercial content.

4.2.4 Other Governmental Agencies

Other governmental agencies play a minor role in broadcasting control. The Federal Aviation Agency establishes limits on the heights of antennas and any proposed tower construction in the vicinity of airports and/or flight patterns must be approved by the FAA. Wage and hour laws, at both the federal and state levels, as well as other Department of Labor and HEW rules and regulations have a modest influence on broadcasting, as do certain state laws, such as those dealing with broadcast coverage of trials, the so-called "blue laws," and zoning regulations. However, none of these, with the possible exception of the FAA tower restrictions, places any real burden on the broadcaster, and certainly no greater burden than is placed on any other business enterprise by the same or similar laws and regulations.

Regulation efforts by some states have resulted in a number of law suits, and there has been editorial comment regarding the role of states in regulating broadcasting in light of the federal role. For example, one article questioned the authority of states to regulate broadcasting when the Federal Government has the ultimate power. The article was discussing the relative merits of a suit brought against a radio station in Ohio for violation of the state's obscenity laws.[4] Here again, the changes in federal control over program content will certainly have an impact upon the effectiveness and validity of state laws.

In Utah the Cable Television Programming Decency Act was passed over the Governor's veto. Similar legislation had been enacted in 1981, but Home Box Office sued the state before the law went into effect. In that action, the court held that the state's definition of pornographic or indecent material was overly broad. (HBO v. Wilkinson, 8 Med. L. Reptr. 1108 (1982)).[5]

Also related to the attempt by individual states to regulate broadcasting is Oklahoma's law that bans the broadcast of advertising for alcoholic beverages. The 10th Circuit Court of Appeals upheld the constitutionality of the Oklahoma statute, ruling that the statute was "reasonably related to the state's goals of limiting alcohol consumption and reducing the problem of alcohol abuse." The Court acknowledged implication of First Amendent rights in the case. The U.S. Supreme Court had agreed to review the constitutionality issue (in the Fall, 1983 session) and will address the question of whether Oklahoma can limit broadcast ads that originate outside the state. (Oklahoma Telecasters Assoc. v. Crisp, 699 F.2d 490 (CA 10, 1983); Capitol Cities Cable, Inc. v. Crisp, Docket No. 82-1795)[6]

The Supreme Court ruling may well have a very important impact on the entire question of the extent to which states may go in regulating an industry that is already regulated by the Federal Government.

4.2.5 Special Problems with Cable Television

Because of the constantly changing posture of the FCC toward cable television, it is not safe to go into detail about specific regulations. Suffice it to say that, concerns of broadcasters to the contrary, there appears to be a potential for greater harm than benefit if the FCC continues or institutes regulations that are too stringent upon the cable television industry. The most important consideration is to provide the best possible service to the general public with less regard for the more parochial interests of the broadcasters and the cable television operators. Naturally, due consideration needs to be given to the financial viability of these two groups, but this must be balanced against the need and desirability of providing the most extensive program content and choice to the greater number of people.

The FCC must consider several important questions: Is it in the public interest to restrict the availability of "distant signals?" Is the public best served by requiring, or not requiring, "public access?" Should the regulation of CATV be left entirely to the states (which might regulate the industry as a public utility)? Should competition be encouraged or discouraged, in the public interest?

These are only some of the questions that have been debated over the years with very few positive results. Should the FCC, or any other federal agency, have control over the cable industry, or should this be left to the states, or local governments, that might have a better realization of what the people in their particular areas need and want? Of course, some of the local governments have enacted franchise ordinances which have been so restrictive as to prevent the successful operation of the franchise. In spite of this, it seems on the surface that the local governments would have a better feel for the "pulse" of their communities, and that therefore, better public service would result. But again, this and the other questions cannot be answered with any great degree of validity because the confusion at the national level has been deleterious to realistic research and study. This type of control appears to be detrimental to the best interests of the viewing and listening public.

In 1983, there were several indications that the FCC was addressing some of these same issues, and some lightening of restrictions and requirements for cable has taken place. Also, in 1983, as a further example of reducing restrictions, the Commission indicated willingness to entertain comments on the proposal to modify the Fairness Doctrine requirements for cable.

4.2.6 General Considerations

In summary, with some admitted prejudice, it appears that there may be too much federal regulation of what we see and hear on television and radio. As will be explored in more detail near the end of this chapter, the public might be better served if serious attention were given to alternative methods. Again, it should be emphasized that some control of commercial content seems advisable, but that program content control may be ill-advised, or at least should be re-examined carefully. It would seem that the primary goal should be to serve the interests of the people who are the recipients of the vast amount of program material. These interests may well be best served by a decentralization of control. At the same time we concern ourselves with government control, we must also address ourselves to all of the other current and potential influences that help determine what is broadcast. These latter concerns, in a variety of forms, will be addressed in the following section, and it is important to keep in mind government's actual or potential control as these other factors are considered.

4.3 CONTROL BY OWNERSHIP

It would be naive to assume that ownership interests do not have a considerable influence on broadcast content, (e.g., the KTTL-FM situation discussed in 4.2.1) whether it be at a station level — perhaps with just a one-family operation — or a national network — with a myriad of stockholders and interlocking directorates and management. Obviously, these various owners have a responsibility to make a profit, as with any business venture, but is it possible that too much attention is paid to profit to the potential detriment of programming?

> ... the greatest control and restricting force on the broadcasting industry (is) GREED. Because broadcasters are a greedy brood, they seek to maxi-

> mize their profits by exploiting the underlying foundation of the commercial broadcasting business. In other words, they are constantly trying to accelerate the monetary intake they receive from the sale of air time to advertisers.[7]

Harsh words? Perhaps, but read what Fred Friendly, well-known journalist, director, and producer had to say after he resigned as head of CBS News.

> Whatever bitterness I feel over my departure is toward the system that keeps such unremitting pressure on men like Paley and Stanton that they must react more to financial pressure than to their own tastes and responsibility. Possibly if I were in their jobs, I would have behaved as they did. I would like to believe otherwise, but I must confess that in my almost two years as head of CBS News, I tempered my news judgment and tailored my conscience more than once. . . *the fact that I am not sure what I would have done in these circumstances, had I been chairman or president of CBS, perhaps tells more clearly than anything else what is so disastrous about the mercantile advertising system that controls television, and why it must be changed* (emphasis added).[8]

As we pointed out earlier, there is absolutely nothing wrong with the profit motive. According to leading economists, this is one of the factors that has made this nation great. But the almost religious dedication to profit to the exclusion of consideration of the public good has spread throughout all aspects of our society (automobile manufacturers are alleged to be more concerned for profit margin than they are for passenger safety), and examples can be cited in all areas of business. Therefore, does it seem reasonable to believe that those who control broadcast properties will forsake the profit motive in order to bring the public more meaningful programming at a fiscal loss? Thus, the owner, or his agent who makes program decisions states that he is "giving the public what it wants," a statement based not on scientific research but on the ratings, which at best indicate what members of the audience have selected from among the offerings available at that time. There is no attempt to evaluate the worth of the program to the audience. *The Beverly Hillbillies* might well outdraw the competition if the competition has even less appeal to the audience than *Hillbillies*.

The owner, or his agent, therefore makes judgments based on very little, and perhaps false, information about what the audience likes. This relates back to the matter of financial control and what type of programming will result in the greatest profit. Professors Gurevitch and Elliott report that in commercial broadcasting the shape of the beliefs and understandings of audience tastes and tolerances perceived by those in financial control of the industry play the largest part in setting the situation within which the program maker works.[9]

Another issue which must be considered is the matter of cross-ownership of various media outlets. For example, a few years ago, in a relatively small city in the Rocky Mountain-Great Plains region, a single family controlled the only television station in town, the only newspaper, the only cable system, and the leading AM and FM radio stations. In such a case (and there are others with similar implications if not quite so startling in impact) it seems reasonable to ask whether the public should expect to receive diversified media output, especially so far as news, public affairs, and comment are concerned.

In recent years, the FCC has taken a number of steps to alleviate some of the potential dangers of cross-media control. Regulations have been promulgated to require certain owners (for example the owners of the media in the aforementioned Rocky Mountain-Great Plains community) to divest themselves of certain media properties. Another regulation requires that, in major markets, the owner(s) of certain broadcasting combinations (radio-TV, AM-FM) must separate the properties at the time of sale. These and other steps taken by the Commission have tended to break-up some of the major media combinations.

Of course, in fairness it should be pointed out that cross-media ownership might be the only way some small communities could gain the benefits of both broadcast and print media, but each case needs to be examined with care to determine whether or not such operation is in the public interest. Again, it is important to avoid the danger of profit motivation overriding service to the community.

4.4 ECONOMIC CONTROLS

While the foregoing section deals with one aspect of economic control, it is important to examine other factors of economic influence if we are to have a fairly comprehensive understanding of the financial pressures exerted upon and by the broadcasting industry, and the role they play in determining what the public sees and hears.

4.4.1 Advertisers

In the introduction, it was pointed out that broadcasting "both reflects and molds the mores of society, yet each has a definite effect upon the other, shaping and developing attitudes, opinions, and values."

This statement is even more applicable to the commercials we see and hear daily. Patrick Walsh says the commercial is the avant-garde of change and reflects and accelerates societal trends more than the rest of television programming.[10] We are seemingly addicted to emulating attitudes, actions, and buying habits of those we see in television commercials. For example, when cigarette commercials were still allowed on the air, most of them showed the smoker as being romantic, manly, good looking, or whatever, in an effort to convince the would-be smoker that taking up the habit would result in the new smoker having the same desirable attributes as displayed by the person in the commercial.

We seem to have the concept that there is an average American, and it appears that it is he that advertisers and broadcasters attempt to reach. This has allowed the creation of certain images by advertisers and their agencies:

> ... there are probably five categories of easily recognizable females in today's media advertisements. We start with the bouncy-haired, acne-faced, pre-puberty adolescent, or a Pepsi generation member. We then proceed to the young, single woman who is hell-bent on marriage and trapping an adequate male. She is the consumer who is dousing herself with enticing perfumes, gargling with minty mouthwashes, brushing with sexy toothpastes, and (pardon me) getting all the confidence she can from her feminine hygiene spray. Next is the young mother who espouses the benefits of certain baby-soft shampoos and dovey detergents. She miraculously transforms into the older housewife, whose voice is reminiscent of a power saw, and whose nose is either stuck in an oven or someone else's business. Finally, we cannot forget the constipated grandmother who relates tales of laxatives and denture creams.[(11)]

There is indeed danger that we may warp the viewer's outlook, not only on life and society as a whole, but also on one's own self and values. Is it not possible for the following scenario to be played out in real life (in fact, it is probable), and result in self-shattering trauma? A rather plain, not too bright person, almost completely devoid of any likable personality traits, is generally ignored by his peers and feels generally miserable about his apparent failure as a human being. But, wait, there is hope! All one need do is use a particular deodorant, toothpaste, aftershave lotion, cologne, or shampoo, or perhaps wear certain clothing and drive a certain car, and all troubles will vanish overnight and popularity will zoom and friends will come running. So our victim follows all of the tips given by the commercials, but nothing happens. It certainly cannot be the fault of the products, because the commercials have told us they will work wonders; therefore, the saddened party must be at fault and impossible to improve. What does that do for the ego? Although the foregoing may seem to be a fantasy, it is not; there is mounting evidence of the influence broadcast commercials have on our actions. Witness the controversy over advertisements during children's viewing hours for sugar coated cereals. Manufacturers (the advertisers) claimed the product created no problem, yet health groups, Action for Children's Television, parents, and others refuted such statements, suggesting that the product is harmful to children. There are many such examples of the effect of advertising on consumer behavior, and not just on children.

If broadcast commercials do, as we have indicated, affect our general habits, values, and actions, as well as convince us to buy certain products, then it behooves the advertiser to reach as large an audience as he can. How does he do this? By buying into the most popular shows. Such "buying in" becomes very competitive, with various potential sponsors attempting to outbid one another for the privilege of having their commercials appear on the most popular programs. Commercial time for such programs, thus, becomes extremely expensive. With such a significant investment, it follows logically that the sponsor wants to make certain that the program retains its popularity, so he assumes as much control as possible over program content. Therefore, the sponsor not only controls what we see and hear during the commercials, but he also has considerable clout insofar as programming decisions are concerned, because the network or station does not want to give up the income derived from that sponsorship.

Sponsor influence, or even control, over program content is generally exercised through the advertising agencies. This control is not something new; it existed from nearly the beginning of sponsored broadcasting, and became more apparent during the post-World War II boom in broadcasting, as evidenced in a column by *New York Times* television critic Jack Gould, reporting on the 1959 FCC hearings into practices and policies of TV networks: "In the case of shows in which they were active, for instance the agencies said that they review all scripts in advance, scrutinize dialogue and story lines, and have their 'program representatives' on hand to check each day's production work."[12]

4.4.2 Networking and Syndication

Television stations, because of the tremendous expense involved, do not do a great deal of local production. Therefore, to fill the hours they program each day, they must turn to the networks or those companies which syndicate programs. (Radio stations depend to a very modest degree on such sources because the greatest share of the programs are locally originated.) Program content control, therefore, may be in the hands of the program producer, influenced of course by the other forces discussed in this chapter, rather than by station ownership or management.

"Stations have the right to refuse to carry networks programs they find objectionable, and, of course, they do not need to purchase those syndicated programs they do not like." Such is the response one gets when the question of program control by networks and syndication distributors is raised. Of course, such a response dodges the real issue of how much program control is relinquished by the stations. Except for local news programs and occasional production regarding some matter of local concern, the television programming you see is determined by decision-makers at networks or in syndication companies.

Such organizations are big business, with responsibilities to their stock-holders which far outweigh any responsibility they might feel toward the general public, the broadcast audience. Because the major motive of the businesses controlling the programming agencies (networks and syndicators) is to show a profit, profit must be the overriding factor in programming decisions. Thus, programs must appeal to the widest possible audience, to a mass audience, not to individuals nor to minorities (and this includes all types of minorities, not just ethnic minorities). Therefore, the greatest number of programs (there are exceptions, of course, like CBS's *Sixty Minutes*) are geared to the lowest common denominator; they are purely entertainment programs with little, if any, attention to public affairs. This is due to the low cost-efficiency of public affairs programs. Neilson ratings are readily translated into numbers of viewers so that we arrive at so many viewers per rating point. In general (again, excepting those rarities like *Sixty Minutes*), network public affairs programming costs three to four times as much per rating point as any other form of program.

The large number of prime time programs on the air geared to entertainment, compared to the low number of public affairs programs, gives rather strong indication that the networks have made the choice for economic viability over public service programs. Less profitable and less popular with the audience, they are not favored by advertisers if they introduce any kind of controversial subject.

For the same reasons, syndicators must sell those programs with the greatest mass appeal so that more stations will buy them. Thus, we get re-runs of such old network programs as *Gilligan's Island*, *I Love Lucy*, and *Hogan's Heroes*. Not that there is anything wrong with these programs, per se, but some viewers might eventually tire of them and prefer something that stimulated their minds a little more.

In recent years, there has been a trend in syndicated programs toward nature programs, such as *Wild Kingdom* and *Last of the Wild*. These programs offer an attractive alternative to the re-runs mentioned earlier. They are educational as well as informative and interesting, and the producers are certainly to be complimented on making this type of program available. In fact, these indications of a greater public awareness on the part of the syndicators, plus the growing number (though still terribly small by comparison) of network public affairs programs is encouraging. It is hoped that this movement will continue and networks and syndicators will realize even more that there are, after all, large audiences for such stimulating programming.

More important than the content itself, however, is the simple fact that local station management, which by law is supposed to be responsible for program content, does *not* have that control, and decisions as to what we see and hear are made by persons far removed from local concerns and interests.

The increase in the number of syndicators and syndicated programs resulted from an FCC ruling issued in 1970, entitled "Financial Interest and Syndication." The financial interest rule provides that no network shall

> ... acquire any financial or propriety right or interest in the exhibition, distribution, or other commercial use of any television program produced wholly or in part by a person other than such television network,...

The syndication rule provides that networks cannot:

> ... sell, license, or distribute television programs to television station licensees within the United States for non-network television exhibition or otherwise engage in the business commonly known as "syndication" within the United States; or sell, license, or distribute television programs of which it is not the sole producer for exhibition outside the United States; or reserve any option or right to share in revenues or profits in connection with such domestic and/or foreign sale, license, or distribution....[13]

In August, 1983, the FCC issued a tentative ruling reversing their previous stand and calling for the financial interest rule to be eliminated. If that tentative decision is adopted and made final, the networks will be able to participate in syndication except for prime time entertainment shows. The networks assert that the rule should be eliminated to give them more programming freedom, and, as might be expected, producers, who now basically control syndication, advocate retention of the rule, because they say that the adoption of the tentative rule will make the three major networks stronger at the expense of the public interest. They also claim that the proposed rule change will permit the networks to interfere with the creative process. The networks want free competition, claiming that the small independent producers cannot get financing for their projects without network assistance. The networks claim that they have suffered as a result of the rule in that network domination has dropped from 90% to 80%.

It is important to note that since the rules went into effect, independent stations have stopped losing money (according to the FCC statement in support of the proposed decision) and started earning a profit. Judging from the increase at the network level during the same period of time, the present rules do not seem to have affected the networks conversely. Anti-repeal supporters contend that if the networks are allowed to become involved in the syndication process, programs will be "warehoused" to prevent independent stations from purchasing them. They also claim that repeal will permit the major networks to prevent the entry of other networks into the market because of an ability to outbid smaller organizations.[(14)] In September, 1983, the House of Representatives Appropriations Committee attempted to prohibit the FCC from spending money to repeal the rules. The Committee was told to rewrite its funding legislation without that special interest provision.[(15)] (at the time of publication any change in the entire syndication matter seems unlikely)

Some observers of the broadcast industry do not take kindly to the FCC's proposal. Tom Shales, television writer for the *Denver Post*, indicated that he felt the FCC, especially the Chairman, "thinks the American viewing public can best be served by absolute and unchallenged tryanny over what gets on television, even on non-network stations. . . . Although the compromise puts token limits on network participation in syndication (an $800-million-a-year enterprise), essentially it hands the networks a goldmine on a silver platter." Shales quotes Jack Valenti, president of the Motion Picture Association of America: "It is an outrage, an absolute rape. This decision represents a smashing, callous disregard for the public interest. It's a 'compromise' that gives everything, totally, to the networks."[(16)] It must be noted that many motion picture production companies are currently the producers and syndicators of some of America's most popular television programs. At any rate, the ultimate decision will likely have a very great impact in dollars, programming, public interest, and production influence. Here again we face the same question that has been raised throughout this chapter: "Is government control in any form (other than electronic) beneficial or harmful to the broadcast industry, in its broadcast sense, and is the public interest served?"

4.4.3 Profession vs. Business

Is it possible for a group of persons, broadcast programmers, who consider themselves members of a profession, to reconcile that posture with a constant drive for large audiences in order to bring in more money and make more profit? It can be said that doctors, dentists, and lawyers are also "out to make a buck," but they appear to have a greater dedication to serving the public, and their profession has a licensing system to prevent pure charlatans from practicing. No such professional licensing exists for individuals engaged in broadcasting, except for the technical licenses for which broadcast engineers must qualify, and those have nothing to do with program content.

Various proposals have been made over the years regarding systems for licensing, but so far none has appeared very workable nor have any been received with any degree of enthusiasm by broadcasters. Here again, unfortunately, it appears that economics takes precedence, just as was the case regarding the public good. However, it might be appropriate to look at generally accepted criteria for a professional so that some future thought might be given to how broadcasters "stack up." According to Yoder, distinctive attributes of a professional are:

1. Acquisition of a specialized technique supported by a body of theory.
2. Development of a career supported by an association of colleagues which may have developed a code of ethics, with suitable disciplinary action imposed by colleagues.
3. Establishment of community recognition of professional status.
4. May have specific requirements for formal educational preparation.
5. May have specific requirements for admission to practice, usually established by law.[17]

4.4.4 General Considerations

The foregoing discussion of the various economic controls affecting broadcasting is by no means exhaustive. It is a brief overview of *some* of the present and potential problems which, it is hoped, will serve as a basis for discussion and further study. However, as a final word, it should be pointed out that there is really no justification for the assumption that broadcasting is "free" in this country, and that commercially based broadcasting is the only system which will work.

The public has three-to-four times as much invested (in receivers and antennas) in broadcasting equipment as does the entire broadcasting industry. We spend over one billion dollars a year just in repairs and parts for our receivers, to say nothing of the extra dollars we spend for products made more expensive because of the dollars the manufacturer has spent on advertising through the broadcast media.

The nations of Western Europe, for example, have provided sufficient evidence that forms of broadcasting, other than commercially oriented ones, not only can exist but can prosper and provide a greater degree of actual public service. In many cases, these long-existing public broadcasting systems have been joined by commercial systems. However, these latter systems generally are very closely supervised by the government so that there is absolutely no connection between sponsors and program content. The still struggling public broadcasting system in this country may provide, eventually, an alternate system, dedicated to enlightening and informing the public.

4.5 AUDIENCE INFLUENCE

4.5.1 "Giving the Public What it Wants"

We often hear spokesmen for the broadcasting industry, responding to critics, say that the broadcasters give the public what it wants. While this may be an easy answer to the problem of mediocre programming, it cannot be defended if one takes the time to examine the statement.

Our society is composed of a number of different publics, with compositions which vary from time to time, and each of those publics has its own dreams, desires, opinions, and ideas, which may change. Audiences are constantly changing. So it is impossible to program for "the public."

Also, people are not able to tell just what they want. Can you want something that doesn't exist, or that you know nothing about? Could the radio listeners of the 1920s demand broadcasts that featured Benny Goodman or Orson Welles? If you were offered a choice between a steak dinner or a widget, could you make a reasoned choice? How about if you were offered $10 or a widget?. Of course you could not choose wisely because you have no idea what a widget is. This same problem exists regarding broadcast programming. Would the audience rather watch "*Dallas*" or a widget? There is simply no sound research method which will answer that question. One of the major problems may be that there are very few widgets in the broadcasting milieu; not enough new and different programming ideas are tried, so there is no criterion by which to judge. Broadcasters often claim that they cannot afford to program good music, good literature, and serious discussion programs because the audience would not be large enough to make the programs economically viable. While it is true that such programs might not draw top ratings, that does not mean that at least certain "publics" are satisfied with the present offerings. Furthermore, while it is highly unlikely that there are large audiences out there awaiting such programs, neither can it be claimed by the broadcasters that many people would reject the programs; they never have had the opportunity to choose. In addition, people improve their tastes, but only if their taste buds are challenged by new and different things.

4.5.2 The Rating Game

A discussion of ratings is an extension of the discussion in the previous section. While the ratings do indicate a fairly accurate picture of how many people were watching or listening to a given program at a given time, ratings should not be considered a valid indication of actual program preference. Let us consider first some of the more pedestrian shortcomings.

The most generally accepted television ratings, Neilsen, are based on electronic recording of how many sets in the sample (we will not even consider that the sample is too small to be adequate) are tuned to each network. The fact that the set is on and tuned to a specific station does not guarantee that anyone is watching, or how many in the family are. Telephone surveys, most often used for radio, ask, "Is your radio on, and, if so, to what station is it tuned?" The same objections as were indicated for television may be raised here. In neither of these surveys is it possible to determine whether any other activity is taking place, such as conversation while the set is on. However, if these minor objections to the rating systems are overlooked, there is another, more important factor that must be considered.

Let's assume that on a given evening the three networks are carrying programs A, B, and C at seven in the evening, programs D, E, and F at eight o'clock, and programs G, H, and I at nine. Assume a somewhat exaggerated situation for the sake of illustration. For a given audience the actual order of program preference is: A,B,C,D,E,F,G,H. They can watch only one program at a time, based on their preferences, so they will elect to watch A at seven o'clock, D at eight, and G at nine. Thus, the ratings would indicate that those were the three most popular programs which would be far from the truth.

This is an example of what often happens, because networks may well pit their best against one another in a desperate battle for ratings. An equally misleading rating results if two networks happen to schedule poor shows opposite a fair show on the third network. The audience will turn to the third program even if it is not one they particularly like, simply because it is better than the other two. In other words, rather than serving as an absolute indication of preference, the ratings more often indicate what some people (those so addicted that they turn their sets on when they get up and often don't turn them off until they are falling asleep at night) choose to select from what is available at that given moment. This last point re-emphasizes what was said about widgets in section 4.5.1: If unusual programming is not among the choices, how can one judge properly?

4.5.3 Some Observations

Unfortunately, audiences tend to behave like sheep; they will follow almost any trend; they are gullible when it comes to evaluating content of either programs or commercials, and they are not very demanding. Visitors to this country from Western European nations are often appalled at the docile manner in which the American audience accepts what the networks and stations offer. Perhaps it is because, until the relatively recent advent of public television, we simply weren't aware that there could be something else — not necessarily something *better*, but a greater degree of choice.

The broadcast media have too great an influence on us as individuals and upon our society for us not to take some action to alert the broadcasters to our needs. Interestingly enough, most broadcasters will admit that they are often influenced by just a modicum of audience response. When *Star Trek* was originally removed from the air the number of persons complaining was small by network standards (reportedly less than 100,000), but the show was returned to the air because of the *type* of person complaining. The network indicated it was impressed with not only the intensity of the protestors but also with their quality. So, it is not always necessary to meekly accept what the broadcast media offer. A great deal needs to be done to increase the number of *critical* viewers and listeners. This is a responsiblity that needs to be shared by parents, educators, and the broadcasters themselves.

Interestingly enough, broadcasters' reactions to what they preceive as audience desires sometimes result in skewing programming in an unexpected direction. In recent years certain discontented, vocal segments of the audience have complained about under-representation. Typically, broadcasters and advertisers have reacted simplistically and have provided ready-made stereotypes for these various minorities. Consequently, much concern has been expressed over the stereotyped images assigned to certain ethnic and racial minorities by the media. The standard casting of Italians as Mafia extortionists, Jews as penurious penny pinchers, blacks as illiterate household servants, and Chicanos as switchblade-carrying criminals not only reinforced the public's outmoded beliefs and antiquated prejudices, but it also produced a new stereotype — the all-American middle-class, middle-aged WASP bigot.

However, where most whites seemed able to laugh at Archie Bunker's misguided attempts at perpetuating conventional racial and ethnic prejudices (in fact, some viewers undoubtedly regarded him as a hero of sorts), it is doubtful that actual minority members reacted as laughingly to their programming counterparts. In fact, programs dealing with minority situations, that Neilsen has ranked highly, have often been targets of attack from the minority groups themselves. For example, generally blacks did not approve of *Julia*, and pressure from Catholic and Jewish factions forced *Bridget Loves Bernie* off the air. Such a reaction is quite understandable, and it is regrettable that broadcasters often adopt such stereotypes in the interest of pulling large audiences; they don't worry too much about losing some of the minorities if the whites are amused. It is up to combined audiences to bring about changes in this arena.

4.6 PRESSURE GROUPS

Action for Children's Television (ACT) has long engaged in an effort to convince the FCC and/or Congress to ban all advertising on children's television programs. The Committee on Mass Communication and the Spanish Surnamed has been active in Colorado for the past several years in an effort to obtain more broadcast employment for Spanish-Americans and to increase the amount of programming directed at this ethnic group. Atheist groups are demanding equal time to reply to religious programming. Black organizations continually challenge license renewals in an effort to force the FCC to make more broadcast facilities available to blacks.

These are just a few examples of the types of groups putting pressure on broadcasters, the FCC, and Congress. Their efforts vary in viability, tactics, intensity, and specific purpose. In general, these groups are more than justified in their demands. Ethnic minorities have long been deprived of the right to sufficient programming addressing their needs, and have not been given equal employment opportunities. Groups, such as ACT, that are concerned about content, are certainly justified in objecting to those aspects of broadcast programming which are detrimental to any aspect of society. And, in recent years, these pressure groups have been making considerable progress in their struggle.

Considering the entire nation, the struggle by the blacks has perhaps received the greatest attention, and deservedly so. As Whitney M. Young, Jr., Executive Director of the National Urban League wrote as long ago as 1969, Puerto Ricans, Mexican-Americans and American Indians also suffer discrimination, but the black Americans are the only group representing involuntary immigration to this country.[18] In a somewhat broader approach, Payl Ylvisaker wrote in 1964:

> If broadcasting could choose only one community need to serve in this... Second Emancipation, it should be to fan the small fires of self-respect which have been lit in the breasts of the community's neglected and

> disadvantaged citizens. This year (1964) it was the Negro; next year it may well be the Spanish- or Mexican-American; the year after, perhaps, it may be the American Indian or the mountain white.[19]

The pressure by ethnic groups has certainly resulted in some changes in programming, as was indicated to some extent in section 4.5.3. We find many more blacks, for instance, serving as newspersons for both networks and local stations. In July, 1978, ABC announced the use of the first black in an anchor position on network news.

Pressure has also resulted in more programming for and about ethnic minorities. Blacks are now generally fairly well presented as normal human beings (when you discount such early attempts as *Julia* and similar programs), but there is still room for improvement. Foreign language broadcasts are much more readily available, and documentaries are beginning to deal with the real issues underlying social and economic problems of ethnic minorities, whereas formerly, such programs merely dealt with the surface manifestations of such problems and simply showed pictures of slums, rats, poorly-dressed children, etc.

In 1978 pressure of a new kind developed. An adolescent in Florida pleaded innocent to a murder charge on the grounds that television violence had driven him to such a mental state that he was not responsible for his violent actions. The courts found against the young man, but the case was a milestone, because it brings into very sharp focus the mounting concern, in many quarters, about the effect that violence on television may be having on society. Many studies, especially those carried out under the direction of Dr. George Gerbner, Dean of the Annenberg School of Communications at the University of Pennsylvania, lend considerable support to the contention that television violence is affective. Further corroboration may be found in *Television and Social Behavior*, a five volume study done under the general auspices of the Surgeon General's Scientific Advisory Committee on Television and Social Behavior.[20]

Also in 1978 another, and perhaps more consequential, case was filed in San Francisco as a result of an NBC telecast in 1974, starring 15-year-old Linda Blair as an inmate of a state home for girls. In one scene she was artificially raped by four other girls with the handle of a "plumber's helper." Three days later, a nine-year-old San Francisco girl was similarly attacked on a beach by other girls using a bottle. The victim and her mother sued NBC and its San Francisco outlet for $11 million, charging that NBC was negligent in showing the program, particularly during the so-called "Family Hour." A friend-of-the-court brief filed by the California Medical Association asserted that "television is a school of violence and a college for crime," adding that, because the attacking girls "acted out what they saw, we now have a real life victim with real life scars —brought to her by NBC."[21]

In 1981, the California Court of Appeals ruled that NBC was protected by the First Amendment. (Olivia v. NBC, 126 Ca. App. 3rd 488, 178 Cal. Rptr. 888). In a somewhat similar case, the next year, 1982, a Rhode Island Court ruled that NBC was also protected by the First Amendment in the case of the 13-year-old who hanged himself while doing a trick that had been performed on TV by Johnny Carson. (DeFilippo v. NBC, R.I., 446 A2d 1036.) These two cases raise the issue of who, or what, is to ensure that programmers are held responsible for any programs that might conceivably result in serious injury. This matter probably cannot be separated from the overall consideration of program content control, but, again, must be weighed against our basic freedom to the access of all types of information.

In Denver, Colorado, the Center on Deafness filed a complaint with the FCC in July, 1983, charging the four local commercial television stations with violations of the FCC regulations that require emergency programming to be visually informative. Programming during a crippling blizzard and a chemical spill emergency during the winter of 1982 were cited as being insufficient in this regard. Shortly after the suit was filed, station KBTV-TV carried a weekend news story on the complaint, attempting to excuse itself because no one was available to operate the equipment necessary to display the messages visually. During the explanation story they ran titles and explained that it took over an hour and a half to input those stories. According to a spokeswoman at the Center, KOA-TV responded to the charge in writing, stating that the station felt it was complying with the subject regulations. The Center responded with a copy of the complaint which had been sent to the FCC and added affadavits from 12 deaf persons stating they were unable to make sense of the visuals presented during the emergencies. Apparently visuals were used, but the claim was that they were unrelated to the emergency information that the deaf needed to know, particularly those living in the evacuation areas.

The foregoing illustrates the fact that there are groups, other than ethnic minorities who have concerns about equal access to the broadcast media, and thus fall within the general rubric of pressure groups, and there should be no automatic assumption that the term is applied in a derogatory or demeaning sense. Another situation is illustrated by the following complaint on the part of a religious group.

In early 1982, Donald Wildmon launched a boycott against NBC and affiliated companies. Wildmon, of the National Federation for Decency and the Coalition for Better Television, charged that "RCA/NBC has excluded Christian characters, Christian values, and Christian culture from their programming."[22] He urged consumers not to buy RCA products (including televisions, video equipment, etc.), and not to do business with Hertz rental cars, Gibson Greeting Cards, Coronet Carpets, RCA records, and CIT Financial, all owned by RCA. The boycott was to continue until NBC programming (a) downplayed the use of drugs, (b) portrayed violence as "non-approved behavior," (c) presented sex from a Judeo-Christian perspective as "a beautiful gift given by God to be shared by husband and wife," (d) stopped using profanity, and (e) presented minorities and the elderly as "full contributing members of society." NBC was described as being "a little bit worse" than CBS and ABC. The Moral Majority, one of 1,900 groups in the coalition, withdrew from the coalition before the boycott began, with leader Jerry Falwell saying that "the networks are making a sincere effort to improve." Wildmon said he was not advocating censorship, but that he was opposed to the belittling of Christians and their beliefs.

4.7 SOCIOLOGICAL AND PSYCHOLOGICAL INFLUENCES

As indicated earlier, our societal mores are changing quite rapidly, and this fact is reflected in the type of broadcast programming available to us. For example, our concerns about sex are being gradually replaced by our concerns about both violence and showing the manner in which many minorities have to live. The programmatic results have been discussed in earlier sections of this chapter.

Some comments on entertainment programming and the natural psyche are in order. The trend toward the scientific push of the last several years is reflected in programming. *Star Trek*, *Space: 1999*, *Six Million Dollar Man*, *The Bionic Woman*, and motion pictures such as the *Star Wars* trilogy, and numerous other programs give evidence of this general move toward a more scientific society. On the other hand, this movement seems to be inherently resisted by many who feel that the rights, freedoms, and wishes of the individual are being violated by the scientific movement. Their hopes, aspirations, and fears are reflected by such programs as *Grizzly Adams*, *How the West Was Won*, and *Little House on the Prairie*, all examples of man's struggle for individuality.

A somewhat related, but still different, sociological phenomenon is individualism versus collectivism. The idea of group activity, of joining, and working for the good of the whole, is becoming more dominant in our society, and, as with the scientific push, we are finding increasing resistance. In effect, what many people fear so far as collectivism is concerned, is that they will lose the opportunities to do what they want to do with their lives, to feel that they can be what they want without overt pressure from some group. Programming dealing with the 1950s like *Happy Days* tends to offer escape for these persons, as do some of the programs mentioned above.

More important issues arise in the media treatment of public affairs. There has been increasing concern expressed over the years that nationally broadcast election returns of predictions based on computer projections could have an effect on an election outcome, particularly where polls still remained open. For example, in 1976 an 8:00 p.m., EST prediction of a winner would have occurred while 23 states were still voting, with an electoral strength of 216 out of a total 538. This is due not only to difference in time zones, but also to differences in the times at which polls close in the various states. A similar situation existed in 1980.

Psychologically, the question should be raised as to whether the West coast voter, having heard a prediction that his favorite was certain to lose the election, would simply not bother voting because his vote would apparently do no good. There is the companion question about whether or not the projections made by the networks are reliable when they project a winner, perhaps only minutes after the polls in the East close (or may still be open, for that matter), based on only a handful of returns. And yet, those projections might influence considerable numbers of voters, either not to vote at all, or to decide to get on the winning side.

From time-to-time Congress has expressed concern over this problem, and several bills have been introduced with no success. Several years ago, in connection with the consideration of one piece of legislation dealing with this problem, Senator Vance Hartke directed a letter to Chairman Pastore of the Subcommittee on Communications, from which pertinent excerpts follow:

> As you know, (election) coverage by radio and television has reached a high degree of speed and interpretation. Network newsman have computers, researchers, pollsters, and other specialists at their disposal... Yet, the projections based on fragmentary samples may prove incorrect. Viewers and listeners are misled — *often unintentionally* (emphasis added)... The late President John F. Kennedy won the 1960 election by a 112,692 plurality vote. If *one* voter in each of the 173,000 voting precincts in the United States had switched his vote... Nixon would have won the popular vote... Realistically, had there been a switch of one vote in Kennedy's favor in each of the 10,400 precincts in Illinois, plus a switch of nine votes in each of the 5,000

precincts in Texas, Mr. Nixon would have tallied the required 270 electoral votes and would have been our President.[23]

The most frightening fact in regard to this matter is that by his use of the phrase "often unintentional," Hartke is making quite clear his opinion that some of the projections are *designed* to mislead the viewer and listener. Thus, we see the possibility that the pressure of beating the competition in broadcasting (and thus building the confidence and numbers of both audience and advertisers) may lead to erroneous projections of election results, which, in turn, will lead to a psychological impact on the audience which might have a real effect on the final election results.

4.8 TECHNOLOGICAL CONTROLS

Although directly related to governmental controls, discussed earlier, it is necessary to point out that such developments as cable television, pay television, direct broadcast satellites, low power television, and theater television have an effect on what we see because these systems make more program choices available to us (discounting for the sake of this discussion the federal limitations, as discussed in section 4.2.5). Laser beams, fiber optics, and other technological advances have a similar impact.

On the other hand, reticence on the part of the electronics industry, often in concert with the broadcasters, can restrict our programming opportunities. The reluctance of television receiver manufacturers to include UHF tuners in their sets not only restricted our choice, but, perhaps, was the principal cause for the failure of UHF to develop properly. By the time the law required all sets manufactured after April 1, 1964 to contain UHF tuners, the entire matter of frequency allocation and related matters had been determined. It was during those considerations that UHF needed to be made viable.

The same problems exist today with new technologies. Currently, AM stereo is beginning to emerge as a force in radio broadcasting, but the same reticence on the part of the FCC and the EIA that existed 20 years ago is causing confusion, and will result in problems for the listener. There are three technologies which make AM stereo possible, and none is compatible with any other. Thus, a consumer might purchase an AM stereo receiver in one geographic location (selecting the system that will enable him to receive the local stations) and later move to another area where the AM stations, if indeed any exist there, utilize a completely different, and incompatible system, rendering the AM stereo portion of his receiver worthless. As with the delay in the selection of a color system, and the UHF decision, bureaucractic "foot dragging" does a disservice to the consumer.

Similar problems are emerging in the television arena in regard to videotex and teletext. Should certain regulatory policies be established, and, if so, by whom? While it is not possible to provide a definitive answer at this time, nor in this discussion, it is appropriate to indicate some of the variables having an impact on the technology, the policy relating thereto, and the possible need for government or industry guidelines.

In the 1970s, a broadcast information system was developed by the British which utilizes the vertical blanking interval (VBI) of a television signal for the broadcast of characters and graphic information. The term teletext is the generic name for this service. Originally conceived as a method for displaying text on a screen for the deaf and the hearing impaired ("closed captioning"), teletext requires the use of a decoder attached to the user's television set.

Videotex is derived from teletext. The distinction is that videotex is an interactive service which gives the user the capability of requesting or responding to the information. For clarification, it must be pointed out that, so far as teletext is concerned, there should be no problem with compatibility — anyone with a standard television receiver would be able to receive the signal. However, in order to have interactive videotex, computers, at both the transmitting and receiving ends, must be used, and there is the potential of the same problem as mentioned earlier in regard to AM stereo — non-compatibility of the home components of the system with those at the site of transmission.

Videotex, whether intended for private or public network use, promises access to an abundance of information at the touch of a finger. Private videotex networks may usher in the "office at home" or "electronic cottage" concept, direct communication between customers and corporate data bases, in-house training, and other computer network functions advantageous to the business world. Public networks, if proper prior attention is given to the compatibility necessity, can offer interactive services such as home shopping and banking, information retrieval, games, and electronic mail.

For public videotex networks to be a successful market venture, the service must be offered at an affordable price (another aspect of "non-regulatory controls") and with sufficient flexibility to attract a mass **audience. Two telecommunications services are logical entrants into the** field of videotex: the telephone and cable industries. Both have experimented with videotex services in the U.S.; both are proceeding with caution, but, here again, the consumer faces the same recurring problem of the lack of direction from government or industry (e.g., EIA) to avoid incompatibility and confusion. This is due, in part, to the divestiture of AT&T, plus a questionable market. This latter concern would likely be alleviated if there were some industry (preferably) or government leadership.[24]

Which of the aforementioned industries, given technical requirements, the regulatory environment, and the competition perceived between the two, is best suited to offer videotex services? Or can a cooperative/competitive relationship exist in the marketplace? If these two telecommunications giants continue to "doze," it is likely that several, possibly competing, services will emerge in an effort to attract customers. For example it has been reported that a nationwide videotex service will be developed by Sears Roebuck & Co., IBM Corp., and CBS Inc. The service would allow customers to shop, bank, pay bills and receive educational information through their home computers. The service is scheduled to begin in several years. Perhaps such videotex systems will spring up nationwide and worldwide, but will there be satisfactory compatibility without some sort of "outside" leadership and guidance? As indicated earlier, the author does not favor governmental regulation for regulation's sake, but previous experience in the field tells us that a modicum of regulation may be better than having a number of competing systems with differing technologies that are incompatible.

Another area of concern is that there is some variation of definition and technology among the many designers and planners of videotex services, thus heightening the concern expressed above about compatibility. Other generic names like viewdata and videotext are also used, as well as trade names (e.g., Prestel in the United Kingdom or Telidon in Canada).

It is important to establish a universal definition for videotex that is also international in scope. First, many of its features lend themselves to international usage. Second, the primary factor in making videotex affordable is mass production of home terminals; many of the companies entering into videotex will strive for an international market. For this reason, it is important to determine a universally acceptable definition of the term.

"Videotex — or videodata as it is also known — is the name of a new kind of online information service able to use adapted television receivers and suitable for use commercially or in the home."[25] In order to distinguish further between videotex systems and other data communication networks, the definition proposed by Study Group I of the CCITT (Commité Consultatif International Téléphonique et Télégraphique), in cooperation with Study Group VIII, and adopted temporarily by the CCITT VIIth Plenary Assembly in Geneva, November, 1980, is perhaps more helpful:

> 1.1.4 The basic facility provided by the Videotex service is the retrieval of information by a dialogue with a data base. The service is intended to provide this facility for the general public as well as specialist users and should therefore represent a balance between service quality, simplicity of operation, and economic implementation.
>
> 1.1.5 This Videotex service is based on public networks and uses standard television receivers *suitably modified or supplemented* (emphasis added) as the terminal equipment, although the use of other equipment is not excluded. The Videotex service would normally also provide facilities for the creation and maintance of the data bases.
>
> 1.1.6 The Videotex service may also provide the following facilities:
>
> a) the input of information (e.g., response to a questionnaire);
> b) terminal-to-terminal communication and/or store and forward message service between users (further study required);
> c) transactional services (e.g., electronic funds transfer, booking systems and calculation facilities);
> d) data processing services (e.g., calculation facilities);
> e) loading of software from a data base to a terminal;
> f) interconnection with other public switched services;
> g) the management of closed user groups.
> (this list is not intended to be exhaustive).[26]

Having been issued by a recognized international body, this definition is now providing the framework for the developers of videotex. It distinguishes videotex as a unique service while leaving it flexible for the discovery of new applications. Bassically, the CCITT has said that videotex is intended for a mass market, should be economical and simple to operate, and available over a public network. It does not specify what type of network this may be, although it encourages interconnection capabilities among networks. It also presents an overview of foreseeable applications for videotex. What it *does not* do is to confront the issue of compatibility among different systems, and *this* is the concern expressed consistently in these pages.

There has been some concern in the telecommunications field that some of the same constraints enumerated in this section might apply to narrow casting within television. It appears that, while there may well be some problems of nonregulatory "control" as far as narrowcasting is concerned, the danger of any real disruptive issues (as indicated earlier in this section in relation to other telecommunication vehicles) is relatively slight.

Since narrowcasting is basically designed to provide information and services to limited interests, and such limited appeal is not feasible using the current systems, it appears that the result will service those "minority" groups not currently served. An audience that wants to learn more about Navajo rug weaving and dyeing, could be served by cable or lowpower television without harm to the reception of "more popular" programming by the "masses."

One concern that might become aparent involves two facets of a "control" problem. One aspect is that the carrier, be it cable or lowerpower television, has control over the programming carried and may or may not serve the interests of some small interest groups. The other is that more powerful and influential interest groups may put pressure on the carriers to carry or not to carry certain programs.

Although such concerns are very real, it is not unrealistic to suggest that through the balancing process which is part of our Common Law heritage, we can arrive at solutions which best serve the needs of our social communities.

4.9 CONTROL BY OTHER MEANS

The television critic could have a considerable influence on what we see and hear if there were more critics and fewer reviewers. There are only a handful of capable, reliable critics in the country. The rest who may claim that name are merely columnists, often simply reviewing releases they receive from the networks and stations. The public would be well served if there were more critics like Jack Gould of the *New York Times* and Lawrence Laurent of the *Washington Post*, who strive to be truly critical in the hope of improving broadcast fare.

It is undoubtedly true that the broadcast media, and particularly television, have suffered greatly from a lack of motivating criticism. As Robert Landry has said of television; "The only art medium with a universal audience, the one conduit of ideas that must be kept unclogged, is practically without any organized, extensive, general criticism. What little there is, is apt to be offhand, careless, and feeble."[27]

Thus, although broadcast critics do not currently have much influence on what we see and hear, it would seem to be important that valid, learned criticism be increased. This would advise the broadcasting audience in the same manner that theater, book, and motion picture critics advise their audiences. This, in turn would put pressure both on the broadcasters and the advertisers to produce more meaningful programming.

Another group that needs to be considered briefly is the National Association of Broadcasters. This is the national trade association in the broadcasting field, and it is natural to expect that such an organization would not be very harsh in its attempts to police the industry. However, due to a fear of government-enforced programming standards, of stricter regulation of commercial message content, and of general chaos within the industry, the NAB has attempted to undertake self-policing. As far as control of programming is concerned, they established a Radio Code and a Television Code. These were designed to serve as guidelines for the industry by indicating parameters in such matters as the number of commercial messages per hour, type and content of commercial messages, and guidelines for children's programming and general programming.

However, in 1983, a number of television advertisers filed suit in court against the NAB claiming that the Codes amounted to restraint of trade and, in effect, created an illegal monopoly that was harming the firms whose advertising was determined by the Code Authority Board to be in violation of the Code. Subsequently, the courts ruled in favor of the plaintiff, and the NAB was forced to drop the Code. At the same time, efforts are being made by some programmers to have the entire Codes thrown out on the basis that even program restrictions result in restraint of trade. As a result of these two actions, the NAB has been rendered practically impotent in its efforts to bring about some degree of better taste in programming and advertising standards. However, the industry is to be applauded for attempting to prevent excesses.

Will the lack of enforcement authority by the NAB have any *real* impact on what we see and hear? Although there is some slight difference of opinion, the consensus is that the effect will be slight at best. Recent literature bears this out as two journal articles will attest. In one, the history of NAB regulation is traced from 1922 to the dissolution of the Code Authority Board this year. The author says that there won't be any real difference in broadcasting, that the Board had no real authority because membership and rules were strictly voluntary, and the rules were unenforceable, designed only to meet the needs of the broadcasters and not the public.[28]

Another article deals with the group's limitations. It notes that the news item announcing that the NAB would be lifting its ban on ads for contraceptives in a Knights of Columbus publication brought a flood of mail to NAB headquarters, indicating distress at the decision. The article also points out that NAB member stations tended to go along with the

Code's general rules, even though those rules were not really enforceable. Even non-Code and non-NAB members followed the guidelines. It is also mentioned that Morality in Media receives great quantities of mail asking that personal product advertising be stopped.[29]

4.10 REVIEW AND FORECAST

This brief overview has indicated many, but certainly not all, of the various factors that have a controlling influence on broadcasting and what we see and hear. As indicated earlier, space did not permit more extensive dicsussion, but it is hoped that this information will serve as a springboard for further reading, discussion, and research.

Although the information presented in this chapter may seem to be rather pessimistic, it was not intended to discourage anyone from being a broadcast fan, nor was it designed to dissuade anyone from constantly striving to bring about improvements in the broadcasting fare currently available to the American public. Indeed, there have been considerable strides made in recent years, and it is only by continuing careful examination of such factors as have been discussed here that we can hope for this trend to improve and accelerate.

It is to be noted that perhaps the most beneficial influences have been exerted by the various pressure groups — not universally, of course, but for the most part. On the other hand, perhaps the most deleterious influence has been that caused by the commercial nature of the media and the resultant attitude on the part of network management and commercial advertisers and their agencies. Network managers tend to substitute commercial considerations based on circulation for considerations of service to advertisers and the public, in the plan and design of network schedules. In other words, network television is an advertising medium motivated by a commercial concept. Perhaps more critical viewing and listening, causing the ratings (in spite of their weaknesses) to reflect the public's discontent with overcommercialism, would result in a lessening of control by advertisers. Direct protest to sponsors is another avenue for those who would like to see broadcasting serve a larger number of audiences and serve them much more meaningfully.

As indicated earlier, radio deregulation instituted by the FCC in 1983 has been rather far reaching, and the prospect of similar deregulation of television is brighter than at any time in the past. Congress has also shown an interest in this area, but has not acted swiftly. In fact, there is evidence of continuing reservations on the part of some representatives in Congress and the Senate, echoing the sentiments of Rep. John Murphy (D-N.Y.) as expressed in 1978, because he felt that freeing broadcasters from regulation would have a predictable result:

> For most broadcasters, the ratings alone will prevail. In the end, there will be fewer public affairs programs, if that is possible. Our children and young people will continue to get short shrift. The cheap programming that fills the Saturday morning ghetto period — which earns high profits at low cost — will still be their standard fare. And in adult entertainment, banality will continue to rule.[30]

He said further that the new provisions miss the target because they "leave untouched network control over American television."

The last point is the strongest point made; however, the balance of his argument is not necessarily constructive. One of the main tenets of American life has been "a free exchange of ideas in a free market place." Why should broadcasting be exempted from that privilege? It seems reasonable that freedom from government regulation just might result in greater diversification. Of course, government deregulation will not prove to be a panacea, what with all of the other controls and influences currently having an impact on what we see and hear, but it would be a step in the right direction, and that freedom might lead to freedom from other pressures. There is no point in hoping for better programming through freer choice if the first and easiest step, government deregulation, is too slow in becoming a total reality, particularly when some believe that government regulation is there to protect us! That is too much like "Big Brother" watching over all of us. Government deregulation should be instituted soon in the area of cable television and the other new telecommunication processes as well, thus permitting more local control and more access to diversified programming by the public.

In spite of the concerns expressed in this chapter about the great number of controlling factors present in broadcasting, the future appears bright — at least brighter than the past. The broadcasters and sponsors are providing us with more hours of meaningful programming than was available a few years ago (admitting that there is still a great deal of pap on the air), and the government is at least achieving some degree of deregulation. And, perhaps most important, it appears that the public is beginning to demand more of a voice in what programming is presented.

(Note: Research assistance for this chapter was provided by Ms. Dianne Raley, graduate student in the School of Journalism, University of Colorado-Boulder)

REFERENCES

1. Koppel, *Nightline*, ABC, May 18, 1983.
2. Shapiro, *Nightline*, ABC, May 18, 1983.

3. *Communications and the Law*, Spring, 1981.
4. *Comment*, Fall, 1982.
5. *Video Law Monthly*, April, 1983.
6. *Ibid.*, May, 1983.
7. Aaholm, Leslie and Carol Williams, "Majority-Minority," an unpublished graduate research study, University of Colorado-Boulder, 1974, p. 1.
8. Friendly, Fred, *Due to Circumstances Beyond Our Control* (New York: Random House, 1967), p. 265.
9. Gurevitch, Michael and Phillip Elliott, "Communication Technologies and the Future of the Broadcasting Profession," in *Communication Technology and Social Policy*, edited by George Gerbner, Larry P. Gross, and William H. Melody (New York: John Wiley & Sons, 1973), p. 505.
10. Walsh, Patrick, "Commercials and Our Changing Times," in *Television*, edited by Barry G. Cole (New York: The Free Press, 1970), p. 238.
11. Aaholm & Williams, *op. cit.*, p. 2.
12. Gould, Jack, "Control by Advertisers," in *Problems and Controversies in Television and Radio*, edited by Harry J. Skornia & Jack William Kitson (Palo Alto, Ca: Pacific Books, Publishers, 1968), p. 418.
13. Report and Order in Docket 12782, 23F 2d 382 (1970), aff'd sub nom Mount Mansfield Television, Inc., v. FCC 442 F. 2d 470, (2nd) (in 1970).
14. *Broadcasting*, August, 22, 1983.
15. Ibid., September 26, 1983, p. 28.
16. Shales, Tom, *Denver Post*, August 11, 1983.
17. Yoder, Dale, *Personnel Management and Industrial Relations* (Englewood cliffs, NJ: Prentice-Hall, Inc., 1970) p. 396.
18. Young, Whitney M., Jr., "This Is Where We're At," in *Broadcasting and Social Action*, edited by the editors of *Educational Broadcasting Review* (Washington, D.C.: National Association of Educational Broadcasters, 1969), p. 5.

19. Ylvisaker, Paul, "Conscience and the Community," *Television Quarterly*, Winter, 1964, p. 11.

20. *Television and Social Behavior*, A Technical Report to the Surgeon General's Scientific Advisory Committee on Television and Social Behavior (in 5 volumes), G. A. Comstock, E. A. Rubinstein, and J. P. Murray, (ed.), U.S. Dept. of HEW, Health Services and Mental Health Administration, National Institutes of Health (Washington, D.C., DHEW Publication No. HSM72-9060, 1972).

21. Seib, Charles B., "Broadcasters under Attack Again," a column in the *Washington Post*, reprinted in the *Denver Post*, July 24, 1978.

22. *Christianity Today*, April 9, 1982, pp. 74-75.

23. Hartke, Senator Vance, *Senate Report*, August 29, 1967.

24. The foregoing information on teletext was based in part on material drawn from an unpublished M.S. Thesis at the University of Colorado: Stalhut, Janis, "Comparison of Videotex Services over Cable and Telephone Networks," Boulder, 1984.

25. Woolfe, R., *Videotex, the New Television-Telephone Information Services*, Heyden & Son, Ltd., London, England, 1980, p. xi.

26. International Telegraph and Telephone Consultative Committee (CCITT), Study Group I, final report to the VIIth Plenary Assembly, *Document AP VIII, No. 91-E*, September, 1980, pp. 80-81, as cited from Prunier, Patricia Sylviane, *Potential of Videotex on Cable Television Networks*, Master's Thesis, University of Colorado, Boulder, 1981.

27. Landry, Robert, "Wanted: Critics," *Public Opinion Quarterly*, December, 1940.

28. *Communication and the Law*, Spring, 1983.

29. *Federal Communication Law Journal*, 34 Winter, 1982, pp. 49-91.

30. Murphy, Congressman John, in "Deregulation Would Increase TV 'Banality,' Lawmaker Says," a UPI news story, datelined Washington, D.C., appearing in the Boulder *Daily Camera*, July, 26, 1978.

chapter 5

ENGINEERING ECONOMY: EVALUATING THE ECONOMICS OF SYSTEM DESIGN ALTERNATIVES

Robert J. Williams
Department of Mechanical Engineering
University of Colorado
Boulder, Colorado

5.1 ECONOMIC ANALYSIS — A PREREQUISITE TO ACTION IN THE TELECOMMUNICATIONS INDUSTRY

The technologies of the various communication processes are in the midst of changing society much as printing did five centuries ago, or as broadcasting did in the first half of this century. Not only are there many types of processes, but each has been developed to a sophisticated point undreamed of a short time ago. Then, too, processes are being joined in unique ways, and through coupling with the logic of computers, the collection, storage, manipulation, and transmission of information literally knows no bounds. Voice signals, video signals, reference data, and control signals are typical of information that can be reduced to the common denominator of electronic digital data. There are rarely significant technical limitations. The limits are now those of the imagination, of regulation, and of economics.

To design today's complex telecommunications systems, to buy even "off-the-shelf" equipment, and to keep the entire system operating with trained personnel, adequate inventory, and within the regulatory climate, require a centralized function in most organizations. The manager of this function is usually a highly skilled technical individual with formal training. Technical decisions by this individual as to procurement of equipment through purchase or lease generally result in long term financial commitments by the organization. Once such commitments are made and facilities or equipment are built or purchased, the decisions are frequently irreversible. In other words, it does no good to change one's mind once this point of no return has been reached.

5.1.1 Co-Determinants for Decision-Making

Each project, system, or major individual item of equipment under consideration for procurement is normally evaluated in two distinctly different ways — whether it meets the required technical specifications and whether it is economically viable. Each technically acceptable alternative generally presents a different set of economic consequences, requiring an economic feasibility study. Economic evaluations may be intuitive, may be by the "squeaky wheel" method, by the "necessity" process (wait 'till the old one breaks down), or by a comprehensive life cycle cost analysis. The latter method considers all of the costs and all of the anticipated savings over the economic life of the proposed investment. It responds to the question of what are the future economic consequences of a decision to be made now. At one time in history the engineer was content with doing the technical design and the technical feasibility studies. The physical environment was considered to be well-ordered, and the complexities of the economic aspects were left to "someone else." Surely, management could always find the funds to pursue a "hot" idea. The result was that an effective optimization of resources of time, facilities, and money was seldom achieved.

The concepts of what has come to be called *engineering economy* have been developed to assist engineers in presenting the economics of technical alternatives. Other terms, including managerial economy, have been used, but the generally accepted one is engineeering economy. This chapter is intended to provide some understanding of the principles, techniques, and reasoning by which a technical person may be guided in his economic analysis relating to long-term investments in capital equipment. A number of references are cited for those who wish to pursue the topic in more detail. In teaching his courses in engineering economy, the author has found it difficult to keep current with the rapid technical and cost changes in the broad telecommunications field. He has found it more desirable to present general examples or simplified models for reality as a way of focusing on a conceptual understanding of the basic principles. This approach avoids the question of whether the data in the examples is technically and economically up-to-date and permits more emphasis on learning the concepts. Students readily adapt the concepts to their own special fields. This report emphasizes the same approach.

5.1.2 Economics as a Design Parameter

While economic analysis frequently follows a technical feasibility study (Will it work? Will it deliver the promised output?), we now find that economy has become a criterion of design. This means that the engineer and the communications manager, from the first conceptual efforts, must actually consider economics in parallel with function and reliability. Or, to express it another way there is now a growing tendency in design to emphasize the joint achievement of performance objectives and product or service costs. It is apparent at the present time, for example, that telephony is no longer a unique art. Groups outside the telephone business with experience in electronics and information theory are reaching in to specified narrow segments of the business because they recognize a profit potential. They have also been assisted by the 1968 Federal Communications Commission (FCC) Carterfone decision, which permitted non-AT&T equipment to be connected to telephone company lines. Cost parameters within these segments are first established, and then a system is specifically designed to operate within these parameters. *Rocky Mountain Industries* magazine in October, 1977, examined the growth of the interconnect business, stating that "in the Denver area about 2,000 private interconnect systems have been installed," with "the cost of owning a private telephone system as much as 15 to 35 percent less than the monthly rental of a comparable system from Mountain Bell, depending on size and type of equipment."[1] The net result of such cost-effective designs is to drive down unit costs. Many functions of interactive communications media, however, tend both to require and to generate increasing quantities of information, so the capital investment necessary to produce lower unit costs is at the same time increasing.

In the current environment involving high-technology companies, the objective of a company's capital budgeting process may not be simply to maximize a safe, stable return on a capital intensive investment, but rather to secure a dominant share of unique segments of a market with the potential for high-growth rates. Technology intensity replaces capital intensity. The perception of the process is thus changed. Planning horizons are often drastically shortened. Evolutionary improvements are no longer sufficient. Intangible political (including regulatory) and social factors frequently play a much more significant role. The difficulty of expressing these factors numerically increases the attendant risk. But economics still drives the design of the segment to be exploited. The potential for higher returns on limited capital resources by successfully managing the higher risks becomes the prime motivator and criteria.

5.1.3 Application Independent of Economic System

There is a growing acceptance of the concepts of engineering economic evaluation in both market oriented economies and centrally planned ones. Objectives of a public project, of a private market-oriented corporation, and of a business enterprise in a centrally planned economy may each be different. In addition, the definition of the problem to be addressed by the plan will vary because societal needs are valued differently from country to country. These definitions will be a decisive factor in the final solution. In any event, regardless of the economic system, resources are limited, and before any long-term commitment is made, some assessment of the benefits and costs of an engineering project must be done as a part of the decision process.

While more attention might have been given in the past to quantifying the technological factors (preformance specifications, for example), there now exists a systematic method of treating the economic consequences of one project design over another.

5.2 INTEREST AND RETURN

Effective use of capital as a limited resource is essential to the profitability of any company. While the "bottom line" is viewed as the real measure of the effective use of capital, it cannot be an optimum unless each of the individual units of invested capital is optimized. Timing, in terms of the moment, as well as the duration of commitment of individual units, is an essential part of this optimization. Capital expenditure decisions generally involve an investment of funds in anticipation of greater future receipts. An evaluation of the effectiveness (or return on investment) of the expenditure requires comparisons of future sums of money, or receipts, at various points in time. Often the capital expenditures for a particular project also come at various time intervals. An understanding of the time value of money (interest) is essential in order to measure the effectiveness of the investment.

5.2.1 Economy Must be Measured over the Life of the Capital Investment

The cheapest equipment to buy or build, the yardstick of the purchasing agent, is not always the cheapest equipment to own. Savings in construction materials, components, and complete units of equipment may be wiped out by continuing high operating and maintenance costs. What may appear to be expensive refinements as reflected in a higher first cost may prove over the life of the equipment to actually reduce operating and/or maintenance costs. Operating people who have to live with the equipment have a constant fight against costly down-time and always push for the best in the original procurement. They find it difficult to express their justification in economic terms, but know they want the higher priced alternative. Tools of engineering economy use figures to determine which of several possible alternatives costs the least in the long run. These tools are not used to prove that a preconceived solution or alternative is the only correct one. The trend toward consideration of costs over the long run is recognized by the term *life-cycle costing*. Another relatively new term with essentially the same meaning is *planning horizon*. Both emphasize that all costs and all revenues of a proposed project will be considered over its economic life. The useful, economic life forms the planning horizon.

5.2.2 Interest

The term interest is used as the cost of capital, or the rental amount charged for the use of money. It represents a practice as old as the recorded history of man. Time is one of the key variables. It is a measure of the financial gain anticipated by the owner through his lending and of the benefit which the borrower or user anticipates by having capital available. The economic gain through the use of money by either party is what gives money its time value. "The growth of money in time must be taken into account in all combinations and comparisons of payments,"[2] so adequately stated in the early development of engineering economy, still holds today. Regardless of the source of funds, whether equity or debt, it is necessary in economy studies to consider that invested capital must gain compensation for its owners. Measuring the amount or degree of compensation of invested capital forms the basis for economic decision analysis. Idle funds are not gaining compensation and thus are actually costing money, commonly known as the cost of foregone opportunities. (The effects of inflation/deflation will be considered later in this chapter.) Thus, an investment of one dollar that shows an earning of $0.15 at the end of the year would be said to yield a return investment of 15%. The earning may be due to an increase in sales or in savings in operations or maintenance.

5.2.3 Simple and Compound Interest

Interest may be either simple or compounded. In simple interest, the principal and the interest become due only at the end of the loan period. Since this application is seldom found in typical business situations, it will be discussed no further here. All applications in the remainder of this chapter will assume compound interest. When money is borrowed for a length of time equal to several interest periods, interest is calculated at the end of each of these periods. There are a number of loan repayment plans, ranging from paying the interest when due at the end of each period to paying no interest until the end of the total loan period. Interest due but not paid thus becomes an addition to the loan, and interest on this addition is charged at the end of subsequent periods. Interest in this case is said to be compounded. Assume, for example, that $1,000 is borrowed to be paid in 3 years with interest at 9%. Table 5.1 shows the year-by-year activity when interest is paid at the end of each period of one year.

TABLE 5.1
Loan of $1,000 on which Interest is Paid Annually

Year	Amount owed at beginning of year	Interest due at end of year	Amount owed at end of year	Amount paid by borrower at end of year
1	$1,000	$90	$1,090	$90
2	1,000	90	1,090	90
3	1,000	90	1,090	90

The other extreme, payment of no interest until the end of the total duration of the loan, is illustrated in Table 5.2.

TABLE 5.2
Loan of $1,000 on which Accumulated Interest is Paid at Expiration of Loan

Year	Amount owed at beginning of year	Interest due at end of year	Amount owed at end of year	Amount paid by borrower at end of year
1	$1,000.	$ 90.	$1,090.	$ 00.
2	1,090.	98.10	1,188.10	00.
3	1,188.10	106.93	1,295.03	1,295.03

The interest is compounded in Table 5.2, based on an interest charge payable on interest previously due but not paid. The force of compound interest is considerable. At the 9% illustrated, the principal amount doubles in just eight years.* A third arrangement, shown in Table 5.3, is more frequently used and is typical of mortgage payments. Interest is paid when due, and the sum of the payment of the interest due together with a portion of the principal is constant each year.

TABLE 5.3
Loan of $1,000 on which Year-End Payments are Equal

Year	Amount owed at beginning of year	Interest due at end of year	Amount owed at end of year	Amount paid by borrower at end of year
1	$1,000.	$90.	$1,090.	$395.06
	694.94	62.54	757.48	395.06
	362.44	32.62	1,395.06	395.06

5.2.4 Equivalence

Each of the above illustrations satisfies the payment of a loan of $1,000 for 3 years at 9% compounded annually and is said to be equivalent to each other. The out-of-pocket amounts do vary from one to the other because they are made at different dates. The concept of equivalence is the key to understanding engineering economy. It provides a means by which elements of an investment, each perhaps made at a different time, may be compared with earnings that may occur periodically over the total life of the investment. The evaluation of any resulting gain over the

*The number 72 can be used as the product of the interest rate and the time in years to determine the doubling time. For instance, money at interest of 8% will double in value in 9 years. Similarly, a country whose population is increasing at a rate of 4% per year will have its population doubled in 72/4 or 18 years.

life may be thus expressed as a "return" on the investment.

5.2.5 Interest Symbols and Terms Defined

Several symbols and terms are used in describing interest relationships:

i represents an interest rate per period. Interest rates may be for a one-year period, a six-month, or quarterly, or other lesser periods.

n represents the number of interest periods.

P represents a present sum of money at time zero.

F represents a sum of money that occurs at some time, n periods, other than time zero.

A represents a uniform end-of-period payment or receipt made each period. It is known as an annuity.

G represents a uniform increase or decrease per period in the value of A, and is known as a gradient.

The concept of equivalence, with its implication of the time value of money at some interest rate, permits translation or conversion of any amount of money at any date into an equivalent amount at any other date. Sums of money at different dates can only be added or subtracted when they are translated or converted to a common date. The simplest translation would be to determine the future value, F, n periods hence, of a present sum of money, P, placed in a savings account with an interest rate of i per period. Or, knowing an amount, F, desired n periods in the future, one could determine the present amount, P, which must be deposited, at interest rate i per period. These two translations are illustrated in Figure 5.1.

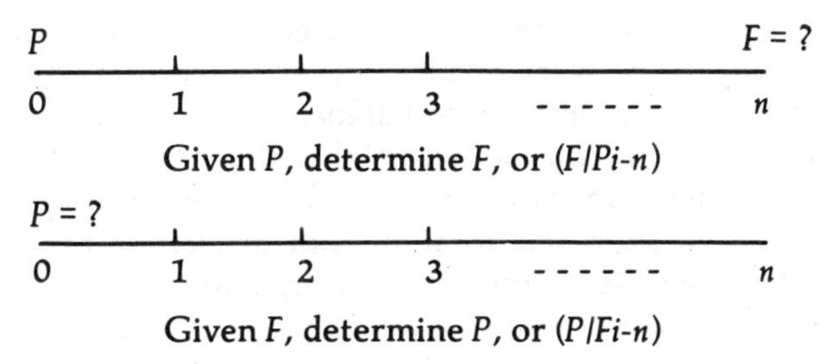

Figure 5.1: Translation of Present and Future Values of a Single Amount

The notation $(F/P\ i\text{–}n)$ is called the future from a present amount, at some rate of interest, i, per period for n periods, Similarly, $(P/F\ i\text{–}n)$ is called the present from a future amount, at some rate of interest, i, per period for n periods. These notations, together with those for a uniform series derived later, were endorsed several years ago by the Engineering Economy Division of the American Society for Engineering Education, and are now used in most writings in engineering economy. They follow very closely the system designed for the 1962 edition of *Engineering Economy*, published by the American Telephone and Telegraph Company. This *Green Book*, [3] as it has been commonly called, is now in its third edition and is a comprehensive work on engineering economy.

5.2.6 Nominal and Effective Interest

Reference has been made several times to the role of interest per period. A period may be any length of time up to and including a year. An interest rate is commonly specified in terms of a year and is known as a nominal rate. However, when the compounding intervals are less than a year, the nominal rate must be divided by the number of compounding intervals, or periods, to determine the rate per period. A nominal rate is always qualified accordingly. Thus, a rate stated as 6% compounded quarterly, would actually be 1-1/2% per quarter. Six percent would be called the nominal rate, and the qualifying phrase following it would indicate the frequency of compounding, or the number of periods.

When the compounding intervals are less than one year, interest is then earned on earlier interest, thus exceeding the nominal amount. The actual interest earned or paid may be expressed as the effective interest rate. If a person deposited $100 at a rate of 6% compounded semi-annually, the amount of accrued interest at the end of one year would exceed $6. At the end of the first six months, or the first period, interest of $3 would accrue as 3% of the $100. During the second interval, the original $100 plus the $3 interest, or $103, would earn 3% or $3.09. The total at the end of the year, or at the end of the two periods, would thus be $106.09. The actual interest rate, or effective rate would be 6.09%. A dramatic distinction between nominal and effective interest rates can be drawn in the case of a typical personal loan where it turns out that the true cost of the loan is considerably higher than the nominal or stated rate indicates. Such loans have a nominal rate of interest applied to the amount borrowed, with this total amount divided by the number of months of the loan to determine the monthly amount. The effective rate can be readily calculated, and for any nominal rate is almost double.

5.2.7 Uniform Series

Other notations are used to describe the four basic translations of a uniform series. A uniform series, *A*, may be translated into either a present single amount, *P*, or a future single amount, *F*. Conversely, a present single amount, *P*, or a future single amount, *F*, may be converted into a uniform series, *A*. The six translations, two for single amounts and the four for a uniform series, become the basis for determining equivalent amounts which in turn are the key to determining the effectiveness of capital investments. The notation, $(P/A\ i\text{–}n)$ is called the present value of an annuity, or the present value of a uniform series, A, at some rate of interest for n periods. The single future amount, F, equivalent to a past annuity, is called the future worth of an annuity, $(F/A\ i\text{–}n)$. A present single amount may be translated into a uniform series using the notation $(AP\ i\text{–}n)$ or annuity from a present amount. A future single amount is translated into a uniform series using the annuity from a future amount notation, $(A/F\ i\text{–}n)$.

5.2.8 Examples of the Series Notations

The present value of an annuity, $(P/A\ i\text{–}n)$, may be used to translate a series of potential annual savings which are expected to result from the installation of a new unit of equipment. The present worth of the savings may then be compared with the investment, with i representing the minimum return expected on the investment. Such a procedure is also known as the discounted cash flow method.

The future value of an annuity, $(F/A\ i\text{–}n)$, is helpful in calculating the amount, such as a sinking fund, one would have saved over a number of years by setting aside uniform amounts every month or year.

An annuity from a present amount, $(A/P\ i\text{–}n)$, may be used to determine the periodic amount one may withdraw over a given period of time from the accumulated savings in the previous paragraph. A common application in engineering economy studies is to determine the annual allocation to costs of an investment in capital equipment. The annuity or uniform amount thus resulting is a combination of interest or return on the investment plus a portion of the original sum or principal. The notation, $(A/P\ i\text{–}n)$, is then called a capital recovery factor. By including the return on the invested capital as a cost, the engineer makes provision in his analysis for a minimum profitability, which the accountant, during and at the conclusion of the investment, will express in dollars of profit on the capital. The telephone industry refers to the portion of principal thus allocated as capital repayment, or the annual apportionment of the original sum of money invested in plant and equipment representing a cost to the firm which must be recovered from the customer.[4]

Assume an investment in a new system costs \$50,000 with an economic life of 4 years. "Capital recovery" costs are established at 10%. No salvage is indicated. The annual capital recovery charge may be readily determined: \$50,000 $(A/P\ i\text{–}n)$ = 50,000 (0.31547) = \$15,774. This amount includes both "capital recovery" or a portion of principal, and a "return" or interest on the unrecovered amount which can be shown on a year-

TABLE 5.4
8% Interest Factors for Annual Compounding Interest

n	Single Payment		Equal-Payment Series				Uniform Gradient-Series Factor
	Compound-Amount Factor	Present-Worth Factor	Compound-Amount Factor	Sinking-Fund Factor	Present-Worth Factor	Capital Recovery Factor	
	To Find F Given P $F/P\ i, n$	To Find P Given F $P/F\ i, n$	To Find F Given A $P/A\ i, n$	To Find A Given F $A/F\ i, n$	To Find P Given A $P/A\ i, n$	To Find A Given P $A/P\ i, n$	To Find A Given G $A/G\ i, n$
1	1.080	0.9259	1.000	1.0000	0.9259	1.08000	0.0000
2	1.166	0.8573	2.080	0.4808	1.7833	0.5608	0.4807
3	1.260	0.7938	3.246	0.3080	2.5771	0.3880	0.9487
4	1.360	0.7350	4.506	0.2219	3.3121	0.3019	1.4038
5	1.469	0.6806	5.867	0.1705	3.9927	0.2505	1.8463
6	1.587	0.6302	7.336	0.1363	4.6229	0.2163	2.2762
7	1.714	0.5835	8.923	0.1121	5.2064	0.1921	2.6935
8	1.851	0.5403	10.637	0.0940	5.7466	0.1740	2.0984
9	1.999	0.5003	12.488	0.0801	6.2469	0.1601	3.4909
10	2.159	0.4632	14.487	0.0690	6.7101	0.1490	3.8712
11	2.332	0.4289	16.645	0.0601	7.1390	0.1401	4.2394
12	2.518	0.3971	18.977	0.0527	7.5361	0.1327	4.5956
13	2.720	0.3677	21.495	0.0465	7.9038	0.1265	4.9401
14	2.937	0.3405	24.215	0.0413	8.2442	0.1213	5.2729
15	3.172	0.3153	27.152	0.0368	8.5595	0.1168	5.5943
16	3.426	0.2919	30.324	0.0330	8.8514	0.1130	5.9045
17	3.700	0.2703	33.750	0.0296	9.1216	0.1096	6.2036
18	3.996	0.2503	37.450	0.0267	9.3719	0.1067	6.4919
19	4.316	0.2317	41.446	0.0241	9.6036	0.1041	6.7696
20	4.661	0.2146	45.762	0.0219	9.8182	0.1019	7.0368

21	5.034	0.1987	50.423	0.0198	10.0168	0.0998	7.2939
22	5.437	0.1840	55.457	0.0180	10.2008	0.0980	7.5411
23	5.871	0.1703	60.893	0.0164	10.3711	0.0964	7.7785
24	6.341	0.1577	66.765	0.0150	10.5288	0.0950	8.0065
25	6.848	0.1460	73.106	0.0137	10.6748	0.0937	8.2253
26	7.396	0.1352	79.954	0.0125	10.8100	0.0925	8.4351
27	7.988	0.1252	87.351	0.0115	10.9352	0.0915	8.6362
28	8.627	0.1159	95.339	0.0105	11.0511	0.0905	8.8283
29	9.317	0.1073	103.966	0.0096	11.1584	0.0896	9.0132
30	10.063	0.0994	113.283	0.0088	11.2578	0.0888	9.1896
31	10.868	0.0920	123.346	0.0081	11.3498	0.0881	9.3583
32	11.737	0.0852	134.214	0.0075	11.4350	0.0875	9.5196
33	12.676	0.0789	145.951	0.0069	11.5139	0.0869	9.6736
34	13.690	0.0731	158.627	0.0063	11.5869	0.0863	9.8207
35	14.785	0.0676	172.317	0.0058	11.6546	0.0858	9.9610
40	21.725	0.0460	259.057	0.0039	11.9246	0.0839	10.5699
45	31.920	0.0313	386.506	0.0026	12.1084	0.0826	11.0447
50	46.902	0.0213	573.770	0.0018	12.2335	0.0818	11.4107
55	68.914	0.0145	848.923	0.0012	12.3186	0.0812	11.6902
60	101.257	0.0099	1253.213	0.0008	12.3766	0.0808	11.9015
65	148.780	0.0067	1847.248	0.0006	12.4160	0.0806	12.0602
70	218.606	0.0046	2720.080	0.0004	12.4428	0.0804	12.1783
75	321.205	0.0031	4002.557	0.0003	12.4611	0.0803	12.2658
80	471.955	0.0021	5886.935	0.0002	12.4735	0.0802	12.3301
85	693.456	0.0015	8655.706	0.0001	12.4820	0.0801	12.3773
90	1018.915	0.0010	12723.939	0.0001	12.4877	0.0801	12.4116
95	1497.121	0.0007	18701.507	0.0001	12.4917	0.0801	12.4365
100	2199.761	0.0005	27484.516	0/0001	12.4943	0.0800	12.4545

by-year basis as follows:

End of year	Unrecovered investment	Capital recovery charge (AP 10-4)	Interest on unrecovered portion	Amount to apply to recover investment
1	$50,000	$15,774	$5,000	$10,774
2	39,226	15,774	3,923	11,851
3	27,375	15,774	2,738	13,036
4	14,339	15,774	1,434	14,340
			13,095	50,001*

*Difference due to rounding error.

Each year, as some of the capital is recovered, the amount of interest or return due is thereby decreased. The same concept of changing proportions between interest and principal takes place in the schedule of a time-payment plan where the amount of periodic payment for a mortgage, auto loan, etc., is established by applying the capital recovery factor.

The annuity from a future amount, $(A/F\ i\text{–}n)$, may be used to determine the amount which must be set aside periodically to pay the principal amount of a bond issue due at maturity. The notation is then called a sinking fund factor. An individual who wishes to accumulate some predetermined future amount may use the notation to determine the necessary periodic amounts which must be deposited.

Mathematical formulas and tables are available to translate any amount occurring at any particular time into an equivalent amount at some other time. A representative table is shown in Table 5.4.

5.2.9 Derviation of the Time Value Factors

A cash flow diagram aids in understanding the necessary translations by providing a graphical description of each problem situation. A horizontal line represents a time scale divided into intervals or periods. Arrows signify cash flows, with downward arrows representing receipts or a positive (+) cash flow, and upward arrows representing expenditures or a negative (–) cash flow. Perhaps it is easier to distinguish the direction of the arrows if one considers that funds that flow downward into the cash "pot" enrich it, and funds that move out (upward) decrease it. A vector-type plot, perhaps more easily understood by some, would reverse the arrows (arrows pointing upward are positive and proportional to cash inflows, while those pointing downward are negatives representing cash outflows). Cash flow diagrams are not always necessary in simple problems, and become rather unmanageable in complex situations. However, for the neophyte the diagrams are helpful in visualizing the various receipts and disbursements.

If an amount, P, is deposited now at an interest rate i per year, how much principal and interest will be accumulated at the end of n years? The cash flow diagram, Figure 5.2, illustrates the time line.

P F = ?

0 1 2 3 --- n

Figure 5.2: Single Payment Present Amount and Future Amount

The future value, F, or compound amount, of an investment of P dollars may be developed as shown in the Table 5.5. The derived *single payment compound amount* factor is designated (F/P i–n).

TABLE 5.5
Derivation of Single Payment Compound Amount Factor

Year	Amount at beginning of year	Interest saved during year	Compound amount at end of year	
1	P	Pi	$P+Pi$	$=P(1+i)$
2	$P(1+i)^2$	$P(1+i)i$	$P(1+i)\ P(1+i)i$	$=P(1+i)^2$
3	$P(1+i)^2$	$P(1+i)^2\ i$	$P(1+i)^2+P(1+i)^2\ i$	$=P(1+i)^3$
n	$P(1+i)^{n-1}$	$P(1+i)^{n-1}\ i$	$P(1+i)^{n-1}+P(1+i)^{n-1}\ i$	$=P(1+i)^n=F$

The *single payment present worth factor*, designated (F/P i–n), may be solved from the derivation in Table 5.5.

$$P = \frac{F}{(1+i)^n} \quad \text{or} \quad P = F\left[\frac{1}{(1+i)^n}\right]$$

As stated earlier, there are four basic translations involving a uniform series. Figure 5.3 illustrates the future value, or the accumulation at some future date, of a uniform series of year-end payments or annuities.

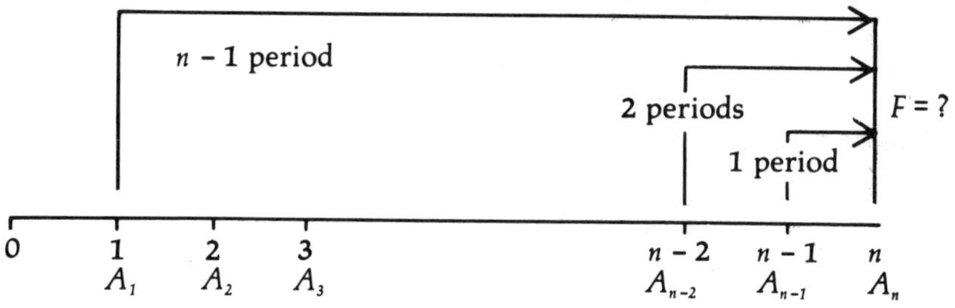

Figure 5.3: Uniform Series Compound Amount

Each single payment, A, draws compound interest for a different number of periods.

Payment A_n, at the end of the period n draws no interest.

Payment A_{n-1}, at the end of the period $(n-1)$ draws interest for 1 period.

Payment A_{n-2}, at the end of the period $(n-2)$ draws no interest for 2 periods.

Payment A_1, at the end of the period 1 draws interest for $(n-1)$ periods.

Each A payment thus corresponds to a single payment present worth amount, and each individually will amount to a corresponding single payment compounded amount, F.

The sum of each of the individual F amounts becomes:

$$F = A(1) + A(1+i) + A(1+i)^2 + \text{------} + A(l+i)^{n-1} \quad (5.1)$$

Multiplying both sides by $(1+i)$:

$$F(1+i) = A(1+i) + A(1+i)^2 + A(1+i)^3 + \text{-----} A(1+i)^n \quad (5.2)$$

Subtracting (5.1) from (5.2):

$$F(1+i) - F = A(1+i)^n - A$$

or

$$F = A \quad \frac{(1+i)^n - 1}{i}$$

The factor $\left[\frac{(1+i)^n - 1}{i}\right]$ is known as the *uniform series compound amount factor,* and is designated $(F/A\ i\text{–}n)$.

If each A value in the time line in Figure 5.3 is considered as a future worth F in a single payment present worth factor, then the present worth values may be summed to derive the *uniform series present worth factor.*

$$P = A\left[\frac{1}{(1+i)^1}\right] + A\left[\frac{1}{(1+i)^2}\right] + \text{---}$$

$$+ A\left[\frac{1}{(1+i)^{n-1}}\right] + A\left[\frac{1}{(1+i)^n}\right]$$

This can subsequently be expressed as:

$$P = A\left[\frac{(1+i)^n - 1}{i(1+i)^n}\right]$$ which is designated $(P/A\ i\text{–}n)$.

Through rearranging, A can be expressed in terms of P:

$A = P\left[\frac{i(1+i)^n}{(1+i)^n - 1}\right]$ which is designated (A/P i–n) and is also known as the *capital recovery factor*.

Similarly, A can be expressed in terms of F to derive the annuity from a future amount factor, (A/F i–n), also known as the *sinking fund factor*.

$$A = F\left[\frac{i}{(1+i)^n - 1}\right]$$

There are frequent times when annual charges are not uniform, as for instance maintenance charges as equipment becomes older, or when receipts increase as an investment becomes more profitable. When such amounts increase by a relatively constant amount each period, such arithmetic progression is called a gradient. A gradient conversion factor may be used to find an equivalent uniform annuity, and to determine the present or future worth of a gradient series.

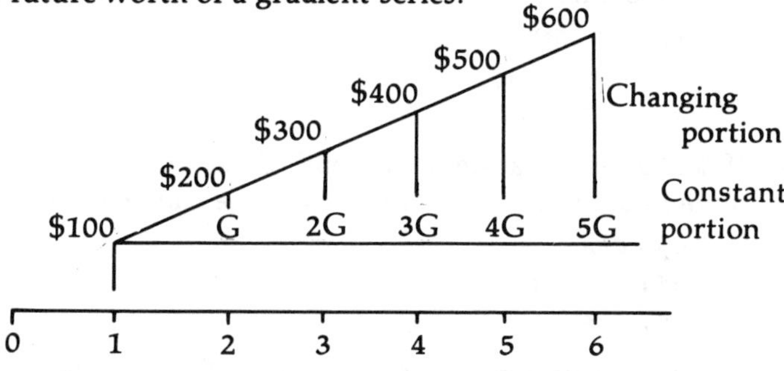

Figure 5.4: Cash Flow Diagram for Disbursements Increasing at a Uniform Rate

The future worth of the complete series illustrated in Figure 5.4 is the sum of the constant portion and the changing portion. Similarly, the present worth of the series may be determined, and also the annuity over the total period which is equivalent to the changing portion of the arithmetic progression. More complete books on engineering economy may be consulted for the development of the respective gradient factors as well as tables for each of the three at varying rates of interest.

A careful review of the various interest factors derived together with a display of the available data on a time scale will indicate the following points which must be kept in mind:

1. P is at the beginning of a year at a time considered as the present, or time 0. F is at the end of the nth year.

2. An A occurs at the end of each period. The P of a series is always one period prior to the first A. Exceptions are termed an annuity due, where the first A is at time 0. An appropriate P or F may be calculated by a shift in the time line.

   The F of a regular series occurs at the same time as the last A.

5.2.10 Examples of Applications of the Time Value Factors

Assume in each of the examples, unless otherwise noted, that an i of 8% is used.

1. A deposit of $1,000 will accumulate to what amount at the end of 10 years?

   $F = (F/P\ 8\text{-}10)$
   $= \$1,000\ (2.1589)$
   $= \$2,158.90$

2. What deposit must be made now if a sum of $1,000 is desired 6 years from now?

   $P = F\ (P/F\ 8\text{-}6)$
   $= \$1,000\ (0.6302)$
   $= \$630.20$

3. If $1,000 is set aside at the end of each year, what will be the amount accumulated in the fund at the end of 10 years?

   $F = A\ (F/A\ 8\text{-}10)$
   $= \$1,000\ (14.487)$
   $= \$14,487$

4. A community has successfully sold a $1,000,000 bond issue to develop a cable TV system. The bonds mature in 20 years. What amount must be set aside at the end of each year to (a) pay the interest due the bondholders, and (b) to eventually pay off the bonds at maturity? Assume the sinking fund earns interest at 5%.

   (a) $I = Pi$
   $= \$1,000,000\quad (0.08)$
   $= \$80,000$ *interest due each year*

   (b) $A = F\ (A/F\ 5\text{-}20)$
   $= \$1,000,000\ (0.03024)$
   $= \$30,240$ deposited each year to retire bond issue at *maturity*

   Total annual cost of bond issue:
   $= \$80,000 + \$30,240$
   $= \$110,240$

5. What would the total annual cost of the bond issue in problem 4 be if the community desired to retire the bonds in 10 years rather than 20, and was able to earn 6% on the annual funds set aside? Interest due each year would remain the same.
   Amount to be set aside in a sinking fund would be:

   $A = F\ (A/F\ 6\text{-}10)$
   $= \$1{,}000{,}000\ (.07587)$
   $= \$75{,}870$

   Total annual cost, 10 year basis:
   $= \$80{,}000 + \$75{,}870$
   $= \$155{,}870$

6. A new piece of equipment, with an installed cost of $65,000, is proposed to augment an existing system. What would be the equivalent annual allocation or charge for this equipment if its economic life is estimated to be 10 years?

   $A = P\ (A/P\ 8\text{-}10)$
   $= \$65{,}000\ (0.1490)$
   $= \$9{,}685$

7. A cable TV company is considering purchasing a new van costing $10,000 for use in installation and repair work. It would like to finance the purchase by borrowing money from a local bank at 12%. What would be the monthly payments if the loan is to be paid off in 3 years?

   Since the payments are monthly, the interest rate per period would be 12/12, or 1%, per month for 36 months.

   $A = P\ (A/P\ 1\text{-}48)$
   $= \$10{,}000\ (0.02633)$
   $= \$263.30$

8. Assume that in problem 7 the company purchased the van and later found it had a larger than anticipated cash flow in its operations during the year following the purchase. Immediately after making the 12th payment the company desired to pay off the remaining amount due. What single payment would be equivalent to the remaining 24 payments? Assume no pre-payment penalty.

   $A = \$263.30$ from problem 7

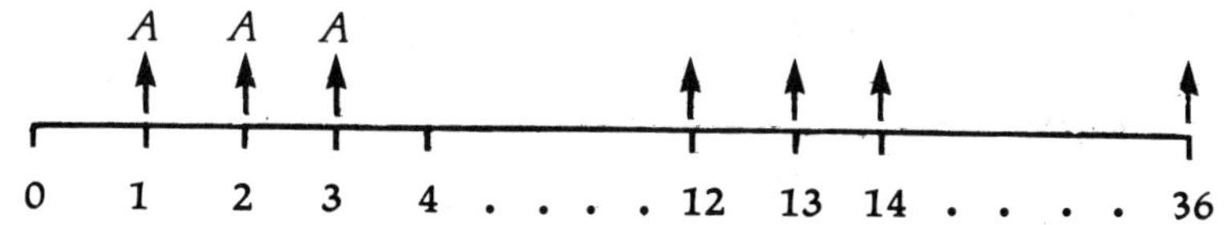

The end of month 12 becomes point 0 *for the 13th through the 36th payments. The equivalent single payment thus is the present worth, at the new 0* point, of the remaining 24 payments. Since each of the $263.30 payments is part interest and part principal, (refer to section 5.2.8), the present worth at the end of the 12th month is the sum of the principal portions of the remaining payments.

$$P = A\,(P/A\ 1\text{–}24)$$
$$= \$263.30\ (21.2433)$$
$$= \$5593.36.$$

Note that the remaining payments are not multiplied by 24, because to do so would ignore the time value of money and equivalence concepts.

5.2.11 Continuous Compounding

The payments and receipts described so far have all been assumed to occur discretely at the end of an interest period. The fact is that in any business, funds generally flow continuously — whether in or out. Some analysts insist that continuous compounding must therefore be used in order to achieve greater accuracy. On the other hand, the estimates of cash flow over the life of a projected investment are approximations which will not be made more accurate with continuous compounding. Readers are directed to some of the references for a further analysis of continuous compounding.

5.3 EQUIVALENT UNIFORM ANNUAL COST COMPARISONS

The time-value translations introduced earlier emphasized that cash flows can be converted into equivalent amounts at any time using a predetermined interest rate. This equivalency concept provides the basis for the three principal methods of economic analysis which may be defined as follows:

1. *Equivalent uniform annual cost,* or annual cost method. Cash flows are converted to an equivalent uniform yearly amount over the economic life of n periods at interest rate i. This can be thought of as representing a leveled, year-by-year amount.
2. *Present worth,* or discounted cash flow. Cash flows over the economic life are "discounted" or converted to a single equivalent amount, P, at time 0, at an interest rate i.
3. *Rate of return.* The negative and positive cash flows are equated. The rate of interest which makes the two equivalent represents the extra compensation gained, or the "return."

Another method commonly used is the benefit-cost analysis. Operationally, it is very similar to the present worth method and is used primarily in evaluating public projects.

Older methods, frequently involving averaging, have been replaced because they do not properly account for the time value of money, and thus frequently lead to incorrect investment decisions. The three most widely used methods indicated above are readily amenable for consideration of after-tax cash flows and after-tax rates of return.

A typical investment in plant and equipment involves a present expenditure, occasionally a supplementary capital expenditure, a series of savings or receipts generated by the equipment, and ending with the disposal for some net realizable salvage value. The cash flow pattern is illustrated in Figure 5.5.

Investment

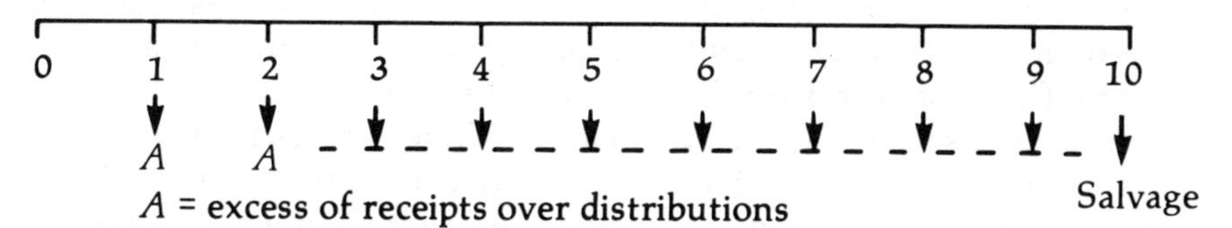

Figure 5.5 Cash Flow Diagram of Typical Equipment Investment

Example 1. Assume that in Figure 5.5 the investment is \$29,000 with an estimated salvage value of \$2,000. The annual excess of receipts over disbursements for the 10-year life is estimated to be \$4,000. Would the investment be justified if the minimum rate of return is 10%?

Using an equivalent uniform annual cost basis, the annuity cost equivalent would be:

$$\begin{aligned} EUAC &= (\$29{,}000 - 2{,}000)\,(A/P\ 10\text{–}10) + 0.10\,(2{,}000) \\ &= (\$27{,}000)\,(0.16275) + 200 \\ &= 4{,}594. \end{aligned}$$

The annual excess of receipts over disbursements of \$4,000 is not sufficient to cover the equivalent uniform annual cost of \$4,594. On the other hand, if the rate of return were 6% rather than 10%, the EUAC would be \$3,668 + 120 or \$3,788, and the investment would be justified. The selection of a minimum acceptable rate of return thus plays an important role in distinguishing which project proposals will return a minimum profit.

5.3.1 Treatment of Salvage Values

In the diagram in Figure 5.5, a salvage value was indicated. This is an estimate of the net realizable value of an asset at the time it is removed from service at the end of its life. The net realizable value is the dollar amount which might be received from the sale or trade of the asset, either as a used piece of equipment or for scrap, less the estimate of the cost of removal. Costs of disconnecting utilities, removal of a base, and the moving of the asset to the point where it may be disposed of, are all included in the cost of removal. Since the net realizable value represents recoverable funds, that portion is not allocated on an annual basis. Thus, in Example 1, the $2,000 is subtracted from the total investment. However, the $2,000 is part of the investment as far as the earning of a return is concerned, and for 10%, this would amount to a necessary earning of 0.10 ($2,000) or $200 per year. It is frequently extremely difficult to estimate net salvage at the end of an economic life. Technological obsolescence plays a major role in contributing to the difficulty in determining what the used market might be in 10, 15, or 20 years. For this reason, many companies, including those in any way related to investment in telecommunication equipment, are disregarding any salvage values. If there is finally some realizable amount, it is added to income that year as a gain on the sale of an asset. Similarly, if there is a loss because of the high cost of removal, that is charged against current income. Such adjustments in accounting are allowable when item, but not group accounting is used. The corresponding tax treatment provides for the benefit of taxation at half-rate if there is a net gain, and full deduction for a net loss. Since gains and losses of a given year must be offset against each other, it is generally to the taxpayer's advantage to group losses one year and gains the next.

For income tax purposes, the 1981 Economic Recovery Tax Act makes no provision for including salvage value of assets when calculating depreciation. It is permissible to have separate books of account for income tax purposes and internal record-keeping.

5.3.2 Comparison of Alternatives on an Annual Cost Basis

Example 2. It has been determined that the present manual method of handling a certain expanding operation is no longer adequate. Two proposals for mechanization have been submitted. Each will perform the required operations within the established technical limits. Plan A has a first cost of $20,000. Annual disbursements for labor and labor extras (social security, paid vacations, various other employee fringe benefits) are estimated to be $9,000. Annual payments for maintenance, power, property taxes, and insurance are estimated to be $1,800. Salvage value is estimated at $1,000.

An alternate proposal, Plan B, has a first cost of $28,000, with an estimated salvage value of $2,000. This plan is considered to be more efficient and more reliable, resulting in lower labor and maintenance costs. Labor and labor extras are estimated to be $7,000, and other payments are estimated to amount to $1,500. It is expected that the need for this operation will continue for 10 years. The minimum required rate of return is 10%. Which is the more economical plan?

Capital recovery:	Plan A	Plan B
= ($20,000 – 1,000) (A/P 10-10) + 0.10 ($1,000)		
= ($19,000) (0.16275) + 100	$ 3,192	
CR = ($28,000 – 2,000) (A/P 10-10 + 0.10 ($2,000)		
= ($26,000) (0.16275) + $200		$ 4,432
Labor	9,000	7,000
Maintenance, power, etc.	1,800	1,500
Equivalent Uniform Annual Cost	$13,992	$12,932

The equivalent uniform cost method is commonly used when the individual or agency to be "sold" is accustomed to considering financial matters in annual blocks of time. All three basic methods will select the same preferred alternative. The choice of method depends upon which might be more easily understood by the person or agency receiving the analysis and also on the form of the data. In public projects, for example, there are frequently large initial investments with relatively smaller annual amounts, usually for maintenance, and the present worth method may be more straightforward.

5.3.3 Asset Life

Mention has been made of asset life in this chapter without any definition. There are several ways to describe the life of an asset, and the following are among the more common.

Physical life implies a period of time over which an asset remains physically sound with reasonable care and maintenance. This period may extend beyond the time for which there is a useful function for it to perform.

Accounting life is based primarily on tax considerations as regulated by the Internal Revenue Service, and is the basis for depreciation charges. It may or may not correspond to its physical life or its economic life. The Economic Recovery Tax Act, of 1981 establishes depreciation lines generally shorter than the useful or economic life.

Economic life is an optimum one over which the costs of the asset are either minimized or its benefits are maximized. It is strongly influenced by technological obsolescence. The economic life is the most appropriate one to consider in engineering economy studies.

An asset such as a solid state device may be as physically sound after several years of use as the day it was purchased. On the books of account it may have a number of years yet remaining before it is fully depreciated.

However, its economic life may be in jeopardy if a more profitable device becomes available as an alternative.

5.3.4 Evaluations of Assets with Unequal Lives

It frequently happens that alternatives under consideration in a study do not have the same economic lives. Some adjustments are thus necessary to avoid skewing the results. The primary consideration is the anticipated life of the plant or larger unit of which the alternatives represent some component. What is the continuing requirement? In no case would it be valid to consider a component as having a longer life than its larger system. If the continuing requirement is for 35 years, for example, and the life of one alternative *A* is seven years, and that for alternative *B* is five years, then five identical alternative *A* units and seven units of *B*, would be considered. The assumption is that each identical replacement would have the same stream of costs. There may be other situations in which the life of the shorter-lived alternative may well be taken as the study period, or conversely, the life of the longer-lived alternative. There may be other study limits, but whatever the analysis, it remains important to carefully examine the consequences of the varying lives, possible differences in salvage value, and the options available subsequent to the study period.

5.4 COMPARISON OF ALTERNATIVES BY THE PRESENT WORTH METHOD

The same data presented in Example 2 in section 5.3.2 may be used to determine the preferred investment by the present worth method.

Example 1:

PW Plan A = *\$20,000 + 9,000 (P/A 10–10) + 1,800 (P/A 10–10) – 1,000 (P/F 10–10)*

= *\$20,000 + 10,800 (6.144) – 1,000 (0.3855)*

= *\$85,969*

PW Plan B = *\$28,000 + 7,000 (P/A 10–10) + 1,500 (P/A 10-10) – 2,000 (P/F 10–10)*

$= \$28,000 + 8,500\ (6.144)\ 2 - 2,000\ (0.3855)$

$= \$79,453.$

Plan B has the lower present worth cost and thus is the preferred alternative as previously determined in Example 2. The equivalent uniform annual costs from Example 2 in section 5.3.2 can readily be converted into present worth equivalents:

PW Plan A $= \$13,992\ (P/A\ 10\text{–}10)$

$= \$13,992\ (6.144)$

$= \$85,967$

PW Plan B $= \$12,932\ (P/A\ 10\text{–}10)$

$\$79,454.$

(Such slight differences are due to the rounding of variations in individual calculators including those used in constructing the tables.)

The choice of using the present worth method, as in the case of the equivalent uniform annual cost method, depends on who will be receiving the study or the preference of the analyst. Either method gives the same results.

5.4.1 Present Worth Represents a Financial Commitment Over Project Life

The use of the present worth, sometimes called present value, reduces the difficulty of comparing expenditures at varying periods of time by converting all such expenditures into a single value for comparison with other single values. Such single amounts represent the *total* commitment, at time *0*, of all future expenditures. This method frequently finds acceptance in large public projects involving deferred investments, in a variety of valuation situations, and in capitalized cost considerations. The latter is not to be confused with the accountant's "capitalizing" an expenditure in order to record it as an asset rather than a current expense. In engineering economy studies the concept of capitalized cost represents a single amount, a present worth, which at some given rate of interest will be equivalent to a cash flow of equal annual payments extending to infinity. The subsequent interest earned from the present worth would provide the required annual annuity indefinitely.

The present worth amount may be that of an excess of receipts over disbursements as well as expenditures alone. It is important that all alternatives be expressed in a similar fashion and that they all cover the same study period.

5.4.2 Importance of Proper Rate of Return

The specified interest rate, or the rate of return, is a crucial element, particularly when comparing expenditures whose timing may vary considerably from alternative to alternative. Careful consideration must be given to use a rate which is consistent with others used in the organization, considering the risks involved. Projects which will show a favorable excess of receipts over disbursements under one return may well show a negative present worth under another return. It is beneficial to remember once again that in calculating a present worth of a series of negative and positive cash flows, a positive present worth means that the stipulated rate has been exceeded. Similarly, a negative cash flow means that the excess of receipts over disbursements has failed to equal the specified rate. Consider the following example.

Example 2: A firm is exploring the possibility of replacing a present piece of equipment with one of two alternatives. Option A has an initial cost of $35,000. It is anticipated that receipts with Option A will exceed disbursements by $6,500 the first year, and will gradually increase by $400 per year over its economic life. Option B has an initial cost of $50,000 with an excess of receipts over disbursements estimated at $7,800 the first year with a gradual increase of $500 per year. Neither will have a net salvage value. With a minimum rate of return of 10% over a 10-year study period, which alternative is the more economical?

(In calculating a present worth of a combination of expenditures and receipts, it is advisable to use a minus (–) sign to denote all expenditures, and a plus (+) sign for all receipts.)

For Option A:

$$PW_A = \$35{,}000 + 6{,}500\ (P/A\ 10\text{–}10) + 400\ (P/G\ 10\text{–}10)$$

$$= \$35{,}000 + 6{,}500\ (6.144) + 400\ (22.8913)$$

$$= +\ \$14{,}093.$$

For Option B:

$$PW_B = -\$50{,}000 + 7{,}800\ (P/A\ 10\text{–}10) + 500\ (P/G\ 10\text{–}10)$$

$$= -\$50{,}000 + 7{,}800\ (6.144) + 500\ (22.8913)$$

$$= +\$9{,}369.$$

Both options will provide the minimum required 10%, and thus either might be acceptable. However, Option A has a larger positive present worth. This means that not only will the $35,000 investment provide a 10% return, but $14,093 above that amount — an amount greater than the excess with Option B.

Now assume that a 15% return had been specified rather than 10%:

For Option A:

$$PW_A = -\$35,000 + 6,500\ (P/A\ 15\text{–}10) + 400\ (P/G\ 15\text{–}10)$$

$$= \$35,000 + 6,500\ (5.019) + 400\ (16.9795)$$

$$= +\$4,416.$$

For Option B:

$$PW_B = -\$50,000 + 7,800\ (P/A\ 15\text{–}10) + 500\ (P/G\ 15\text{–}10)$$

$$= -\$50,000 + 7,800\ (5.019) + 500\ (16.9795)$$

$$= -\$2,362.$$

Option B has now failed to meet the minimum required rate of return. Option A still meets the minimum plus an additional sum. In the calculations using 10%, it is obvious that Option A is the preferred alternative. The interpretation of the amounts may not always lead to an obvious choice. In that case the question may be raised as to the earning power, expressed as a rate of return, of the extra $15,000 initial investment required by Option B. That investment will earn $1,300 per year more in addition to a gradient of $100 per year. This may be expressed by setting the present worth equal to zero. Some rate of interest will then make the receipts equivalent to the initial investment.

$$PW = 0 = \$15,000 + 1,300\ (P/A\ i\text{–}10) + 100\ (A/G i\text{–}10)$$

The rate which makes these amounts equivalent is 2.67%. This calculation shows that the additional investment required by Option B fails to meet the minimum required rate of return. It would be better to invest the extra $15,000 elsewhere in the firm where it could earn at least the minimum.

5.4.3 Comparison between Immediate and Deferred Alternatives

Example 3: A company is planning for a new operation which will require its own PBX. Model A, a small unit with a first cost of $5,000, will handle the work for four years, after which the volume is expected to increase such that a second unit will be required. Model B is a larger unit and would meet the anticipated increase with no difficulty. Its price of $7500 appears to offer some economy, even though not all its capacity would be needed right away. The annual operating and maintenance costs of A are estimated at $800 per year, while B's costs are estimated at $1,150. Model A has an estimated salvage value after 8 years of $450, the Model B at 8 years an estimated $750. Use the present worth method to determine whether it is best to buy the larger unit now or to select the smaller one and then purchase a similar additional unit four years hence. Use a 10% minimum rate of return. The study period is 8 years, after which it is likely the operation will be abandoned. In that case the second unit of A would have an estimated salvage value of $2,500.

PW of Model A = –$5,000 – 800 (P/A 10–8) – $5,000 (P/F 10–4)
–800 (P/A 10–4) (P/F 10–4) + 450 (P/F 10–4)
+ 2,500 (P/F 10–4)
= –$5,000 – 800 (5.335) – $5,000 (0.6830)
–800(3.170) (0.6830) + (450 +2,500) (0.3855)
= $13,278.

PW oF Model B = –$7,605 – 1,150 (P/A 10–8) + 750 (P/F 10–8)
= –$7,500 – 1,150(5.335) + 750(0.3855)
= $13,351.

The present worth equivalents of all the estimated costs for each PBX are within $73 of each other, and thus either one could be selected. The final determination will depend upon a close examination of the intangible factors. The difference in outcome of present worth studies are generally considered of no consequence if the net present worth between the two best plans is less than 5%.

Example 4: Radio relay stations in areas of extreme heat usually have high battery failure rates. One proposal is to provide air conditioning for the stations and thereby increase battery life from the present maximum of 6 years to a more normal life of 15 years. Assume a station's complement of batteries costs $16,000. Equipment to air condition the stations to maintain recommended temperatures has a first cost of $15,000. Annual testing, repairs, and ad valorem taxes on the batteries are estimated at 10% of first cost. Maintenance and ad valorem taxes on the air conditioning equipment are estimated at 3% of first cost. Assuming a 30-year life for the air conditioning equipment, and an 8% minimum rate of return, compare the present worth of the two alternatives. Salvage on batteries and air conditioning equipment is assumed to be negligible.

Note: To reduce some calculations, it may be easier to find some equivalent uniform annual costs and convert these to present worth.

No air conditioning:

PW of batteries = *16,000 (A/P 8–6)(P/A 8–30)*
= *16,000 (0.21632)(11.258)*
= *$38,965*

PW of annual charges = *0.10 (16,000)(P/A 8–30)*
= *0.10 (16,000)(11.258)*
= *$18,013*

Total PW = *$56,978*

Air Conditioning:

PW of batteries = *16,000 (A/P 8–15)(P/A 8–30)*
= *16,000 (0.11683)(11.258)*
= *$21,044*

PW of annual charges for batteries = *$18,013* (same)

PW of air conditioning equipment = *$15,000*

PW of annual charges for air conditioning = *0.03 (15,000)(P/A 8–30)*
= *450 (11.285)*
= *$5,066*

Total PW = *$59,123.*

The difference in the present worth of the two alternatives is $2,145 or 3.7%. Since the difference is also less than 5% as in the previous example, the consideration of intangibles will play a significant role. Intangible benefits of air conditioning the stations include improvement transmission qualities because of longer battery life and reduced equipment deterioration. Access to the remote areas would also be a factor.

5.4.4 Use of Present Worth in Determining Valuations

Another common use of the present worth method is in determining valuations. Problem 8 in section 5.2.10 is one example. Another is in determining a single figure equivalent to a series of payments on the use of a patent. Still another common application is in establishing the amount one would pay for a bond to achieve a particular yield. The term *yield* is the same as a rate of return and is commonly used in the commercial stock and bond market. The value of a bond, or the amount necessary to achieve a given return, equals the worth of all interest payments plus the face value of the bond.

Example 5: Assume that the bond market appears somewhat depressed, meaning that a number of bond issues are selling at less than their value at maturity. An investor is willing to risk investment if he can achieve an 8% yield. What price should he pay for a $1,000, 6% bond with interest payable semi-annually and which matures in 15 years?
The interest payments will be 3% per period or $30. Thus:

$$\begin{aligned} PW &= 30\ (P/A\ 4\text{–}30) + 1{,}000\ (P/F\ 4\text{–}30) \\ &= 30(17.292) + 1000\ (0.3083) \\ &= 518.76 + 308.30 \\ &= \$827 \end{aligned}$$

The 6% bond rate is only used to determine the annuity, or amount of interest, per period.

5.4.5 Use of Future Worth in Determining Valuations

While it is frequently convenient to make cost comparisons based on the present time (close to the decision time), there are other occasions when it is more helpful to know the future worth or future value. Assume, for example, that a company is comtemplating purchasing a parcel of land as a possible future site for an installation. Since the site installation is speculative, the company might want to calculate the future worth of its investment to determine how it would compare with a realistic selling price. The future worth of the investment is the sum of the compounded amount of the original cost and the compounded amounts of any other costs including annual property taxes and any improvements.

Example 6: Assume that in 1975 a small cable company established a 10 year development plan which included a major acquisition seen as necessary in early 1984. Interest rate forecasts available in 1975 indicated a strong possibility of increases by that year. The company decided to set aside annual deposits of $50,000, beginning at the end of 1975, into a special account rather than face borrowing at high interest rates in 1984. However, the company ran into some financial difficulties in 1979 and 1980 and was unable to make any deposit those two years. It is now the end of 1983, and a company official wishes to check the accuracy of the accumulated amount indicated to him by the accounting office. If the fund was to earn 10%, what should the total be immediately after the 1983 deposit?

Solution: Because of the interruptions in 1979 and 1980, the payments do not represent a uniform series throughout the life. However, the first 4 payments may be treated as a series through the end of 1978 and as a single sum the remainder of the time. Thus, the accumulated amount at the end of 1978 will be:

$50,000 (F/A 10–4) = 50,000 (4.641) = $232,050.

That portion will accumulate as a single sum so that by the end of 1983 its value will be:

$232,050 (F/P 10–5) = $232,050 (1.6105) = $373,716.

The payments resumed in 1981 will constitute a series ending with the last in 1983, and will accumulate to:

$50,000 (F/A 10–3) = $50,000 (3.310) = $165,500.

The total amount, or future worth of the fund at the end of 1983 will be:

$373,716 + 165,500 = $539,216.

The future worth may also be determined by assuming that deposits were made each year, and then to subtract out the accumulated amounts of the assumed 1979 and 1980 payments.

If deposits had been made each year, the future worth would be $50,000 (F/A 10–9) = 50,000 (13.579) = $678,950. The accumulated equivalents of 1979 and 1980 payments which were not actually made would be:

$50,000 (F/A 10–2)(F/P 10–3) = $50,000 (2.100)(1.3310) = $139,755.

The net accumulated worth:

$678,950 – 139,755 = $539,195

(Note: The small difference, $21, is due to rounding errors in the calculator).

5.5 COMPARISON OF ALTERNATIVES USING THE RATE OF RETURN METHOD

In the two prior methods, the equivalent uniform annual cost technique, and the present worth process, some minimum rate of return was assumed as a cost, or the necessary return. In the rate of return method, the measure of efficiency, or the rate, is the unknown factor. It is that interest rate which makes a series of expenditures and a series of receipts equivalent to each other.

5.5.1 Analysis of the Concept

The use of the rate of return as a measure of performance is frequently confusing because the term describes two different concepts. In the widespread accounting applications, it translates the net income or profit realized during the accounting period as a percentage or index of the "investment" or book value necessary to achieve that profit. In the decentralization of profit responsibility it is a useful measure of management stewardship. This application of the phrase "rate of return" as a measure of management stewardship is not satisfactory as a planning tool, however, because it is viewed as an overall return on the overall investment in the profit center. In planning, it is important to examine the profitability of each increment of investment such as each additional amount proposed for investment in specific units of equipment, process facilities, communications systems, buildings, and other assets. In this sense, "rate of return" reconciles the financial or the accounting viewpoint with the measurement of efficiency logic of the engineer.

The return on investment concept, as generally expressed in financial statements, does not give weighted consideration to the time period over which the investment earns its return. The time value of money concept differentiates very clearly the significance of immediate and near-future receipts from those some years away.

Some authors and some companies have used the term "profitability index," describing such an index as a number which equals the annual percent of compound interest which money invested in facilities and the like will yield over the life of the project. Thus, the index is usually a synonym for rate of return or return on investment.

The rate of return is also known as the *internal rate of return,* because it is devoid of external influences, and is completely dependent upon the internal cash flow consequences of the proposed investment.

5.5.2 Calculating the Rate of Return

The three principal elements are the investment, the savings or excess of receipts over disbursements, and the duration of the study period. The interest rate is the unknown, and the usual way is to solve the appropriate equation by trial and error. An estimate of the rate of return is made first to achieve some approximation.

Example 1: Company X is investigating the possibility of purchasing System Y with an installed cost of $84,000 and an estimated economic life of 6 years. A careful examination of the potential net savings estimates them at $26,500 per year. It has also been determined there will be no net realizable value at the end of the six years. What is the rate of return before income taxes?

Note that such problems may be set up on either the annual cost or present worth basis.

$$PW = 0 = -\ \$84{,}000 + \$26{,}500\ (P/A\ i{-}6)$$

$$(P/A\ i{-}6) = \frac{\$84{,}000}{26{,}500} = 3.169$$

Examination of the appropriate interest tables indicates that the factor of 3.169 falls somewhere between that of 20% (3.326) and of 25% (2.951). While the derivation of the interest formulas earlier in the chapter indicates a geometric progression, straight line interpolation may be used between relatively small intervals for all practical purposes. The error resulting from linear interpolation will usually be much less than the errors in the estimates of cash flows over a period of years. Thus, the rate, *i*, may be found:

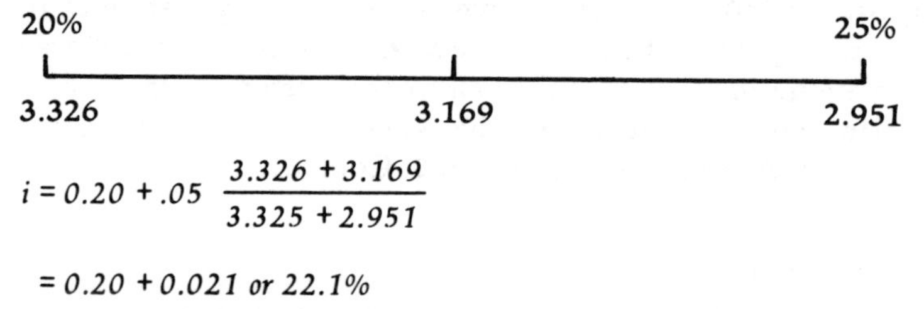

$$i = 0.20 + .05\ \frac{3.326 + 3.169}{3.325 + 2.951}$$

$$= 0.20 + 0.021 \text{ or } 22.1\%$$

5.5.3 Interpreting the Results

Now the question is the significance of the 22.1%. If Company X has experienced a before-tax return on investment of 10% it might be tempted to give this proposal a high priority. Similarly, Company X may be tempted to rate this proposal high if it considers this return in terms of the cost of the various kinds of capital available to it. Both temptations may lead to difficulty. Investment proposals should compete with one another on comparative merit alone — independent of the company's "return on investment" and the source of funds. A basic concept of economics states that all resources, including monetary funds, are limited, and it is therefore important to use those funds which will be most productive. System Y must be compared with other demands for corporate funds before a decision can be made. The technical person with a "hot" idea frequently cannot understand why funds are not immediately available to implement his request. He must recognize the need for understanding profitability calculations as a prerequisite to action.

5.5.4 Income Tax as a Factor in Rate of Return Calculations

Previous examples have avoided consideration of income tax. Section 5.8 will deal briefly with the impact of taxes. The complexities of the Internal Revenue Service requirements, as experienced not only in the United States but also in most other countries of the free world, are best left to the accountants. However, the technical person must understand income tax as a cost of doing business, and must include this cost in his calculations. The most common consideration occurs when a choice is to be made between a high-cost investment in an asset and a lower cost alternative. Generally, the only reason to invest in the higher-cost alternative is when the resulting savings "earn their way." In terms of individual projects, the lower annual costs of the higher-cost alternative may well result in lower deductions for tax purposes, and thus have a higher income tax cost. This fact should be kept in mind regardless of which of the evaluation methods is used.

Example 2: Company Z is considering two proposals for improved operations. Alternative A involves a present investment of $100,000 with anticipated savings of $25,000. Alternative B is more expensive, with a first cost of $140,000, but will have lower operating costs such that the annual savings are anticipated to be $29,500. It is expected there will be no remaining value with either alternative after 20 years. Estimated annual income taxes with alternative A are $10,000 with the corresponding tax with Alternative B as $11,250. Compute the rate of return on each.

For Alternative A:

$$PW = 0 = -\$100{,}000 + \$25{,}000(P/A\ i\text{–}20) - \$10{,}000(P/A\ i\text{–}20)$$

$$= -\$100{,}000 + \$15{,}000(P/A\ i\text{–}20)$$

$$i = 13.9\%$$

For Alternative B:

$$PW = 0 = -\$140{,}000 + \$29{,}500(P/A\ i\text{–}20) - \$11{,}250(P/A\ i\text{–}20)$$

$$= -\$140{,}000 + \$18{,}250(P/A\ i\text{–}20)$$

$$i = 11.6\%$$

Assume that the minimum after-tax rate of return for any preliminary consideration by the company is 10%. Which alternative should be selected? One would select Alternative A.

Both alternatives exceed the minimum of 10%. However, a further step, utilizing the rate of return approach, dramatizes one of the values of this method and clarifies the differences. The incremental rate of return is a useful device whether comparing two alternatives or a number of them. Does the extra savings of $3,250 with Alternative B justify the extra $40,000 investment?

$$PW = 0 = -\$40{,}000 + \$3{,}250\ (P/A\ i\text{-}20)$$

$$i = 5\%$$

Without hesitation now, one can clearly state that in no way is the higher cost alternative justified.

Example 3: The two previous examples were straightforward and required no trial and error calculations because there was only one unknown.

Assume that an investment of $50,000 with an estimated salvage value of $8,000 at the end of 10 years results in a year-by-year excess of receipts over disbursements as follows. Calculate the expected rate of return.

End of Year	*Cash Flow*
0	$50,000
1	+7,500
2	+8,000
3	+8,000
4	+8,000
5	+8,000
6	+8,000
7	+7,600
8	+7,600
9	+7,600
10	+7,600
10	+8,000

The resulting equation would be set up as follows:

$$PW + 0 = -\$50{,}000 + \$7{,}500(P/F\ i\text{-}1) + \$8{,}000(P/A\ i\text{-}5)\ (P/F\ i\text{-}1)$$
$$+7{,}600(P/A\ i\text{-}4)(P/F\ i\text{-}6) + \$8{,}000(P/F\ i\text{-}10)$$

This may also be written as:

$$PW = 0 = -\$50{,}000 + \$8{,}000(P/A\ i{-}10) - \$400(P/A\ i{-}4)(P/F\ i{-}6)$$
$$-500(P/F\ i{-}1) + \$8{,}000(P/F\ i{-}10)$$

Before attempting a trial and error solution, a brief examination indicates an essentially level rate of about $7,700 in addition to the salvage. A perusal of the tables indicates that the present worth of an amount 10 years hence varies between 1/3 and 1/2 over a wide range of interest rates.

The equation to give an approximate rate of return is then:

$$PW = 0 = -\$50{,}000 + \$7{,}700(P/A\ i{-}10) + \$3{,}000(P/A\ i{-}10)$$
$$= \$47{,}000/\$7{,}700 = 6.10;\ \text{try } 10\%$$

for 10%

$$PW = 0 = -\$50{,}000 + \$7{,}500(0.9091) + \$8{,}000\ (3.791)(0.9091)$$
$$+ \$7{,}600(3.70)(0.645) + \$8{,}000\ (0.3855)$$

$$= +\$1{,}073$$

for 11%

$$PW = 0 = -\$50{,}000 + \$7{,}500(0.9009) + \$8{,}000\ (3.696)(0.9009)$$

$$+ \$7{,}600(3.102)(0.5346) + \$8{,}000(0.3522)$$

$$= -\$1{,}184$$

$$i = 10.48\%$$

While the calculations for the above trial and error solution may seem somewhat tedious, the use of computer programs tailored to individual company parameters permits a number of very rapid computations and comparisons. There are clues of where to begin as illustrated above.

The rate of return method has a definite advantage when used to evaluate incremental investments, because it readily indentifies the efficiency of additional amounts over the required minimum. Comparisons are thus readily facilitated.

A major problem, perhaps, is interpreting the rate of return which has been calculated. It should certainly be no lower than the cost of capital to be minimally acceptable. This condition, unfortunately, implicitly assumes the risks of the proposal are minimal. When studies are made for the purpose of rationing capital to the most profitable proposals, lower limits are usually quickly established among competing alternatives.

5.6 BENEFIT-COST ANALYSIS

The benefit-cost (B/C) analysis method is most widely used in the public sector. It requires essentially the same data as the other evaluation techniques. Presumably, mention of an excess of benefits over costs is more palatable to the taxpayers of the world than to speak of a rate of return. The latter implies that the government is earning a profit on money extracted through taxes. Taxpayers are either in revolt or on the verge of it in many countries, resulting in an increased need for some better measure of the use of government funds. Taxpayers, the World Bank, similar agencies, and foreign country assistance, do not represent an endless source of funds, and thus the use of limited funds presumably must be optimized as in the private sector. The benefit-cost analysis is not a new method, but is enjoying widespread and accelerating popularity because governments and development institutions, as well as private investors, are requiring a much more sophisticated analysis before committing funds. Even so, there are many related problems, primarily because the people receiving the benefits are frequently not those paying the costs, and *vice versa*.

5.6.1 Quantification of Benefits and Costs

Each proposal must first attempt to identify those who are to receive the benefits (also known as the users) and the agency which is to pay for them. All of the benefits minus any disbenefits (additional costs to the users) are included in the numerator and all of the agency costs minus any savings to the agency are placed in the denominator. A proposal should have a ratio of at least 1.0 to justify expenditures of funds. Again, since public funds are limited, competition for funds would favor those projects with a much higher ratio.

As in the other methods, all benefits and all costs must be expressed in equivalent terms — either in an equivalent uniform annual basis, or in the single figure present worth mode. Occasionally benefits can only be expressed as differences in the costs to the user.

Expressing benefits and costs in equivalent terms implies using some rate of interest. Public projects frequently suffer from a wide variation in interest rates, expected project lives, and operational charges. Variations are particularly widespread when a public agency is attempting to prove its case vis-a-vis a privately financed competitor. Lower interest rates tend to favor projects which might otherwise be undesirable. The present worth of benefits on a long term project is very sensitive to the interest rate used. On federal government projects a minimum rate has been established based on the current rate for long-term treasury obligations as reported in the *Treasury Bulletin.*[5] The use of this rate carries the assumption that there is no capital rationing, or, in other words, money for all purposes can be borrowed at the same rate of interest. While this is a common practice, it does not consider the degree of risk and the opportunity cost (the interest rate on the best opportunity foregone) as is commonly followed in the private sector.

5.6.2 Advantages of the Benefit/Cost Method

There are no inherent advantages to the benefit versus cost approach as compared with rate of return, present worth, or annual cost methods. However, the use of the B/C approach, particularly with regard to local governments, offers some gains which might otherwise not be obtained:

i. It requires the agency to think in terms of alternatives rather than advocating a single possibility.

ii. It helps to separate unprofitable programs or activities from those with a greater potential for productivity.

iii. It focuses attention on long-range costs and consequences rather than on only the immediate control for productivity.

iv. Measurable benefits can be weighed against measurable costs. Then unmeasurable benefits can be examined separately to see whether they are impressive enough to justify the necessary measurable costs.

5.6.3 Calculations

Before a ratio can be computed, all of the measurable benefits and all of the measurable costs must be identified and expressed in monetary units, either on a present worth or equivalent annual basis. There are two methods of establishing the numerator and the denominator. One is expressed in section 5.6.1. An alternative is to subtract annual agency costs from the benefits. There is no decision problem with either method as long as the analyst is consistent.

Example 1: An emergency phone system is proposed for a certain section of remote highway where breakdowns, for some unknown reason, seem to occur more frequently than elsewhere. The cost of installing the system is $100,000, and annual operating and maintenance charges are estimated to be $5,000 during its 10-year useful life. After that time a major re-routing and improvement program should be completed, and the emergency system will not be needed. If the system is installed, it is estimated that annual road user costs due to breakdowns will be reduced to approximately $7,000 from the $30,000 average at present. Savings will result primarily from reduction in down time for commercial vehicles, and the reduction of service and towing charges. If an 8% cost of capital is used, what is the benefit to cost ratio for the proposal?

Benefits = Road user savings = *$23,000*
PW 10 yrs. = *$23,000 (P/A 8-10)*
= *$23,00 (6.710)*
= *$154,330*
Costs = *$100,000 + $5,000 (P/A 8-10)*
= *$100,000 + 33,500*
= *$133,500*
B/C = *$154,330 = 1.15*
$133,500

Benefit cost analysis provides a quantitative basis for making decisions in relatively subjective areas which otherwise might be treated by intuition alone. In the above example, other alternatives might be provided, and could be compared. The method lends itself to incremental analysis similar to that used in the rate of return method.

5.7 OTHER YARDSTICKS

Different methods of evaluation are frequently used for each type of investment proposal in order to prove a previously formed opinion. Personal preferences may be viewed with some suspicion when that preference switches with each new situation.

5.7.1 Payout Period

One of the common "yardsticks" is the payout or payback period, commonly defined as the length of time required to recover the intial investment from the net cost flow produced. A frequently used maximum period is three years. A proposed investment is not considered justifiable if the investment divided by the annual net cash flow exceeds three. This is the period of time after which the project presumably could be abandoned with the investment fully "paid back."

This method has shortcomings. It fails to consider the time value of money and the failure to recognize the effect of any achievements past the early years. No salvage values are included. Although there are some modifications of the simple formula, .none addresses the two major shortcomings.

The payout period probably enjoys widespread use for two reasons. It is simple to use, requiring a minimum of calculations. For those who are anxious about risk and the liquidity of their positions, it seems to favor quick-profit projects.

5.7.2 MAPI Formula

An elaborate analysis procedure has been devised by George Terborgh of the Machinery and Allied Products Institute (MAPI). The package provides a variety of assumptions, graphs, and worksheets for use in replacement investment proposals. The trade association represents firms doing business in the replacement market. This bit of knowledge may make it easier to understand the emphasis on factors for "accumulated inferiority" of present equipment. Many of the assumptions are unrealistic, and the MAPI formula provides no basis for direct, quantitative comparison with other alternatives. The MAPI method has consequently fallen into general disfavor.

5.8 DEPRECIATION AND INCOME TAX

Obsolescence and physical wear and tear are two major concerns of the owners of physical assets. The former may be regarded as functional depreciation, and is due to changes in technology, demand, or requirements. It is of particular concern in high technology equipment where new developments may lead to retirement of many units practically overnight, and is of more concern than physical depreciation.

5.8.1 Definitions of Depreciation

Considerable confusion surrounds the use of the term *depreciation*, especially when various "official" definitions are examined. The Federal Power Commission System of Accounts for electric power companies and the FCC System of Accounts for telephone companies are very nearly alike in interpretation:

> "'Depreciation, ...means the loss in service value not restored by current maintenance, incurred in connection with the consumption or prospective retirement of. . . .plant in the course of service from causes which are known to be in current operation and against which the utility is not protected by insurance."

This is the "wear and tear" definition adopted by the commissions following hearings on depreciation of railroads and telephone companies during the 1930's. This definition was essentially affirmed by the Internal Revenue Code of 1954 which said: "General Rule. . . .There shall be allowed as a depreciation deduction a reasonable allowance for the exhaustion, wear and tear (including a reasonable allowance for obsolescence). . . .(1) of property used in the trade or business, or (2) of property held for the production of income."

The effects of inflation have given rise to other interpretations. Following World War II the U.S. Steel Company found it could not replace worn out facilities except at greatly inflated prices compared to the original cost and viewed depreciation as a process where funds are raised to pay for the replacement of plant and equipment. In its 1964 Annual Report the American Telephone and Telegraph Company urged this same interpretation:

> "We continue to urge also that both taxing and regulatory authorities should allow depreciation charges that recognize the lower purchasing power of today's dollar. Although in recent years the decline in the dollar's value has slowed considerably, much of the total investment in industry was made when each dollar was worth much more. Hence depreciation allowances limited to the number of dollars originally invested cannot recover the true cost of the investment being consumed. In our judgment the tax law should permit recovery, through depreciation, of the full purchasing power of each dollar invested."[6]

In a case involving telephone rates, the Indiana Public Service Commission authorized the accrual of depreciation upon the cost of property repriced in current dollars.[7]

An interpretation found in *Accounting Terminology Bulletin 1,* issued by the American Institute of Accountants, gained wide approval, and until 1981 was the interpretation for U.S. federal government income tax purposes.

> "Depreciation accounting is a system of accounting which aims to distribute cost or other basic value of tangible capital assets less salvage (if any), over the estimated useful life of the unit (which may be a group of assets) in a systematic and rational manner. It is a process of allocation, not of valuation."

The key term is *allocation* in which a periodic charge is made to allocate or amortize as a cost of current business, for income tax purposes, a previously incurred capital expenditure.

The word "depreciation" has thus been variously defined, with the allocation concept finally gaining widespread acceptance. This was the case until the Economic Recovery Tax Act of 1981 essentially replaced the word with the phrase "accelerated cost recovery system" (ACRS) called "acres".

5.8.2 The Accelerated Cost Recovery System (ACRS)

Depreciation deductions for income tax purposes have always been a part of the United States federal tax system. The need for funds for public works programs initiated during the Great Depression caused the

U.S. government to sharply curtail the deductions by requiring unrealistically long lives. The resulting smaller deductions meant increased taxes paid. Over the years since then there has been pressure to liberalize the depreciation rates, and changes have gradually taken place. A significant change was the 1954 Code by which Congress granted permission to taxpayers to use accelerated methods of depreciation. The two primary methods were the double-rate declining balance and the sum-of-the-years digits.

The Economic Recovery Tax Act of 1981 with ACRS is the latest in this series of changes and is the most extreme. Congress felt that additional incentives were needed to stimulate capital investment. Prior to the passage of the new law, taxpayers were permitted to depreciate the cost of an asset over its useful life. ACRS provides for recovering the cost of the assets by depreciating them over pre-determined recovery periods which are generally shorter than their useful life. The recovery periods are 3 years, 5 years, 10 years, or 15 years, depending on the classification of the assets. While there are a few exceptions, ACRS is a mandatory system which must be used by all taxpapers. Properties placed in service prior to 1981 are not changed to the new schedule. Since ACRS is not based on estimated useful lives, but on a cost recovery schedule, it is not, strictly speaking, a depreciation system. The Code does, however, classify it as system of depreciation.

Salvage is no longer a factor. Recovery percentages are applied to an unadjusted basis. A determination is made as to which of the four classes of property the asset belongs, and a statutory recovery percentage is assigned for each year of the recovery period.

The recovery percentages for the ACRS deduction for 1981-1984 are as follows:

If the recovery year is:	The applicable percentage for the class of property is:			
	3 year	5 year	10 year	15 year public utility
1	25	15	8	5
2	38	22	14	10
3	37	21	12	9
4		21	10	8
5		21	10	7
6			10	7
7			9	6
8			9	6
9			9	6
10			9	6

11	6
12	6
13	6
14	6
15	6

There are some small changes built in for 1985. The recovery percentages for property placed in service in 1986 and thereafter are as follows:

If the recovery year is:	The applicable percentage for the class of property is: 3 year	5 year	10 year	15 year public utility
1	33	20	10	7
2	45	32	18	12
3	22	24	16	12
4		16	14	11
5		8	12	10
6			10	9
7			8	8
8			6	7
9			4	6
10			2	5
11				4
12				3
13				3
14				2
15				1

Three year property includes special tools, autos, light trucks, and equipment used in research and development.

Five year property includes most machinery and equipment not used in reasearch. Business equipment and furniture are in this category.

Ten year property includes some public utility property (with a previous class life of more than 18 years but not more than 25 years), manufactured homes, and railroad cars.

Fifteen year property includes buildings and certain public utility property.

5.8.3 Investment Tax Credit

An investment credit, or credit against income taxes, was first introduced by Congress in 1962. It has undergone a number of changes in the attempt to encourage business spending for certain classes of depreciable property. Under ACRS, taxpayers are given a choice between reducing the credit or reducing the basis for recovering the investment. This tends to limit, but not eliminate, a double benefit.

5.8.4 Previous Depreciation Methods

Although ACRS is a mandatory system, it does prohibit properties placed in service prior to 1981 from being changed to the new schedule. The earlier methods will thus continue for some time. Like the new ACRS, they are each based on time and thus are independent of use, as were those specifically permitted by the U.S. Treasury Department. Each of the three methods —straight line, sum-of-the-years digits, and double rate declining balance —has unique features which appealed to different management philosophies. The latter two, known as accelerated methods, provided for the allocation of the bulk of the investment during the early years of the asset's life. This provided some hedge against obsolescence and other sudden changes which could render the assets less valuable. Regardless of the method, including the new ACRS, taxes are not avoided, but only the timing is changed. The timing, however, usually has significant input on the rate of return.

5.8.5 Straight Line Depreciation

The straight line method is the simplest to understand and to apply. A constant depreciation charge is made each year. The total amount to be depreciated, installed cost minus salvage, if any, is divided by the estimate of useful life in years. The book value is the difference between the installed cost and the product of the number of years of use and the annual depreciation charge. It was essentially the only method used between the major changes in tax policy of the U.S. Treasury Department in 1934 and the accelerated methods allowed beginning in 1954. Many assets experience rapid decrease in value during the early portions of their lives, and straight line depreciation does not recognize this condition. Although its use was declining, many firms, expecially public utilities, continued to follow it. It remains one of the allowed exceptions under ACRS.

Example 1: Consider an interconnect device with a first cost of $35,000, an estimated life of 10 years, and an estimated salvage value of $3,000.

Straight line rate = *($35,000 −$3,000)/10* = *$3,200* **depreciation per year**

5.8.6 Sum-of-the-Years Digits

Both the sum-of-the-years digits (SYD) method and the double rate declining balance (DRDB) method are known as accelerated methods.

Both permitted allocation for tax purposes of most of the asset's value during the first third of its life. The SYD method takes its name from the calculation procedure. The annual allocation is the ratio of the digit representing the remaining years of life to the sum of all the digits for the original entire life. The sum-of-the-year digits from 1 to n is

$$SYD = \sum_{j=1}^{n} j = \frac{n(n+1)}{2}$$

Example 2: Using the same data as in Example 1 in section 5.8.5, determine the first year depreciation charge using SYD:

$$SYD = \frac{10(11)}{2} + 55$$

$$D_1 = \frac{10}{55}(\$35{,}000 - \$3{,}000) = \$5{,}818$$

The book value after five years with the straight line method would be:

$$\$35{,}000 - 5(\$3{,}200) = \$19{,}000.$$

The unallocated amount at the same time by the SYD method would be:

$$\begin{aligned} BV_5 &= \$35{,}000 - (40/55)(\$35{,}000 - \$3{,}000) \\ &= \$11{,}727. \end{aligned}$$

5.8.7 Double Rate Declining Balance Method

The third major method is the declining balance. A constant depreciation rate is applied to the remaining book value of the asset. The rate allowed by the Internal Revenue Service depended on the type of property. Most industrial and business assets qualified for the double rate which is twice the straight line rate. The salvage value is not subtracted from the installed cost before applying the depreciation rate.

Example 3: If the double rate declining balance method were to be used for the interconnect device in Example 1, Section 5.8.5, the first year depreciation charge would be 2 × 1/10 or 20%:

$$\begin{aligned} D_1 &= .20 \times \$35{,}000 \\ &= \$7{,}000. \end{aligned}$$

The new book value at the beginning of the second year would be

$$\$35{,}000 - \$7{,}000 = \$28{,}000.$$

The depreciation allocation at the end of the second year would be

$$D_2 = .20 \times \$38{,}000 = \$5{,}600.$$

The book at the end of the 5th year is

$$BV_5 = \$35{,}000(1 - .20)^5 = \$11{,}469.$$

A comparison of selected depreciation amounts and book values for Example 1 is of interest to note the differences in write-off:

	Straight Line	SYD	DRDB	ACRS *(1983)
Installed cost	$35,000	$35,000	$35,000	$35,000
Depreciation, 1st year	3,200	5,818	7,000	5,250
Depreciation, 2nd year	3,200	5,236	5,600	7,700
Book value, end of 5th year	19,000	11,727	11,469	0

*Under ACRS, the device would be in the 5-year category

It may be noted that about two-thirds of the installed cost has been depreciated by the end of the first half of the asset's life when the accelerated methods are used. The relative size of the salvage value will cause some variations with the SYD method. The more rapid allocations result in a higher rate of return on invested capital. High depreciation charges and consequent low taxes in the early years provide an advantage because of the time value of money. When the accelerated methods, including ACRS, provide for smaller allocations than the straight line method in the later years, the income tax charge will be higher. By the end of the asset's life, the amount of the tax paid will be the same. The ability to defer federal income taxes from the use of accelerated depreciation provides interest-free funds for additional investment. This is especially true under ACRS.

5.8.8 After-Tax Evaluations

In many engineering economy studies it will be found that an alternative which appears to be most economical in a before-tax cash flow analysis will also be the choice after income tax considerations. However, a clear picture of any differentiation of proposals can only emerge when all significant costs have been gathered, including income tax. Section 5.5.4 pointed out that more efficient alternatives will frequently have lower deductions for tax purposes, and thus must include a higher tax charge as an appropriate cost.

The calculation of corporate federal income taxes is complicated and requires specialists. Computations are further complicated by investment tax credits, permitted under the Internal Revenue Code for specified investments in new plant facilities. These credits have been revised under the Economic Recovery Tax Act of 1981.

Consider the following example of an after-tax analysis (old schedule):

Example 3: An $85,000 investment in new equipment is proposed. Preliminary studies indicate a potential saving in labor and other charges of $21,500 a year over the next 10 years. The investment will be depreciated for income tax purposes by the sum-of-the years digits method assuming a 10-year life and zero salvage. The effective income tax rate is 50%. A 10% investment tax credit will be taken at year zero. What is the prospective rate of return after taxes?

A year-by-year tabular cash flow display will be necessary.

$$SYD = \frac{10(11)}{2} = 55.$$

End of Year	Cash Flow Before Income Tax		Depreciation	Taxable Income	Income Tax	Cash Flow After Income Tax
0	-$85,000		$ —	$ —	+$8,500	-$76,500
1	+ 21,500	(10/55)	-15,455	6,045	- 3,023	+ 18,477
2	+ 21,500	(9/55)	-13,909	7,591	- 3,796	+ 17,704
3	+ 21,500	(8/55)	-12,364	9,136	- 4,568	+ 16,932
4	+ 21,500	(7/55)	-10,818	10,682	- 5,341	+ 16,159
.	.	.	.	.	.	.
.	.	.	.	.	.	.
10	+ 21,500	(1/55)	- 1,545	19,955	- 9,978	+ 11,522

The after-tax rate of return may be determined by equating the net outlay ($76,500 with the tax credit) to the net annual savings after income taxes. Note also that the difference between the year-by-year after tax cash flows appears to be a constant ($773), and thus a gradient factor may be used.

$$PW = 0 = -\$76,500 + \$18,477\ (P/A\ i\text{–}10) - \$773\ (P/G\ i\text{–}10)$$

The unknown i may be estimated by temporarily ignoring the gradient amount:

$$(P/A\ i\text{–}10) = 76,500/18,477 = 4.14.$$

From the tables, $(P/A\ 20\text{–}10) = 4.192$. Thus 20% may be used as the first assumption:

$$\begin{aligned} PW = 0 &= -\$76,500 + \$18,477\ (P/A\ 20\text{–}10) - \$773\ (P/G\ 20\text{–}10) \\ &= -\$76,500 + \$18,477\ (4.192) - \$773\ (12.8871) \\ &= -\$9,007. \end{aligned}$$

The savings failed to generate a 20% return by falling short \$9,007.

Assume $i = 15\%$

$$PW = 0 = -\$76{,}500 + \$18{,}477\ (5.019) - \$773\ (16.9795)$$

$$= +\$3{,}111. \text{ Thus 15\% falls too low.}$$

Interpolating between 15 and 20%

$$ROR = 15\% + (.05)\ \frac{3{,}111}{3{,}111 + 9{,}007}$$

$$= 16.28\% \text{ or } 16.3\%$$

Example 4: Consider the same problem as in Example 3, but with straight line depreciation.

End of Year	Cash Flow Before Income Tax	Depreciation	Taxable Income	Income Tax	Cash Flow After Income Tax
0	-\$85,000	\$ —	\$ —	+\$8,500	-\$76,500
1-10	+ 21,500	- 8,500	-13,000	- 6,500	+ 15,000

$$PW = 0 = -\$76{,}500 + \$15{,}000\ (P/A\ i{-}10)$$
$$(P/A\ i{-}10) = 76{,}500/15{,}000 = 5.1$$

From the tables, $i = 15.5\%$

Example 5: Consider the same problem as in Example 3, but under the 1983 ACRS schedule. The equipment would now shift to the 5-year recovery period, and assume the investment tax credit would have to be reduced two percentage points to 8%. Also that the equipment would remain in service for 10 years.

End of Year	Cash Flow Before Tax	Capital Recovery Percentage & Amount	Taxable Income	Income Tax	Cash Flow After Income Tax
0	\$-85,000	—	—	+ 6,800	\$78,200
1	+21,500	(15%) 12,750	8,750	4,375	17,125
2	21,500	(22%) 18,700	2,800	1,400	20,100
3	21,500	(21%) 17,850	3,650	1,825	19,675
4	21,500	(21%) 17,850	3,650	1,825	19,675
5	21,500	(21%) 17,850	3,650	1,825	19,657
6-10	21,500	—	21,500	10,750	10,750
		(100%) 85,000			

The after-tax rate of return is calculated as in Examples 3 and 4.

$$PW = 0 - \$78{,}200 + 17{,}125\ (P/F\ i{-}1) + (20{,}100)(P/F\ i{-}2) + 19{,}675\ (P/A\ i{-}3)\ (P/F\ i{-}2)$$

for $i = 18\%$, $PW = -\$3{,}838$
for $i = 16\%$, $PW = +1100$

Interpolation between 16 and 18% gives a $ROR = 16.57\%$.

It is interesting to note this result (16.57%) under the new ACRS system as compared with the old schedule used in Example 3 (16.28%). The net outlay under ACRS is greater because of the reduction in the investment tax credit. That has a more immediate impact than offsetting savings occurring later. The net cash flow for the first year under ACRS is less than under the old schedule. Subsequent cash flows under ACRS have a slight advantage, and the rate of return advances but 0.3 of 1%.

In the early years, the larger after-tax cash flows resulting from the accelerated methods (including ACRS) as compared with the straight line method, have a greater impact than the lesser ones toward the end of the life. The result is greater efficiency in the use of the investment, or a better return.

The impact of using accelerated depreciation and thus deferring Federal income taxes was highlighted in 1968 during the California Public Utilities Commission rate hearings of Pacific Telephone and Telegraph Company. Prior to that time the Bell System used a straight line depreciation for both book and tax purposes. The California PUC held the view that the larger tax deductions for depreciation in the early years of a depreciable asset would produce tax savings, and that these savings should be treated as a reduction in expense and allowed to flow through to earnings. Pacific Telephone and Telegraph Company argued that while this method would temporarily improve earnings, the taxes that were deferred would ultimately have to be paid. The PUC said as long as the company's growth continued, there would be no cause for concern. Other regulatory agencies, including the Federal Power Commission, had ordered the flow through principle as early as 1964.

5.9 MULTIPLE ALTERNATIVES

Preceding sections have dealt primarily with two alternatives. Frequently there are more than two possibilities, or multiple alternatives. When the selection of one alternative precludes the selection of others, the alternatives are termed mutually exclusive.

There are a number of approaches, including the basic equivalent uniform annual cost and present worth methods. They are straightforward, once it has been determined which rate of return to use. Both the rate of return and benefit/cost approaches lend themselves more readily to the incremental concept where each additional portion of investment above the minimum required can be tested for its efficiency. Since more capital can be invested in other acceptable alternatives, the incremental investment must be justified. For the benefit/cost method of analysis, the alternative to be selected is the one requiring the largest investment and whose incremental investment over another acceptable alternative meets or exceeds a B/C ratio of 1. Similarly, for the rate of return approach with mutual exclusive alternatives, the one to select is again that requiring the largest investment and whose incremental investment over another acceptable alternative meets or exceeds the minimum required rate of return.

5.9.1 Incremental Rate of Return

The rationale for investing in any alternative greater than the minimum necessary to do the job is that the extra investment is productive in the sense that it provides at least the minimum required rate of return. If not, only the basic investment is justified, and the extra amount is utilized elsewhere, where it can be at least minimally productive.

The specific steps to be used when evaluating mutually exclusive multiple alternatives may be summarized as follows:

1. Rank the alternatives according to increasing size of intitial investment.
2. If potential savings or cost reductions are indicated for each, calculate the rate of return for each investment, using either the EUAC or present worth methods over the economic life. Eliminate any which do not meet the minimum required rate of return. If only operating and similar annual costs are specified, determine the incremental rate of return of the second alternative with the least investment. If the incremental rate of return *does not* meet or exceed the minimum, eliminate that alternative and go on to the next highest and check its incremental return, comparing it with the lowest. If the incremental rate of return of the second alternative *does* meet the minimum requirement, use the second alternative as a new base and compare the third alternative with it.
3. Repeat the comparison of pairs until all alternatives have been evaluated. It is important that the correct pairs be compared. An alternative whose incremental rate of return fails to meet the required minimum is no longer a valid alternative and must be eliminated.
4. Select the alternative requiring the largest investment whose incremental rate of return meets or exceeds the minimum required.

Example 1: Consider the following five alternatives which have been found to meet the required technical specifications, including a 10-year useful life. If the minimum rate of return is 12%, which single alternative should be selected? (Income tax considerations are omitted to simplify the demonstration of the concept).

Alternative	A	B	C	D	E
Installed cost	\$4,000	\$3,500	\$8,000	\$7,500	\$6,000
Annual Operating and Other Expenses	1,100	1,200	450	500	750

The first step is to rearrange the alternatives in terms of increasing size of initial investment.

Alternative	B	A	E	D	C
Installed cost	\$3,500	\$4,000	\$6,000	\$7,500	\$8,000
Annual Operating and Other Expenses	1,200	1,100	750	500	450

This rearrangement portrays a typical situation in which frequently the more expensive alternative does require less maintencance, or less labor, or has longer intervals between overhauls or major outages, etc. The real issue becomes the justification of the added expense.

In comparing A with B, an extra \$500 initial investment results in savings of \$100 annually. Over the 10-year period the rate of return is:

$\$500 = \$100\ (P/A\ i\text{–}10)$
$(P/A\ i\text{–}10) = 5$ *or* approximately 15% from the tables.

There is justification in spending the extra \$500. The next question is whether alternative E is viable.

In comparing E with A:

$\$2{,}000 = \$350\ (P/A\ i\text{–}10)$
$(P/A\ i\text{–}10) = \$2{,}000/\$350 = 5.7$ or between 12% and 15%.

Alternative E is a valid one.

Next, compare the extra investment of \$1,500 required by Alternative D with its savings of \$250 annually as compared with E.

$\$1{,}500 = \$250\ (P/A\ i\text{–}10)$
$(P/A\ i\text{–}10) = 6$ or between 10% and 11%. The extra investment required by D is not justified and D is discarded as an alternative.

Alternative C is now compared with E, the largest *valid* alternative so far:

$\$2{,}000 = \$300\ (P/A\ 1\text{–}10)$
$(P/A\ i\text{–}10) = 6.33$ or between 9% and 10%. Alternative C is not acceptable.

The largest investment whose incremental rate of return meets or exceeds the required 12% is the \$6,000 represented by E. Extra funds that might be spent for alternatives D and C would be better utilized elsewhere for other projects meeting the 12% requirement.

The incremental concept provides a valuable tool in evaluating multiple alternatives through selective pairing. Care must be exercised in following the correct procedure, or the wrong alternatives may be selected. The largest rate of return does not optimize the selection. A relatively small investment with a very high rate of return may not generate savings or earnings of the magnitude of a larger investment which earns the required rate.

There are many other applications of the incremental concept. In evaluating the economics of leasing versus purchasing, the same technique may be applied. The question becomes: "Does the extra investment now in ownership, with its presumably lesser annual charges, provide the required minimum return over the annual costs of leasing?" Subtracting the leasing costs from the ownership costs provides a basis for calculating the rate of return. The rate must then be interpreted as to how it compares with other requests for funds.

Example 2: A company can either buy a small warehouse or lease it on a 20-year lease. The purchase price is $85,000. The lease cost is $5,000, payable at the beginning of each year. In either case the company would have to pay all maintenance and property taxes. It is estimated the building could be sold at the end of 20 years for $105,000. What rate of return before income taxes (including capital gains tax) would the company receive by purchasing rather than leasing?

The difference between the two alternatives represents the extra investment in ownership and the consequent savings each year. These incremental amounts form the basis for determining the rate of return.

$$PW = 0 = \$85{,}000 - \$5{,}000 - \$5{,}000(P/A\ i\text{–}19) - \$105{,}000\ (P/F\ i\text{–}20)$$

$$i = 6.9\%$$

Whether or not a 6.9% rate is acceptable depends upon the minimum return this particular company expects to receive on its investments.

5.9.2 Decision Criteria

Whether choices are to be made among mutually exclusive capital investment alternatives, or from those which are not, there are tools and techniques of analysis which assist the technical person in making a decision. Pure judgment and intuition are only somewhat helpful. Capital is limited, and capital can be productive. The essential element in decision making is to choose that alternative which will be the most profitable in the long run.

5.10 REPLACEMENT

An article in the *Wall Street Journal* in May 1983 was entitled, "To Exploit New Technology, Know When to Junk the Old".[8] While the writer was referring primarily to the technological changes in the product produced, the same rationale of "knowing when to junk the old" may be applied to the evaluation of existing capital assets. The replacement question can be analyzed utilizing the same approaches outlined in this chapter. Each alternative, including that of keeping the old, can be assessed through establishing estimates of how much money will be spent and how much will be received (or saved) as a consequence of its selection.

However, there are some special considerations which deserve attention. The need for reviewing the possibility of replacing existing capital assets may be due to any one of several changing conditions. Assets tend to deteriorate with age and use and thus may experience higher operating or maintenance costs. They may lack the capacity to meet increasing demands and then become functionally obsolete. Advances in technology may result in an asset becoming economically obsolete. There is frequently a reluctance, however, to move beyond the mere awareness of changing conditions. The existing asset may be unconsciously favored because it is already in place, because there is still some unrecovered investment, or because it is still operational and producing a benefit. Additional reasons for delaying the analysis include the risk or uncertainty of the costs of new equipment, and the difficulty of forecasting the need for any replacement.

However, once a decision has been made to seriously consider the possibility of replacement, there are some special aspects which have to be kept in mind. As for the existing assets, sometimes referred to as the defender, there are questions such as what remaining economic life should be considered, what capital value should be used, and what are the possible income tax considerations. The one-year remaining life test is frequently used for the defender because the resulting higher capital recovery cost (as compared with a longer period for the defender) partially compensates for the higher risk. If the costs for the one-year test are less than that of the challenger, then it may be well to keep the existing asset for one more year, particularly if there is a great deal of uncertainty about technological advances, possible changes in system requirements, or economic conditions.

The question of what asset value to use is a different one. The "book" value, or undepreciated amount is frequently not a realistc figure. There is a reluctance to accept the idea of a "sunk" cost, which is the difference between the undepreciated value and the actual or net realizable value. It is helpful in this case to take an outsider's view. Consider that the equipment or system in question is the only fixed asset owned by the company. Assume also that an outsider is about to start a similar business requiring only that same asset. He is unencumbered with an existing asset, and so has a choice between buying a used piece exactly like yours with its similar operating and other costs, or buying a new piece. His cost of the used piece would be the market price, not an inflated undepreciated "book" value amount.

Income tax considerations include possible tax write-offs for the sunk costs. The tax write-offs do not decrease the cost of a replacement (must differentiate between the "cost" of an asset and its "financing"), but do represent an opportunity cost which may be foregone if the asset is kept.

Replacement decisions are very critical. The newest is not always best. One Denver area firm is successfully using some World War II vintage lathes which have been rebuilt and used for certain applications where the tolerances are less precise, and speed of operation is not important. A comprehensive economic study was made with the company using a minimum after-tax return of 25%.

Consider also the quick switch by some car owners when more fuel efficient models came on to the market several years ago. It was not uncommon to pay an additional $5,000 on a trade-in in favor of a car whose fuel efficiency was estimated at 25 mpg as compared with the defender's 15 mpg. Assume a car is driven 15,000 miles per year, and the gasoline costs $1.25/gallon. The savings in gasoline per year would thus be 600 gallons or $750. The monthly saving would be $62.50. At 12% financing over 36 months, the monthly payment for the $5,000 would be $166. There are, of course, other considerations, but perhaps the calculations do illustrate the comment that many people favor economy, and they are willing to pay a lot for it!

Engineers in a firm are responsible for recognizing when an asset may not be economically employed, and for planning economically feasible alternatives.

5.11 LEASING

Several examples have already been given illustrating typical lease versus. purchase comparisons. Despite the advertising which stresses the advantage of leasing over ownership, there are no automatic, intuitive superiorities. Leasing is a method of financing, and lease-buy decisions should be determined by an actual cost comparison analysis. Leasing does not have the fifty cents on the dollar cost after taxes any more than ownership. Under ownership, depreciation charge, labor, and material costs, for instance, are deductible for tax purposes. Given the same 50% tax rate, those costs may also be considered approximately fifty cents on the dollar. Investment tax credits usually accrue to the leasor and generally are not passed on to the leasee.

5.12 INFLATION

The reality of inflation (as well as occasional deflation) does affect future periods during the life cycle. Consideration of it has been omitted thus far in this chapter in order to emphasize other important aspects.

During the mid 1970's, inflation became a national as well as a worldwide concern, and thus is an important factor in any planning. Inflation factors should, as nearly as possible, be estimated on a year to year basis, and on an individual cost basis. Labor costs, for instance, may not inflate at the same rate as certain material costs. Different materials inflate at different rates. Manufacturing labor may not inflate at the same rate as customer service labor. Escalation rates of a geometric nature may be treated as a compound amount factor in estimating future costs. Others may increase on an arithmetic basis and future costs may be estimated using a gradient factor.

Economists speak of actual dollars (or rubles, pesos, rupees, etc.) and real dollars (or rubles, pesos, rupees, etc.). Actual dollars are the actual numbers of dollars at the point of expenditure, and the kind of dollars in terms which people usually think. People anticipate their future salary in actual dollar terms. Real dollars are dollars of the same purchasing power as of some point in time, regardless of when the corresponding actual dollars occur. Actual dollars as of any time, n, can be converted into real dollars of purchasing power at any time, k, using the relation:

$$R\$ = A\$ \left(\frac{1}{1+f}\right)^{n-k} = A\$\,(P/F f, n-k)$$

where

f is the inflation rate per period over n periods.

Most minimum annual rates of return currently used by industry are a combination (the sum) of real interest rates (increase in real purchasing power) and the inflation rate.

5.13 ACCOUNTING AND ENGINEERING ECONOMY

Although the accountant and the engineering economy analyst may work with the same financial data, they do so for different purposes. The major difference lies in the way each views a firm's profitability. The accountant is primarily concerned with evaluating the results of past decisions and operations to determine how profitable the firm has been. The engineering economy analyst attempts to predict what the profitability of a current or future decision might be. The time line used in the earlier portion of this chapter might serve to illustrate the differences by assuming that both the accountant and the engineering economy analyst are standing on it at time zero and with their backs to each other. As they move along in the progression of time, the accountant is continually classifying, recording, measuring, summarizing various financial transactions that have already occurred. The accountant looks for the dollars of profitability. On the other hand, the engineering economy analyst, using some of the accountant's data, quantifies the expected future differences in costs and returns of alternative engineering proposals. He includes a capital recovery factor as a cost, which, if everything proceeds satisfactorily, the accountant will eventually find as a profit.

Accounting records provide a valuable source for the engineering economy analysis, but the technical person must be alert to the fact that accounting data is frequently collected, measured, and summarized, etc., in a different form. For example, average costs may be summarized whereas the engineering economist may need to determine incremental costs. It is essential that the technical analyst have sufficient knowledge of the accounting process so he is able to handle the requisite data intelligently.

5.13.1 The Accounting System

Basic documents for the financial management of a business of any size are the income statement (or profit and loss statement) and the balance sheet. The income statement may be expressed as a basic formula:

Profit = Revenue – Costs

The income statement reports how well a company has done over a specified period of time such as a month, quarter, or year. With the aid of computers, some firms make this determination as frequently as weekly, or in a few instances, daily, to aid internal decision-making.

The balance sheet may be likened to a snapshot of the company at any one time in which the assets, liabilities, and net worth are displayed. The corresponding basic formula is:

Assets = Liabilities + Net Worth

or:

Assets – Liabilities = Net Worth.

The change in a company's position over a period of time may be determined by comparing a succession of balance sheets. More will be said later about some significant financial ratios which are used in such comparisons.

Accounting uses a double-entry system for every transaction, and this system may be viewed as a series of equations, debits = credits, which must always be kept in balance. The terms *debit* and *credit* are conventional, and have no particular significance. One is advised to merely memorize them rather than reasoning through to a definition.

The purchase of a piece of equipment costing $6,000 with a down payment of $2,000 and the rest charged, would result in entries as follows:

debit equipment	$6,000
credit cash	2,000
credit accounts payable	4,000

The debits = credits equation continues to balance, as does assets = liabilities + net worth. The accounts, equipment, cash, and accounts payable are simply storage terms for a particular asset or liability.

Assets are everything of value owned by a firm or owed to it. All cash assets plus those resources which might be converted to cash within the normal operating cycle (no more than a year) of the firm are termed current assets. Fixed assets are those such as land, buildings, and equipment which are not intended to, nor can normally be quickly converted into cash.

Liabilities are debts or obligations of the firm which must eventually be paid. Current liabilities consist of those debts which must be paid within the normal accounting period (a year or less). Long-term debt or long-term liabilities extend over a period longer than a year and are normally represented by bonds.

Net worth is the total of all assets minus all liabilities. This is really the ownership portion of a firm, or equity. Equity is made up of the preferred and common stock, the surplus from the sale of stock above its par value (capital surplus), and the earned surplus or retained earnings. Earned surplus is the accumulation of past earnings not paid out in dividends but reinvested in the business.

In the equation, assets = liabilities + net worth, the right hand side may be viewed as the source of all funds within the firm. Some funds are thus available through short and long-term borrowing (a credit card charge is a short-term loan), some through the sale of stock, and others are available through previous earnings which have been "plowed back" into the company. The left side of the equation indicates the distribution or the form in which the funds are to be found. Thus, some funds are in cash and marketable securities, others in inventory, equipment, land, buildings, patents, etc. It can be readily seen that in the double entry system any transaction does affect at least two accounts.

5.13.2 The Balance Sheet

The balance sheet for the Telecom Company shows the major accounts as of a particular time, December 31, 19xx, and conforms to the fundamental equations, assets = liabilities + net worth.

Telecom Company
Balance Sheet
December 31, 1984

Assets		Liabilities	
Cash	$151,800	Accounts payable	$ 5,500
Accounts receivable	8,300	Notes payable	26,000
Inventory	49,200	Accrued taxes	3,400
Land	15,000		$ 34,900
Building	91,000		
Equipment	39,600	Net worth	
		Capital Stock	$200,000
		Earned surplus	120,000
			$320,200
	$354,900		$354,900

The amount of detail displayed depends on the statement's purpose.

5.13.3 The Income Statement

An understanding of the financial situation of a company also requires an income statement. For the Telecom Company, the following represents the major activity during the preceding year.

Telecom Company
Income Sheet
Year ending December 31, 19XX

Gross income from sales		$360,080
Cost of goods sold		189,160
Net income from sales		$170,920
Operating expense:		
Salaries	$ 87,100	
Depreciation	9,300	
Advertising	8,200	
Insurance	2,400	
Utilities	19,500	$126,500
Net profit from operations		44,420
Interest expense		2,300
Federal and state income tax		21,000
Net profit		$ 21,020

Profit must come from operations, which are the producing and selling of goods and services. The first item on the statement represents the amount of money taken in from the sale of goods and services during the year. Next, all the costs and expenses of doing business must be subtracted. In the Telecom Company these include "cost of goods sold" which represents all the production, material, and labor costs associated with making the goods and services, salaries, promotional expenses, etc. After subtracting all the operating costs, a profit from the operations activities can be determined. The interest on the notes payable (see balance sheet) is a tax deductible expense, and is deducted before income taxes are calculated. After determining the income taxes, the remaining amount is the "net profit" which is the amount available for dividends and/or retained earnings.

Many of the items of expense are difficult to determine in an ongoing firm. There are problems in determining what costs are appropriate for a particular period of time and how they should be charged. Depreciation is one of these, and as noted earlier, can be calculated by several different methods over some estimate of time. "Overhead" costs present another difficulty. The light, heat, and insurance expenses usually cannot be related to a particular product or service. Accountants concern themselves with ways of allocating such costs. There are also company-wide costs such as engineering, research, personnel, procurement, and the like. They each make a contribution to the production of goods and services, but such contributions are often difficult to measure and apply to the appropriate good or service.

5.13.4 Fixed and Variable Costs

Everything a firm does costs money. Some charges occur whether anything is done or not. Different expenses have different effects on a firm's profitability, depending on the variations in operations and the timing of the operations.

Fixed costs are those which are independent of any production or sales activity. Taxes, insurance, security expenses, interest on debt, and administrative salaries, are among those costs which continue to accumulate regardless of the level of activity.

Variable costs vary directly with output and comprise, for the most part, the two major items of direct labor, and material. The charges for each of these two categories are made directly to the product.

An engineer frequently fails to understand these costs. An error commonly made is to believe that a reduction in labor costs will result in corresponding proportionate decrease in overhead costs.

Because the major fixed and variable cost accounts are summaries of a wide variety of lesser accounts, it is frequently difficult for the engineering economist to ascertain certain labor, maintenance, tax, and other expenses required in a study. Considerable care must be exercised to search the accounting records until the necessary information is found. Only then can true incremental costs and incremental benefits be determined. Cost data that gives average values are frequently not adequate and may be seriously misleading.

5.13.5 Financial Ratios

There are a number of key financial ratios which may be determined by analyzing the two basic financial statements. Some are of significance to the outside investor. Others are of concern internally. In either case, it is the trend of such ratios which provides indicators of the strengths and weaknesses of a firm.

Ratios are generally classified into several fundamental types:

1. Liquidity ratios measure the ability of a firm to meet its short-term obligations.
2. Leverage ratios measure the contributions of the owners' equity as compared with the financing provided by borrowing.

3. Activity ratios measure how effectively the firm utilizes its resources.
4. Profitability ratios measure management's effectiveness as shown by the profit (or return) on sales and investment.

Each of the ratios can be compared with previous ratios in the same organization and also with those from the same type of industry. Considerable data is available for the latter comparisons and may be found in the financial reference sections of many libraries.

Another way of classifying the ratios to utilize the information which is supplied by the accountant may be for the engineer to view them as follows:

1. Scorecard ratios — Am I doing well or badly in the areas for which I'm responsible?
2. Attention-directing ratios — What problems should I be looking into?
3. Problem-solving ratios — Of the several ways of doing the job, which is best?

There is a growing trend, as stated initially in section 5.1.2, toward utilizing costs as a joint parameter in design. An analysis of one's own firm's costs, or an analysis of others' costs in specified segments of business, frequently indentifies the area in which a new or an improved technical design is needed. The engineer must also recognize cost as an active factor in the design process rather than something that is calculated after the design has been completed.

The ability of the engineer or related technical person to read and analyze financial statements can be a distinct advantage, both professionally and personally. Perhaps the advice should be presented even more strongly, for the ability to read and interpret financial statements is almost essential to survival, both professionally and personally.

5.14 SUMMARY OF CONCEPTS IN ENGINEERING ECONOMY

Eugene L. Grant, recognized as a pioneer in engineering economy, and now Professor of Econmics of Engineering, Emeritus, at Stanford University aptly summarized the concepts of engineering economy at a conference some years ago. Although the conference dealt with highway planning, the concepts summarized are sufficiently general in nature as to have wide application.[9]

1. All decisions are among alternatives; it is desirable that alternatives be clearly defined and that all reasonable alternatives be considered.

2. Decision-making should be based on the expected consequences of the various alternatives. In comparing investment alternatives, it is desirable to make the consequences commensurable with the investments insofar as practicable. Money units are the only units that make consequences commensurable with investments.
3. Only the differences between alternatives are relevant in their comparison.
4. It is necessary to have a criterion for decision-making (or possibly several criteria). The criterion for investment decisions should recognize the time value of money and related problems of capital rationing.
5. In looking at the predicted consequences of various alternatives and in establishing criteria for decision-making, it is essential to decide whose viewpoint is to be adopted.
6. Insofar as possible, separable decisions should be made separately.
7. In organizing a plan of analysis to guide decisions, it is desirable to give weight to the relative degrees of uncertainty associated with various forecasts about consequences. In this connection, it is helpful to judge the sensitivity of the decision to changes in the different forecasts.
8. Decisions among investment alternatives should give weight to any expected differences in consequences that have not been reduced to money terms as well as to the consequences that have been experessed in terms of money.
9. Decisions among investment alternatives must be made at many different levels in an organization. The implementation of rules aimed at rational decision-making may appropriately be different at different levels.

REFERENCES

1. Thomas, V., "Interconnect: Anti-competitive?", *Rocky Mountain Industries,* Vol. 10, No. 8, October 1977, p. 6.
2. Fish, J. C. L., *Engineering Economics,* 2nd ed. (New York: McGraw-Hill, 1923), p. 20.
3. American Telephone and Telegraph Company, *Engineering Economy,* 3rd ed., (New York: McGraw-Hill, 1977).
4. *Ibid,* p. 149.

5. Circular No. A-76, Executive Office of the President, Bureau of the Budget, March 3rd, 1966, p. 5.
6. American Telephone and Telegraph Comapny, *Annual Report*, 1964 (NewYork: American Telephone and Telegraph Company, 1965), p. 6.
7. Indiana Telephone Corporation (Ind. 1957) 16 PUR 3d, 490, 497.
8. Foster, Richard, "To Exploit New Technology, Know When to Junk the Old", *Wall Street Journal*, May 2, 1983.
9. Grant, E. L., *Concepts and Applications of Engineering Economy*, HRB Special Report 56, Workshop Conference on Economic Analysis in Highway Programming, Location and Design (Washington, D.C.: Highway Research Board, 1959).
10. Engineering Economist, *The Quarterly Journal of the Engineering Economy Divisions of the American Society for Engineering Education and the American Institute of Industrial Engineers*, Richard S. Leavenworth (editor), American Institute of Industrial Engineers, 25 Technology Park, Norcross, Georgia 30092.
11. Au, Tung, and Thomas, P., *Engineering Economics for Capital Investment Analysis*, (Boston: Allyn and Bacon, 1983).
12. Grant, E. L., W. G. Ireson and R. S. Leavenworth, *Principles of Engineering Economy*, 7th ed., (New York: John Wiley and Sons, 1982).
13. Newman, Donald G., *Engineering Economic Analysis*, 2nd ed., (San Jose: Engineering Press, Inc., 1983).
14. Riggs, J. L., *Engineering Economics*, 2nd ed, (New York: McGraw-Hill Book Comapny, 1982.

chapter 6

TELEPHONE RATES: ECONOMIC THEORY AND CURRENT ISSUES

Wesley J. Yordon
Department of Economics
University of Colorado
Boulder, Colarado

6.0 INTRODUCTION

Discussion of the pricing policies of a public utility is customarily divided into treatment of the rate level (or revenue requirement) and treatment of the rate structure (i.e., the relationships among various rates for different services). This division results from practical procedures: the agency responsible for setting such prices often finds it expedient to regulate the margin between revenue and expense by changing all prices by some approximately uniform percentage. In the U.S., where many utilities are privately owned, the objective of such adjustments is to set rates at a level which will enable the firm to earn a fair rate of return on its investment, judging fairness by comparison with the rate of return in competitive industries.

Rate-level hearings are often prolonged, but the issues are conceptually simple. In contrast, questions about rate structure are profoundly difficult, especially in telephony. Economic analysis of rate structure begins with the proposition that the price of such service should equal its marginal cost, as discussed below. Two major difficulties in applying this principle to telephone rates are the great difficulty of estimating marginal costs, and the fact that setting each price equal to apparent marginal cost would probably result in deficits for the telephone companies. The first problem exists because many of the costs are "joint" between one or more services. (This concept is treated in some detail below, but, to illustrate, consider the difficulty of ascertaining how much of the cost of a cow should be attributed to beef and how much to hide.) The second problem might be ignored in some nations where the telephone system is operated by a government with a treasury ample enough to sustain deficits, but in most nations it must be faced whether the system is public or private.

In recognition of the fact that many readers lack training in economic theory, the exposition here is at an elementary level, but in view of the importance of joint costs in telephony this complexity cannot be avoided even though it is difficult enough to be excluded from most introductory and intermediate economics texts. Discussion of events in the U.S. aims at answering questions such as: Have telephone rates been set so as to discriminate against some users and favor others? In what way has there been subsidization of local rates from excess revenue obtained from high long-distance rates? Why is this subsidization being eliminated? Why are telephone companies moving to change their pricing structures by introducing charges for services which often have been "free," such as directory assistance, installation, and unlimited local calls for a flat monthly fee? Why are rates being reduced for bulk users of long-distance service while local rates are increasing?

To analyze such issues requires some knowledge of economic theory and of telephone technology, costs, and data-gathering systems. This chapter attempts to provide this in a general fashion, but cannot describe existing practices in detail because they vary widely even within the U.S. The circumstances which have brought these issues to prominence in the U.S. are largely associated with the nearly unique American policy of providing telephone service through privately-owned enterprise, but the essential issues also exist, albeit with less fanfare, where telephone service is provided through government enterprise. Readers from nations where the latter practice prevails should be able to profit from the American experience.

6.1 THE LOGIC OF MARGINAL COST PRICING

Almost everyone intuitively understands the functions of a price system as it customarily operates, but a more analytical approach is usually necessary in order to reason about the merits of changing the *status quo*. For example, most readers probably feel that it is logical that long distance phone rates are lower at night than during the day, but few will have wondered why taxicab fares are not lower when the weather is pleasant than when it is inclement. The exposition of economic theory here will start at an elementary level, skim over some material which is customarily covered in introductory and intermediate theory courses, and extend to material which is not usually treated in such courses because it involves the awkward case of a multiproduct enterprise with a high proportion of joint and common costs.

Consider a person who is picking apples to satisfy his hunger. The first apple or two is intensely desired, but as the hunger pangs are eased the value of subsequent apples declines. Conversely, the first few apples are within easy reach, but subsequent ones are attained only with greater effort. How many apples should he pick and eat?

Not so many that the pain of picking the last one exceeds the satisfaction which it yields. Not so few that the nearest unpicked apple would yield more satisfaction than the pain of picking it. In the jargon of economics, the optimal quantity is reached when marginal utility equals marginal cost. (The word "utility" is used in economics to designate the satisfaction achieved from consuming a good, but economic usage does not imply usefulness as opposed to pleasurability. The word "marginal" is used to designate the borderline unit; it is the economists quaint term for what mathematicians call a first derivative.)

To express this proposition mathematically, let U equal the utility achieved from eating apples, let D equal the disutility of picking them, and let A equal the number of apples picked and eaten. The person wants to maximize net benefit (U-D), and the necessary condition is $d(U-D)/dA = 0$, or $dU/dA = dD/dA$.(See Appendix A.6.1 on mathematics if needed.)

In an economy where there is division of labor, the consumer of a commodity is generally not its producer. Suppose that the society consists of a picker, an eater, and a benevolent central planner whose objective is to maximize social welfare. Assuming that the planner can compare the utility of consumption with the disutility of production, he should direct that the number of apples to be picked and eaten be such as to satisfy the condition $dU/dA = dD/dA$. If instead of a planner who directly controls quantities there is a regulator whose task is to set an optimal price, he will maximize social welfare by setting a price such that $(dU/dA)/\lambda_1 = P_a = (dD/dA)/\lambda_2$ where P_a is the price of an apple, and λ_1 and λ_2 are the marginal utility of money to the consumer and to the producer, respectively. (As a mathematical formality, it is evident that λ_1 and λ_2 provide comparability among the three terms of the equation; the economic significance of this will be discussed below.) According to the theory of consumers' behavior, a rational consumer will choose to purchase the quantity which satisfies the left-hand equality, so the regulator's task is to bring about the right-hand equality. The value $(dD/dA)/\lambda_2$ may be viewed as the marginal cost of providing an apple, so the regulator will maximize social welfare if he sets price equal to marginal cost and orders the producer to supply the quantity which the consumer wishes to purchase at that price. (If marginal cost varies with output, as in the initial illustration, the regulator must take this effect into account, but since the complications associated with rising or falling marginal cost will be treated at length below, it may be helpful at this point to simplify by temporarily assuming that marginal cost is constant, unaffected by the volume of production.)

The discussion above sketches in simplest form the logic behind the important proposition that maximization of social welfare calls for the price of a commodity to be equal to its marginal cost of production. A number of qualifications have been glossed over for the moment, but before turning to these, some readers may appreciate a slightly more elegant version of the proposition which does not rely so heavily on the concepts of utility and disutility. It can be shown that a more rigorous expression of the necessary condition for maximization of social welfare is

$$\frac{dU_1/dA_1}{dU_1/dB_1} = \frac{dU_2/dA_2}{dU_2/dB_2} = \frac{V_A}{V_B} \tag{6.1}$$

where A and B represent any two commodities, the subscripts designate any two persons, and V_A and V_B may be thought of as the respective marginal costs of A and B. (Thus dU_1/dA_1 designates the marginal utility which person #1 receives from his consumption of A, etc; properly the derivatives should be partials but we dispense with this refinement.) This form involves only ratios of utilities and costs and thus avoids interpersonal comparisons of pleasure and pain; in the language of economic theory the ratios on the left are termed marginal rates of substitution and the one on the right is called the marginal rate of transformation.

For present purposes the latter concept is important. Many would have reservations in thinking about the cost of providing telephone service in terms of the disutility of someone's labor, but monetary cost is not necessarily a correct measure either. Marginal rate of transformation is a better concept. It is an expression of the *opportunity cost* of commodity A in terms of commodity B; i.e., V_A/V_B indicates how many units of B must be foregone by society in order to obtain a unit of A. Thus, if it requires 10 minutes of labor to produce a unit of A and 20 minutes to produce a unit of B, and if labor is the only resource used, $V_A/V_B = 1/2$; this indicates that production of a marginal unit of A requires society to forego the production of the ½ unit of B, all other things being equal.

Again there is a presumption that each consumer will purchase goods or services in quantities such that $(dU/dA)/(dU/dB) = P_A/P_B$, so the regulator's task is to set prices such that $P_A/P_B = V_A/V_B$. If the monetary costs of A and B reflect the real cost of the resources (labor, land, and capital) used up in the production of commodities A and B, then the monetary cost ratio is the marginal rate of transformation, but as we shall see much of the problem of rate structure concerns this "if." The point of this subsection has been to sketch the logic of the proposition that for optimal allocation of resources the price of a good or service should be set equal to its marginal cost of production, and to suggest that in this context "marginal cost" means the opportunity cost of the resources used, the cost of which may be difficult to measure since it is not necessarily identical with monetary expense as reported by an ordinary accounting system. Fuller expositions of the logic of marginal-cost pricing may be found in most intermediate microtheory texts, but for present purposes elaboration of the logic of the rule is less important than a number of limitations on its validity and applicability, namely questions about consumers' rationality, externalities, and equity.

6.2 ADJUSTMENTS FOR IRRATIONALITY AND EXTERNALITIES

The argument for marginal-cost pricing sketched above assumes that a consumer purchases goods in quantities which maximize his or her individual utility (in the face of given income and prices), and that a change in a consumer's utility changes social welfare by an equivalent amount (since the individual is part of society). In theory, the first assumption is treated as concerning externalities, but in practice the two issues often are difficult to separate. For example, it may be argued that if gin were sold at a price equal to its marginal cost some consumers would "irrationally" drink more than the amount which maximizes their utility, hence that it is desirable to offset this tendency by imposing a high tax and thus holding the price of gin well above its actual cost. Others may think that the drinker does maximize his own utility, but object to the fact that his intoxication imposes disutility or costs on others.

The question of consumer rationality does not appear to be a serious issue in telephone ratemaking. Reservations may be held concerning a subscriber's knowledge of technical matters, rates, and options, or about his susceptibility to advertising and salesmanship, but the appropriate remedy would be provision of better information rather than manipulation of rates.

Externality issues have been important. In 1910, Theodore Vail (President of AT&T, 1885-7 and 1907-19) argued that the addition of a subscriber brought external benefits to existing subscribers, and thus justified a rate structure featuring low rates for residential service. [(1)] To illustrate numerically, suppose that Mr. Smith would gain $5/mo. in utility from phone service, but declines to subscribe because the monthly cost is $10. But 10 of his acquaintances would each gain $1/mo. in utility if they could call Smith, so the total increment in social utility would be $15 compared to a cost increment of $10. In this circumstance marginal-cost pricing fails to maximize social welfare, and society would gain if service were offered to Smith at $5/mo., with each of his 10 acquaintances contributing $.50/mo. to offset the telephone company's loss, say by leasing an extension cord from the company at a rate which exceeds its actual cost.

It has been argued that a similar externality exists with respect to telephone calls, because the callee as well as the caller may gain utility from the call. If outgoing calls are priced at their marginal cost, this will inhibit calls for which the marginal utility to the caller is less than the charge, but it is possible that the sum of the caller's and callee's marginal utilities exceeds the marginal cost of the call and that social welfare would rise if the call were made.[(2)] This argument is less persuasive than the first because the caller-callee relationship is likely to be close enough so that one party can take into account the benefit to the other, and reciprocity would tend to even out the financial burden of a series of calls.[(3)]

Another externality issue of great importance has concerned the attachment of "foreign" equipment to the system; i.e., equipment not provided by the telephone company. The companies (especially those in the Bell System) have argued that such attachments are likely to impose harm on other subscribers by jeopardizing the quality of the telephone network. On the basis of this reasoning telephone companies prohibited attachment of foreign equipment until recently, and thus were able to exclude competition from the market for station equipment (see discussion in section 6.5.2 below).

6.3 THE PRICING DILEMMA IN DECREASING-COST ACTIVITIES

The arguments above concern efficiency, not equity; marginal-cost pricing (in the absence of irrationalities and externalities) is a necessary but not a sufficient condition for maximizing social welfare. We now turn to considerations of equity for buyers and sellers.

With respect to a buyer, it was stated above in section 6.2 that economic theory postulates that a rational consumer purchases the quantity of commodity A which makes $(dU/dA)/\lambda = P_A$. But if the consumer is poor, λ (the marginal utility of money to the consumer) will be large and he will be forced to restrict his purchase of A even though the utility from an additional unit would be large. If society judges such a situation to be unjust there are two possible remedies: grant him more money income, or reduce the price of A below its marginal cost. The latter is preferable only if irrationality and/or externalities are acting in conjunction with his poverty to limit his consumption to less than the socially optimal quantity; such an argument is commonly applied to education, housing, preventive medicine, and even to food. Telephone service is not viewed in this light; on the contrary, prior to 1968 the New York City Welfare Department regarded telephone service as an unwarranted expense in the budget of families on welfare. Opinion has shifted so that concern is now expressed over the inability of low-income families to purchase telephone service (especially the ill or the elderly for whom it may be vital in case of emergency), and a form of inexpensive ("lifeline") service has been proposed to meet this need. In general, however, such proposals are not intended to set price below cost but only to provide an option which links low price with low usage (and hence low cost) so as to avoid forcing everyone to buy the type of service which is appropriate for the average family. In sum, general opinion in the U.S. does not regard consumption of telephone service as an activity which should be promoted by subsidizing it from general taxation.

The discussion above may seem to belabor the obvious, but in fact certain technical characteristics of the telephone industry generate a pricing dilemma; apparently the principle of marginal-cost pricing can be applied only if accompanied by subsidization from general tax revenues, but social considerations apparently do not justify such a subsidy. This dilemma arises because telephone service appears to be a decreasing-cost activity rather than an increasing-cost activity, and since most presentations of economic theory focus on the latter, it is desirable to elaborate the distinction and the implications thereof. Consider the extraction of petroleum as a good example of an increasing-cost activity. For clarity, imagine that petroleum may be extracted from any one of 10 sites and that the rate of production obtainable from any given well is a constant of nature, say one million barrels per year. Due to differences in the depth of the oil, the cost of drilling and extraction vary: oil may be obtained from site #1 at a cost of $1/bbl., from site #2 at a cost of $2/bbl., etc. Suppose that the demand for oil is such that 8 million bbls/yr. will be sold if the price is $8/bbl., then demand and supply will be equated if sites #1 to #8 are exploited and the oil is sold at $8/bbl., as depicted in Figure 6.1.

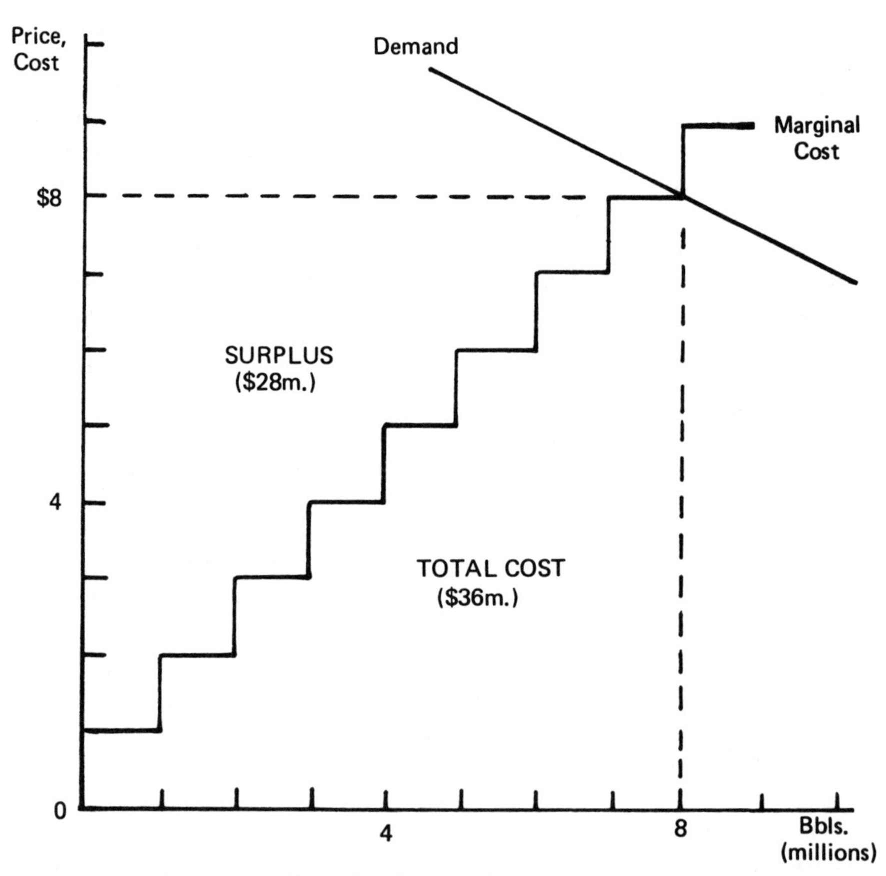

Figure 6.1: MC Pricing Where Cost Increases by Steps

Given these conditions, site #8 is the "marginal" well, and the cost at site #8 ($8/bbl.) is the marginal cost of producing a barrel of petroleum. With a uniform price equal to marginal cost, the total revenue generated by selling 8 million barrels is $64 million, while the total cost is $36 million ($1 million at site #1 plus $2 million at site #2 plus . . .), so there is a surplus of $28 million. If the oil wells are owned by the government, this surplus provides the government with funds which may be used for various purposes — among them perhaps to provide fuel oil for needy families. If the wells are privately owned the surplus is nevertheless available to the government through taxation. Oil from site #1 may be taxed at the rate $6.99/bbl., oil from site #2 at the rate of $5.99/bbl., etc., without causing any diminution of production, and the total tax take will be $27.9 million, ignoring collections costs.

If production costs increase smoothly rather than by discrete steps, the task of capturing the surplus through taxation may be more difficult, but the essential point remains: marginal cost exceeds average production cost, and marginal-cost pricing will generate total revenue in excess of total production cost. The case of smoothly increasing cost is depicted in Figure 6.2. If price is set equal to marginal cost at the output level which satisfies demand (P*,Q*) total revenue will be P*Q* or the area of rectangle OP*EQ*. Total cost of production will be the area of rectangle OBFQ*, and the surplus will be the area of rectangle BP*EF, or the area of triangle AP*E.

Parenthetically, to avoid confusion for those who have been exposed to an introductory course in economics, it should be noted that most economics texts assume the the surplus is captured by the land owner; i.e., that it becomes a rent or royalty which the producer must pay. Accordingly it appears as a financial cost to the producer, so that production cost plus rent equals total revenue, and average *total* cost equals price. A graph of average total cost would be represented by a U-shaped curve which reaches a minimum at point E, indicating that the producer just breaks even when price is P* and output is Q*. However, the standard treatment tends to obscure what is essential for present purposes; namely, that selling a product at a price which equals its marginal cost of production generates a surplus in activities where marginal cost rises with output.

For decreasing-cost activities the analysis is symmetrical, and the opposite conclusion follows, namely that selling a product at a price equal to its marginal cost of production will generate total revenue less than total production costs, resulting in a deficit. This situation is depicted in Figure 6.3, where total revenue is the area of rectangle OP*EQ*, total production cost is area ORSQ*, and the deficit is area P*RSE.

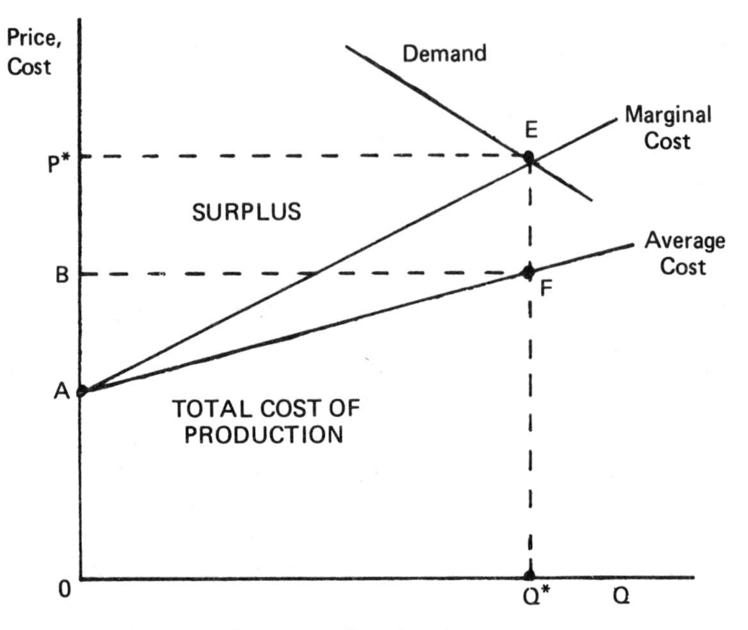

Figure 6.2: MC Pricing in Increasing Cost Activity

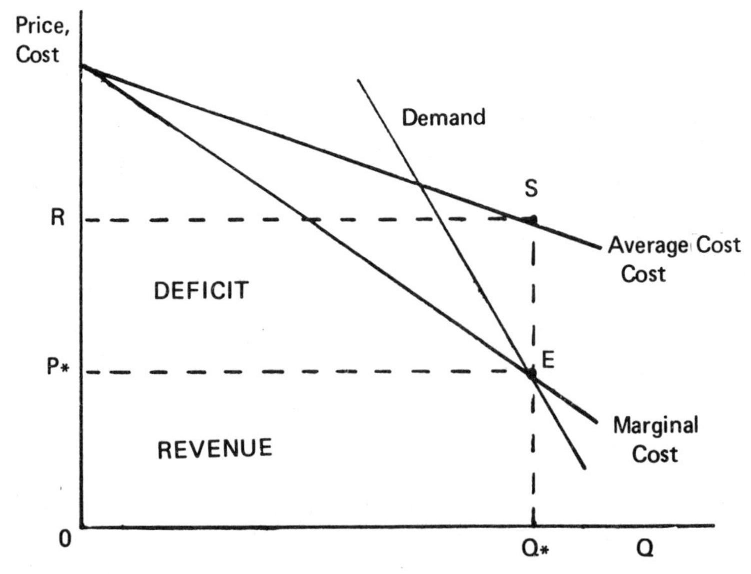

Figure 6.3: MC Pricing in Decreasing Cost Activity

In the common intermediate case where average cost is constant and equal to marginal cost, marginal-cost pricing generates exactly the revenue needed to cover total cost.

Local telephone service was viewed as an increasing-cost activity by some analysts writing in the first half of the century. This view is based on the law of combinations, which states the number of paired connections between N telephone stations is given by the formula $N(N-1)/2$; e.g., 6 for 4 subscribers, 15 for 6 subscribers, etc.[4] This was a sensible argument when connections were made manually, but its validity became dubious with the spread of mechanical switching (dial systems). A recent study shows that access investment cost per subscriber falls with the number of subscribers in a given area (about $450 for 1,000 subscribers, $290 for 50,000 subscribers, and $260 for 100,000 subscribers; Figure 6.4 roughly reflects this relationship.[5]

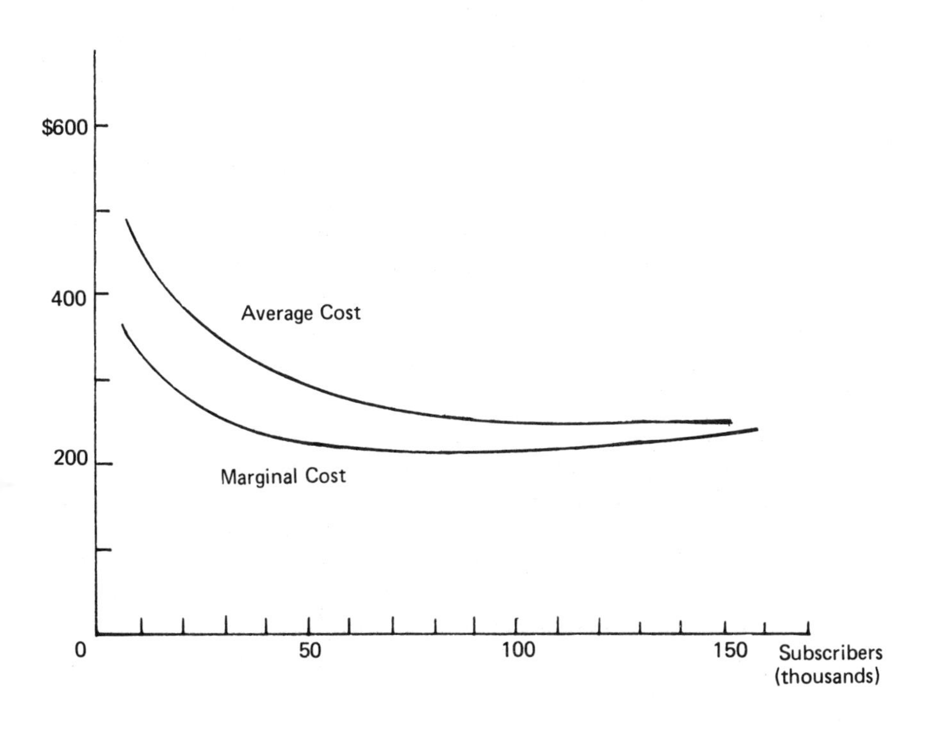

Figure 6.4: Investment Per Telephone Subscriber (Access Equipment)

Where average cost is falling, marginal cost must be less than average cost; so if marginal-cost pricing were applied to access investment, revenue would not cover total cost, and the deficit would have to be offset with surplus revenue obtained from other activities.

For "long lines" the presence of decreasing costs has long been evident. The costs of right-of-way and poles (or trenching) are about the same for 2,000 circuits as for 2, and the cost of cable does not increase proportionally with the number of circuits. But with expansion of the market for long-distance service and the development of microwave transmission, decreasing costs no longer prevail for service between large cities.

It must be stressed that analysis of telephone costs is a formidable task. Until recently the concern of citizens and regulatory agencies in the U.S. focused on the overall rate level rather than on rate structure, so there was little motivation to identify disaggregated costs. The Uniform System of Accounts (use of which is required by law) is out of date by half a century, and while reform is in progress, it will be some time before data become available in the quantity and form needed for a tolerably accurate estimation of the cost functions for the numerous dimensions of telephone service. Consequently the cost functions described should be viewed only as illustrations of the probable tendency in some facets of telephone service, and it is possible that some other aspects are subject to increasing costs. It will be argued below that a refined rate structure, which prices various dimensions of telephone service according to their actual costs, would substantially mitigate the problem described here. However, it seems probable that the overall tendency toward decreasing costs would remain a problem even with refined cost analysis and ratemaking, and in any case it is important to understand that the telephone rate structure has historically been shaped by the assumption that decreasing costs predominate.

Some economic theorists have argued that the ideal of pricing at marginal cost should not be abandoned simply because it would result in deficits in decreasing-cost industries; they argue that the deficit should be offset by a subsidy from the fisc.[6] But where will the fisc obtain the necessary funds? Possibly by borrowing in the short run, but sooner or later only from taxation of other activities. If there are increasing-cost activities of sufficient magnitude, it might be possible to extract from them surplus enough to subsidize the deficits in the decreasing-cost activities; but if not, then the attempt to subsidize in order to maintain marginal-cost pricing in the decreasing-cost activity will cause prices to exceed marginal costs in others, and therefore be self-defeating. The former condition may exist in a few nations such as those whose income arises mainly from extractive industries such as petroleum, but the latter prevails in most industrial nations such as the United States where less than 5% of national income originates in agriculture and extractive industries. Given the magnitude of other demands on government funds, which in the U.S. have pushed taxes to about 40% of national income, it is unrealistic to assume that there is an untapped reservoir of non-distorting tax revenues which could be used to subsidize telephone service. Consequently the discussion will proceed using the assumption that telephone service as a whole is to be unsubsidized, with only a reminder that the alternative approach is not ruled out for nations where the extractive sector dominates and provides distortion-free tax revenues in excess of high-priority demands for government spending.

6.4 VALUE-OF-SERVICE PRICING

The question of how to structure prices in situations where simple marginal-cost pricing fails to generate sufficient revenues came to prominance in the 19th century with the growth of roads, canals, and railways. It was evident that the *use* of such facilities imposed small real cost, and that a pricing system which charged fees equal only to the cost of use would not generate enough revenue to cover construction cost. In 1844 Jules Dupuit, a French engineer, provided an excellent analysis of the problem and pointed out the merit of basing prices on value of service rather than on cost of service:

> Thus when a bridge is built and the state establishes a tariff, the latter is not related to cost of production: the heavy cart is charged less than the sprung carriage even though it causes more wear to the timber of the carriageway. Why are there two different prices for the same service? Because the poor man does not attach the same value to crossing the bridge as a rich man does, and raising the charge would only prevent him from crossing. Canal and railway tarrifs differentiate between various classes of goods and passengers, and lay down markedly different rates for them although the costs are more or less the same. In drawing up these tarrifs in advance the legislator merely defines certain features and characteristics which seem to him to indicate a greater or lesser degree of utility in the same service rendered to different people.[7]

The similarity between value-of-service pricing and marginal-cost pricing with respect to efficiency in resource allocation should be evident; in both cases the aim is to set a price which permits a person to consume a commodity up to the point where the marginal utility to the person equals the marginal opportunity cost to society. If a facility exists and use of it imposes negligible cost in terms of wear or congestion, use of it should not be parsimoniously rationed. On the other hand, if it is possible to segregate users who will not be deterred by a high fee, they can be made to bear the financial burden of the construction cost without curtailing the use of the facility. Such a pricing system is sometimes described as "charging what the traffic will bear," but it may also be viewed as a system of "*not* charging what the traffic will *not* bear."[8]

It is also commonly described as price discrimination, but for reasons which will become clear, this term is likely to generate more heat than light. At this point it is important to understand that value-of-service pricing by no means ignores the cost of service. For convenience of exposition, writers commonly choose examples where actual cost appears to be negligible, but it is clear that they do not envision a charity program which renders services at prices lower than the opportunity costs of providing them. In fact, with the aid of modern economic theory it becomes evident that in some cases value-of-service pricing coincides with a sophisticated form of marginal-cost pricing, but more generally it is a system where cost of service establishes the minimum charge and a variable markup is added to collect additional revenue from those who value the service highly.

Up to this point the terms "cost" and "marginal cost" have been used without specifying whether these refer to the short run or to the long run. It is my belief that the popular introductory and intermediate economics texts provide more confusion than help on this matter for present purposes. The problems are that they focus on a single-product enterprise, assume nearly perfect information, and usually illustrate with reference to wheat farming (which is an increasing-cost activity with no long-lived or lumpy investments). Samuelson has argued that the rule for optimal pricing is to equate price with short-run marginal cost, which will be equal to long-run marginal cost if and only if the capacity of the facility is optimal.[9] But in a world where there is uncertainty and where projects are lumpy rather than infinitely divisible there is ambiguity in the concept of optimal capacity.

Where need varies stochastically what capacity is required to handle the load? Does a roof have excess strength if no known snowfall has ever been enough to break it? If a facility is deliberately built oversize in relationship to current need because it is uneconomical to rebuild it as need increases, does such a facility have excess capacity in any meaningful sense? Some economists would answer affirmatively and argue that price should be held down while the capacity is excessive, then raised as growing demand makes it necessary to ration use. But if long-term plans are based on current prices such a policy will include erroneous planning,[10] and for public utilities such as telephone service it would be administratively awkward to vary rates from month to month in accord with changing relationships between demand and capacity.

Consequently it is impractical to base telephone rates on short-run marginal costs, and the relevant cost base more nearly resembles long-run marginal cost. However, the concept of long-run marginal cost in abstract theory is derived on the assumption that *all* inputs are variable (i.e., that there is no existing plant or equipment) whereas in the telephone industry some inputs have such long lives (e.g., a right of way) that the time will never come when the industry will have a fresh start with a clean planning slate. Hence the relevant cost base is not quite the same as theoretical long-run marginal cost, but with this qualification the discussion here will refer to the relevant cost base as long-run marginal cost because we envision a planning period which is long enough to permit substantial variation in plant and equipment.

For some pricing decisions, however, it is appropriate to assume that plant and equipment are fixed. For example, the relevant cost of placing a call at night is low because the facilities to handle it are already in place, and are under-utilized. Some economists view this as a situation where short-run marginal cost is low and is the correct base for price. It is more accurately viewed as a situation which involves joint cost; i.e., the same facilities provide two essentially different services, and since the demand for daytime service is greater than the demand for nighttime service, the latter may be viewed as a byproduct of the former. This is not really a short-run phenomenon because the difference between daytime and nighttime demand presumably will endure for the foreseeable future, but in any case the conclusion is valid that the relevant cost is lower for nighttime calls than for daytime calls, and that the rate structure should reflect the difference. Generally this will imply that calls during the offpeak period should be priced at less than *average* cost, so there is a strong similarity between the problem of joint cost and the problem of decreasing cost. Either case calls for value-of-service pricing; in the former, value of service is used to identify the relevant costs and the resulting price structure may be viewed as marginal-cost pricing; in the latter, value of service is used to vary the margin between price and marginal cost so as to satisfy the revenue requirement as efficiently as possible. In practice it is virtually impossible to distinguish between these two cases, but we proceed with a simplified example which starts with a pure case of joint cost.

Imagine that a municipality contemplates construction of a fully automated telephone exchange. (For simplicity we assume that each subscriber will provide his own station equipment and line; the municipality will provide only the service of switching calls among the lines.) A referendum is held to ascertain whether residents favor such a project, but it is unanimously defeated because each resident deems the cost to be excessive. Then 1,000 merchants contract to pay $1 per month apiece for service (a sum sufficient to fully compensate the municipality for the cost of constructing and operating the exchange) and the exchange is built and placed in service. At this point in time the situation is similar to the theory described in section 6.1 — one "consumer" has declined to purchase something because its marginal cost exceeds its marginal utility, while another "consumer" has agreed to the purchase because marginal utility exceeds marginal cost, but here the consumers are groups rather than individuals and the indivisible nature of the project makes the term marginal somewhat inappropriate since the purchase involves a large rather than an infinitesmal increment.

After the exchange is constructed it becomes evident that the merchants use it only during the day, and that it might be used at night for residential service. Since construction cost has already been covered, the marginal cost of nighttime use is low, so the municipality can allow residential subscribers to connect with the exchange for a fee much lower than $1 per month if they place calls only at night. The extra revenue will permit a reduction in the fee charged to business subscribers (which may induce more merchants to subscribe, a complication treated below) but they still pay more than the residential rate. Nevertheless, they are better off than they would be if there were no residential subscribers, so the arrangement seems eminently fair.

The above example all too conveniently assumes that the decisions took place in a certain sequence which revealed that the telephone exchange was more intensely desired by the merchants than by the residents. Let us now modify the assumptions by postulating that the decisions are to be made simultaneously, in the light of perfect information about the demand intensities of the two groups. This and other assumptions are admittedly unrealistic, but we gain generality and rigor. The analysis may seem tedious and abstract, but there is no other way to convey an understanding of the main issues in telephone ratemaking.

Suppose that telephone subscribers conveniently fall into two groups: business users who make calls only during the day, and residential users who make calls only at night. It is not feasible to meter calls, but experience indicates that one switching unit is needed in the exchange for each 100 subscribers in order to avoid catastrophic failure of the system. The monthly capital recovery expense (interest and depreciation) of each switching unit is $100; i.e., $100 per month per switching unit provides a "fair return" on the investment in the exchange. If an exchange were built for the exclusive use of either business or residential subscribers, it is evident that the appropriate monthly fee would be $1 per subscriber, but since they will jointly use the same facility, the $1 is a joint cost which must somehow be split. It might seem fair to split it 50-50 but if (as is almost certain to be the case) there is a difference in the intensities of demand for the two types of service, an even split would be unfair and inefficient. Figure 6.5 illustrates the problem with hypothetical demand functions $B = 2000 - 1000P_b$ and $R = 1450 - 1250P_r$, where B and R are the numbers of business and residential subscribers and P_b and P_r are the respective monthly fees. A common fee of $.50 would attract 1500 business subscribers but only 825 residential subscribers, generating revenue of $1,162.50 and requiring an exchange with 15 switching units (at a cost of $1,500) having excess capacity at night. Hence a uniform fee of $0.50 is neither optimal nor feasible, but this amount may be used as a reference point in calculating the optimal price structure, so it is plotted in Figure 6.5 and labeled AVC (Average Variable Cost — but not variable in the short run). Optimum quantities and prices may be ascertained by finding a quantity such that P_b and P_r are equidistant from AVC, the former above and the latter below. This point occurs where $S = 12$, $P_b = \$.80$, and $P_r = \$.20$. This shows that fees should be $.80 per month for business subscribers and $.20 per month for residential subscribers; with this rate structure 12 switching units will provide proper capacity both day and night and the revenue requirement of $1,200/month will be exactly satisfied by collecting $960 from businesses plus $240 from residences.

This price structure appears to be a case of price discrimination, but in reality it is not.(11) The prices are exactly equal to their respective marginal costs, but because the costs are not easily indentifiable some further explanation may be needed to make it clear that the fees are indeed equal to the respective costs of service. Consider the short-run marginal cost of adding another subscriber. The cost will be catastrophic system failure (or less dramatically, major inconvenience to phone users) unless an existing subscriber's service is simultaneously terminated. If the latter action occurs, the decrease in social utility and the revenue loss to the company is $.20 or $.80 (depending on whether the new subscriber is residential or business). Note that this reasoning applies only if the exchange is used to capacity (otherwise the opportunity cost of adding a subscriber is zero), and that capacity will be fully utilized only when the fees are $.80 for business subscribers and $.20 for residential subscribers. The long run marginal cost is $1.00 per pair of subscribers (one business, one residential) and the 80-20 split is needed in order to match

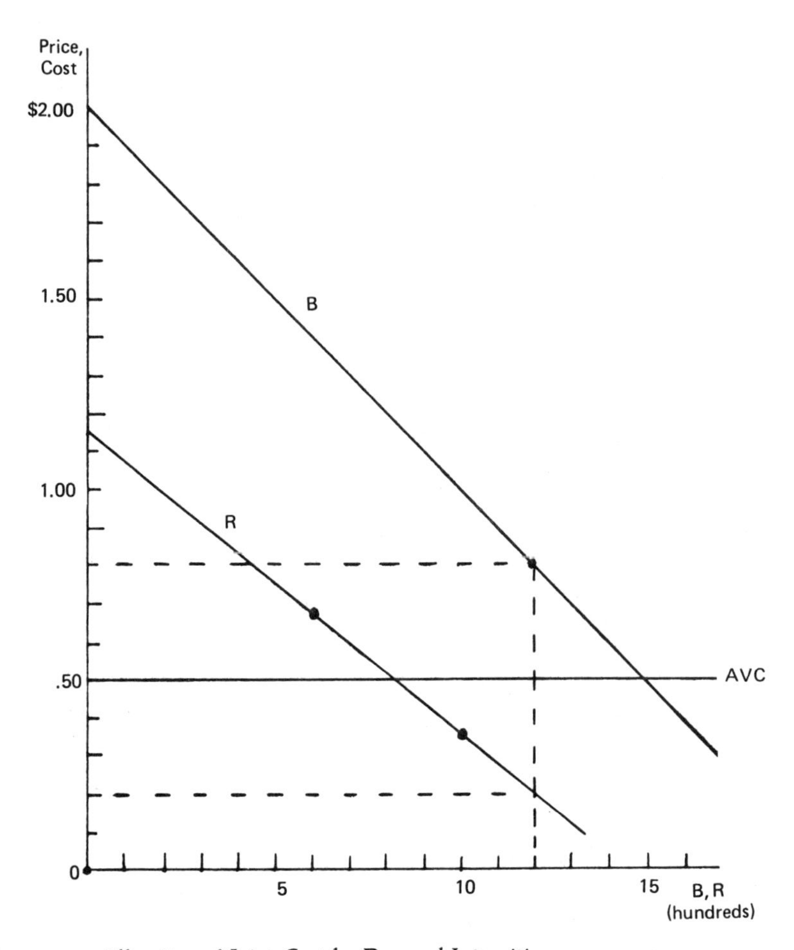

Figure 6.5 Allocation of Joint Cost by Demand Intensities

the number of residential subscribers with the number of business subscribers. Accordingly, this price structure can be labeled value-of-service pricing because it does take into account the intensities and sensitivities of business and residential demand for exchange service, but it is not a case of price discrimination. Note also that the joint cost cannot rationally be allocated according to relative use because the aim of differential pricing is to equalize use.

Unfortunately even the simplest example becomes substantially more complex if we introduce decreasing long-run costs in place of the convenient constant-cost assumption used above; i.e., that the monthly cost of an exchange was simply $100 per switching unit. Let us now take a small step toward realism by assuming that the cost is $100 per switching unit plus a lump sum which is unaffected by the size of the exchange; e.g., the salary of a custodian. This does not affect long-run *marginal* cost, but it makes long-run *average* cost exceed long-run marginal cost. Consequently the optimal price structure (P = LRMC) no longer generates sufficient revenue to cover costs, and we are forced to consider a price structure which will be quasi-optimal or second best.

The logic of quasi-optimal pricing was developed rigorously by F.P. Ramsey in 1927 and has been discussed more recently by Baumol and Bradford.(12) Here the formula will be presented and illustrated, with no attempt to describe the reasoning behind it. It is important to appreciate, however, that the formula reflects maximization of social welfare, not maximization of profit. The measure of social welfare is the money value of consumers' utility minus the money cost of production. It thus assumes that a dollar has equal value to rich and poor alike, and the only question of income distribution it handles is the one of how to generate the revenue required with the least possible harm to consumers as a group.

In the case at hand (where cross-elasticities are zero) the Ramsey rule for quasi-optimal price structure is

$$\frac{P_b - MC_b}{P_r - MC_r} = \frac{MR_b - MC_b}{MR_r - MC_r} \tag{6.2}$$

where P is price, MC is marginal cost, and MR is marginal revenue. (marginal revenue is the first derivative of revenue; when the demand function is linear MR has twice the slope of the demand function.) The demand functions displayed in Figure 6.5 are $P_b = \$2 - .001B$ and $P_r = \$1.16$ $\$1.16 - .008R$, where B and R are the number of business and residential subscribers, respectively. Substituting the numerical values for the terms MR and MC, we obtain

$$\frac{P_b - .80}{P_r - .20} = \frac{2 - 2\,(.001B) - .80}{1.16 - 2\,(.0008R) - .20} \tag{6.3}$$

Examination of this equation will reveal that the relative magnitudes of the margin between price and marginal cost vary with the vertical intercept of the demand function (i.e., $2 for business demand, $1.16 for residential demand) and with the slope of the demand function (i.e., -.001 for business and -.0008 for residences). If one thinks of the intercept as a measure of the intensity of demand, and the slope as a measure of the sensitivity of demand, we may say that the margin between price and marginal cost should vary positively with intensity and negatively with sensitivity. A commonly used concept in economics which combines these two aspects (in inverse form) is the price-elasticity of demand, and this concept may be used to express the Ramsey rule as

$$\frac{(P_b - MC_b)/P_b}{(P_r - MC_r)/P_r} = \frac{E_r}{E_b} \tag{6.4}$$

where E_r and E_b are the price-elasticities of demand for residential and business subscriptions. However, since elasticity varies along a given demand curve, one cannot speak of *the* elasticity of demand function, and confusion is likely to result if one refers to the rule as prescribing a lower markup in the more elastic market. (Column (g) in Table 6.1 is included to illustrate this point.) Consequently, most readers are advised to think in terms of intensity and sensitivity (height and slope) of demand rather than in terms of elasticities.

Table 6.1 displays two quasi-optimal or "second best" price structures together with the optimal structure and the monopoly structure. Column (c) shows the net revenue; i.e., total revenue minus the cost of switching units (@ $100 month). It is this net revenue which is available to cover lump-sum fixed costs such as the expense of a custodian, amortization of design fees, or (in a more complex situation) differences between average cost and marginal cost due to decreasing marginal cost. The optimal or first-best solution provides zero net revenue, and hence is not feasible. Two second-best solutions are shown, generating net revenue of $360 and $576, to cover two possible levels of fixed cost. Finally, for comparative purposes, the monopoly solution is shown which generates net revenue of $648 which is the maximum possible.

Examination of the table will reveal why many discussions of price discrimination generate more heat than light. Row (1), the optimal solution, displays a situation which most people would regard as "discriminatory" simply on the basis of the differential prices shown in column (a), or with more sophistication, on the basis of the relationships to average total cost as shown in column (e). But according to economic theory the situation is really non-discriminatory, with prices for both types of service equalling their respective marginal costs as shown in column (f). The rate structures shown in the other rows may legitimately be described as discriminatory, but it is not clear which group is

TABLE 6.1
Examples of Value-of-Service Pricing

	(a)		(b)	(c)	(d)	(e)		(f)		(g)	
Rate Structure	Monthly Fee		#of subscribers in each class (also variable cost in dollars)	Revenue minus variable cost	Average total cost	Fee/ATC		Fee/MC		Elasticity	
	Bus.	Res.				Bus.	Res.	Bus.	Res.	Bus.	Res.
Optimal	$ .80	$.20	1200	$ 0	$ .50	1.60	.40	1.00	1.00	.67	.19
Quasi-optimal	1.00	.36	1000	360	.68	1.47	.53	1.25	1.80	1.00	.45
Quasi-optimal	1.20	.52	800	576	.86	1.40	.60	1.50	2.60	1.50	.81
Monopoly	1.40	.68	600	648	1.04	1.35	.65	1.75	3.40	2.33	1.42

favored. A comparison between the fees and average total cost (column (e)) suggests that there is discrimination against business subscribers, but from the viewpoint of economic theory average total cost is a meaningless concept in this context. The relevant concept is marginal cost, and comparison of the ratios of fees to marginal costs in column (f) reveals that the rate structures shown in rows 2, 3, and 4 involve discrimination *in favor* of business subscribers even though their rates are higher.

One of the points of this example is that no normative significance should be attached to the term "discrimination" in the context of public utility economics; one must make an effort to avoid the common language connotation of wrongdoing. In the language of public utility economics the semantic difficulty is overcome by use of a double negative — an economist would testify that the rate structures displayed are "not unduly or unreasonably discriminatory." But such language fails to convey the optimal properties of the rate structures shown, and it should be emphasized that they maximize the sum of the utilities of business and residential subscribers, subject to the constraint that total revenue cover the cost of service. On the other hand, it must be stressed that the term "value-of-service pricing" also applies to many price structures which are neither optimal nor quasi-optimal and which may exhibit price discrimination in the pejorative sense of the term.

The major points which should be evident from this illustration of ideal value-of-service pricing are:

1. The ideal method of allocating a joint cost is on the basis of relative intensities and sensitivities of demand; the resulting price structure may appear to be discriminatory, but actually it is a sophisticated form of marginal-cost pricing.
2. The pricing dilema created by decreasing costs may be resolved by using value-of-service pricing; the resulting price structure will be discriminatory but if it follows Ramsey's rule for quasi-optimality the discrimination is socially beneficial.
3. To ascertain whether a rate structure exhibits price discrimination and to determine whether the discrimination is socially beneficial or harmful requires information concerning the cost and demand functions for the services in question; gross errors result if judgments are based only on a comparison of the prices or rates.

In order to provide clarity, this illustration unrealistically assumed that the cost and demand functions were perfectly known, and that there was a hard and fast distinction between the two categories of subscribers. In reality the relevant functions are largely unknown and the distinctions between categories are more-or-less arbitrary for reasons to be discussed in the following section.

6.5 TELEPHONE RATES IN THE U.S.

6.5.1 Overview

The vocabulary used in discussions of telephone economics is specialized and has been largely shaped by the Bell System. Telephone service is classified into two broad categories, *basic* and *non-basic*. (When specialized terms in this section first appear they are italicized and defined explicity or implicitly.) Basic service is local exchange service for residential and business subscribers; non-basic service encompasses everything else: message toll service (MTS, or ordinary long-distance calls), Wide Area Telephone Service (WATS), private-line service (PLS), special terminal equipment such as private branch exchanges (PBX), key telephone systems (KTS), teletypewriters, and conveniences for the ordinary subscriber such as extra-long cords, extension phones, deluxe instruments, and answering devices.

A necessary and significant part of the cost of basic service is *access cost*, the cost of the equipment needed to connect a subscriber to the local exchange (the subscriber loop, drop line, inside wiring, and the basic telephone). The typical residence or small business needs no more than one line, and in the usual case of a single-party line it is used only for calls to and from the subscriber, so use of the line imposes no significant congestion costs on others. (Incoming calls may be blocked by use, but presumably the subscriber takes this into account for calls which he desires to receive. Also "call waiting" service now makes it possible to receive calls while the line is in use.) Use of access equipment imposes negligible wear. Consequently, access cost is the cost of a lump-sum investment which is not *traffic sensitive* because the amount of investment needed is unaffected by the number of calls, their duration, or their distance. Access thus resembles my example of the custodian (above, 6.4.3) or Dupuit's bridge (above, 6.4.0), but a difference is that access equipment is not shared so it would be feasible to require each subscriber to pay the entire cost of the initial investment. However, as noted above (6.3) the average cost of access investment decreases with the number of subscribers, so a rate structure which charged each subscriber with the marginal cost of his access investment would usually fail to generate enough revenue to cover the average cost. Furthermore, the argument concerning externalities (above, 6.2) justifies charging even less than marginal cost. For these reasons local telephone companies in the U.S. customarily paid the lump-sum investment in access equipment and charged subscribers fees for local service which were not large enough to generate a fair rate of return on that investment.

Nevertheless the telephone companies had to earn a fair rate of return on their total investment, so the deficits generated by pricing access at less than cost had to be offset by pricing other services above cost. The stated policy of the Bell System was to make up the difference by pricing non-basic services so as to obtain maximum profit margins on them, and thus permit low rates for basic service.[13]

If one examines the examples of non-basic service listed above, one may be reminded of Dupuit's comment on the logic of charging a higher toll for the rich man's carriage than for the poor man's cart; the Bell System has portrayed its pricing policies as discriminating against big business subscribers in favor of small business and residential subscribers. To practice such discrimination requires monopoly status; if competitors are permitted to enter these markets they will *cream skim*— i.e., they will undercut prices and take away business in the profitable segments without serving the unprofitable segments. The erstwhile monopolist then must raise prices in the favored markets or face bankruptcy.

Until the 1950s, the Bell System's portrayal of its pricing policies, its argument in favor of monopoly status for local telephone companies, and for AT&T's Long Lines Department were widely accepted by economists, regulators, and legislators. But in 1956 a U.S. Circuit Court of Appeals ruled that the Bell Companies could not prevent a subscriber from attaching a Hush-A-Phone to the telephone, and in 1959 the FCC decided to grant licenses for private microwave transmission above 890 MHz and abandoned its previous requirement that such licenses demonstrate exceptional circumstances or lack of common carrier facilities. These two rulings opened the door (though only by a crack) to competition in the markets for telephone equipment and long-distance service. A chain of subsequent rulings widened the openings, and the associated investigations into the Bell System's rate and cost structure generated doubts about the accuracy of the System's portrayal of its *cross-subsidization* (i.e., supporting basic service with contributions from non-basic service) and about the desirability of monopoly in telephony. In 1980 an FCC ruling removed the last regulatory barrier to entry into interstate telecommunications markets, and in 1982 a settlement of a seven-year-old antitrust lawsuit against AT&T stipulated divestiture by 1984 of AT&T's subsidiary local telephone companies.

The divestiture is dramatic, but significant changes in the U.S. telephone rate structure would have been inevitable even if AT&T's corporate structure had been left intact due to the impact of technological changes.

The development of microwave converted long-distance service from a natural monopoly to a market in which there is room for competition and the development of electronic switching makes it feasible to price local telephone service on the basis of cost of service, instead of the traditional flat rate. These forces, in conjunction with legal rulings which increased competition in the industry, have made usage-sensitive pricing for local service inevitable.

6.5.2 The Terminal Equipment Market

Ancillary terminal equipment other than the basic telephone instrument has been classified by the Bell System as *vertical services*. According to the traditional pricing policies advocated by the Bell System prices (historically rental charges because such equipment was leased, not sold) for such equipment were to be set above cost in order to generate revenue to cross-subsidize basic service. To maintain prices which generate supranormal profits requires protection from competition, and for nearly a century telephone companies insulated themselves from such competition by discontinuing service to any subscriber discovered using a foreign attachment — i.e., equipment not supplied by the telephone company.

These restrictions were defended as necessary to prevent harm to the network or to telephone company employees (from high voltage which might leak to the lines from defective equipment), but in some cases the restrictions applied to equipment which could not possibly cause harm. Such was the "Hush-A-Phone," a purely mechanical cuplike attachment which snapped onto the handset. The FCC approved prohibition of this attachment, but in 1956 a U.S. Circuit Court of Appeals overruled the FCC.(14) In 1968 the FCC, following guidelines set by the court in the Hush-A-Phone case, ruled that the Bell Companies could not prevent a subscriber from using his telephone in conjunction with Carterfone, a device which acoustically coupled the handest to a two-way radio and thus made it possible to link the telephone network with private mobile radio radio units.(15) The gist of the Carterfone decision was that telephone companies could not prohibit harmless foreign attachments, and so it became possible for subscribers to buy from other sources specialized terminal equipment such as computer terminals, facsimile machines, and teletypewriters, and connect them to the telephone network through a protective interface device. In 1977 the Supreme Court upheld an FCC ruling which extended the opening to competition by making it possible for subscribers to purchase equipment from independent suppliers and connect them directly to the telephone network provided that they have been certified for quality by the FCC.

These decisions eliminated the ability of telephone companies to set prices for vertical services in excess of their costs, and to the extent that they were in fact earning supranormal profit on some of these services, in consequence, they had to raise prices for other services in order to earn a fair rate of return in the aggregate. In their arguments for protection from competition, the telephone companies portrayed the old pricing system as one which discriminated against large businesses (which leased much special terminal equipment) in favor of small business and residential subscribers who made do with basic service. The FCC did not accept the accuracy of this picture, and countered: "... we find the studies of contribution loss submitted by the telephone industry without merit. Among other things, the studies submitted generally utilize inappropriate cost and allocation methodologies, thereby failing to show whether there is, in fact, any direct contribution to basic local telephone service including residential exchange to be lost due to interconnect service."[16]

During FCC and court proceedings on these issues, thousands of man-days were devoted to investigation of the extent and direction of cross-subsidization between vertical services and basic service, but no clear-cut findings emerged. It is therefore impossible to provide an estimate of the impact of competition in the market for vertical services except to say that it eliminated any possibility for telcos to earn excess profit in that market.

From a broad perspective, questions about the existence and magnitude of price discrimination against business customers are unimportant. Businesses must charge prices which cover their costs and so would pass on to their customers whatever burden they were paying for telephone service. If, in fact, there was discrimination against business customers, the elimination of it benefits the average consumer by reducing many prices by a small amount, and these benefits exceed any associated increase in the average residential phone bill. But the benefits are imperceptible and the increase is evident, so public resentment is to be expected.

6.5.3 The Long-distance Market

The Bell System's monopoly in telephone was initially established on the basis of the patent of the telephone granted to Alexander Graham Bell in 1876. When this patent expired in 1894 a number of independent (non-Bell) telephone companies began offering local service, but in the interval AT&T's Long Lines Department had achieved a monopoly in the long-distance market due to its control of patents on necessary equipment and economies of scale. AT&T refused to connect independent companies with its long-lines network unless they became subsidiaries of AT&T; when this practice ended in 1913 (in response to the threat of an antitrust suit) Bell affiliates were providing 85% of local telephone service in the U.S.

In the period from 1907 to 1919, under the leadership of Theodore Vail, AT&T urged persuasively that the telephone industry was a natural monopoly which should be regulated by government rather than by competition. It may well be debated whether it was socially desirable for the Bell System to control so many of the local telephone companies, but it was efficient that local service in each municipality be provided by a single company and that long-distance service be provided by a single, nation-wide network. This was so because economies of scale were so significant when the network consisted of wires or cables strung on poles or laid underground (above, 6.3). The development of microwave altered this condition and made it feasible to move toward competition in the long-distance market, but to appreciate the magnitude of the impact of such competition it is necessary to understand the system of *separations payments*.

To obtain access to AT&T's Long Lines Department, a telephone subscriber uses equipment provided by the local telephone company, and is billed for the long-distance call on the basis of its distance, duration, and time of week. The local company turns over to Long Lines a portion of the toll to cover the cost of the service provided by Long Lines, and retains a portion to cover the cost incurred by the local company; the latter is termed the separation payment. A portion of the local company's cost is traffic sensitive (i.e., the cost varies according to volume), and this portion may be estimated by ordinary accounting techniques. The other portion of the local company's cost is the access cost which is a non-traffic sensitive joint cost which cannot be scientifically allocated between long-distance and local service (above, 6.5.1). At first, long-distance tolls did not include any coverage of access cost, but in 1943 (in response to a ruling of the Supreme Court) the FCC changed its regulations so as to include some coverage of access cost in long-distance tolls.[17] In the absence of a scientific method of allocation, the formula for splitting access cost between interstate (long-distance) and intrastate was determined in negotiations between the FCC and the Utility Commissioners of the States. (From a theoretical perspective the important distinction is local versus long-distance service, but in practice the focus is intrastate versus interstate because states regulate the former and the FCC regulates the latter.) A series of changes raised the fraction apportioned to interstate from 3% in 1943 to 19% in 1974.[18] The changes resulted because local service costs were rising with general inflation while technological advances were reducing the cost of long-distance service; the State regulators felt pressured by their constituencies to hold down the cost of basic service, but the FCC felt less pressure to reduce long-distance rates. In 1981 approximately 37% of each dollar collected for interstate MTS and WATS service was used to support basic exchange service; i.e., on the average MTS and WATS rates exceeded their long-run marginal costs by 37%. (To avoid misinterpretation I restate that because access cost is not traffic sensitive, it is not part of the long-run marginal cost of long-distance service, but of course it must be incurred in order to gain access to such service.)

This substantial departure from cost-of-service pricing could not be maintained in the face of free competition, and such competition was in fact emerging due to the development of microwave technology and a series of legal decisions.

In World War II the U.S. Army constructed a radio microwave communications network in North Africa, and in the late 1940s AT&T used microwave to provide carriage of network television which required a bandwidth of 600 voice-grade circuits and would have strained the capacity of the cable network. In 1950 the FCC licensed TV networks to build microwave relay stations in remote areas which lacked common carrier facilities but warned that this permission was temporary pending provision of service by the telephone common carriers. Developmental licenses were also granted to railroads, pipelines, and lumber camps in remote locations.[(19)] In 1959 after three years of investigation (one of the important questions being whether the available spectrum could accommodate numerous users) the FCC decided to make available frequencies for private microwave for business use.[(20)] The FCC stated that such licenses no longer require a demonstration of exceptional circumstances or lack of common carrier facilities, but the applications would not be approved if there was a reasonable likelihood that adverse economic effects would result. Expansion of private microwave was limited, however, because the FCC did not permit separate organizations to share a system, and the telephone companies did not permit interconnection of private microwave facilities to the telephone network.

The FCC relaxed its prohibition against shared private microwave in 1966 and took another step toward increased competition in telephone communication in 1969 when it approved the application of Microwave Communications, Inc. to offer common-carrier service between Chicago and St. Louis and nine intermediate points.[(21)] Up to five subscribers could share a single channel for the carriage of voice, facsimile, and data; rates were significantly lower than those charged by the existing common carriers in part because MCI offered a channel whose bandwidth was only 2KHz, instead of the customary 4KHz. Moreover, the FCC recognized that MCI's success would depend on its ability to link its microwave system with its subscribers through local distribution lines leased from telephone companies, and stated that such interconnections would be ordered in the absence of a showing that they were not technically feasible. Over the next few years a series of judicial rulings required the telephone companies to provide interconnections with microwave common carriers thus enabling the latter to compete more effectively with AT&T's Long Lines Department in the market for private line interstate service. By 1975 the new specialized carriers had gross revenues of $35 million, compared to telephone company gross revenues of $1,613 million from private line service. This was only a small percentage of the market (2%), but the economic effect of the new competition was greater than might be inferred from this statistic because AT&T responded to the competitive pressure by lowering its own rates for bulk services.

The telephone common carriers had opposed authorization of microwave common carriage on the grounds that such service could succeed only by practicing creamskimming; i.e., they charged that the new specialized carriers would provide service at lower rates between those points where traffic was heavy and which therefore could be served at low cost, but would leave the low-density, high-cost markets to be served only by the telephone common carriers (mainly AT&T's Long Lines Department). But since the FCC had not been persuaded to exclude competition, AT&T responded to meet it by lowering its rates for heavy users of Long Lines service. In 1960 it introduced Wide Area Telephone Service and TELPAK, which was a bulk lease service offering packages of 12, 24, 60, or 240 voice-grade circuits at rates 51% to 85% lower than existing rates for private lines.

Motorola (a manufacturer of microwave equipment) and Western Union argued before the FCC that the TELPAK rates were unduly discriminatory, that they were set below the cost of the services in order to drive out competition in the market for bulk communications, and that AT&T was using profit from its monopoly markets (basic exchange service and MTS) to cross-subsidize TELPAK. After investigation the FCC ruled in 1964 that 12 and 24 circuit packages were unduly discriminatory but allowed the rates to stand for 60 and 240 circuit packages pending additional cost analysis. An investigation of comparative rates of return on Bell System's interstate services (commonly referred to as the Seven-Way Cost Study) showed that the rate of return on investment in 1963-4 was 10% for MTS and WATS, 4.7% for voice-grade private line, 2.9% for teletypewriter exchange service, 1.4% for telegraph-grade private line, and 1.3% for TELPAK.[22] It then appeared that AT&T's rate structure benefited those who were heaviest users (i.e., subscribers to TELPAK and private line service) in comparison to moderately heavy users (WATS) or light users (MTS).

AT&T has countered that the Seven-Way Cost Study erred in apportioning costs on a fully distributed cost (FDC) basis, and argued that a more appropriate method would use long-run incremental cost (LRIC). The dispute over the proper method of cost allocation has persisted; in 1976 the FCC rejected AT&T's proposed LRIC method and ruled that future rate filings must be justified on a fully distributed cost basis. The commission also ruled that existing TELPAK rates were unjustly discriminatory and ordered termination of the special bulk rates in 1977 (except for existing customers).[23] The FCC also ruled that customers who leased PLS could resell unneeded capacity to others; this reduced the separation between the markets for PLS and WATS because a cooperative group of smaller users could obtain the lower rate intended for heavy users. In 1981 permission for reselling was extended to WATS, thereby drawing away business from MTS, and TELPAK was completely terminated. (The delay resulted from a series of legal contests initiated by TELPAK customers who wanted the service continued.)

During the same period, court rulings and FCC responses permitting the specialized carriers to branch out from bulk services toward services which were similar to and competitive with AT&T's WATS and MTS. The FCC had rejected MCI's application to offer Execunet (a switched service), but court rulings overturned the rejection and ruled that AT&T must provide interconnections for Execunet service.[24]

The pro-competitive decisions of the courts and the FCC could have been superseded by Congressional action. In 1976 Congress seemed close to enacting the Consumer Communications Reform Act which would have required potential entrants to prove that their services would not duplicate existing or potential services of established carriers and would have directed the use of incremental cost accounting. If enacted, this bill would have closed the door to new entrants and perhaps would have enabled AT&T to drive out the specialized carriers who had already entered. However, by 1978 the "Bell bill" had been rejected, and Congress had demonstrated its preference for competition over regulation.

With the elimination of uncertainty about their future, the new long-distance carriers grew in number and size, and although AT&T continues to dominate the long-distance market, it now faces competition in all markets.

In sum, by 1983 the markets for bulk PLS, single PLS, WATS, and MTS could not be separated and were no longer protected against competition. Price discrimination was not viable, and the pressures were inexorably toward cost-of-service pricing for all long-distance service.

6.6 TOWARD COST-OF-SERVICE PRICING FOR LOCAL SERVICE

The events described in 6.5.3 have made it impossible to keep a telephone rate structure which prices long-distance service above cost in order to subsidize local service. In response the FCC ordered that separations payments be phased out over a five-year period starting in 1984, and that the attendant reduction in revenue be offset by explicit charges for interstate access to be billed to each subscriber.

The plan called for initial fees for interstate access of $2 per month for residential customers and $6 per line per month for businesses, with periodic increases until they became high enough to fully offset the eliminated separations payments. However, Congressional opposition to access fees for residential customers has been so strong that imposition of them has been deferred indefinitely. Apparently Congressional sentiment favors continuance of the cross subsidy to residential subscribers, using revenue derived from access fees imposed on all long-distance carriers in the magnitude of $150 per line per month, far above the cost of the line. Such a large gap between price and cost creates strong incentives for the long-distance carriers to create their own local networks, bypassing the regular telcos. Possibly taxes could be imposed on bypass schemes, but the feasibility of enforcement is questionable.

The threat of bypass applies not just to long distance, but to any pricing system which attempts to discriminate substantially against business customers. The technology of microwave and broad-band cable makes it possible for business to bypass the telcos completely, and they will do so if confronted with telco rates which are much above cost.[25] The threat is not confined to large businesses; a large office building or industrial complex could provide a communications network for all of its tenants.[26]

The pressures described above compel telephone companies to eliminate the traditional deficits in local service; they face bankruptcy if they are unable to find ways to raise more revenue and to reduce costs. Fortunately, technological advances in switching make it possible to achieve enormous gains in efficiency by moving toward a more rational pricing system, a system such that telephone users are forced to take into account the costs which their usage imposes.

The cost imposed by telephone use varies according to the number of calls, their distance, duration, and timing. (The cost is mainly a capital cost, but it is properly viewed as variable, as argued in section 6.4.) As outlined above (6.1), it is economically logical that prices reflect costs, but until recently most telephone companies in the U.S. charged for local service only a flat rate which did not vary with use. This practice arose because with the earlier technology it was expensive to record the necessary information. Before 1921, calls were manually switched, and although it was possible for the operator to record distance, duration, and time for each call, it would have been inordinately expensive to do so for local calls. The introduction of dial systems in the 1920s slightly altered the situation; mechanized message registers could cheaply tally the *number* of local calls, but without operator handling it became more expensive to record the other information.

It was a matter of judgment whether the recordkeeping was worth the effort, and opinions varied widely. A few telephone companies in the U.S. set charges for local calls in accord with an elaborate metering system, a few offered pay-as-you-call coin-box service for residential subscribers, some adopted a rate structure which used a simple count of the number of monthly calls in one way or another, but most continued to charge a flat monthly rate regardless of the number of local calls.(27) Outside of North America most telephone companies counted local calls and billed accordingly.

But while telephone costs vary with the number of calls, they are also affected significantly by the duration and timing of calls; the latter is especially important because at times when congestion is no problem, the cost of a call is essentially zero if it does not require operator handling. Therefore, a billing system based on simple message counts is far from cost-of-service pricing, and it may be debated whether such a system is superior to a flat rate.

The pricing system for long-distance calls has always been *usage sensitive*; the price of a call varies according to duration, distance, and timing. In fact, the classification of a call as "long distance" reflected a judgment that it was expensive enough to be worth special metering and billing, whether manually in the early days or with special equipment later. But the development and installation of electronic switching systems is eliminating the difference between long-distance and local calls by making it feasible to apply usage-sensitive pricing for all calls.

With electronic switching it is possible to record the duration, distance, and time of week of each call at a cost of less than $1.00 per month per line, making it feasible to price local calls in the same way that has always been used for long-distance calls. The evident benefit of such a pricing system is that it discourages wasteful use and reduces the amount of equipment needed to provide satisfactory service, and given their financial problems telcos cannot ignore this opportunity to economize.

There will be strong opposition to the establishment of usage-sensitive pricing from customers who have become accustomed to the benefits of heavy usage without any need to calculate the cost, for example, firms that practice mass solicitation by phone. One may therefore anticipate that unmeasured service will long continue to be an available option, but the flat rate for such service will rise to such an extent that many customers will switch to a measured service option. As this happens there will be an increasing public acceptance of the logic of cost-of-service pricing and decreasing political support for a flat-rate option which benefits only a small number of heavy users. One might expect that eventually all calls will be priced on the basis of duration, distance, and time of week. Urban and suburban customers who respond by limiting their calls and by timing them to get off-peak discounts should experience only a modest increase (beyond general inflation) in their local bills, but some rural customers will experience substantial increases. Long-distance rates will decline.

Other moves toward cost-of-service pricing are less dramatic, but provide significant reduction in waste. For example, directory assistance was provided at no charge; 80% of the requests for information came from 20% of the subscribers but the cost was imposed on all subscribers, adding about $.30 per month to the average bill.[28] Most companies now charge for local directory assistance; the change improves both equity and efficiency. Service connection charges formerly were not differentiated adequately according to conditions, and the average charge was less than the average cost. Most companies now adjust such charges according to the amount of labor and encourage customers to install their own phones if the wiring and jack are in place, thereby conserving resources.

In summary, reform of telephone rate structure in the U.S. is placing the burden of costs on those who cause them. Within the limits of practicality and revenue requirements, the objective is to encourage use when marginal utility exceeds marginal cost and to discourage use when marginal utility falls short of marginal cost. Many customers will be upset by the changes, but telephone companies can no longer provide costly services for free. The movement toward cost-of-service pricing is well underway, and it will bring significant gains in economic efficiency.

REFERENCES

1. Horn, C. F., "Factors Affecting Telephone Pricing," in J. T. Wenders (ed.), *Pricing in Regulated Industries: Theory and Applications* (Denver: The Mountain States Tel. & Tel. Co., 1977), p. 6.
2. Squire, L., "Some Aspects of Optimal Pricing for Telecommunications," *Bell Journal of Economics and Management Science*, Vol. 4, Aug. 1973, pp. 515-25.
3. Littlechild, S. C., "The Role of Consumption Externalities in the Pricing of Telephone Service," in J. T. Wenders (ed.), *op. cit*., pp. 44-5.
4. Clemens, Eli W., *Economics and Public Utilities* (New York: Appleton-Century-Crofts, Inc., 1950), p. 328.
5. Mandanis, G. P. *et al*., *Domestic Communications and Public Policy* (Systems Applications Inc., 1973) as reported in J. H. Alleman, *The Pricing of Local Telephone Service* (Washington, D. C.: U.S. Dept. of Commerce, Office of Telecommunications, April 1977), p. 114.
6. Hotelling, H., "The General Welfare in Relation to Problems of Taxation and of Railway and Utility Rates," *Econometrica*, Vol: 6, July 1938, pp. 242-69.
7. Dupuit, J., "On the Measurement of the Utility of Public Works," English translation by R. H. Barback in *International Economic Papers*, Vol. 2, p. 89; original in *Annales des Ponts et Chaussees*, 2d series, Vol. 8, 1844.
8. Hadley, Arthur T., *Railroad Transportation* (New York: G. P. Putnam's Sons, 1886), p. 76.
9. Samuelson, Paul A., *Foundations of Economic Analysis* (Cambridge: Harvard University Press, 1955), p. 242.
10. Lipinski, Jan, "The Correct Relation between Prices of Producer Goods and Wage Costs in a Socialist Economy," in D.C. Hague (ed.), *Price Formation in Various Economies* (New York: St. Martin's Press, 1967), p. 124.
11. Kahn, *op. cit*., pp. 89-109; Baumol, William J., *Economic Theory and*

Operations Analysis (Englewood Cliffs, N. J.: Prentice-Hall, Inc., 1977) pp. 173-5.

12. Baumol, W. J. and D. F. Bradford, "Optimal Departures from Marginal Cost Pricing," *American Economic Review*, Vol. 60, June 1970, pp. 265-83.
13. Testimony of Frank J. Alessio, New Mexico S.C.C. Docket No. 673, Hearing on Remand, Mountain Bell Exhibit B, *In the Matter of Rates and Charges of the Mountain States Telephone and Telegraph Co.*; Corman, W. F., "The Pricing of Telephone Service," *Telephone Engineer and Management*. Oct. 1, 1971, pp. 93-110.
14. Kahn, op, cit., Vol. II, pp. 140-2.
15. *Ibid*., pp. 142-5.
16. *In Re: Economic Implications and Interrelationships Arising From Policies and Practices Relating to Customer Interconnection, Jurisdictional Separations and Rate Structures*, First Report, FCC Docket No. 20003, Sept. 23, 1976, pp. 123-5.
17. Horn, *op. cit*, p. 19.
18. *Ibid*., p. 20.
19. Irwin, M.R., "The Communications Industry," in Walter Adams (ed.), *The Structure of American Industry* (New York: The Macmillan Co., 1971), pp. 391-3; Kahn, *op. cit*., Vol. II, pp. 129-36.
20. *Allocation of Frequencies in the Bands Above 890 Mc*, 27 FCC 359 (1959), 29 FCC 825 (1960).
21. *In Re: Applications of Microwave Communications, Inc*., 18 FCC 953 (1969).
22. Reported in Irwin, *op. cit*., p. 394.
23. *In Re: American Telephone and Telegraph Co. Long Lines Department*, Docket No. 18128, FCC Report No. 12343, Sept. 23, 1976.
24. MCI v. FCC. 561 F.2d 365 (1977) and MCI v. FCC, 580 F.2d 590 (1978).
25. Netschert, Bruce, "The Bypass Threat. . .," *Telephony*, July 18, 1982, pp. 112-6.
26. "Gambling with Long Distance Resale. . .," *Telephony* , April 18, 1983, pp. 34-6.
27. Alleman, *op. cit*., pp. 11-12.
28. *Ibid*., p. 103.

chapter 7

TELEPHONE SYSTEMS

Floyd K. Becker
Department of Electrical and Computer Engineering
University of Colorado
Boulder, Colorado

7.0 INTRODUCTION

The title of this chapter is disarmingly simple and therefore, probably misleading. The telephone system in the 48 contiguous states is the grandest technological achievement ever. It is ubiquitous and its excellence causes users to treat it commonly. The following pages will not thoroughly describe the telephone system because such an attempt is futile; the technological and corporate changes occurring in the structure of the system outpace the descriptive powers of authors. What will be attempted here is a brief sketch of the voice communication portion of the system. Included in this description of the voice system will be its use for data communications.

Voice and data comprise more than 90% of the usage of all systems based upon almost any measure one may care to apply. This chapter is divided into four main topics:

1. Speech and hearing; i.e., "voice"
2. Voice transmission systems
3. Switching systems
4. Signaling systems

The emphasis of this chapter is not on people or companies. Also, most parts of this topic are highly technical and require precise treatment, while the approach here is of necessity, brief. Further caveats and apologies for this treatment will be avoided.

Early telephony was concerned with the development of subsystems which "worked" and standards or measurements of excellence were unavailable or ignored, but as telephone technology matured it was obvious that such standards and measures were imperative. It was an economic necessity that systems developers knew what was good and what was only acceptable or unacceptable, and that brings us to the first topic.

7.1 SPEECH AND HEARING

The transmission requirements of a voice system were determined by conducting tests on a very large number of subjects in order to ascertain the pertinent parameters of human speech and hearing. Statistical averaging of the data shows that the "average" talker emits only 32 microwatts of power and the spectral content of average speech; i.e., the sound power per hertz of bandwidth, is as shown in Figure 7.1.

Specific talkers vary significantly from this "average talker," and it is clear that a designer must include the capability for all talkers to utilize the system. Instantaneous speech power may vary over a range of nearly one million to one and average speech power may vary over a range of one thousand to one. Figure 7.1 shows that the majority of this speech energy lies in the low frequency region, but does the ear utilize all this energy in a direct relationship? Testing the aural response of many subjects and again applying statistical methods indicates that the response of the ear extends from 20 Hz to 20 kHz and is somewhat dependent upon the level (pressure at the ear) of the sounds. This multidimensional response of the ear is shown in Figure 7.2 as equal loudness contours. Simply stated, Figure 7.2 is the result of detecting a

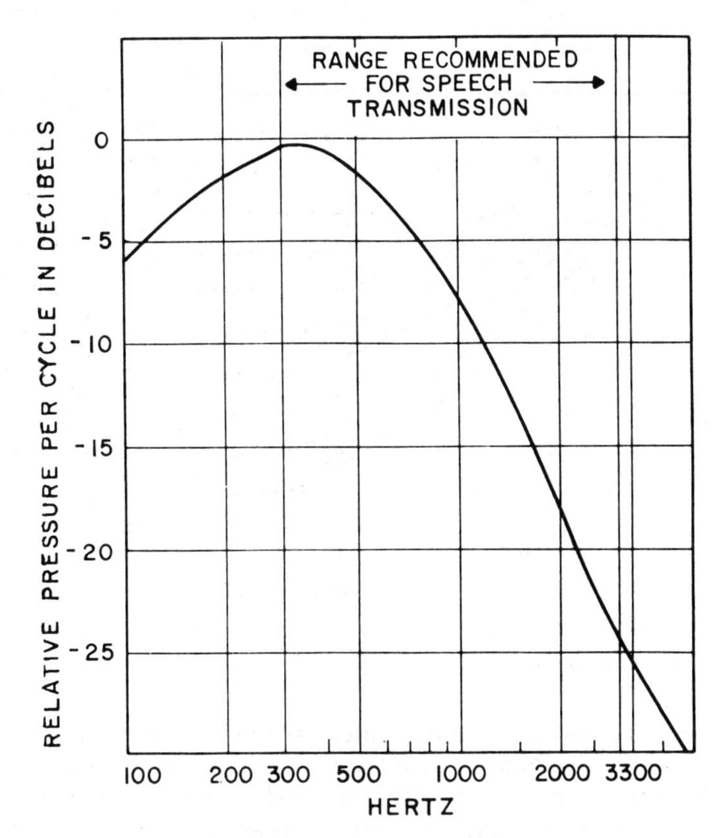

Figure 7.1: Intensity Level of Speech versus Frequency

subject's threshold of hearing, increasing the power level an arbitrary amount to a second level of sound pressure, and then increasing the power a second time until a subject perceives the loudness *increase* to equal the first. Several times, the power is again increased and the same criterion applied. On any equal loudness contour, a tone different from 1000 Hz is adjusted until the subject perceives it to be equally loud as the 1000 Hz tone.

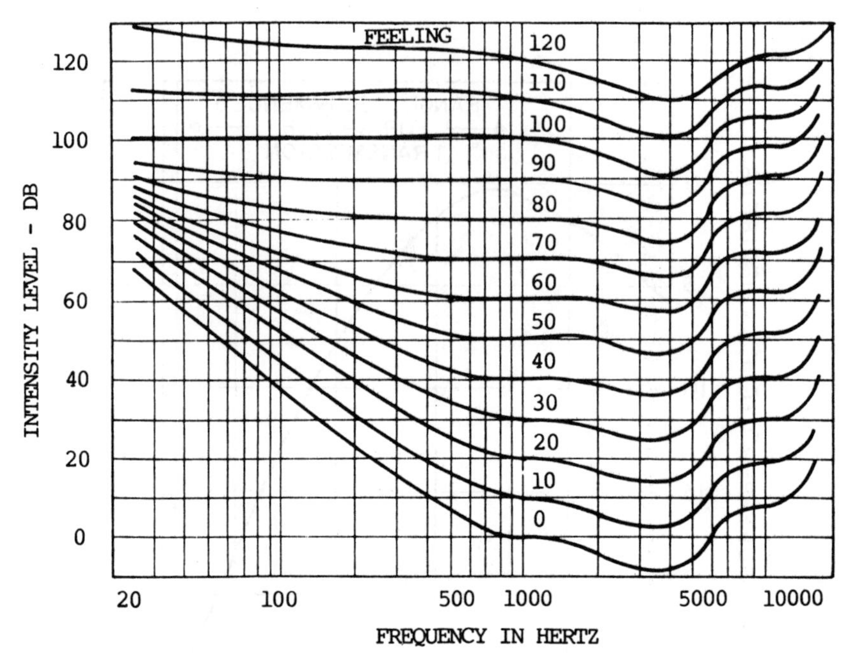

Figure 7.2: Sensitivity of the Ear as a Function of Frequency — Loudness Level Contours versus Intensity Level

The ear perceives equal loudness increases when power ratios are constant rather than when power increases are equal. This response of the ear gave rise to the use of the BEL or decibel as the standard of measurement in transmission systems. Definitions and methods of use of decibels is covered in an appendix of this book. People prefer speech and music at levels between 80 and 90 dB on this curve, where the response of the ear is relatively uniform with frequency.

There is an obvious disparity between the bandwidth requirements for speech and the ability of the ear, which again raises questions of the important elements of the speech spectrum and what effects reduced transmission bandwidth will have upon a user's perception of quality and the intelligibility of a voice passed through the network. Further tests indicate that a transmission band extending from 300 to 3 kHz is highly intelligible for the spoken word and is adequate regarding loudness, voice recognition, and quality. Restriction of this band rapidly reduces intelligibility or articulation and perceived loudness. These results are shown in Figures 7.3 and 7.4.

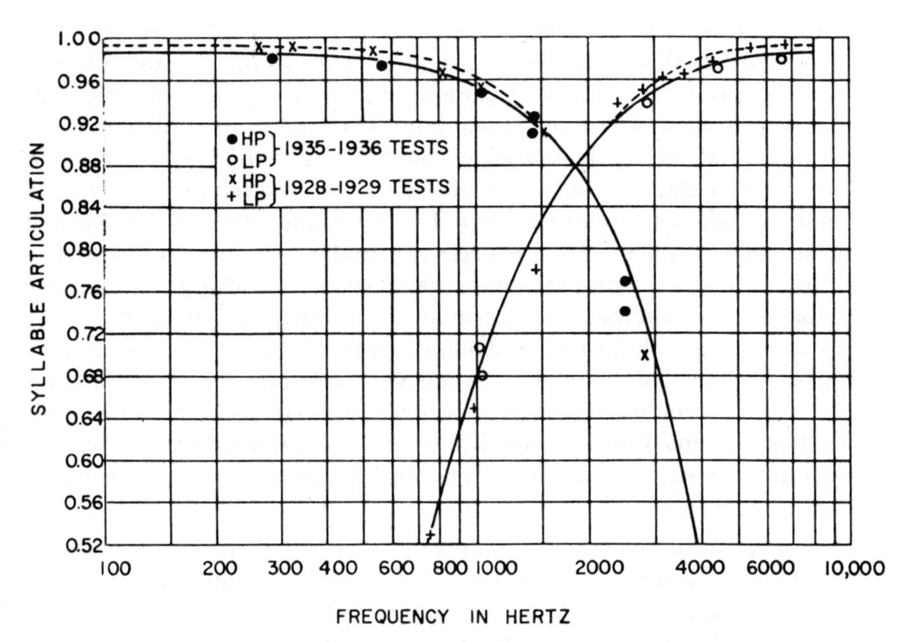

Figure 7.3: Speech Syllable Articulation versus Dual Filter Cutoff Frequency

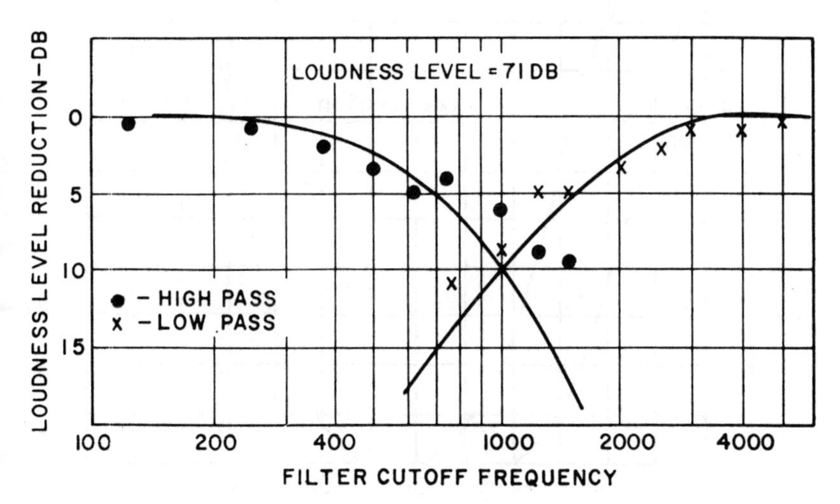

Figure 7.4: Loudness Level Reduction Test on the Coupled Tone Having 26 Components in the Range 150 to 4000 Hz.

Unfortunately, voice systems are susceptible to impairment from sources other than bandwidth restriction; such as noise, crosstalk, and echo. Noise is the induced or generated sounds in a system which are identifiable as "hiss" or occasional impulses. The sources of such noise are manifold and largely unpreventable. Hissing sounds are generated in electrical conductors and are dependent upon the temperature of the conductor and its ohmic resistance. This is usually referred to as thermal noise. There are many other sources for similar sounding noise. Impulse sounds are due to natural phenomena such as discharge of induced voltages on conductors and man-made noise due to switching transients and the like. Steady background noise may reach levels which are annoying and affect understanding. Impulsive noises may become annoying when their level or frequency of occurrence becomes high, but usually neither happens. While the amplitude and rate are insufficient to bother the human listener, they normally determine the error rate in voiceband data transmission.

The measurement of noise is an effort to determine the amount of annoyance a complex signal causes the listener. Listener satisfaction is based on volume and noise and, hence, a ratio. Figure 7.5 shows the results of tests in which the industry-standard 500-type telephone was used.

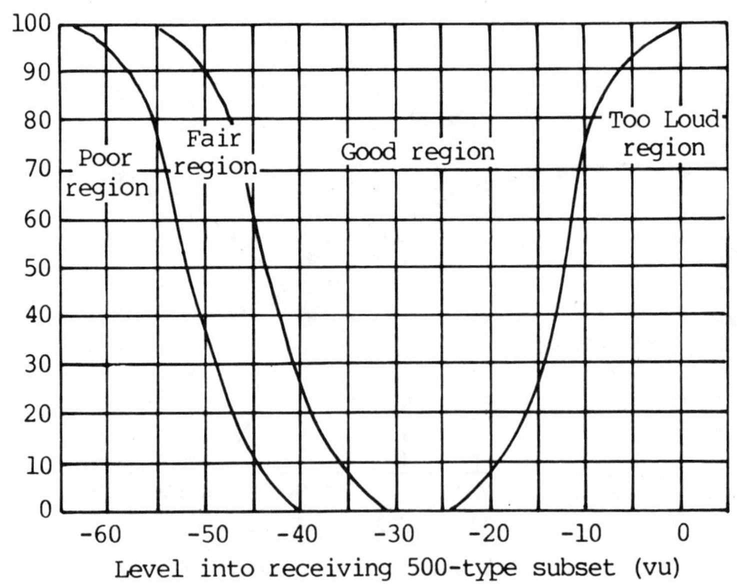

Figure 7.5: Customers Judgement of Grade of Telephone Service versus Received Volume

System objectives place the volume such that 95% of the users perceive the volume as good and that not more than 5% perceive the volume as only fair. Similar tests, displayed in Figure 7.6, indicate that 95% of the users of the 500-type set will judge the system as good or better if the noise level at the set is 27 dBrnc. Note that Figure 7.5 expresses volume in VU and that Figure 7.6 expresses noise in dBrnc. This is consistent with measurement standards and techniques. The reference noise level is -90 dBm and the C-message weighting curve is shown in Figure 7.7. Such measurements are usually made with a true-rms meter with about 200 millisecond averaging.

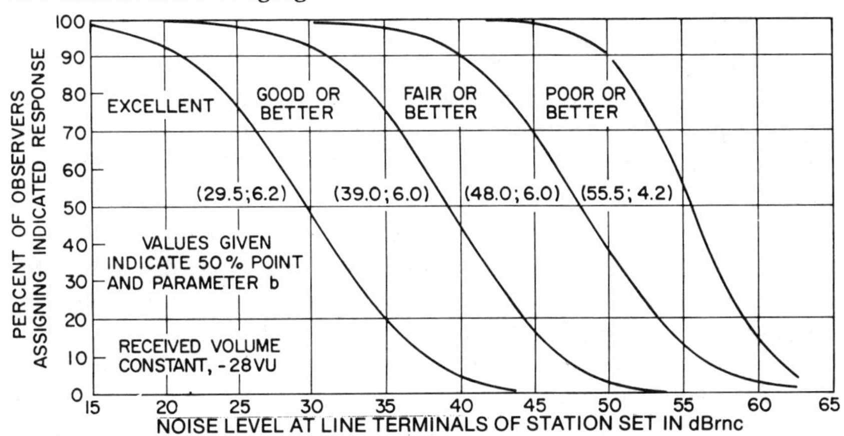

Figure 7.6: Noise Opinion Curves

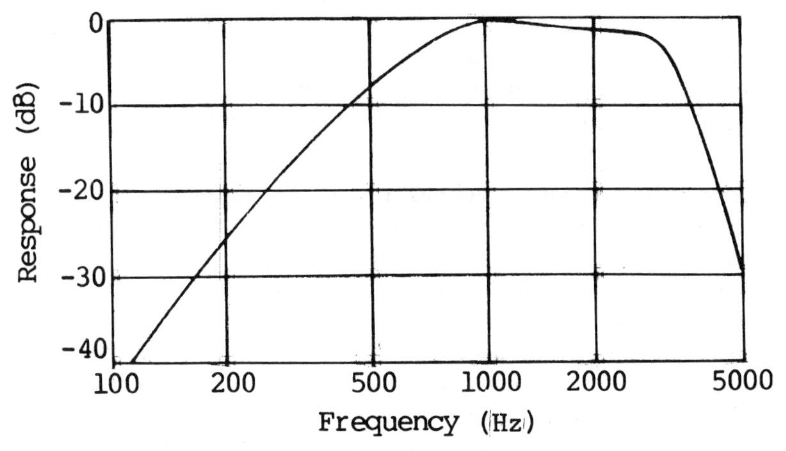

Figure 7.7: C-Message Frequency Weighting

The second source of impairment in voice systems is cross-talk. Cross-talk is the unwanted coupling of other talkers' signals into one's circuit path. Cross-talk may be due to direct inductive or capacitive coupling between conductors, or devices or crossmodulation in carrier systems, or coupling between time slots in time-division systems, or one of many other causes. Cross-talk is discernible by its babbling sound and may be divided into intelligible or unintelligible categories. When only a few circuits couple together, intelligible cross-talk may result and violate privacy while unintelligible cross-talk may be lumped with other noise sources as in Figure 7.7.

A unit often used in cross-talk computations is the dBx, which was invented, like the dBrn, to allow cross-talk to be expressed in positive numbers. Like the dBrn, its base is −90 dB; thus, if the coupling between two circuits is −60 dB, this fact is expressed by quoting the crosstalk coupling as being +30 dBx. The smaller the number of dBx, the better the circuit.

In summary, if intelligible cross-talk results from coupling between circuits, the loss between circuits should be −75 dB or +15 dBx. If unintelligible cross-talk results either because of great numbers of interfering channels superimposed or frequency inversion due to nonlinearities in carrier systems, the cross-talk may be viewed as noise, and the results in Figure 7.6 generally apply.

Echo is the final impairment to voice systems to be considered here. Subjective testing shows that propagation times between the originating and terminating end of a circuit may be as high as 300 milliseconds without user dissatisfaction. Longer times than this cause the user to experience difficulty, at least initially, in conversing. Fortunately, this upper limit is just within the propagation time of a stationary satellite and is the reason double satellite hops should be avoided where possible.

As stated, propagation time is not a problem. It is the effect of propagation time and the presence of echoes that impair transmission systems. Discussion of echoes on transmission systems deals with user reaction to echoes as well as circuit stability against "singing" or oscillating. Echoes are generated at any amplification or termination point in a two-wire circuit. Also conversion of four-wire to two-wire circuits causes echoes. Nevertheless, most echoes are caused by improper or near impossible impedance matching. Examples are negative impedance converter amplifiers and voice frequency amplifiers used in the exchange plant and four-wire to two-wire conversion for switching. Whatever the cause, echoes are a prime concern in circuit layout.

The magnitude of an echo which may be tolerated by a talker is a function of the amount of delay encountered in round trip from the talker to the point causing the echo and return to the talker. Table 7.1 shows the magnitude of the tolerable echo versus roundtrip delay.

TABLE 7.1

User Reaction to Echo Delay

Round-Trip Delay	Ratio of Echo to Transmitted Signal
0 ms.	85%
20	30
40	15
60	7
80	4
100	3

Values shown in Table 7.1 will result in about 1% of the talkers expressing dissatisfaction with the circuit. The measure of echo is called Return Loss. In this instance loss is used in order to again deal with positive numbers. Return Loss is defined as 20 $\log_{10}$ *impressed signal / reflected signal* and is usually measured at 1,000 Hz. Table 7.1 shows the ratio of the reflected signal to the impressed or transmitted signal.

The above discussion has dealt with what is appropriately termed talker echo. As shown in Figure 7.8, a listener can hear a re-echo in his direction. A listener's reaction to this echo is again dependent on the amount of delay, but a circuit which is adequate for the talker's satisfaction is, in general, satisfactory for the listener.

A circuit which is badly misterminated will oscillate or "sing." In order to sing, the echo must be as great or greater than the transmitted sound and will occur only under trouble conditions. But a useful quantity for a systems designer is the margin against singing and is found by increasing gain until the circuit breaks into oscillation at any frequency. This amount of gain increase is expressed in dB.

The average return loss at two-wire switching points is 27 dB for properly balanced offices but is as low as 15 dB for the final four-wire terminations. Because of this low return loss, circuits over about 45 milliseconds round-trip delay (800 miles in length) are equipped with echo suppressors. Virtually all such long-haul circuits are four-wire circuits which will be described further. Echo suppressors are voice operated devices located at each end of long circuits which insert as much as 50 dB loss in the return path of a four-wire circuit, thus effectively removing all

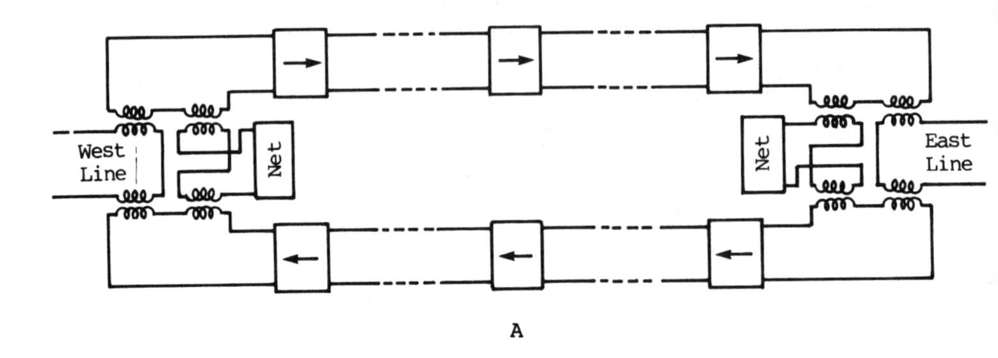

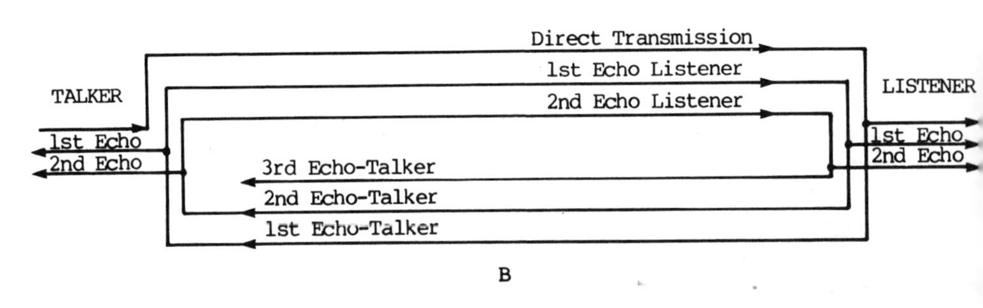

Figure 7.8: Echo Paths in a Four-Wire Circuits

echoes but allowing sufficient transmission for the opposite end to interrupt the talker. Such devices do effectively reduce echoes but they introduce other impairments by clipping the beginnings and endings of words. More serious problems arise when two circuits, each containing echo suppressors, are linked together. Also, it is possible for each end to gain control by talking simultaneously and render the circuit momentarily useless.

Circuits which utilize satellites have round trip delays of about 0.5 seconds and echoes have been a persistent problem. Echo suppressors may not operate in a satisfactory manner because of their interactions with the users. It is possible for users to get into a "ping-pong" conversation; each end beginning to speak and then abruptly ceasing, because of such problems.

In order to alleviate the echo problems on satellite links echo cancelers have been introduced. An echo canceler is also located at each end of the

circuit and subtracts a delayed image of the talker's signal from the echo in order to reduce the echo to acceptable levels. Large-scale integrated circuits have made such devices reasonably priced for wide-spread introduction into satellite systems.

Before discussing the next three topics, namely: transmission, switching, and signaling, some introductory comments are appropriate. Figure 7.9 is a representation of the telephone voice system. It consists of telephones or data sets on the customer's premises, connections to the central office, the switching equipment in the central offices, intra- and inter-office transmission systems, and the associated signaling equipment. Telephones, data sets, PBXs and key systems contain signaling equipment as well as the central offices.

Referring to Figure 7.9, calls are originated at the customer location by removing the handset from its cradle, or other appropriate action which accomplishes the same function. This going "off-hook" signals the central office that service is requested. The central office returns dial tone when it is ready to receive the address signals. The user, using the dial, signals the address of the called party. The originating central office selects a trunk to the appropriate central office which may be itself or another perhaps a thousand miles distant, and signals the address of the called party. If the terminating central office is the same office as the originating office, the trunk connecting the calling party to the called party is not much more than a pair of wires. If the terminating office is from several miles to thousands of miles from the originating central office, transmission will be by means of a carrier system. Closely located central offices will be reached through short-haul carrier systems and more distant central offices will be reached by means of long-haul carrier facilities. A description of the switching, transmission, and signaling functions occurring during this period of time will be described subsequently. The terminating central office tests the called line for idle/busy conditions. If the line is idle an audible tone is returned to the calling party to signal that the called party is being rung. Simultaneously, the called party is signaled of the incoming call by a 20 Hz ringing signal on that line. The called party goes "off-hook" thereby signaling the terminating central office that the incoming call has been answered and this information is returned to the originating central office. At this point automatic message accounting begins and a voice path is established between the originating central office and the terminating central office. As can be seen from this description of a simple voice call, central offices

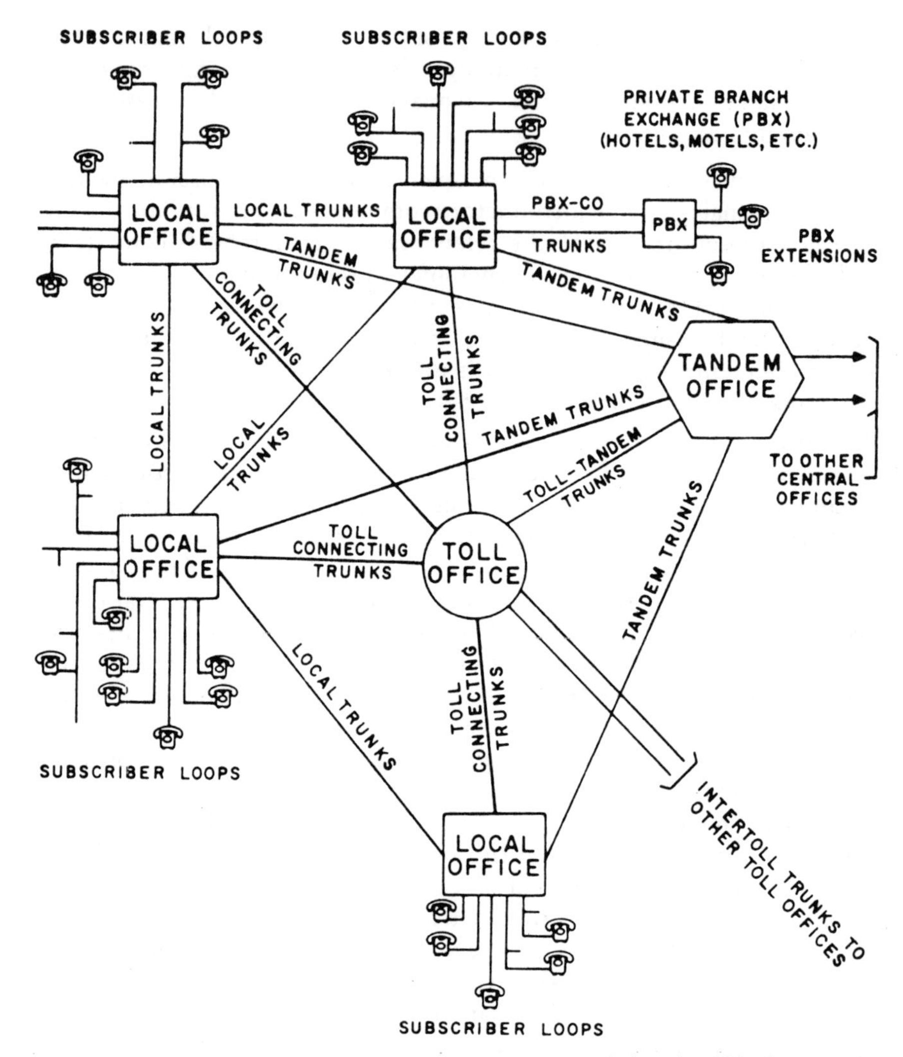

Figure 7.9: Typical Pattern of Interconnections Among Subscriber Telephones and Switching Offices in a Telephone Network

are required to receive, translate, and transmit information, one to another, in order to establish calls. Breaking down connections is equally complicated. We shall now address transmission, switching, and signaling in greater detail.

7.2 VOICE TRANSMISSION SYSTEMS

This section deals with the various voice transmission systems common in the U.S. today. Such systems may be divided into three transmission techniques. These divisions are:

1. Space division multiplex
2. Frequency division multiplex
3. Time division multiplex

A further refinement in the descriptions of the various transmission methods concerns their application as short-haul or long-haul voice systems.

Space division multiplex is a complicated way of categorizing a system which uses separate pairs of conductors for each voice path. Such systems are used only for short-haul transmission systems; primarily from the user's premises to the first switching point, or serving central office and between closely located central offices. In terms of the telephone company's capital investment, such systems represent the largest investment of the corporation's financial assets. The cost of cables and the attendant installation expenses cause telephone companies to utilize sophisticated Operations Research techniques to optimize the investment.

Pole-mounted open-wire and cable are rapidly disappearing in both rural and urban areas. Buried cable results in greater reliability, lower maintenance costs, and more constant transmission parameters. The term "cable" deserves a definition. Telephone cable intended for outdoor use is generally available in bundles of 6 to 2,700 pairs and in gauges from 26 to 19 gauge. Small numbers of pairs are usually 19 and 22 gauge while the larger cables are 24 and 26 gauge. The small size, heavy gauge cables are used on long routes with a sparsity of users and may provide party-line service in the rural area. 19 or 22 gauge cable is the usual inter-office cable. In both of the above instances, heavier conductors are used for the superior transmission performance, including their lower dc resistance, which extends the talking and signaling range. The customary limiting parameter for cables is 1,300 ohms dc resistance. This limit is dictated by the central office signaling range. 1,300 ohms of 19 gauge cable is approximately 15 miles while only 3 miles with 26 gauge cable. Mixtures of gauges are used to minimize cable costs and keep a cable route within the serving central office's signaling limit. Thus, routes up to 15 Kft. would be served by 26 gauge pairs while routes between 15 Kft and 25 Kft consist of a mixture of 26 and 24 gauge, and so on, until the longest routes, over 75 Kft., use all 19 gauge cable.

Cable pairs without loading have a frequency-dependent loss characteristic as shown in Figure 7.10. The addition of inductance or loading coils in the amount of 88 millihenries every 6,000 ft. to the cable converts the cable to a multi-section low-pass filter with relatively flat attenuation in the passband. This is also shown in Figure 7.10. While the attenuation is

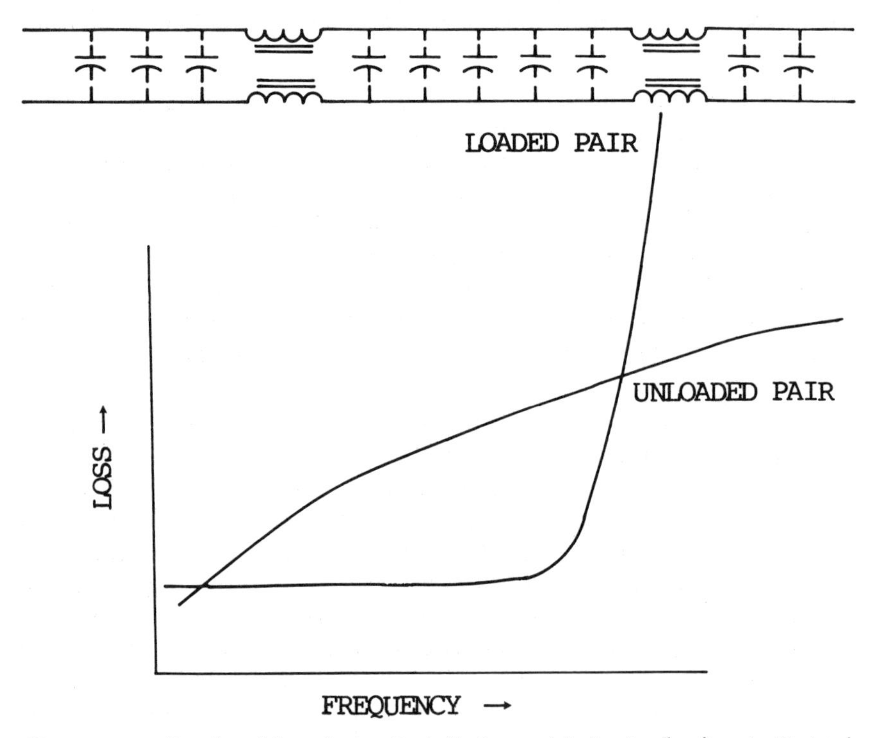

Figure 7.10: Results of Introducing Periodic Lumped Inductive Loading in Twisted Cable Pairs

remaining relatively constant the input impedance is not. Figure 7.11 shows the input impedance of loaded and non-loaded cables on a statistical basis. The general rule is to load all cable pairs over 18 Kft. in length. Economic factors require that cable pairs be accessible at multiple points and when a user is connected to a cable pair, not at the end of the cable, the unused stub is not removed or terminated. Other administrative

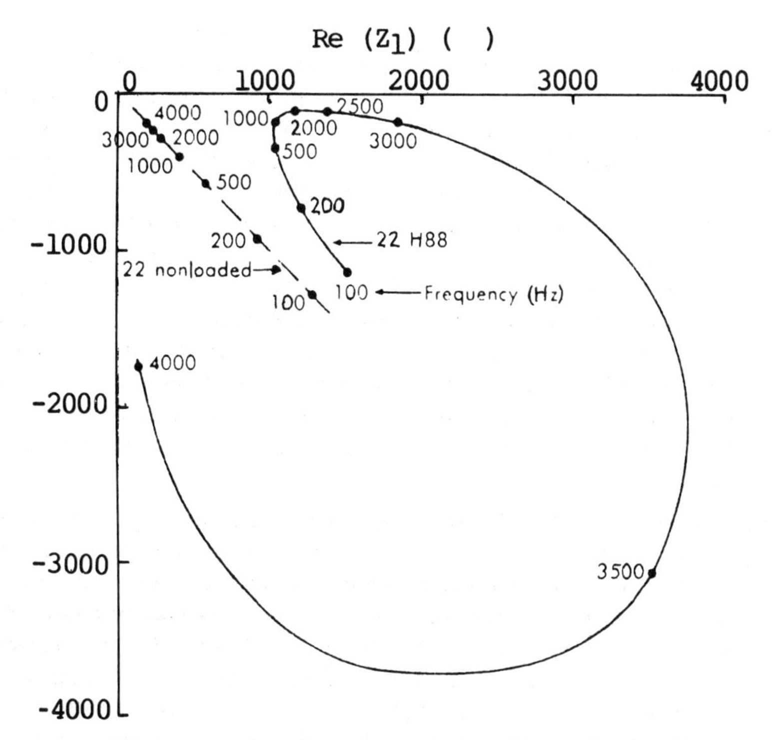

Figure 7.11: Characteristic Impedance, Z_l, in the R-X Plane at Various Frequencies

procedures call for bridged taps on cable pairs. In general, a stub or bridge tap will not exceed 6 Kft., but such an unterminated transmission line will produce a noticeable transmission defect. The effect is similar to the "studio effect" or a "hollow" sound. PBX trunks and other special service pairs do not have stubs or bridged taps on them in order to minimize impairments to speech and data. Ninety-five percent of all cable pairs have 8 dB insertion loss or less at 1 kHz, most having less than 6 dB loss. The noise measured on the same pairs will be 20 dBrnc or less. Impulse counts may vary widely but data sets should experience one error in every 10,000 bits transmitted or less due to impulses.

A trunk is a communication channel within a switching system or between switching systems. Trunks are defined differently for transmission and switching purposes, the differing terminology applying to the same trunk. Since this section deals with transmission we shall use transmission terminology, and in the switching section we shall use that terminology, but noting the transmission means.

Voice and data service in the U.S. require large quantities of voice paths distributed in complex patterns. The system has grown from a few isolated offices without connections, to other offices, to a massive pattern of voice paths allowing interconnection to virtually all points on the globe. This network has evolved as demand and technological opportunities developed. As demand grew and technology progressed, new plants were added to the existing system. The long life of the facilities in use dictates their retention. Also, this diversity along some routes protects services against special hazards. Thus, radio parallels coaxial cable between many points. Catastrophe to one system may not affect the other. Several centers in the U.S. use radio and coaxial cable and, if needed, satellites could be used also; however, presently, the public voice system utilizes almost no satellite channels for domestic service. Large numbers of satellite voice channels are used for overseas calls and those to Canada and Alaska.

The transmission plant is highly complex. Facilities may be open wire, cable pairs, coaxial cable, radio, including satellite, and most recently, glass fibers. All these systems, in varying numbers, are presently in place and providing voice and data service. These circuits may be combined in groups of only one or a very few up to several thousand on a broad-band system. Systems using coaxial cable and glass fibers utilize many cables or fibers as part of the same system and result in hundreds of thousands of voice channels. This is a combination of space and frequency or time dimension multiplex.

For short distances nothing is cheaper than a pair of wires. As material and labor costs increase, the definition of "short" keeps shrinking. As the demand for additional channels develops, it becomes economical to use the existing pairs of wires as the transmission medium for a carrier system. Carrier systems multiplex many voice channels on a single cable pair or coaxial cable but at considerable expense. Both frequency-division and time division multiplex systems are in use in large numbers. Voice frequency systems are needed for both short and long distances. The short-haul systems' end-to-end requirements are the same as the long haul systems'. Short-haul systems, thus, have considerably less stringent requirements per section, span, repeater, or whatever, than those used on a 4,000 mile long-haul system. In addition to the less demanding requirements, the much greater number of short-haul circuits creates a demand for inexpensive carrier systems.

Chronologically, analog frequency division multiplex carriers were developed after the space division multiplex systems became too expensive for further expansion. Frequently, analog carrier systems employ the same wire pairs that previously carried voice frequency circuits. Electronic technology was favorable to expensive copper pairs and the attendant labor costs.

Short-haul frequency division multiplex systems translate several individual voice channels, extending from 300 to 3 kHz, to contiguous bands usually much higher in frequency than the original voice band. 4 kHz is the normal allocation for a voice channel. This channel allocation permits relatively moderate filter requirements.

The key elements in frequency division multiplex systems are modulators and filter networks. Modulators effectively multiply two waveforms to produce the algebraic product. To illustrate, consider a voice signal to be comprised of a number of sine-waves, each of different frequency and amplitude, but located between 300 Hz and 3000 Hz.

This voice signal is one input to the modulator as shown in Figure 7.12.

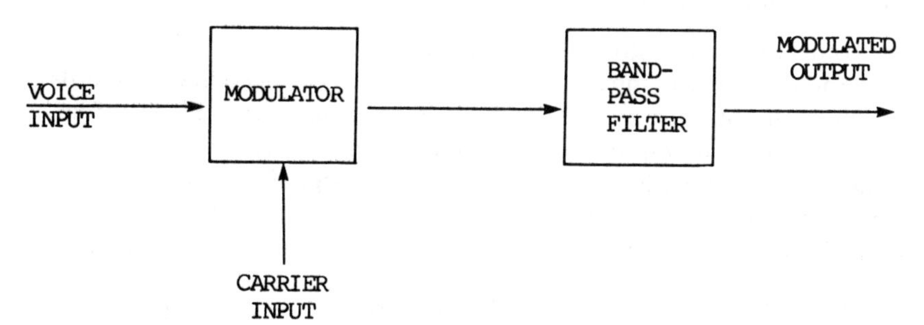

Figure 7.12: Modulator and Filter to Produce Single or Double-Sideband Modulation

The second input to the modulator is a sine-wave at the desired carrier frequency. Recalling the trigonometric identities:

$$\cos(a + b) = \cos a \cos b - \sin a \sin b \tag{7.1}$$

and

$$\cos(a - b) = \cos a \cos b + \sin a \sin b \tag{7.2}$$

and subtracting equation (7.1) from (7.2), and dividing by 2 produces

$$(1/2)\cos(a - b) - (1/2)\cos(a + b) = \sin \cdot a \sin b \tag{7.3}$$

substituting now in equation (7.3), time functions for angles a and b, let $a = w_1 t$ and $b = w_2 t$, then (7.3) becomes

$$\sin \omega_1 t \cdot \sin \omega_2 t = (1/2)\cos(\omega_1 - \omega_2)t - (1/2)\cos(\omega_1 + \omega_2)t \tag{7.4}$$

Equation (7.4) states that multiplication of two sine-waves creates two new sinusoids at frequencies equal to the sum and difference of the original waves. Referring again to Figure 7.12, the voice signal is assumed to consist of several sinusoids of different amplitudes and the carrier signal to be a single sinusoid; hence, the modulator output would consist of the voice components spaced above and below the carrier frequency as illustrated in Figures 7.13a and 7.13b.

The bandpass filter is present in Figure 7.12 because modulators may not be perfect productors, (1) resulting in some residual voice band signals and usually additional frequencies grouped about the odd harmonics of the carrier. The upper and lower sidebands of Figure 7.13b may be transmitted but either of the sidebands contains the necessary signal components to regain the original voice signal, as will be illustrated. Consider first the case where both upper and lower sidebands are transmitted. Figure 7.14 is a demodulator that performs similar functions to the modulator in Figure 7.12.

The translated voice signals of Figure 7.13b are one input to the demodulator and the other is the same carrier frequency as used in the modulator. Now the outputs of the modulator are again sum and difference components of the inputs, hence, the difference components lie in the same frequency positions as the original voice signal, and the sum components are at twice the carrier frequency and eliminated by the low pass filter of Figure 7.14. The output of the demodulators, as just stated, is:

$$\sin \omega_c t \cdot (\cos(\omega_c - \omega_1)t - \cos(\omega_o + \omega_1)t)$$

$$= -\sin \omega_1 t + 1/2 \sin(2\omega_c - \omega_1)t - 1/2 \sin(2\omega_c + \omega_1)t \qquad (7.5)$$

eliminating the double carrier components the output of the demodulator is equal to $-\sin w_1 t$. Note that if $\cos w_c t$ were used at the demodulating carrier, the voiceband frequencies would have disappeared. If both upper and lower sidebands are transmitted, the phase of the demodulating carrier is critical. If in Equation (7.5) only the upper or lower sideband were transmitted, the resultant output of the demodulator would be one-half as great, but its amplitude would be independent of the phase of the demodulating carrier. The first case considered is called double-sideband amplitude modulation and the second is single-sideband amplitude modulation or DSBAM and SSBAM. Modern frequency-division multiplex systems use single sideband amplitude modulation because of the 50% bandwidth reduction. Single sideband AM suffers a 3dB signal/noise reduction compared to double-sideband AM, however.

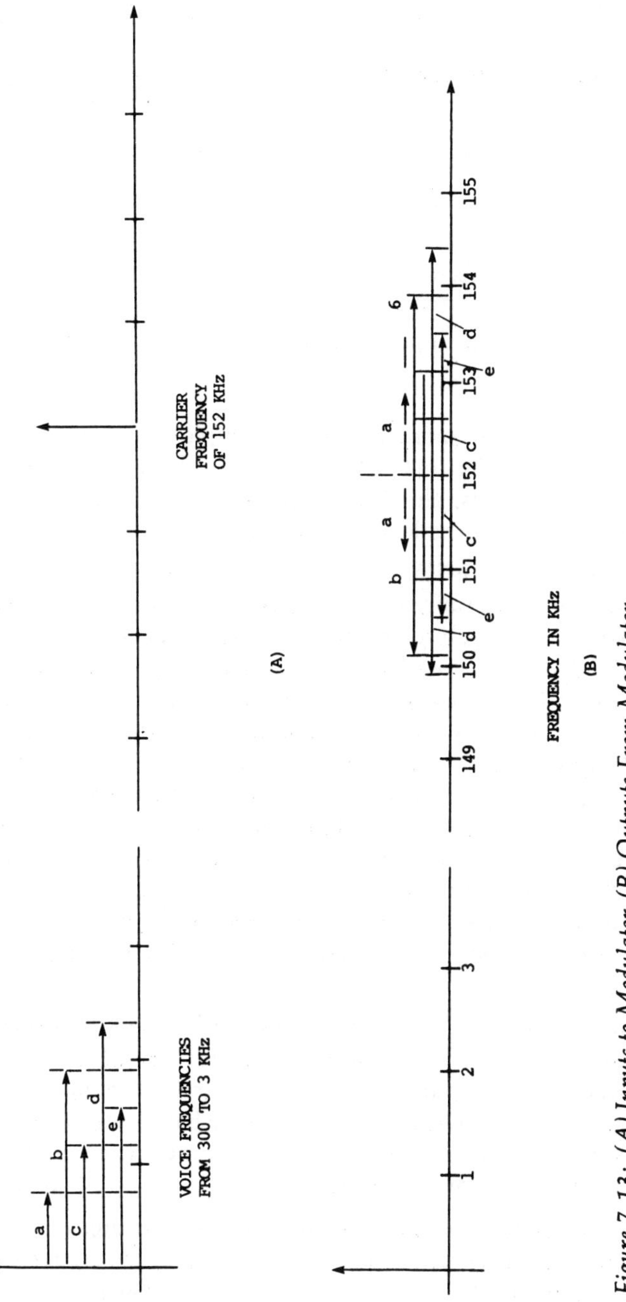

Figure 7.13: (A) Inputs to Modulator, (B) Outputs From Modulator

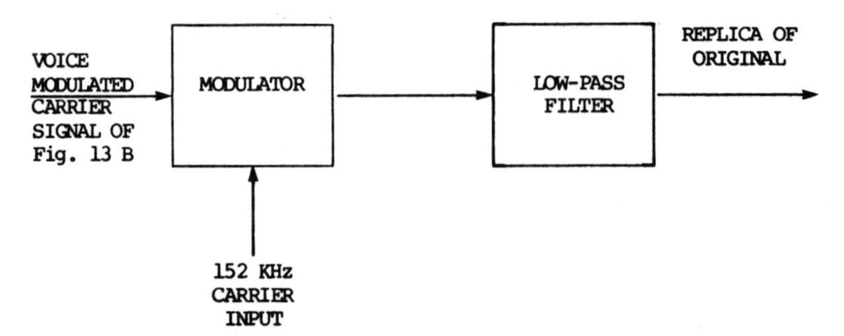

Figure 7.14: Demodulator for Recovery of Voice Signal From Carrier System

To complete the frequency division multiplex descriptions, consider the system shown in Figure 7.15. Prior to modulation, a voice signal may be compressed; that is, the high amplitude speech signals are amplified relatively less than the low level speech.

Such compression is done with an averaging time equal to the typical syllabic rate so that about one-fifth of a second is required to change the gain from low to high for high and low level speech, respectively. At the demodulator, this procedure is reversed, resulting in a reasonably true reproduction of the original speech. Two advantages are gained by this action. First, since several voice channels will share common amplifiers after frequency translation by the modulators, loud signals will be relatively lessened and are less likely to overload the amplifiers and secondly, and much more important, is that noise signals induced into the channels will be reduced during the silent and low speech level intervals and expanded along with the speech during the louder speech periods. The effect upon the listener of such "companding", a contraction of compression and expansion, is a perceived 22 to 28 dB reduction in the received noise level.

The twelve compressed voice channels of Figure 7.15 are modulated by twelve carriers whose frequencies extend from 152 kHz to 192 kHz and are spaced 4 kHz apart. These initial carrier frequencies are chosen because of the technology of the following bandpass filters, which eliminates the upper sideband of each voice channel. The twelve channels are then combined to form a group, as shown in Figure 7.16a.

Two such groups are then combined by modulating a 280 kHz carrier with group 1 and filtering the result to remove the upper sideband, and group 2 is modulated by 232 kHz carrier and again a filter removes the

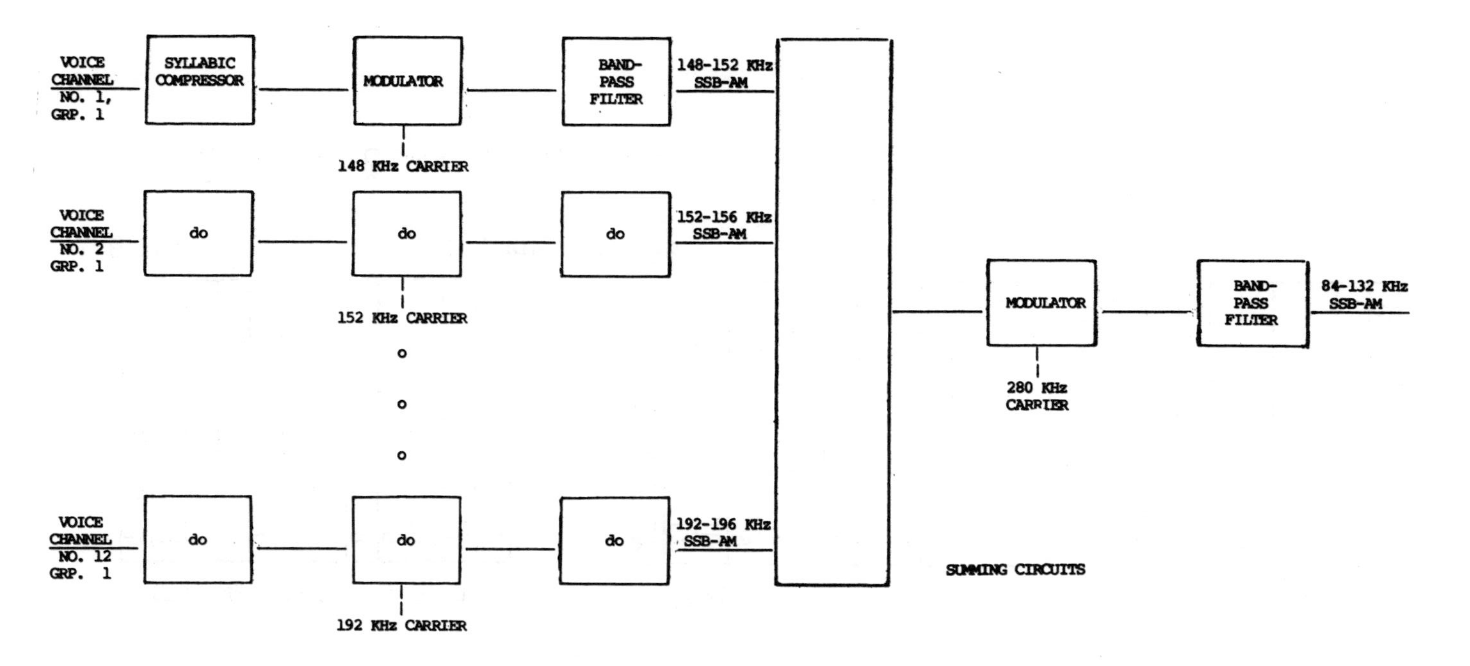

Figure 7.15: Twelve Channel Frequency-Division Multiplexing Arrangement

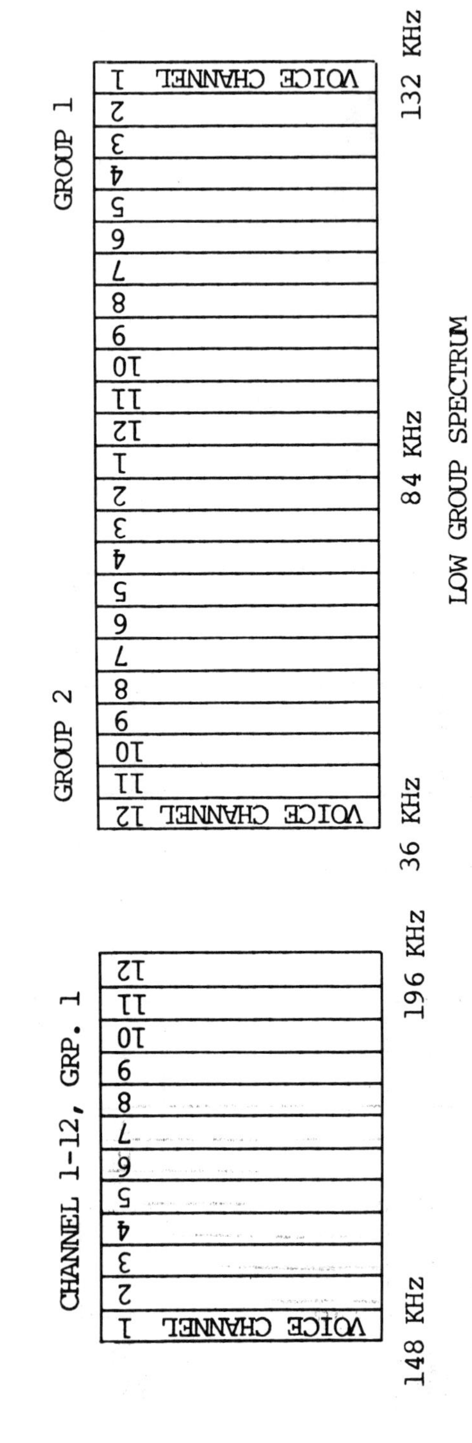

Figure 7.16: Low Group Spectrum of 24 Voice Channel System Utilizing SSBAM, Frequency Division Multiplexing

upper sideband before adding it to the group 1 signal. The result is 24 voice channels from 36 kHz to 132 kHz as shown in Figure 7.16b. A third modulation step using a 304 kHz carrier inverts this band, in frequency, and translates it to the region of 172 kHz to 268 kHz. These two bands are known as the low band and high band.

A distinction is made between the transmission line and its associated repeaters, or amplifiers, and the terminal equipment. The above descriptions pertain to a specific terminal system which provides 24 voice channels in a 96 kHz band extending form 36 to 132 kHz or 172 to 268 kHz. The line system which is used to transmit these signals is a physical four-wire circuit which may transmit the high band and receive the low band at a terminating central office. Intermediate repeater points, spaced 5 to 8 miles apart when paired 19 gauge cable is used, interchange the high and low bands so that weak signals arriving at the first repeater point are translated to the low band, amplified and transmitted to the next repeater point. By this technique, called "frequency frogging", the cross-talk between the high-level transmitted signal and the low-level received signal is eliminated by frequency separation.

The second advantage of frequency frogging is the partial amplitude equalization of the signals. Figure 7.18, which is an extension of Figure 7.5, shows the attenuation characteristic for unloaded cables. Transmitting alternately in high and low bands with a stacking of the voice-bands inverted in each produces partial amplitude equalization, easing the gain control burden at the terminal equipment. As shown in the figure, channel 1 has the least attenuation when transmitted in the low band and the greatest when transmitted in the high band. Also, the slopes of the attenuation characteristic within a voice-band are equalized. Individual channels must be equalized to a fraction of a dB in order to meet transmission performance objectives.

At the receiving terminal of this system, whether the high band or low band is received, the signals are again split into two groups of twelve voice channels in the position of the first modulation, as shown in Figure 7.16a. Single sideband filters similar to those of Figure 7.15 separate the voice channels prior to demodulation and low pass filtering and expansion. The net result is the voice signal occupying its original frequency position with individual component amplitudes intact.

The analog carrier system just described uses the 19 and 22 gauge cable pairs intended for single-voice channels. Since the carrier system uses two pairs, one for each direction, to achieve 24 voice channels the pair gain is 12. Several pairs in the cable will be used for such systems,

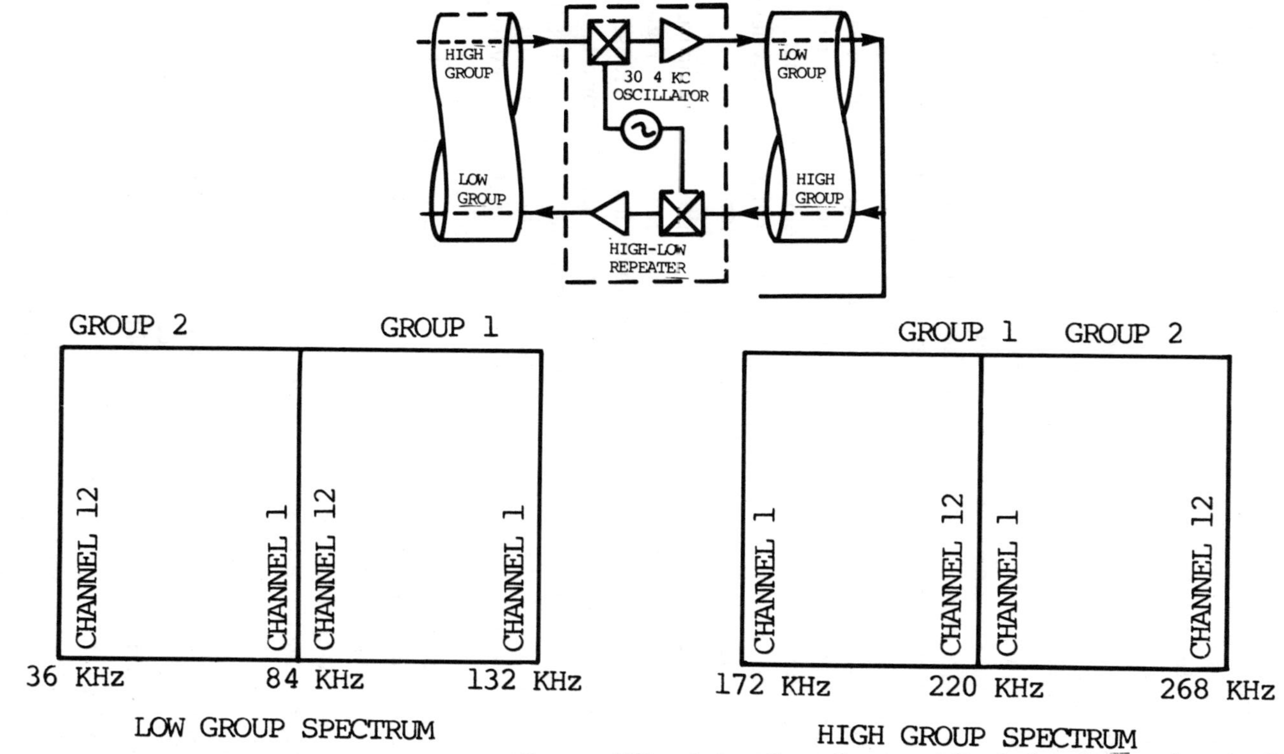

Figure 7.17: Short-Haul Carrier System Spectrum Allocation. When the Low Group Spectrum is Received at a Repeater Location, The High Group Spectrum is Transmitted, and Vice Versa

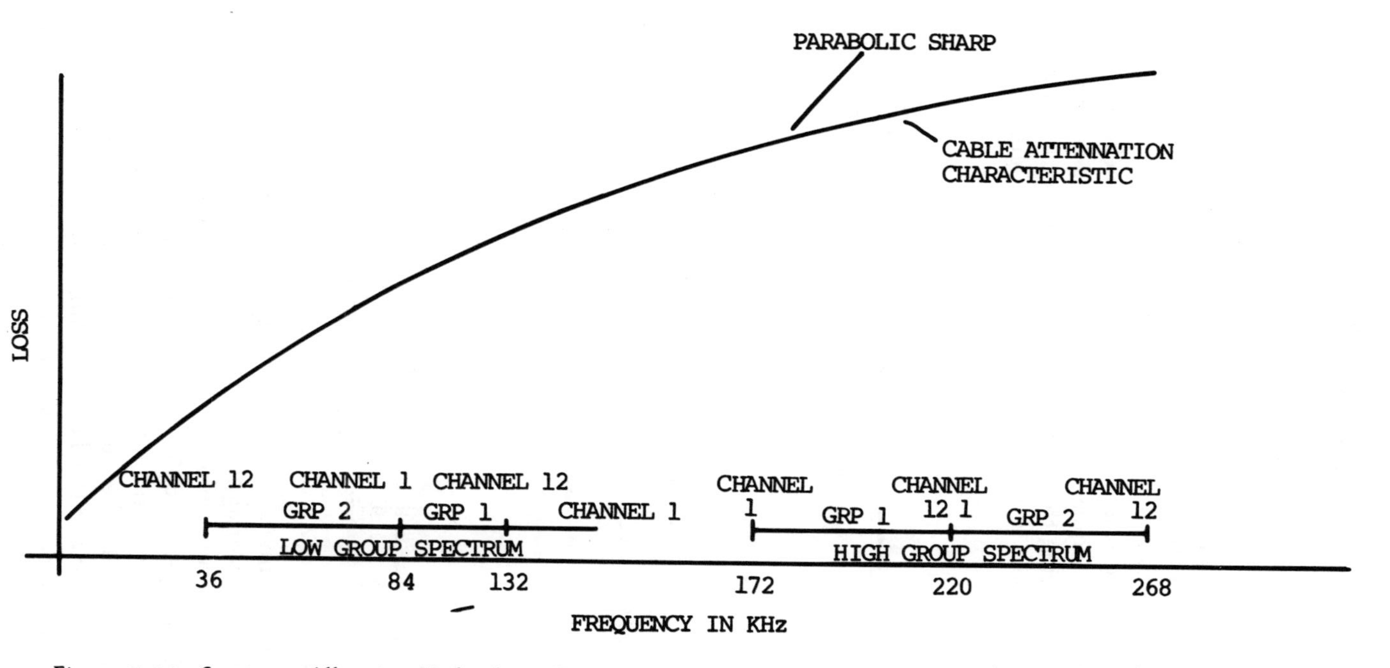

Figure 7.18: Spectrum Allocation Utilized in "Frequency Frogging" in Order to Partially Equalize Cable-Pair Attentuation Effects

effectively increasing the total available voice channels by the factor of 12. Because of the induced cable noise and cross-talk in the systems, they are limited to about 200 miles total length if the present noise objectives are to be met.

Short-haul radio systems were introduced to provide communication services where there were no wire lines and the terrain made such installation uneconomical. Radio also provides temporary service or relief on overloaded routes. Six hundred voice channels are grouped together, using techniques similar to those above for a single radio channel, the difference being that syllabic companding is not usually employed. The usual grouping of voice channels is shown in Figure 7.19.

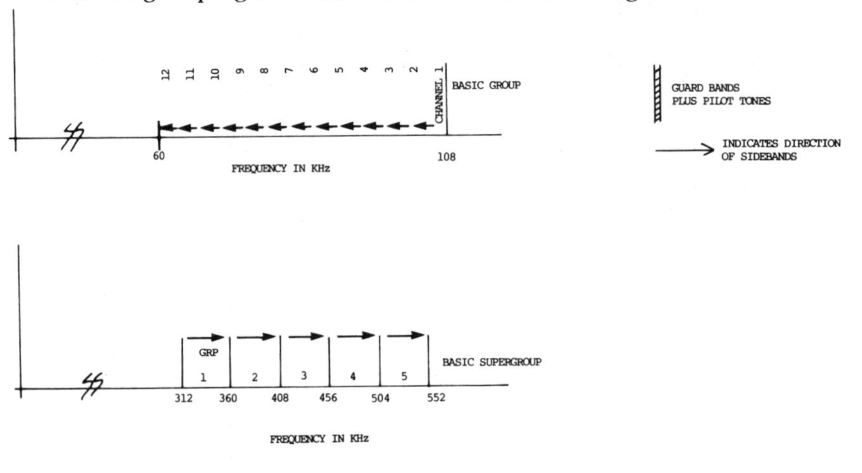

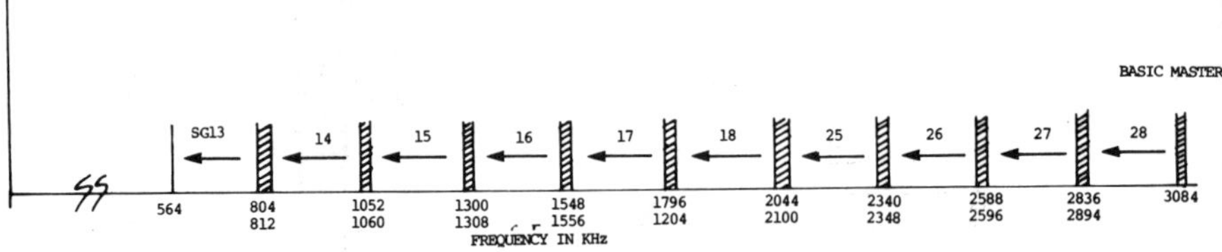

Figure 7.19: Frequency Allocations of Basic Group, Supergroup and Mastergroup

Twelve voice channels are placed in a group; five groups are placed in a supergroup, for a total of 60 voice channels, and 10 supergroups are placed in a mastergroup for a total of 600 voice circuits. These divisions of voice circuits permit 600, 60 or 12 voice channels to be split out of a radio system at any point on the path. The combined 600 voice channels frequency modulate the radio carrier, which may be at a frequency ranging from 2 GHz to 18 GHz, depending upon the FCC licensing. The total transmitted power ranges from 100 milliwatts to about 10 watts.

Several such radio channels may be transmitted on the same system of antennae for a total of about 5,000 voice channels. Short-haul radio systems are generally in the 6 to 18 GHz range. Later advancements utilize syllabic-companded voice channels and single sideband AM radio transmission in place of the frequency modulated radio carrier, thereby increasing the number of available voice channels by a factor between 5 to 10. These advancements have been made possible by new vacuum tube and solid state radio frequency amplifiers. Short-haul radio systems are usually limited to about 10 hops for a total of about 250 miles.

Application of radio to voice telephony has been a very dynamic field since its beginning, and new systems with differing voice signal processing and radio modulation techniques are appearing regularly. One recent addition is the single channel mobile radio system utilizing cellular concepts which permits reuse of radio frequencies and combines sophisticated "handing-off" techniques which track users traveling from one radio cell to another. Technically, this is a form of frequency and space division multiplexing. Switching centers accommodate such movement by "following" the user from one transceiver site to another.

Long-haul analog voice systems use coaxial cable to achieve the necessary bandwidth to transmit thousands of voice channels. Two-way carrier systems require two coaxial cables within a single cable sheath, but usually several such pairs of coaxial tubes are included in the same sheath along with twisted pairs which provide signals to regulate all systems within that sheath. The main source of power is between the two center conductors of the coaxial pair and is on the order 4,000 volts or ±2,000 volts to ground. The cross-section of such a system may show 100,000 two-way voice channels in operation.

The long-haul frequency division multiplexing equipment is similar to that previously discussed and the grouping of voice channels is also the same; i.e., from single voice channels to mastergroups, the important distinction being that companding is not employed in the long-haul analog cable or radio systems except, as will be noted, for satellite systems. Also, present day 3/8" diameter coaxial cable allows operation to nearly 65 MHz giving a capability of about 14,000 voice channels per pair of cables and cable sheaths containing up to 11 pairs of cables. This great bandwidth allows combining 6 mastergroups into a jumbogroup and then multiple jumbogroups are formed. It is technically feasible to extend the bandwidth of these cables already in place and operating up to as high as 150 MHz, which would more than double the present capacity to perhaps as high as 30,000 voice channels per pair of coaxial cables, or 300,000 channels per route.

Long-haul radio systems are not unlike the short-haul systems, but there are notable exceptions. Systems designed for 4,000 mile operation usually operate at 4 and 6 GHz as opposed to the short-haul band of 6 and 11 GHz systems widely used. The lower frequencies experience less moisture attenuation and, in general, the systems operate to much tighter specifications and, hence, they are considerably more expensive to purchase and maintain. Current systems use low index frequency modulation to transmit the voice signals. This modulation is usually done at an intermediate frequency centered at about 70 MHz and then translated to an appropriate point in the radio spectrum, between 3.7 and 4.2 GHz and combined with the 9 other channels. Presently, one radio route may transmit about 15,000 voice channels in each the 4 GHz band and 6 GHz band using the same structure and antennae. Thus the cross-section of any radio route, fully loaded, is considerably less than large coaxial cable routes, but the ease of constructing radio systems with the resultant lower cost per circuit mile has led to about 60% of the long-haul voice traffic to be carried on radio.

The newest radio systems utilize single sideband transmission rather than frequency modulation. This development is made possible by the inclusion of syllabic compandors and greatly improved dynamic equalization to the system and results in 10 radio channels, each accommodating 3,600 voice channels. Hence, at locations using both the 4 and 6 GHz systems the total capacity will be close to 75,000 voice channels per route.

Certainly satellite systems should be included in long-haul voice systems. Satellites, because of their ubiquitous transmission qualities, are being used in a great variety of systems. Present satellites use the same 4 and 6 GHZ band as terrestrial radio. 6 GHz is used from the ground to the satellite and 4 GHz from the satellite to ground. Satellites contain transponders, up to 30, each capable of receiving and rebroadcasting television, data, or voice signals. The choice is made at the up-link system. Each transponder rebroadcasts what it receives so that signals may be frequency modulated or amplitude modulated. The most recent voice and data satellites use syllabic compandored single-sideband transmission and achieve nearly 100,000 voice channels. Satellites in the future will use higher frequencies than the present in order to achieve the narrower transmitting beams. The greatly reduced beam width is required to accommodate satellites which are only three degrees apart at the equator, which is the present proposal for future systems, but consideration is being given to 1½ degree separation. Since satellites have a life of 7 to 10 years, the evolution of such a system will take at least that amount of time.

Satellite communication is used for domestic voice systems and international traffic. Many of the domestic voice networks are provided for corporate telecommunications use and not for general public use. Such networks provide for voice, high-speed data, and full motion video conferencing between corporate centers. The public network uses satellites predominately for international voice, color video, video conferencing, and high-speed data.

The last long-haul analog system to be touched upon is the undersea cable system. The first undersea cable went into service in 1956 and accommodated 36 voice channels over 2 cables and used about 50 vacuum tube repeaters. The latest cable recently completed carries 16,000 voice channels. Cables for undersea use are about 1½ inch diameter coaxial cables with the center conductor being the strength member of the cable. Such cables are powered by very high voltages, about 7,000 volts at each end of the cable. The cable is an analog transmission system, so channels for voice, data, or television are created at the terminals of the cable. Voice systems presently use noncompandored single-sideband transmission.

The next undersea cable is planned to be a three pair fiber-optic system, capable of about 100,000 voice channels. The modulation techniques likely to be used will be discussed under digital carrier systems.

The short-haul frequency division multiplex systems were usually uneconomical under 25 miles and a need persisted for a still cheaper exchange transmission system. This need was apparent at the time the transistor had reached a mature state of development, and it had become clear that transistor digital circuitry was easier and cheaper to implement than analog systems. This coincidence of economic need and technological maturity resulted in a digital carrier system, known universally as T1 carrier. The system is modularized for ease in maintenance and flexibility. Low cost is achieved by the shared use of companding and signal processing equipment. Expensive channel filters are eliminated. Band-limiting filters are required but have proven far less expensive than the channel separating filters required in frequency division multiplex analog systems.

T1 systems consist of the same two parts as analog frequency division multiplex systems. The T1 terminal equipment and the line provide twenty-four, 4-wire voice paths. Terminal equipment samples each band-limited voice channel 8,000 times per second and forms a 7 or 8 bit binary coded signal from each channel. A framing bit is added to each frame of 24 eight bit blocks, resulting in a line bit rate of $8{,}000 \times (24 \times 8 + 1) = 1.544 \times 10^6$ bits per second.

The T1 line transmits this bit rate over 19 to 26 gauge cable pairs which have digital regenerators spaced about every 6,000 ft. The 6,000 ft. is a design criterion intended to coincide with loading coil spacing of the voice-frequency lines. The system was intended to fill the void for systems up to 25 miles in length, but distances of 100 to 200 miles are accommodated without significant impairment. Thus new short-haul systems currently being installed are invariably digital carrier systems.

The terminal equipment is shown in Figure 7.20. Twenty-four voice channels are individually band-limited to remove voice frequencies above 4 kHz. These 24 voice channels are then sampled sequentially and placed on a pulse-amplitude modulated bus. The time sequence of this action is shown in Figure 7.21. Each of the 24 channels is sampled, in turn, 8,000 times per second.

Therefore, this bus is a time division multiplex system. The combined 24 channels are instantaneously compressed, as compared with the syllabic compression used in analog transmission systems, in order to reduce the quantization noise inherent in these systems. The 24 compressed voice channels are then each encoded into an eight bit code. This operation is usually referred to as pulse-code modulation (PCM). This stream of 8x24 bits plus the framing bit is repeated 8,000 times per second.

The signal transmitted is a 1.544 megabit per second pulse stream after conversion to a bi-polar format. The bi-polar format transmits alternate positive and negative pulses to indicate the "one" state and no pulse to indicate the "zero" state. This bipolar signal format is shown in Figure 7.22.

Major advantages of bi-polar transmission are:

1. There is no direct-current component in the line signal avoiding problems associated with ac coupling to the transmission lines.
2. Most of the energy of the signal lies below the bit rate frequency, limiting need for very high frequency response of the line.
3. There is a natural error detecting mechanism in that two consecutive pulses of the same polarity indicate errors, either ones to zeros or zeros to ones.
4. The line signal is self-timing. That is, the timing signal is extracted from the signal at individual repeater points.

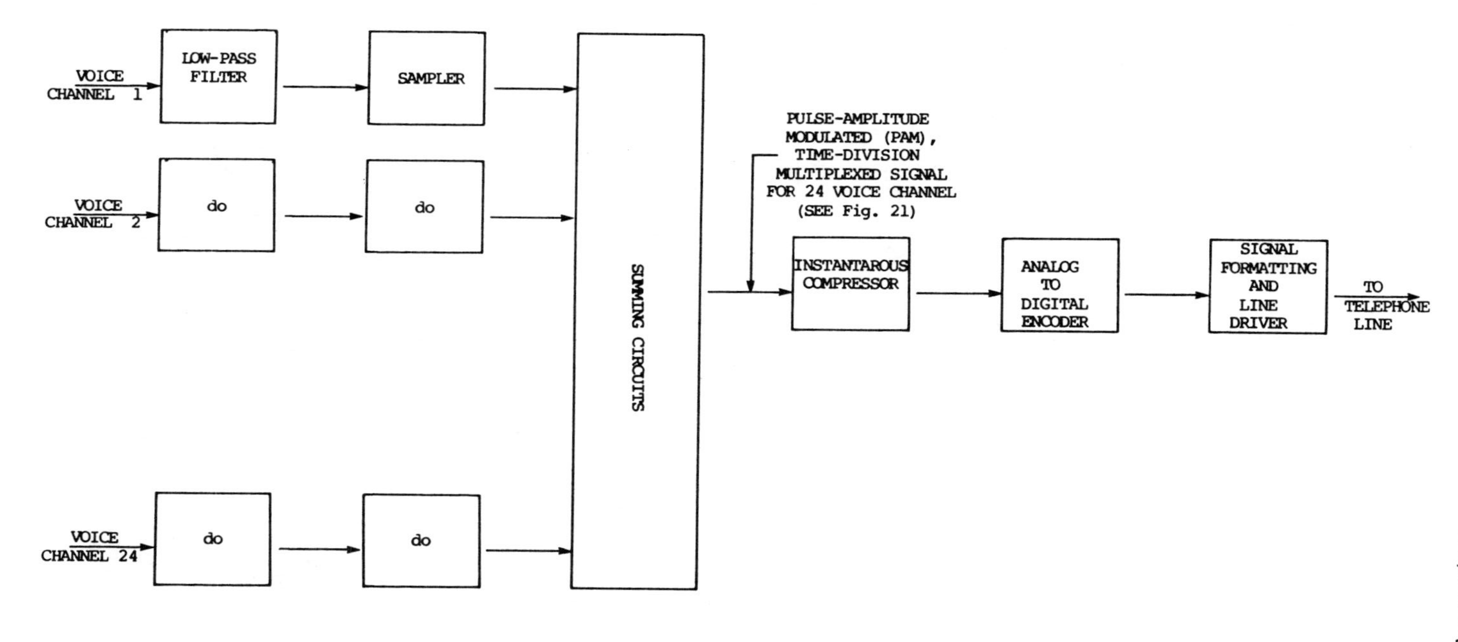

Figure 7.20: Basic Block Diagram or a 24 Voice Channel Digital Time Division Multiplex System

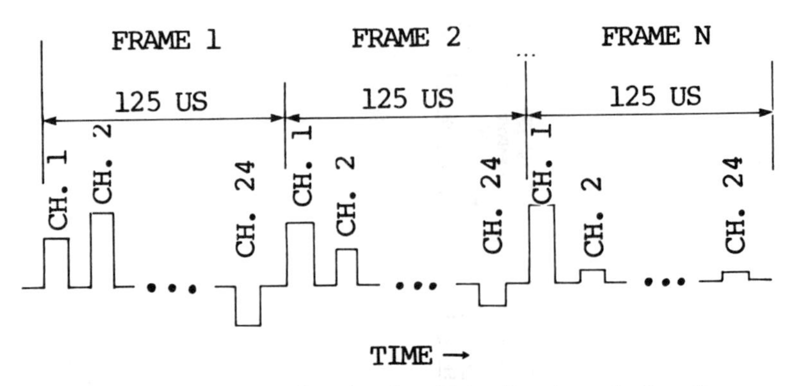

Figure 7.21: Instantaneous Samples of 24 Voice Signals on the Pam Bus

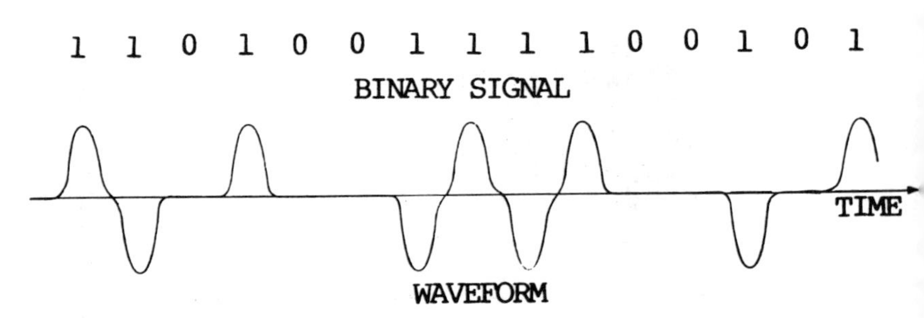

Figure 7.22: Formatted Line Signal for Short-Haul Digital Carrier Systems

The T1 line repeater is a pulse regenerator, rather than an amplifier as used in the analog systems. Figure 7.23 indicates the functions performed by T1 line regenerators. Line equalizers are chosen by measurements on the line. Recovery of the 1.544 MHz timing signal is based on the techniques of phased-lock loops and this timing signal is used to "strobe" the decision circuit for ones and zeros. The bi-polar pulse generator originates a new stream of pulses for transmission on the next section of line.

Error rates in T1 lines are less than one error for each million bits transmitted. This error rate is satisfactory for most data systems and is unnoticeable in speech. Quantization noise is the difference between the continuously variable voice signal and the discrete level which the terminal must reproduce. Eight bit codes produce 256 possible signal levels, 128 each of positive and negative polarity. Figure 7.22 shows the relative level of the regenerated voice signal versus the input level. The unequal

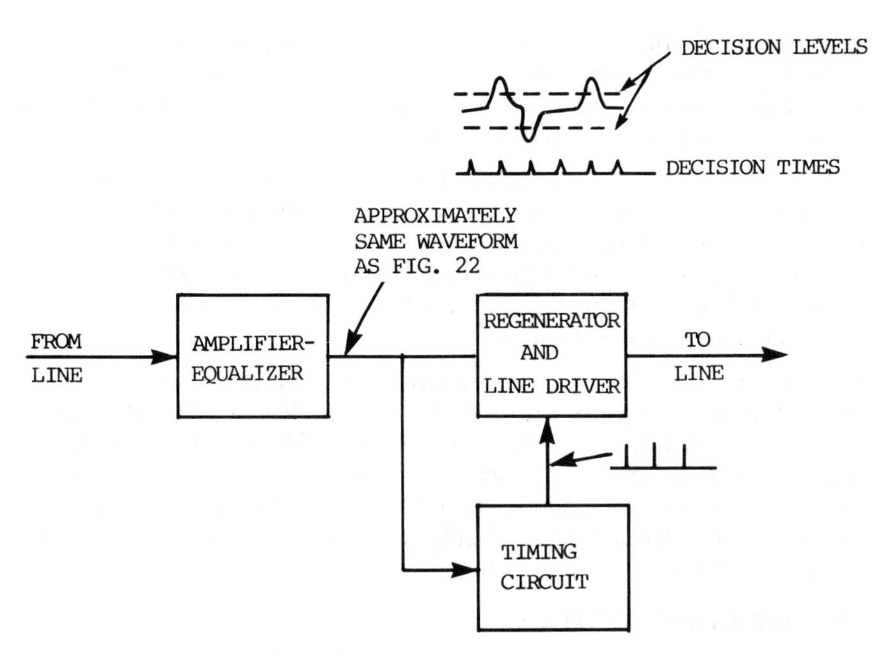

Figure 7.23: Digital Carrier Regenerator

steps, larger at the larger signal amplitudes are a result of the instantaneous compandor action. Nearly equal signal to quantization noise is achieved by this means. A single link of T1 carrier has about 36 dB signal to distortion ratio when the input levels are within their expected limits. Each doubling of the number of links reduces this ratio 3 dB, therefore, four lengths would have about 30 dB signal to distortion ratio. Because of the nature of digital systems no other noise appears in the voice output of the system.

T1 carrier proved so effective that a digital carrier hierarchy was envisioned and is presently being implemented. T1 carrier can serve both the loop and exchange plant, providing service to rural and urban users. T2 digital systems place four T1 carriers on an exchange pair. The four T1 systems need not use common clocks for their bit timing. The T2 system runs at a speed sufficiently above the $4 \times 1.544 \times 10^6 = 6.176$ megabits/sec so that only a trouble condition could cause the T2 rate to be insufficient for the four T1 systems. A bit rate of about 6.3 megabits per second has been found sufficient. The extra bits are used to insert framing information to enable the four T1 systems to be reassembled at a terminal point and to add information about which T1 carrier will have a bit "stuffed" into it and whether such stuffing occurred. The T2 receiving terminal reframes individual T1 carriers and deletes any stuffed bits. Timing recovery in individual T1 systems smooths the received T1 system clock to its original bit rate.

Framing is a challenge in all digital carrier systems. If framing signals are incorrect, channels may be switched and calls momentarily misdirected, but usually bursts of noise are heard by all listeners until proper framing is achieved. The framing pulses form unique sequences which logical circuits in the receiving terminals detect and which are used to re-establish framing when required.

Present extension of the T carrier family is the T4 system which operates at 274.176 megabits/second on ⅜ inch coaxial cable pairs, 18 GHz radio, and most recently on single-mode optical fibers. The T carrier hierarchy is in a state of dynamic turmoil at present because of the rapid advancements in millimeter radio systems and fiber optics. As mentioned earlier, the next transatlantic cable probably will be optical fiber and hence a member of the T carrier hierarchy.

7.3 SWITCHING SYSTEMS

Transmission systems enable voice circuits to span the earth, but switching systems are needed to connect circuits between desired points. A trivial example of a telephone system without central switching is shown in Figure 7.24. Transmission lines must be run between all locations and a user wishing to call another within that system would have some means of alerting the called party; perhaps each user would have a bell on each line terminated. The called party would rotate the switch to the calling line and conversation could take place. In this instance, a one-by-four switch is utilized, the user's own number being in the idle position. Telephone systems started in this fashion, but soon the switching system was centralized and a single transmission line ran from the user to the switching center. The switches shown in Figure 7.24, are located now in the central office and only inches apart.

Figure 7.25 shows a similar network. This time the dots indicate two switches, as in Figure 7.25b. The trunk circuit would supervise the connection, provide talking battery and ringing signals to the two ends, and close the switch when both parties were ready to speak and open the switch when either party disconnected. Such systems were once manually controlled, but presently are controlled by a variety of automatic switching means. In this section representative samples of modern switching systems will be examined.

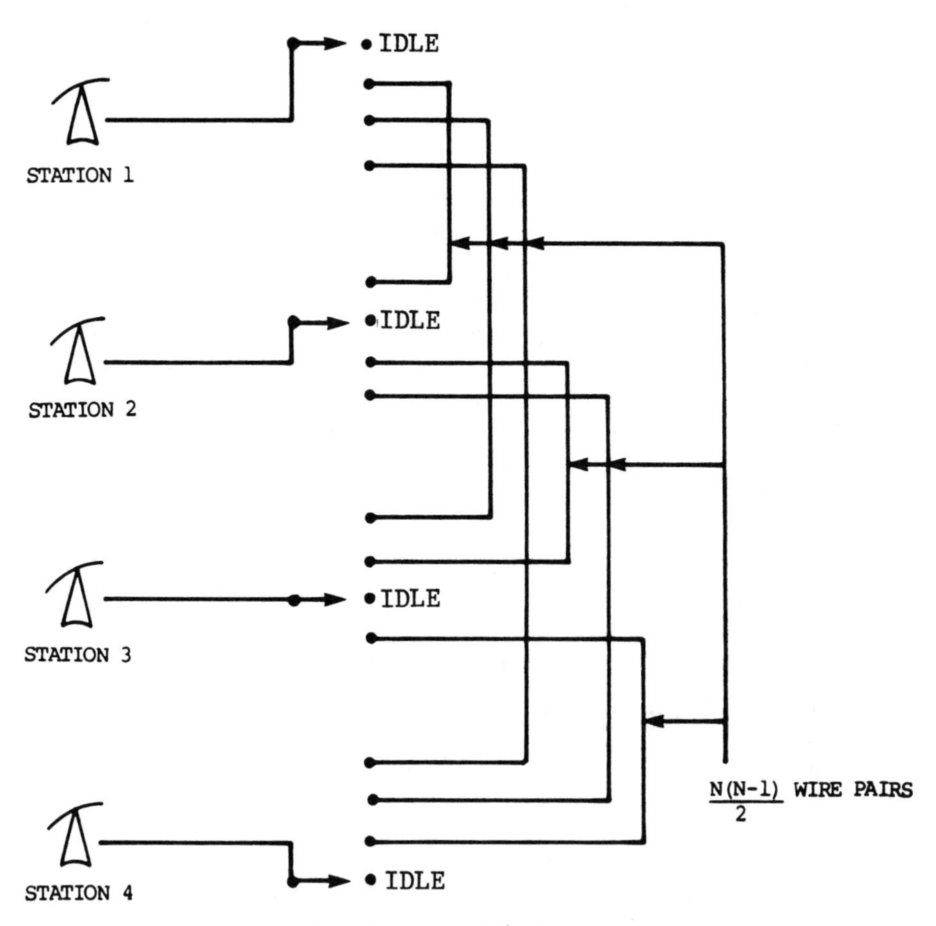

Figure 7.24: Telephone Switching System Without Centralized Switching

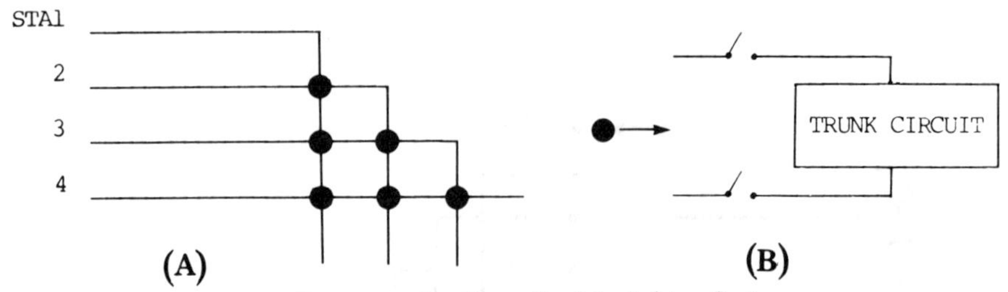

Figure 7.25: Schematic Diagram of a Centralized Switching System

Switching systems may be broadly categorized in the same fashion as transmission systems, that is, (1) space division, (2) frequency division, and (3) time division. The simple system just described is an example of a space division switch, which is the most prevalent. Frequency division has never found widespread use because switching carriers to input and output ports require a switching system of their own. Time division switching, while not the most widely used as yet, is receiving the greatest amount of attention in recently developed systems. Time division systems can take several forms and can be combined with space division to make very large switching centers. These systems will be treated in turn.

Switching systems generally have these common requirements and properties:

1. Any input can connect to any outlet.
2. Many simultaneous paths must be available so that many inputs may be separately connected to outlets.
3. Not all inputs will be used simultaneously.

The first property permits full access between inputs and outlets, the second is an obvious requirement, but the quantity of simultaneous connections governs the traffic capacity of the system and is statistically characterized by the calling requirements of individual users. The switches themselves must provide adequate transmission for the signal to be switched when operated and provide no transmission through the network when not operated. Simple metallic contacts provide adequate transmission and isolation, but the cabling associated with large networks may cause some cross-talk, as discussed in the transmission section.

Space division networks can be made up of a single matrix, or matrices may be grouped in various patterns. Figure 7.26 shows a single matrix with input ports on the left and output ports at the bottom. The dots

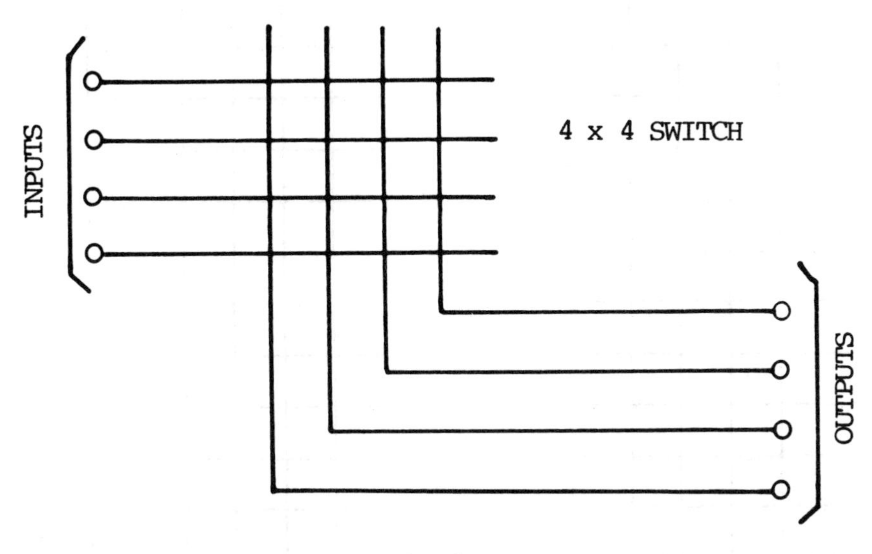

Figure 7.26: A 4 x 4 Single Matrix Switch

represent switches which may operate or not operate, depending upon the connections in progress. This type of matrix requires $n \times m$ switches or cross-points to be able to connect n inputs to any combination of m outputs.

In small systems this single stage is justified because of its simple control requirements. Such single stage networks have a potential problem in that failure of a single cross-point prevents one input from reaching a specific outlet. In large systems, the costs of cross-points surpass the of complicated control means and an arrangement similar to Figure 7.27 may be employed. This is a two-stage network. In this instance the 16 inputs have full access to the 16 outlets, but only 4 inputs may be connected at any given time. Telephone traffic engineering must determine whether this number of links is adequate.

Figure 7.27 illustrates a common practice in switching networks, namely, 16 inputs are *concentrated* on 4 links and these 4 links are *expanded* to 16 outlets. Figure 7.28 illustrates another more versatile arrangement employing 3 stages of switching. The center stage is a *distribution* stage.

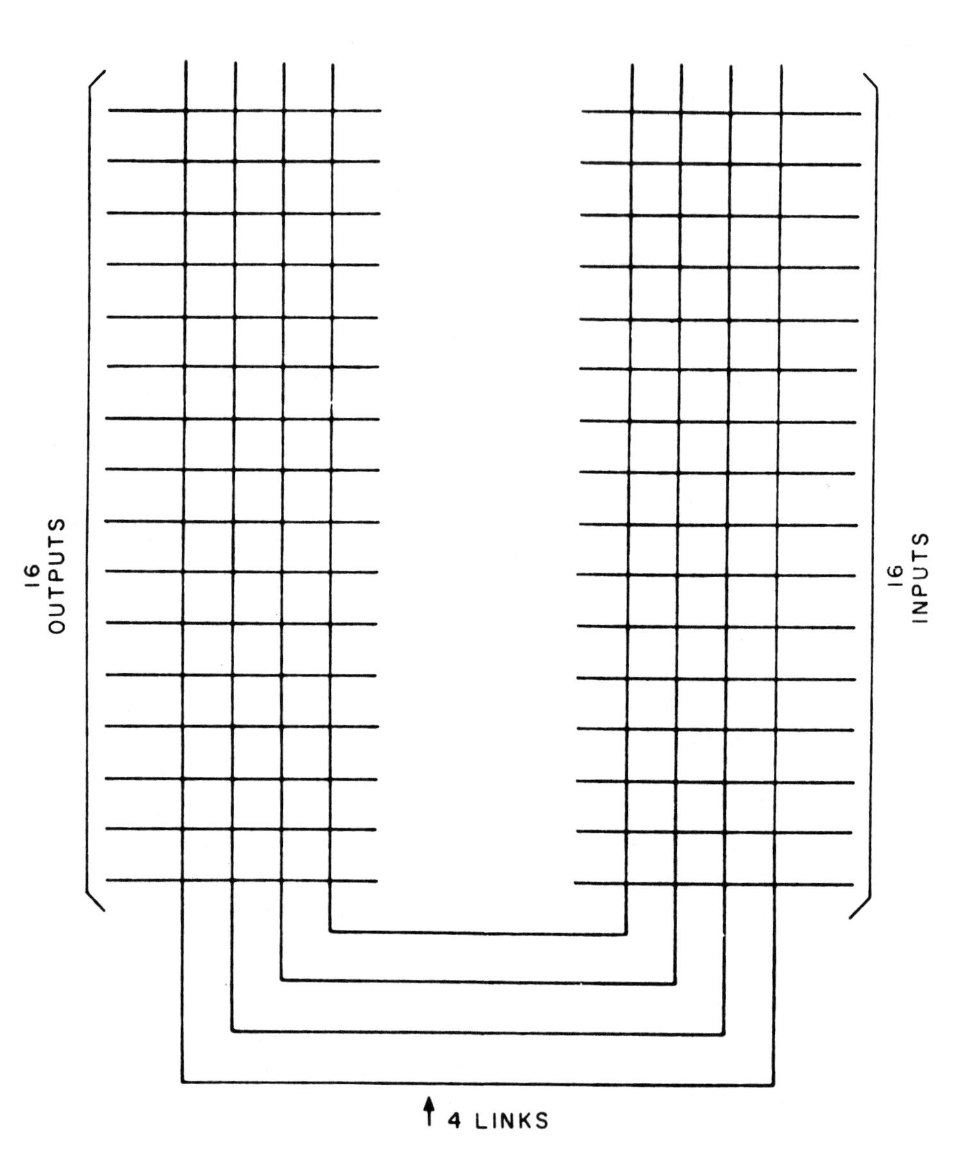

Figure 7.27: A 16-to-16 Network Concentrated onto Four Center Links

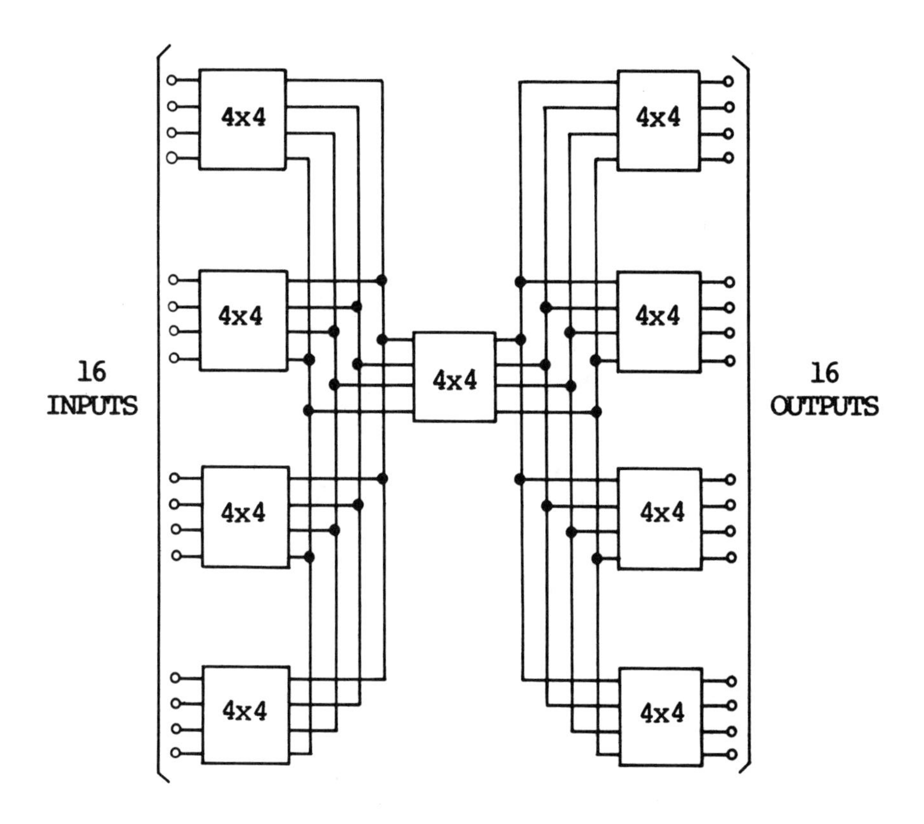

Figure 7.28: A 16-input, 16-output Switching Network Allowing Four Simultaneous Connections

Figure 7.28 combines three functions: concentration, distribution, and expansion. Each of the three functions may consist of several stages of switching. Figure 7.29 illustrates the usual method of interconnecting a three-stage network. Switching systems designed to handle 100,000 users may have 8 or more stages of switching and require complicated means of control.

Time division switching is attractive because of its compatibility with electronic control systems and, when quantized appropriately, it is directly compatible with the T carrier transmission systems, but with some complications.

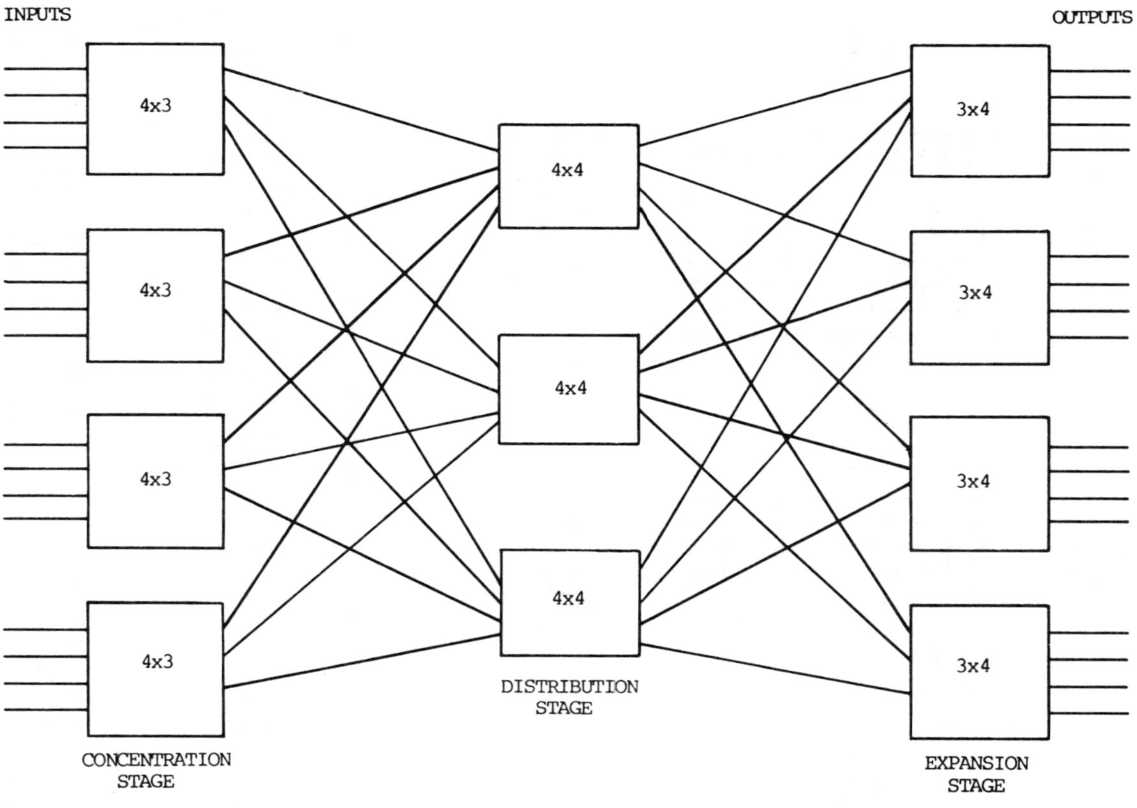

Figure 7.29: Conventional Interconnecting Arrangements in a Multi-Stage Switch

Time division systems are further split into several categories:

1. Pulse amplitude modulation
2. Pulse width modulation
3. Pulse position modulation
4. Pulse code modulation

and there may be others.

There are many similarities between the T carrier system and a time division switch as illustrated in Figure 7.30. As the name implies, time is divided into sampling intervals occurring at twice the rate of the highest frequency to be transmitted. Switches I1, I2 through IN operate momentarily and sequentially at a rate equal to or in excess of 8,000 times per second for voice. The low pass filters on both the input side and output side eliminate all frequencies in excess of 4 kHz.

The resulting series of pulses on the common bus are as shown in Figure 7.31. This system results in amplitude modulation pulses (PAM), one for each channel occurring during each frame. The transmission characteristic of the bus must be adequate to prevent the pulses from being stretched, in time, and resulting in intelligible cross-talk.

If input 1 is to be connected to outlet 4, then switch O4 of Figure 7.6 must operate simultaneously with I1. Other switches on the outlets must similarly operate simultaneously with the desired input switches. The low pass filters in the outlet circuits convert the pulse train into a continuous signal which is a replica of the signal on the input side.

It is possible to have the common bus operate as a two-wire transmission system, carrying signals in both directions in the same manner as a metallic cross-point in a space division switch, but present practice in systems employing PAM favor a 4-wire bus and, therefore, 11 and O4 consist of two switches rather than one.

The bus cannot accommodate signals from all inputs without having severe cross-talk problems so that during a frame there may be a limit of 32 or 64 time slots, and hence only 32 or 64 simultaneous voice connections are possible. The control system determines which switches operate on input and outlet sides, thereby establishing connections.

Pulsewidth, pulse position, and pulse code modulation systems usually employ modulators and demodulators for each input and outlet. A space division switch may precede the time division switch, inserting voice

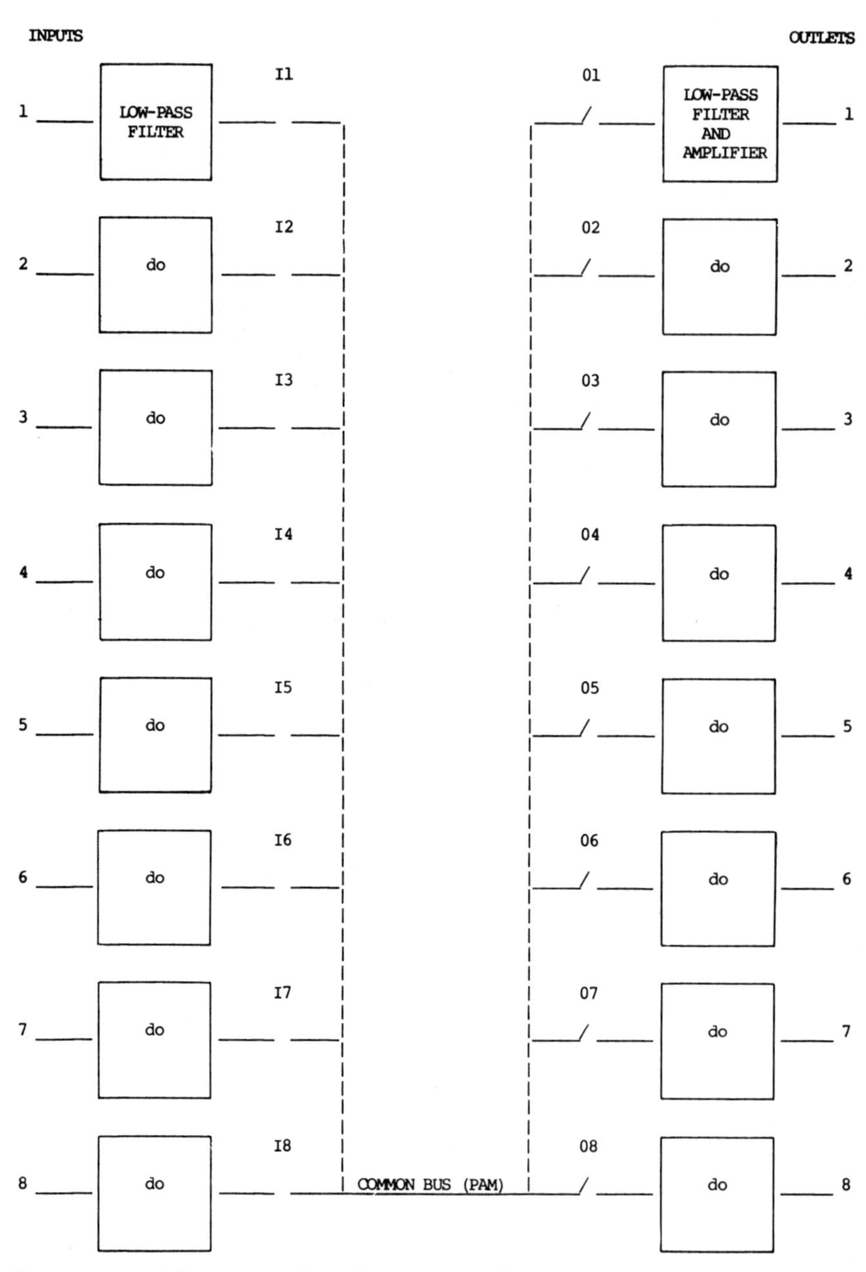

Figure 7.30: A Basic 6 X 6 Time Division Switch

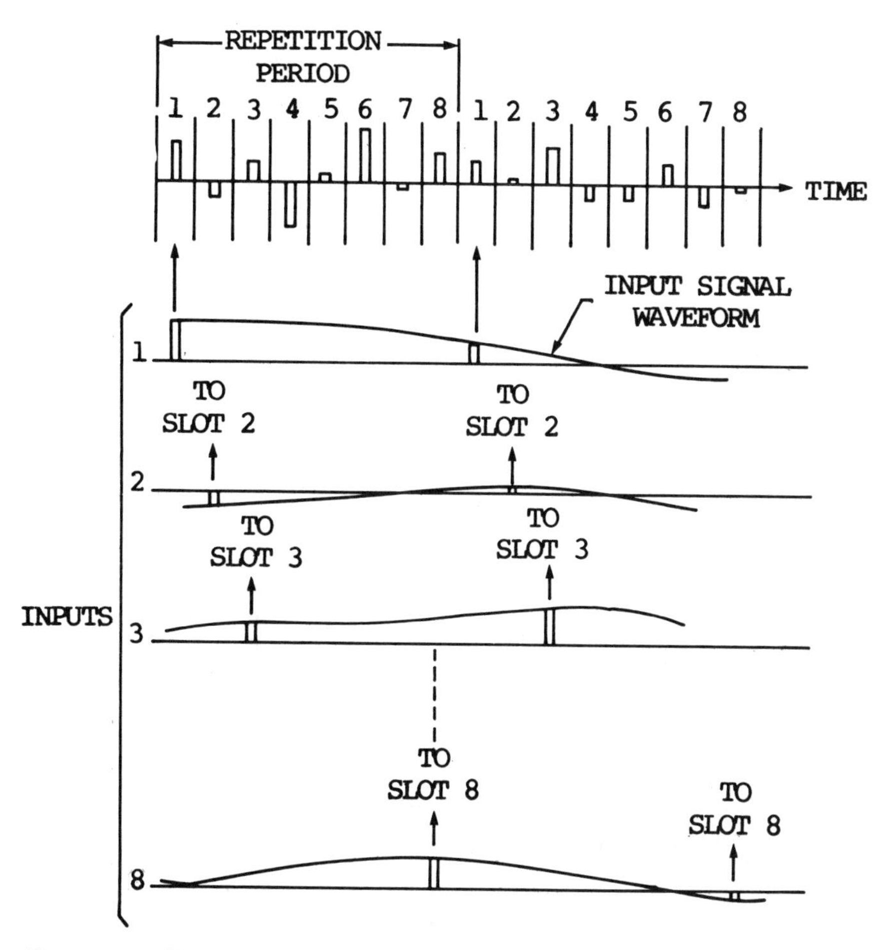

Figure 7.31: Time Division Switching

signals into the time division inputs, and reducing the number of time division circuits. When the switching system needs exceed the capabilities, a single common bus, space division switching between buses is accomplished by time-slot interchange. A suitable analog or digital memory device stores a sample in a given time slot on one bus and inserts it into the appropriate time slot on a second bus. Digital or PCM systems are best suited to this architecture. Multi-bus analog systems usually return to voice and are re-sampled by the second bus system. Examples of an analog system, which includes PAM, PWM and PPM, are shown in Figure 7.32 and the PCM in Figure 7.33.

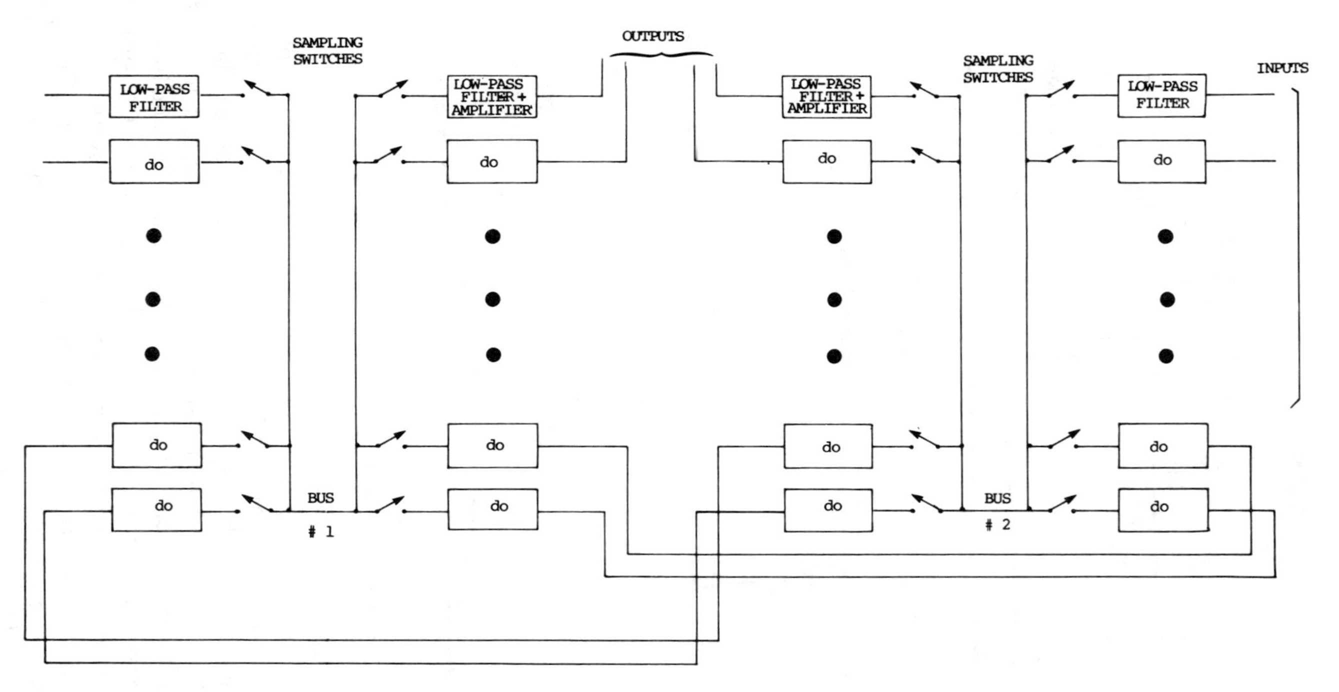

Figure 7.32: A Multiple Bus PAM Time Division Switch

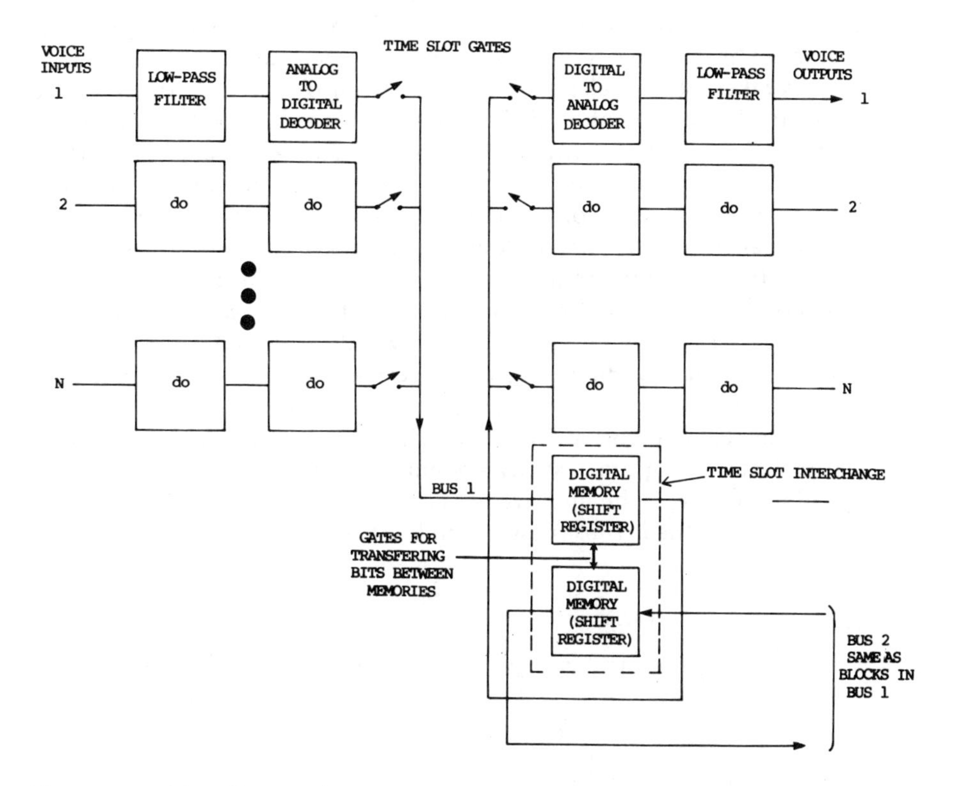

Figure 7.33: Time Slot Interchange for Multiple Bus Digital Time Division Switching

Multiple bus systems use available time-slots in inter-bus connections, and it is clear that at some time all the available time slots are used up in inter-bus communications and there are no time-slots left for inputs or outlets. Such systems must resort to higher numbers of time-slots or combine with space division stages, Presently, the largest time division switches employ PCM, some utilizing optical fibers as the bus.

An electro-mechanical system of space division switching, known best as step-by-step switching or as the Strowger System, after its inventor, was once nearly universal, and it is still a substantial portion of the total system. The step-by-step system was a system design. It translated actions taken at the telephone instrument into switching functions at the central office. A rotary dial added to the telephone set indicates the address digits to the central office by interrupting the central office signal battery at the rate of 10 times per second. Moving the dial "off normal" to the digit 4, for instance, and releasing it, causes 4 interruptions of central office current flow. These interruptions cause a step-by-step switch to count the dial pulses by actions of electromagnets and ratchets. When the dial has completed its travel back to the normal position, a second slow-operating relay indicates that all dial pulses associated with that digit have been received and the system prepares itself to receive the next series of dial pulses. This is a form of *progressive control;* the dial controls switching action at each stage in the network. Figure 7.34 is a diagram of such a system. When the user requests service by going off-hook, the line circuit is activated and the relay chain, not shown, causes the *line-finder* switch to hunt until that line is found.

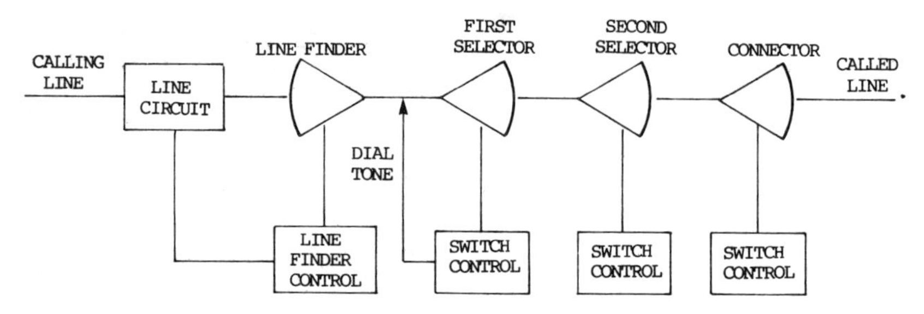

Figure 7.34: A Representation of a 4 Digit Step-By-Step Switching System

The line-finder is directly connected to the *first-selector* switch in the chain and that switch returns the dial tone to the calling party. The first digit dialed causes this switch to step to the proper position, according to the dialed digit. As dialing progresses, each stage of the network responds in the same fashion until, after all the digits are dialed, the called line is reached.

A single switch in a step-by-step office is a 10 x 10 matrix of 3 conductors. Two conductors carry the voice signal and the third is for supervisory purposes. Figure 7.35 is a diagram of such a switch and Figure 7.36 is a physical representation of the switch bank.

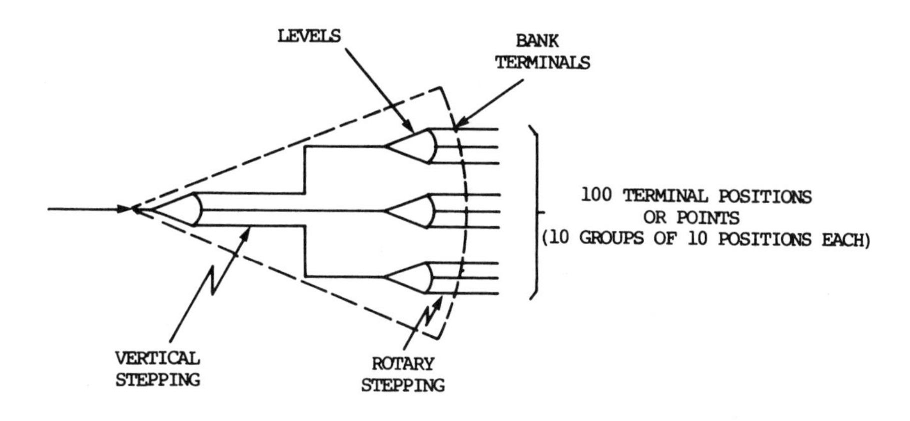

Figure 7.35: Diagram of the Control and Switching Functions of a Step-By-Step Switch

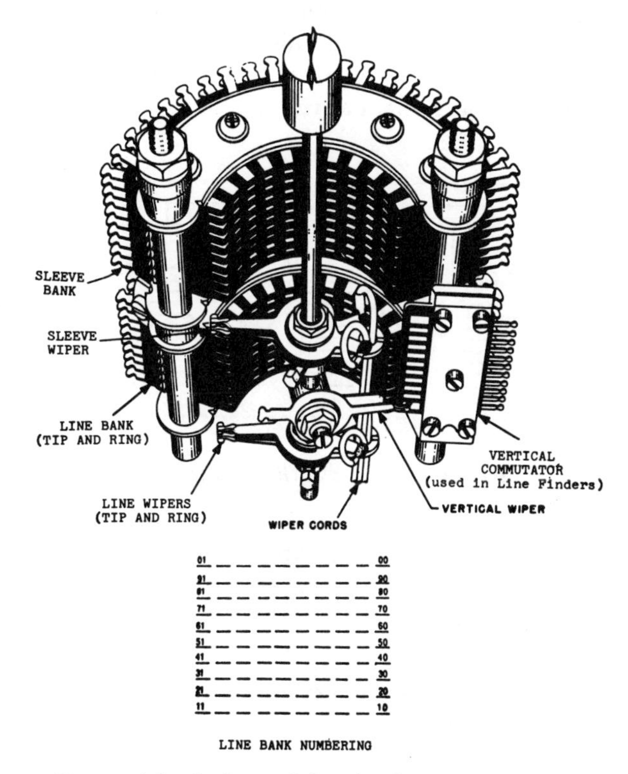

Figure 7.36: Terminal Bank of 100 Subscriber Lines

The vertical and horizontal motion controls are not shown here. The rotary dial pulses cause the central shaft of the switch to rise one step per dial pulse. When that digit is completed the *wipers* are rotated into contacts on the banks until an idle connection to another switch is found. All switches involved in the establishment of the connection are held for the entire call.

Figure 7.37 is a more customary representation of a step-by-step office. Each horizontal line in the diagram represents 10 switch positions of 3 contacts each, as shown in the previous Figure 7.36. The final switch in the train is called a *connector* and both the vertical and horizontal movements are controlled by the rotary dial. If at any point in the dialing procedure a path cannot be found to the next selector or connector, the switch rotates to the end of the horizontal contacts and immediately returns a busy tone to the calling party. When either party disconnects, all switches in the connection are returned to the idle condition. A seven digit call requires a line-finder, first, second, third, fourth and fifth selectors, and a connector. Calls to other offices are accomplished after the first, second, or third digit. Toll calls usually must be indicated by dialing a 1 before dialing the seven or ten remaining digits. Very few, if any, step-by-step systems remain as toll offices.

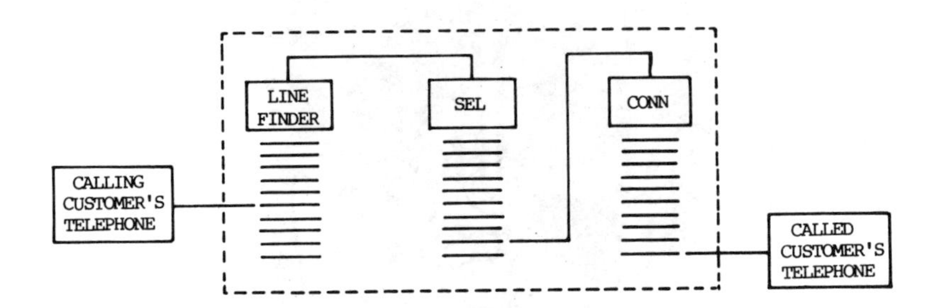

Figure 7.37: A 1000 Line Step-By-Step Switching System

Step-by-step was a popular system in the past because it served basic functions at a time when operators handled most toll calls and all extraordinary types of calls. But the progressive nature of the control does not permit the addition of new features or the automatic re-routing of calls without adding expensive control systems which preprocess calls.

Progressive control systems gave way to *common control* systems with a fundamental organization as shown in Figure 7.38. Local users' lines terminate in line circuits which in turn connect to the first stage of a matrix type switching network. Trunks and service circuits terminate on the outlet side of the network. Service circuits include originating registers which give dial tones and receive address digits from the users' telephones, incoming registers which receive address signals from incoming trunks, busy tone, 20 Hz ringing signal, etc. There is no universal architecture for this type of system. Each system differs from others for reasons not always clear to the casual observer.

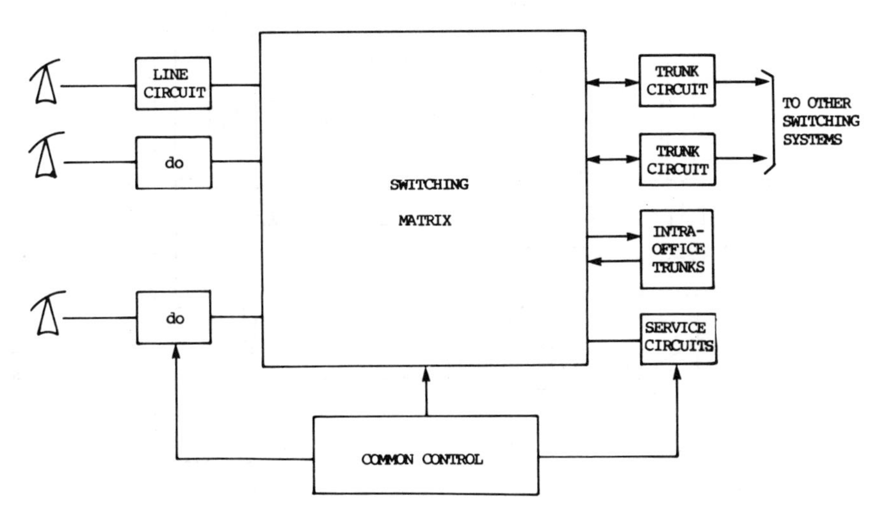

Figure 7.38: Basic Block Diagram of a Common Control Switching System

Common control systems were pioneered because the expensive controls were involved only during set-up and take-down of calls. The cross-points of the matrix, the line circuits and trunks are held during the call. The controls, then, may serve many calls during the duration of a single call. Electro-mechanical systems utilize several controls simultaneously, each processing separate calls. These electro-mechanical controls usually are referred to as *markers* because they *mark* each end of the network indicating the inputs and outlets to be interconnected. A marker usually requires several hundered relays for its logical functions and some systems employ as many as 18 markers. Figure 7.39 shows a diagram of a typical electro-mechanical system, referred to as a crossbar office because of the particular type of matrix switch utilized in the switching network.

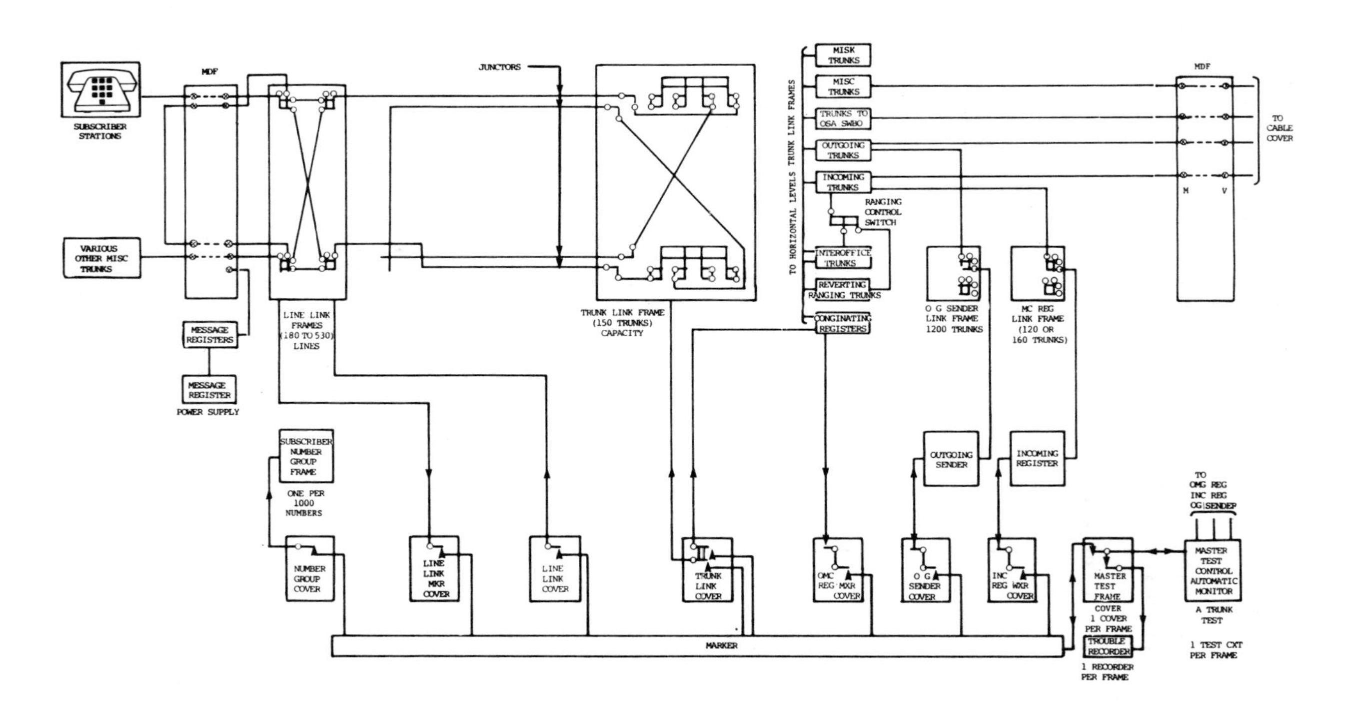

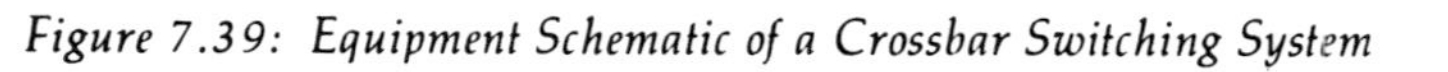
Figure 7.39: Equipment Schematic of a Crossbar Switching System

User lines are cross-connected from the main distributing frame (MDF) to a line circuit in the line-link frame. A great advantage of common control systems is their ability to function with no necessary relationship between an equipment location and a telephone number. A translation memory in the system, in this case the subscriber group frame, identifies the calling party for billing purposes when required on toll calls, otherwise, only equipment locations are required. The called party at the terminating office is identified in the address by a 7 digit number which must be translated to an equipment location. Also contained in the system memory is the *class-of-service* of the calling party. Figure 7.40 is a flow chart of the actions of this particular system in detecting a request for service to the point where the calling party receives dial tone. The steps are as follows:

1. User requests service by going off hook.
2. The line relay and the line link frame operate through the station connection, indicating a service request.
3. Line-link frame connector, consisting of one or several large relays with 100 or more contacts, connects an idle marker to the line-link frame.
4. The marker, in turn, finds an idle trunk-link frame with an idle register. Only one marker at a time may operate with a line-link frame or a trunk-link frame.
5. Marker selects an idle originating register.
6. Marker, through the line link connector, determines the location and class of service of the calling party.
7. Marker stores above information in the chosen originating register.
8. Marker selects and operates a path from the calling party through the line-link frame, a junctor and the trunk link frame to the chosen originating register and disconnects itself to service another call.
9. The originating register may perform some test on the line before returning dial tone to the calling party through the network connection.
10. The user receives dial tone and, in turn, dials the address of the called party.

Assuming no special treatment is required, the address in the originating register allows the call to be completed by many series of operations similar to those just described. The call may be completed within the same office or may be forwarded through several offices to its destination. A single office or wire center of this type may represent many

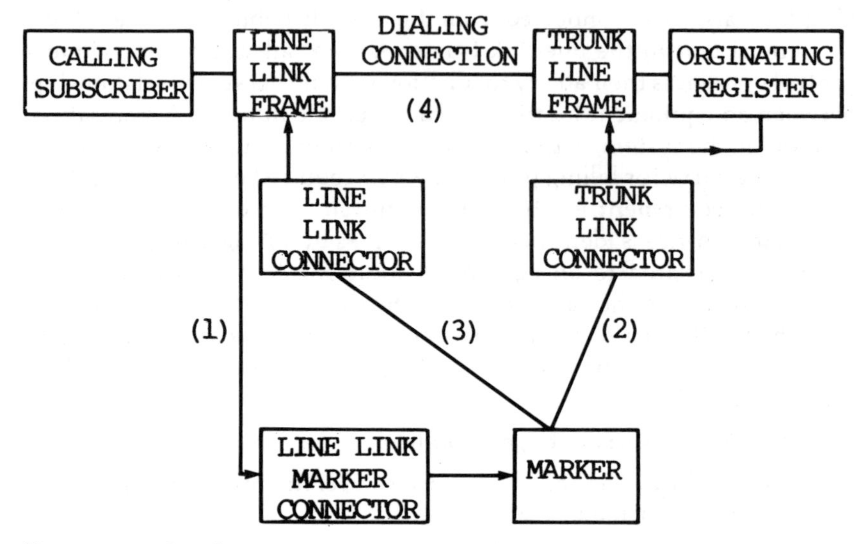

Figure 7.40: Crossbar System Operations to Obtain Dial Tone

exchange codes and be capable of terminating 100,000 lines and several thousand trunks. Most large offices manage traffic for about 30,000 lines, however. Perhaps this is the place to note that in the U.S. there are more than 10,000 such wire centers serving about 90,000,000 lines. In the past, step-by-step systems were placed close to the center of the area served and operated as a single exchange office. Common control systems, with their vastly increased logical power, have reduced the number of wire centers while the total number of lines has increased greatly.

What was just described is an electromechanical, common control switching system. Some memory features; e.g., translation, may be electronic, however. For the most part, billing information from the switch is given to computers and they also give the switching systems routing and signaling information to be passed on to the next office to complete the call. Other systems, using similar switching networks, have utilized electronic control systems.

Electronic controls became desirable when calling rates in large offices could not be handled by the electromechanical controls and markers. Additional markers were not an answer because the slow call-processing by the individual markers kept all the line and trunk link frames occupied. Additional offices also compounded the problem, and the solution was found in high speed call processing using electronics. Single electronic control systems are fast enough to handle up to 500,000 call attempts per hour in busy periods. It is common practice to have duplicate controls and memories to insure reliable service. Reliable service has been defined as two hours of cumulative outages in 40 years! U.S. systems largely meet this objective, except in catastrophic circumstances.

The addition of electronic common control to switching systems led naturally to stored program controls replacing the earlier wired logic systems. Again, a need to meet growing demands for increased calling rates and more versatile services and the maturing technology of stored program computers converged to produce the electronic switching systems.

A great deal of research and development preceded the new generation of electronic systems. Perhaps it was fortunate that their manufacture was delayed for economic reasons, but the delay enabled a comprehensive system design to be completed before its implementation was undertaken. Some forays were made into the stored program control switch field before the technology had matured sufficiently to produce an appropriate economic climate for the widespread acceptance of the concepts involved.

The basic block diagram of a stored program machine is the same as Figure 7.38, the functional requirements of all common control systems being the same. Stored program control gives greater flexibility to switching systems and delays the need to introduce new hardware and technology in order to provide new switching services. New hardware technology was largely dictated by the economics of integrated circuits as opposed to the discreet transistor design of the first systems. Increasing scale of integration and low cost high speed memory systems have changed the architecture of the controls of these systems. Initial systems sought to limit the size of memories and hence instructions, translation tables, parameter tables, etc., were rationed carefully. High speed integrated circuit memories, both permanent (ROM) and temporary (RAM), largely removed the barriers of size of instruction sets or program words and all other memory functions. Present electronic systems have about 1,000,000 bytes of program and memory. Byte lengths of 30 to 50 bits are common.

High speed processors, comparable to high speed main frame processors used in the largest commercial computers, have replaced the original relatively slow discrete transistor systems. Present day electronic systems, as shown in block diagram in Figure 7.41, can terminate 100,000 lines and 50,000 trunks, and simultaneously serve as local and toll offices.

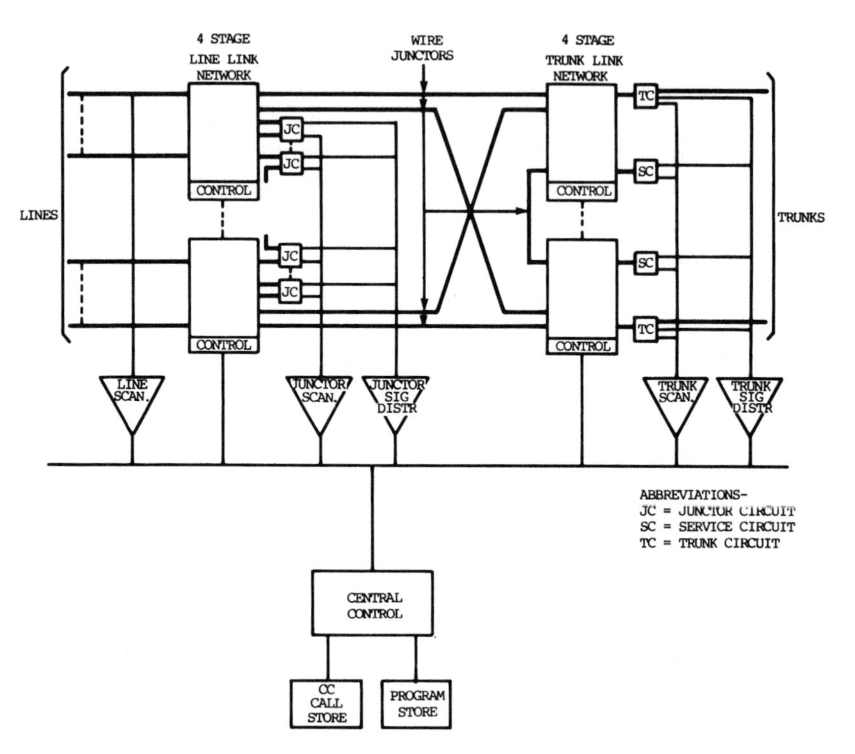

Figure 7.41: Functional Schematic of an Electronic Stored-Program controlled Switching System

This representative system uses an 8 stage switching network for economy of cross-points at the expense of real time of the processor. The functions performed are similar to those previously described. There is some parallel processing employed in most electronic systems. Digit analysis, signal processing, switching network control, memory mapping of the switching network, etc., are frequently handled by peripheral processors. The main processor may become a supervisory system over the peripherals which perform the needed functions.

Switching networks for the bulk of the present systems are electromechanical, space division networks, but the newest systems, led by the largest toll switches, have incorporated digital switching compatible with the T carrier systems. (Some comments about digital switching in relatively small systems in private branch exchanges (PBX) are appropriate and follow later in this chapter.) The trend is toward complete integration of transmission and switching. The large electronic toll offices terminate 100,000 to 200,000 trunks and process 500,000 calls per hour. Analog voice signals are converted to PCM and in many instances reconverted to voice for transmission on the long-haul carrier systems, the bulk of which are still analog, frequency division multiplex as previously discussed.

Led by the hope that T carrier or a similar system could penetrate the bulk of the loop plant in the near future, manufacturers have produced a local/toll office which incorporates all-digital switching similar to the toll office just discussed. These new switches offer the same basic service to voice users as the analog offices, but inherent in the hope of an all-digital loop plant is the universal availability of megabit per second rates for data or mixtures of voice and data. This system at least has a name, Integrated Services Digital Network (ISDN). Substantial inroads have been made in the direction of such a system, however, with the adaptation of T carrier to 26 gauge cable pairs and the installation of optical fibers in some new loop plant.

In some of these systems, switching has been incorporated, either in the form of a simple concentrator, a system which shares 96 voice channels with 200-300 users, or as a data network using packet switching. It is not entirely clear that all users are interested in the high speed data capability that this system offers, but present day signal processing systems, which are increasingly economical through VLSI, can transmit full-motion video at a 1.544 megabits/sec bit rate. Certainly, reduced bandwidth full duplex video systems could provide added services to users as could Teletext systems which offer news, entertainment, interactive games, and various computing capabilities. The boundaries between transmission and switching are becoming increasingly obscure.

Business communication needs traditionally have been met by private branch exchanges (PBX), or key telephone systems (KTS). The introduction of the interconnect industry into business telecommunications and the computerization of the white collar work force or Office of the Future is again producing a need in coincidence with technological capabilities. A number of business telecommunication systems are being manufactured with certain common characteristics:

1. Digital switching, i.e., voice is converted to PCM and switched digitally.
2. High-speed data transmission over inside-wiring twisted pairs.
3. Business oriented calling features.

Some systems feature digitizing equipment in the telephone set and 64 kbit/sec transmission to and from the switch over house pairs in addition to 64 to 100 kbit/sec data over the same pairs. Most systems, however, use 2 to 3 pairs for such purposes. Architectures of these systems bear remarkable similarities, but each new advance in the integrated circuit field produces new capabilities in these relatively small switches. Most such systems serve 100 to 200 lines but some can be extended to 30,000 lines and resemble central offices. Indeed, some are adaptations of central office equipment.

Corporate private transmission networks are being coupled with these new switching systems and are producing private national and international telecommunications systems, small replicas of the public network.

7.4 SIGNALING SYSTEMS

Sophisticated switching and the elegant interconnecting transmission systems allow voice communications after the appropriate switching and transmission functions have taken place. Signaling systems provide the means of conveying information from user to system and system to system to complete a call for the user. Signaling functions can be divided into two broad categories: *supervisory* and *address*. These two types of signals must be mutually appropriate for the switch and transmission systems involved. Because of this intimate relationship between all the elements, there are several forms of signaling systems required to provide compatibility. The salient elements of these will be discussed.

Signaling systems receive information from a source and process the information for delivery to the next step in the chain. This processing may require reception and decoding and subsequent encoding, then signal generation and transmission. Systems may do a few or all of the above functions depending upon the location in the chain. In a restrictive sense, signaling systems transport or convey information; they are not the source of the information but take information form a variety of sources in order to appropriately process and transmit a signal. Address information in the form of a distant telephone number must have routing information appended to it, and perhaps some type of class-of-service information also may be included in a signal transmitted from one central office to another in the process of completing a call. There is a fine distinction between providing a means of carrying a signal between systems and the processing of the signals at intermediate or end offices. The techniques and systems involved in carrying signals are discussed, but not the portions of the systems which provide processing.

North American signaling systems are comprised of individual links between switching offices. Signals are passed in both directions between two offices; repetition of the signals to the next office involves a second link. It is believed that the advantages of link-by-link signaling are greater than end-to-end signaling, as shown in Figure 7.42a and 7.42b. Link-by-link signaling involves signal reception and regeneration, and perhaps modification, at each switching point as opposed to end-to-end signaling which involves originating and transmitting signals sequentially as links of the system are joined at switching points.

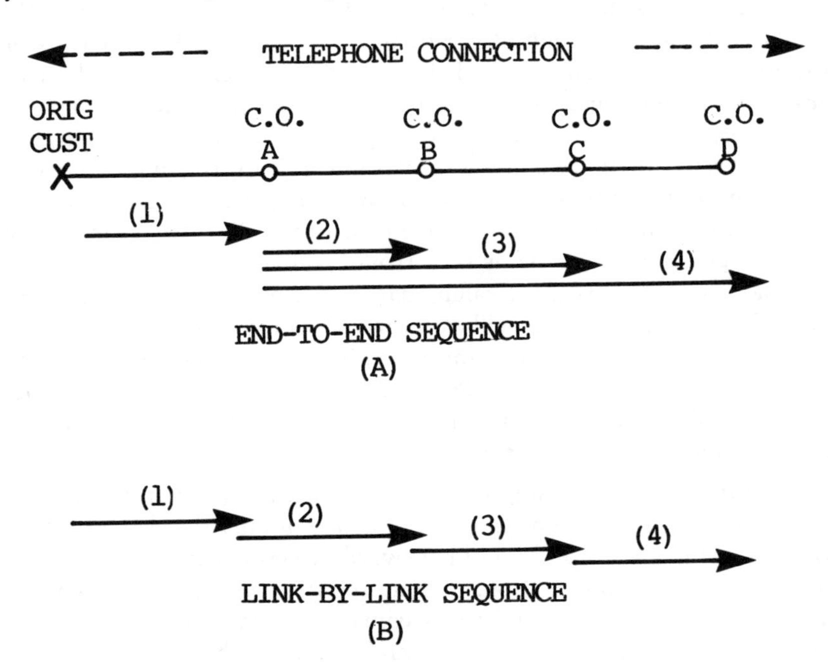

Figure 7.42: End-To-End versus Link-By-Link Signaling

Supervisory signals consist of *control* and *status* signals. Control signals take a variety of forms which may be the position of a switch hook or a bit in a memory system. Most control systems are two-state and continuous, but there are exceptions. *Status* signals indicate the condition of a system, i.e., whether the called party has answered; when an incoming register is ready to receive digits; all-paths busy; etc. Address signals are typified by the telephone number of a called party. Appended to this prime address of a call may be routing information. Additional types of signals are *audible* signals and *network management* signals. Audible signals indicate call progress, i.e., ringing, busy, etc. Network management signals control the routing of calls between any two points so as to maximize the use of a network and avoid delays at congested switching points.

As stated, one type of control signal is generated by a user requesting service by going off-hook. The audible signal returned to the user is a dial tone indicating the central office is prepared to receive the address digits. Address signals are generated by the user's dial and transmitted as timed interruptions by the rotary dial or dual tone multifrequency (DTMF) signals. Until audible ringback is returned, the calling party is unaware of the inter-office signaling which is occurring. When the called party answers, audible ringback is stopped and conversation may ensue. *Answer supervision* is returned to the calling office for control, and charging if appropriate. If the calling party disconnects by returning to the on-hook condition a control signal goes from the local central office forward to all links of the connection to take down the call. If the called party disconnects, a status signal is sent from the terminating office to the originating office, and if the calling party does not immediately disconnect, the originating office will wait for a short interval and then signal forward to take down the call and then give the dial tone, indicating its ready condition to receive a second call. Obviously, switching systems, and hence signaling systems, must cope with all possible actions on the part of calling and called parties.

The capabilities of switching systems affect the requirements of signaling systems. Step-by-step systems can transmit only simple trunk control signals in the forward and backward directions and trains of dial pulses in the forward direction. Common control systems, using senders and incoming registers, send trunk control information and switching system's status signals and address signals using much more sophisticated means than step-by-step systems. Stored program control of switching is having the greatest impact on signaling systems. Data links, separate from the voice network, are increasingly used to transmit trunk control, status, and address signals between central offices. This common channel inter-office signaling (CCIS) system carries information for several hundred trunks between offices. Such information is still conveyed link-by-link.

Another consideration affecting signaling is the use of one-way and two-way trunks. Points close together having heavy traffic usually use trunks on a one-way basis. Traffic is always originating in the same direction on a trunk route. Points at greater distances, especially those between time zones, utilize two-way trunks for increased usage. Both trunk and signal circuits for one-way operations are simpler and less expensive since they are designed to fit the requirements unique to each end. Signaling on two-way trunks is symmetrical. Each input-output function applies identical signals for each direction of operation. This is a great advantage of link-by-link signaling. Individual offices may be changed without disrupting major portions of the network.

The development of exchange and long-haul carrier systems has had a large impact on the co-development of compatible signaling systems. Exchange trunks converted from space division multiplex transmission to frequency or time division multiplex operation have had accompanying developments of new signaling systems. The greatest impact is the conversion from dc signaling, sufficient for metallic and voice frequency circuits, to ac signaling necessary for analog carrier systems, and digital signaling for digital carrier systems.

DC signaling methods include loop signaling systems, such as the telephone and systems that originally were used to multiplex telegraph signals on telephone voice circuits. Simplex, duplex and composite dc signaling enable speech and related signaling to be carried on the same wire facility. Some systems continue to utilize this method of signaling on trunks between PBXs, i.e., tie-trunks, or from PBX to central office operate in a related fashion. Carrier systems necessitated ac signaling techniques or parallel data paths.

AC methods, carrying signal over the same facility as voice, are in widespread use but are being supplanted by CCIS between stored program controlled toll switching and many local switching systems. AC signaling gives a choice of in-band signaling or out-of-band signaling. Most U.S. systems employ an in-band signal, which implies that the signal is in the band of frequencies transmitted for voice and hence can be heard by the user. Normal circuit conditions remove the in-band signal by switching or filters so that a user rarely hears the signal. The advantage of in-band signaling is that no special provisions are required in the carrier system; the signaling equipment is independent of the transmission system. The greatest disadvantage is that speech or other signals in the voice channel may imitate signaling and cause unintended switching to occur.

Present day in-band supervisory signaling uses a single 2,600 Hz tone. A few systems remain which use 1,600 or 2,400 Hz in-band and a few out-of-band signals exist also. Presence of the 2,600 Hz tone indicates an idle circuit or the on-hook condition. When a trunk is seized by a switching system the single frequency (SF) signal is removed, indicating off-hook and dial pulse address signals are indicated by pulses of 2,600 Hz similar to the current interruptions on wire lines.

Another signaling method is compelled signaling. This technique transmits a signal in the forward direction until it is acknowledged by a signal on the return path at which time the forward signal is removed. Upon detection that the forward signal has been removed, the acknowledgement signal is removed. The originating end must detect the end of the acknowledgement before starting another signal forward. For each compelled sequence four transmission times, two round trips are required which reduce signaling speed to an intolerable level by North American standards.

A seizure of a trunk by a central office switch causes it to signal forward to its companion circuit in the next central office an "off-hook." Figure 7.43 shows the user actions as well as the trunk actions between two common control central offices. If one or both offices are step-by-step, the sequences are altered from those of Figure 7.43.

Figure 7.43 illustrates the division of supervisory signals into control and status categories. Control signals include a *connect* signal, a sustained off-hook signal toward the called end. This signal is sent immediately upon seizure to minimize occurrences of simultaneous seizures at both ends on two-way trunks. *Disconnect* signal is a sustained on-hook signal in the forward direction. Disconnect signals are timed at the called end and become effective after a fraction of a second. *Ring-forward* signal is a timed on-hook signal of shorter duration than the disconnect signal. It is used much as a switch-hook flash for the calling end to ask for assistance at the called end.

Status signals are transmitted backward from the called end to the calling end; *Answer signal* is a sustained off-hook at the called end. Automatic charging is initiated when appropriate. *Hang-up* signal is a sustained on-hook signal at the called end. This signal stops the charging equipment at the calling end. *Address* signaling is mostly of two types; *dial pulsing* or decimal trains of pulses correspond to those of a rotary dial. Common control systems may pulse at 20 pulses per second rather than the customary 10 pulses per second. *Multifrequency* pulsing systems transmit each digit of an address by a unique combination of two out of a possible six frequencies in the voice band. Like the dual-tone multifrequency systems of the telephone set, these signals are in-band, making it possible to signal over any voice circuit. In MF addressing there are 15 possible signals, 10 are used for digits, 2 for beginning and ending signals and 3 for operator use. MF tones used for inter-office addressing are different tones than the DTMF two out of seven codes to prevent interaction between telephone sets and inter-office signals. A clear distinction is necessary between trunk supervisory signals and switching systems address signaling. DC or SF signaling provides supervisory signals between trunk circuits and, as will be described, is independent of

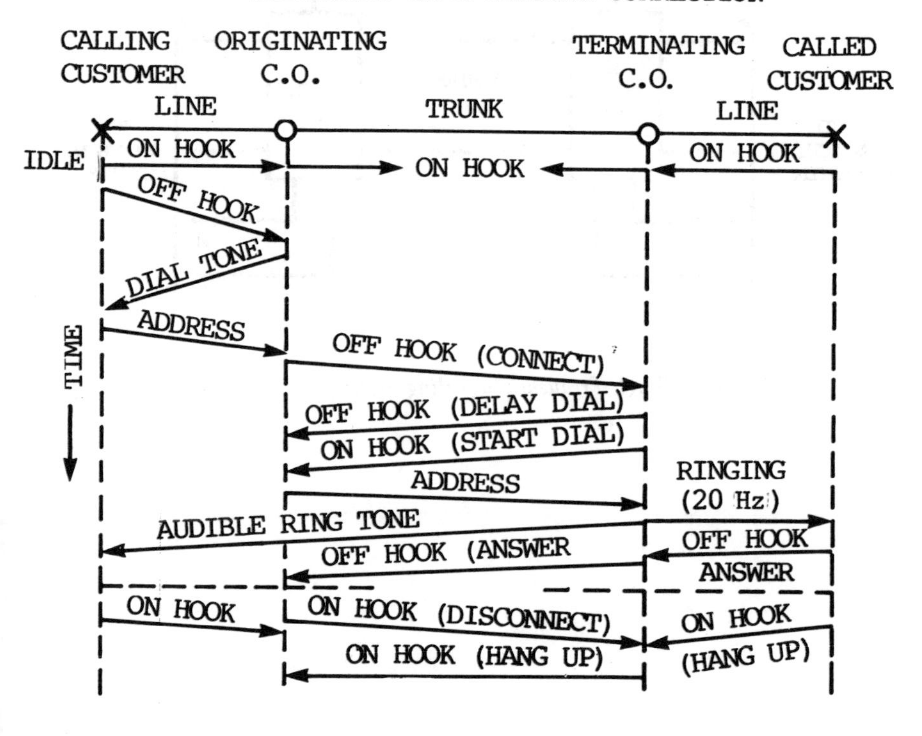

Figure 7.43: Signaling on a Typical Connection

the switching system with which the trunk is associated. Addressing signals are switching system dependent. Step-by-step offices operate only with dial pulses, and if they must be driven by MF or DTMF signals a conversion to dial pulses is required. Common control systems can use dial pulses and/or MF signaling for loop, exchange or toll signaling. Dial pulsing can be DC or SF. Inter-office addressing is carried by MF and loop-signaling can be rotary dial or DTMF.

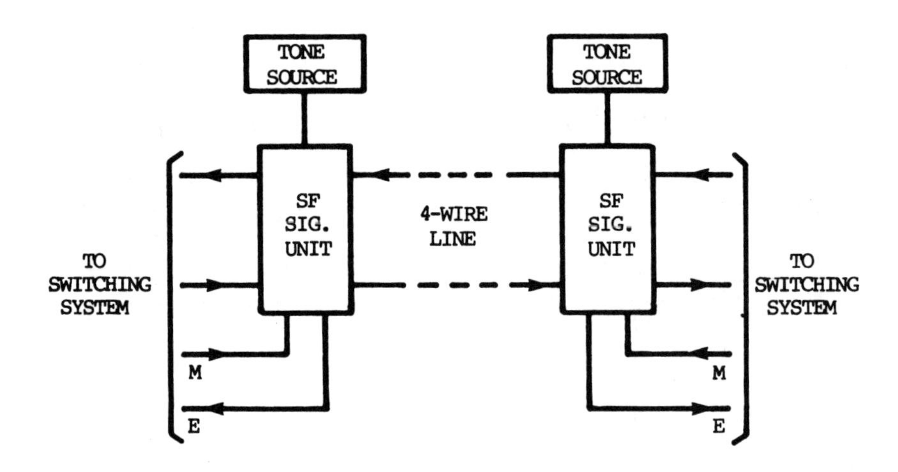

Figure 7.44: Typical Single Frequency Signaling System

Signaling and trunk circuits interface in two distinct fashions. Loops, or lines from central offices to users' premises, and trunks within an exchange area use *loop* signaling; i.e., in a manner similar to the telephone rotary dial, signaling is on-hook and off-hook. Intertoll signaling has developed a different interface in which input and output controls are called E and M respectively. A useful mnemonic is (*E*ar for input and *M*outh for output. Inter-toll signaling systems such as the single frequency systems (SF) convert E and M lead dc signals to ac signals for transmission over the various carrier systems. Figure 7.44 shows the interface between the four-wire trunk circuit and the SF signaling unit. A total of 6 leads forms the transmit and receive pair plus the E and M leads.

SF signaling units must satisfy certain fundamental requirements such as reliable operation over carrier facilities with extreme noise levels, maximum compander tracking error, marginal gain variations, etc. Also, they must be able to function without degrading the quality of voice transmission. In a phrase, they must function ultra-reliably and remain transparent to the carrier system. The second basic requirement is that the system be immune to false operations from speech, data, or noise signals. This false operation is known as *talk-off* and does occur with in-band signaling systems, but system design strives to eliminate it or at least reduce it to acceptable levels. If *talk-off* is sustained for more than about 1/3 of a second, the switching system will perceive the signal as a hang-up or disconnect and terminate the connection. The problem of *talk-off* could be reduced by going to high signal levels for the SF tones, much higher than would be anticipated as the average level for speech and subsequently reducing the SF receiver threshold level. The level of SF tones is limited, however, by cross-talk and carrier overload consideration. Fortunately when large volumes of voice are present, the SF is absent and when a large number of trunks are idle, that is, when the SF tones are present on a majority of trunks there is less opportunity for speech peaks from several channels to coincide and overload the carrier system momentarily. Still the SF level has to be maintained at a level which is not harmful to the quality of the voice carrier system. A final requirement is that the signaling must be made as fast as possible consistent with reliable operation in order to minimize set-up and take-down times. A part of the speed requirement is concerned with user reaction and another with pure economics, since about a half trillion calls are made annually and one or two seconds per call is one half to one percent saving in total call time but about a ten percent saving in set-up time, which is overhead.

To reduce talk-off to acceptable levels, the SF receiver unit is designed to monitor the total energy on the line, whether the energy is voice or noise, and compare that energy with the output of a relatively narrow band filter centered at the SF signal. When these two energies approach equality, a circuit threshold detector operates, indicating that SF tone is present. When the SF unit holds this condition for about one-third of a second, switching systems will respond by taking down the connection. It is understandable why reliable operation is necessary.

Digital transmission systems signals use designated portions of the bit stream. As mentioned in the description of the T carrier system, 24 voice channels are each allocated 8 bits per 125 microsecond frame. Early systems used one bit of the eight for status information and other bits normally assigned for speech during call set-up for dial pulsing or E and M type signaling. Later systems, emphasizing improved voice transmission, have organized the bit streams of 193 bits per frame into 12 frame groups referred to as *super frames*. In each voice channel the least significant bit of the voice signal in frames 2 and 8 of the superframe are used as status signals. The digital channel bank system provides for the signaling required for each channel. There are many arrangements in use, largely dependent upon the system use, that is: loop or exchange carrier and the type of central offices being interfaced. It is clear the digital carrier systems could use SF inband signaling but digital means have proven much less expensive. The type of signaling required over a path is accommodated by the use of plug-in modules in the channel band equipment. Addressing signals normally transmitted inband by MF pulsing may be accommodated in the voice band as in the analog carrier systems or by conversion to dial pulsing as above.

PBX trunks, foreign exchange lines (FX), coin lines and certain others signal the central office by means of a *ground-start* trunk or line circuit. Ground-start trunks and lines are used almost universally in North America for these purposes but other national systems use loop-start, imitating a telephone set at a user's premises. Ground-start trunks are used on PBX trunks and FX lines to avoid *glare*. Glare is a simultaneous seizure of a trunk or line by the PBX and the central office. The central office must recognize that such a condition has occurred and release its end of the connection so the PBX may proceed.

Tip and *Ring* are the titles associated with the two conductors on trunks and lines. In the idle condition, ground-start trunks, looking from the PBX to the central office have –48 volts on the Ring conductor and the Tip is unterminated. The PBX has –48 volts on the Tip and the Ring is unterminated. When the PBX *seizes* the trunk, it does so by grounding the Ring lead and the central office responds by grounding the Tip as an acknowledgement signal. Similarly, if the central office first seizes the trunk, it does so by grounding the PBX Tip conductor and the PBX must mark the trunk busy within about 50 milliseconds. If instead of marking the trunk busy the PBX seizes the trunk already seized by the central office within the 50 millisecond interval, the central office *backs-off,* choosing an alternate trunk to the PBX. One-way trunks eliminate these difficulties and are widely used in larger PBX's. The signals exchanged between PBX and central office must account for *direct-inward-dialing* for completion of calls or *attendant completed* incoming calls and PBX disconnect or hang-up and central office disconnect or hang-up. These signaling sequences are important so that a switch-hook flash will be properly interpreted as an operator recall or dial tone recall and the disconnect or hang-up signals are recognized as quickly as possible so the trunk may be used for ensuing calls with minimum lost time. Again, one-way trunks offer more expedient signaling arrangements.

The growth of stored program controlled offices or stored program controlled signaling systems in electromechanical offices has made it possible to gather the signaling information required for large numbers of trunks and concentrate this information on a single dedicated data channel. This common channel inter-office signaling (CCIS) system is being extended between all of the above types of central offices, which will shortly handle almost all toll calls. A complete CCIS network, using existing voice facilities, is being created. This network transmits and receives signaling information between the local and toll switching centers and between toll centers. Long distance carriers probably will utilize stored program controlled switching systems exclusively and hence all signaling between toll centers will be CCIS. An international standards organization, known best by its initials, CCITT, has issued standards known as CCITT Signaling Standard 6 and CCITT Signaling Standard 7 for the protocols and formats of these signaling messages, so that such signals may be exchanged between long distance carriers nationally and internationally. CCITT 6 are the signaling standards for stored program space-division switching systems and CCITT 7 for the all-digital switching systems. At present only CCITT 6 systems are in operation. Figure 7.45 shows the comparison between the per-trunk signaling systems and CCIS systems.

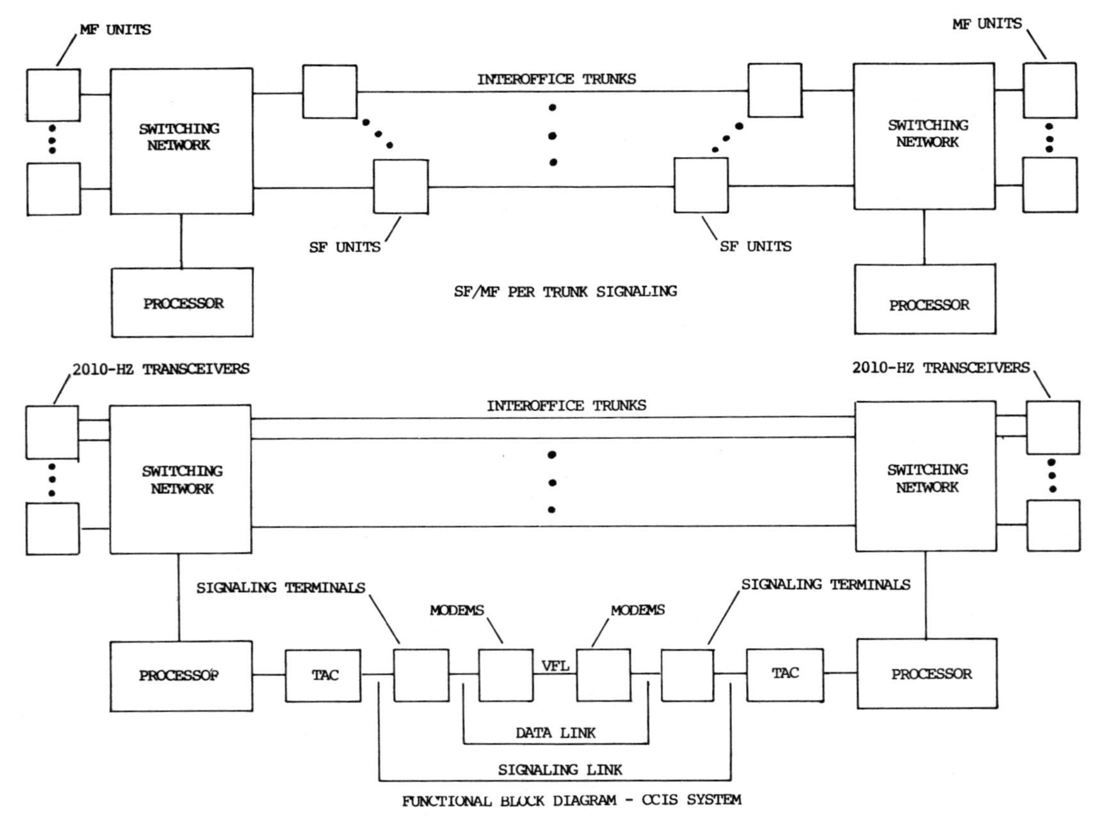

Figure 7.45: Signaling System Block Diagram

CCIS systems transmit 28 bit blocks of data which contain assorted information such as the number of blocks in the message and, of course, and end of message signal in the final block. Trunks are identified and address information, trunk nature information to prevent unintentional double satellite hops, etc., is included. Error correction utilizes error detection and retransmission. The last eight bits of the 28 bit blocks are error detections bits.

Included in the system's design, but not essential to CCIS, is a verification test of the voice transmission path. During call set-up, on a link-by-link basis between switching centers each 4-wire voice link is verified by looping the system back to the link origination point to verify circuit continuity by transmitting and receiving a 2,010 Hz signal. Two-wire facilities transmit a 1,780 Hz tone and receive a 2,010 Hz tone full duplex for circuit verification. Signal levels are also verified so that the transmission level plan is kept effective. Faulty circuits are flagged for repair and marked busy after a circuit has failed a verification test twice in sequence. Whether such circuit verification takes place and the subsequent action is in the policy domain of the long distance common carriers and may vary from carrier to carrier.

The CCIS network reliability is also in the policy area of the carriers but all will probably construct systems containing redundant modems, transmission paths, and data terminals, as well as alternate routing of CCIS signals to isolate faulty nodes, should they occur. Large private networks may use per-trunk signaling or CCIS depending, of course, on corporate needs and capabilities.

CCIS data transmission systems use 2,400 and 4,800 bit/second modems on analog transmission systems, or 4,000 bit/second digital transmission on digital carrier facilities by using alternate framing bits which are not needed when CCIS is employed. Dependent on the data rate, links of CCIS systems control 2,000 or more trunks per data link.

7.5 SUMMARY

The public telephone network consists of many operating companies which provide local service and access to the relatively few long distance carriers. The manufacturer's offerings of transmission, switching, and signaling systems is varied and increasingly state-of-the-art so that the new systems which will be installed in the next few years probably will not conform to the descriptions contained herein. There is no one transmission system which will dominate; cable, point-to-point radio, satellites and optical fibers will each have a place in the new systems. The number of systems which integrate digital transmission and switching will increase but will not totally replace the present analog carrier systems and analog space division switching systems. Satellite transmission is a mixture of analog and digital systems and will remain so, certainly during the life of the present satellite systems. Signaling systems are growing more complex and intelligent at increasing rates. The separate networks of the CCIS systems look less and less like signaling systems and more like computer communications networks, which of course, they are.

If the reader wishes to be informed of technological advances and state-of-the-art systems, there are a wealth of sources of such information varying from corporate technical journals, professional society publications and many fine trade magazines. Persons concerned with managing telecommunications systems will need all these knowledge resources.

chapter 8

COMPUTER NETWORKING AND DATA COMMUNICATIONS

Harvey M. Gates
BDM Corporation
University of Colorado
Boulder, Colorado

8.0 INTRODUCTION AND OVERVIEW

When a computer is connected to a communication system, one enters the realm of data communications or datacom, as some prefer. Data communications is the interchange of digital information between two or more processors using communications. And it is interesting to note that the separation between the digital processor and the communications system is rapidly converging and becoming tightly interwoven. The purpose of this chapter is to provide an overview of data communications in the context of computer interconnection.

Figure 8.1 illustrates a simplified end-to-end data link. In this diagram there are several key components. First, there are the two pieces of data terminal equipment (DTEs) which represent the data source, the data sink, or both. The data terminal equipment options include: a large mainframe computer, a personal computer (PC), a microprocessor, a mechanical or electronic teletype-writer (TTY often referred to as a dumb terminal or a process control device). The list of data terminal equipment is endless. In Figure 8.1, the two DTEs are geographically separated and thus require some form of linking through data communication equipment (DCEs). The DTEs provide the functions required to establish, maintain, and terminate the digital conversation between the two DTEs. A DCE can also convert a digital signal into a analog signal which is often easier to transmit over long distances, and thus the DCE has classically been the communication modulator-demodulator or modem. Modems will be discussed at the end of this chapter.

The data link in Figure 8.1 is shown as the classic telephone line which up until recently constituted the bulk of data linking. There are now many ways to communicate between DTEs. For example, a word processor could be connected to another word processor within the same office

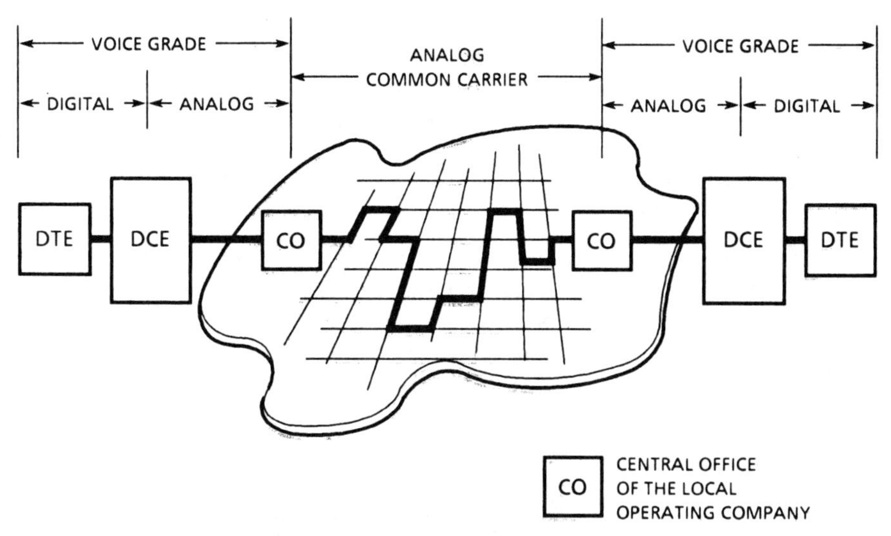

Figure 8.1: The Basic Telephone Data Link Model

building one day, and the next day connected to a system half way around the world. The intra-office communication channel would probably be a local area network (LAN) while the media in the second case would surely be a satellite link. The alternatives are extensive.

This chapter begins by discussing the codes and protocols used between DTEs and passed through the established communications media. There will be an in depth discussion of the communication channels themselves which will first include the conventional narrow-band voice type telephone links followed by selected wide-band systems common to satellites and terrestrial microwave. Discussions of the LAN are reserved for another chapter of this text although there will be natural overlap.

8.1 TRANSMISSION CODES

Data processing terminals compute and communicate in binary code-1's and 0's. Alpha-numeric characters must therefore be represented by binary sequences, and so must various simple commands such as "space" or "carriage return." There are 26 letters in the English alphabet and 10 numeric characters. If one chooses to represent S symbols with a binary sequence, N binary positions or digits are required such that:

$$S = 2^N \tag{8.1}$$

Therefore we could represent our 26 letter alphabet and 10 numeric characters which total 36 with $N = 6$ or 6 binary positions. Notice that if N

= 5, S = 32 which is short by 4 characters; but if $N = 6$, S = 64. We therefore have 28 unassigned binary combinations to represent other useful symbols and commands such as a period, a question mark, quotation marks, a carriage return, a space, and so on. In theory, a data terminal device can now communicate with another terminal device in 6-bit combinations provided both agree on what each binary symbol represents. As we will learn later, this is not necessarily as easy as it may first sound.

There are several codes in use today, although the three most common are: (1) the International Telegraph and Telephone Consultative Committee (CCITT) Alphabet No. 1 called Baudot, which is named after an early French telegraphic pioneer, (2) the American Standard Code for Information Interchange (ASCII), and (3) the Extended Binary Coded Decimal Information Code (EBCDIC).

8.1.1 Baudot Number 2

Baudot No. 2[1] or just Baudot is the oldest of the three mentioned codes. It is one of several five-unit TTY codes. Other Baudot type five-unit TTY codes include the popular international Telex code, the weather code, the TWX code, and a United States military version of Baudot.[2] The Baudot and other five-unit codes allow for only five binary positions ($N = 5$) and therefore only represent 32 symbols ($S = 32$). However, there is a clever scheme which can extend the character set by using shift up and shift down commands. The formal shift up command is called a Figure Shift (FS). The shift down command is called a Letter Shift (LS). Figure 8.2 shows the Baudot code conversion and will help to clarify how these shift commands expand the 5-bit code.

Figure 8.2 has four columns. The first column represents the binary code and bit numbers. The second column represents the Letter Case. The third column is the Figure Case for Baudot, CCITT No. 2. These three colums represent the pure teleprinter Baudot No. 2 code. The fourth Figure Case column is the Telex replacement for the third pure Baudot No. 2 and demonstrates that one can have variations or other code assignments using these Figure Case alternatives.

Consider, for the following example, only the first binary column, the second Letter Case column, and the third Baudot No. 2 Figure Case column. If we transmit the letter *A* in Baudot, we would first initialize the process by transmitting the LS code (11111) followed by *A* (00011). This initialization places us in the letter character column. We stay in this mode until directed to move to the other column with an FS code. Therefore, if we continue our transmission with alpha characters, we do not use either the LS or FS commands. Notice that a space command (SP)

BINARY	LETTERS CASE	BAUDOT NO. 2 FIGURES CASE	TELEX FIGURES CASE
00000	BLANK	BLANK	BLANK
00001	E	3	3
00010	≡	≡ LINE FEED	≡
00011	A	-	-
00100	SP	SP SPACE	SP
00101	S	'	
00110	I	8	8
00111	U	7	7
01000	<	<CARRIAGE RETURN	<
01001	D	WRU WHO ARE YOU?	WRU
01010	R	4	4
01011	J	BELL	BELL
01100	N	,	,
01101	F	%	$
01110	C	:	:
01111	K	(	(
10000	T	5	5
10001	Z	+	"
10010	L	)	)
10011	"	2	2
10100	H	#	#
10101	Y	6	6
10110	P	0	0
10111	Q	1	1
11000	O	9	9
11001	B	?	?
11010	G	$	&
11011	↑	↑ FIGURE SHIFT (FS)	↑
11100	M	.	.
11101	X	/	/
11110	V	=	;
11111	↓	↓ LETTER SHIFT (LS)	↓

Figure 8.2: The Baudot CCITT No. 2 and the Telex Five-Unit Teleprinter Codes.

is in both columns. This duplication saves us from having to transmit the FS or LS commands with each space between words or figures. This is how the transmitted codes between two terminals look for "send 10 cases of 1969 vintage." The parentheses are only used to help separate the commands from the text.

(LS) SEND (SP) (FS) 20 (SP) (LS) CASES (SP)
OF (SP) (FS) 1969 (SP) (LS) VINTAGE (FS)

And if we tapped the data link we would see the following binary sequences. Once again, the parantheses are used only to separate the 5-bit codes for easier reading.

(11111) (00101) (000...............01) (11011) (11100)
LS S S E FS

The Baudot code is very efficient for the natural language transmission — like the transmission of this textbook material, because only about 5.06 bits per character are used compared to ASCII and EBCDIC which require 7 and 8 bits per symbol, respectively.

8.1.2 ASCII

The ASCII code (pronounced ask-key)[3] requires seven bits for alphanumeric characters and control and an eighth bit for parity checking. Parity checking is a means for determining an error in the 7-bit sequence and it is covered in Chapter 10, Telecommunications Systems. The 7-bit sequence allows for 128 unique bit patterns ($S = 128$). These combinations allow for both upper and lower case alpha characters to be transmitted between digital terminal equipment.

There are several versions of ASCII which often assume the personality of the language or particular region in which the code is used, such as the dollar sign in the United States versus the pound sign used in the United Kingdom. The ASCII code shown in Figure 8.3 represents the results of the International Standards Organization (ISO) in which the United States' representative is the American National Standards Institute (ANSI). Figure 8.3 is the ANSI X3.4 version of ASCII. Notice in this figure that the bit positions are numbered B1 to B7. The ASCII ANSI X3.4 code for G is:

B7	B6	B5	B4	B3	B2	B1
1	0	0	0	1	1	1

Notice there are 96 printable or displayable type characters and 32 control characters. These control characters are divided into:

(1) Format Effectors such as Line Feed, Carriage Return, and Backspace

(2) Communication Control such as ACK, NAK, and EOT

(3) Information Separators such as File Separator and Record Separator; and

(4) Miscellaneous Commands such as NUL, Bell, and CAN

BITS					B7	0	0	0	0	1	1	1	1
					B6	0	0	1	1	0	0	1	1
					B5	0	1	0	1	0	1	0	1
BITS	B4	B3	B2	B1	COLUMN / ROW	0	1	2	3	4	5	6	7
	0	0	0	0	0	NUL	DLE	SP	0	@	P		p
	0	0	0	1	1	SOH	DC1	!	1	A	Q	a	q
	0	0	1	0	2	STX	DC2	"	2	B	R	b	r
	0	0	1	1	3	EXT	DC3	#	3	C	S	c	s
	0	1	0	0	4	EOT	DC4	$	4	D	T	d	t
	0	1	0	1	5	ENQ	NAK	%	5	E	U	e	u
	0	1	1	0	6	ACK	SYN	&	6	F	V	f	v
	0	1	1	1	7	BEL	ETB		7	G	W	g	w
	1	0	0	0	8	BS	CAN	(	8	H	X	h	x
	1	0	0	1	9	HT	EM	)	9	I	Y	i	y
	1	0	1	0	10	LF	SUB	*	:	J	Z	j	z
	1	0	1	1	11	VT	ESC	+	;	K	[	k	{
	1	1	0	0	12	FF	FX		<	L	/	l	l
	1	1	0	1	13	CR	GS	-	=	M	]	m	}
	1	1	1	0	14	SO	RS	.	>	N	>	n	~
	1	1	1	1	15	SI	US	/	?	O	-	°	DEL

COL/ROW	MNEMONIC AND MEANING*		COL/ROW	MNEMONIC AND MEANING*	
0/0	NUL	NULL (MS)	1/0	DLE	DATA LINK ESCAPE (CC)
0/1	SOH	START OF HEADING (CC)	1/1	DC1	DEVICE CONTROL 1 (MS)
0/2	STX	START OF TEXT (CC)	1/2	DC2	DEVICE CONTROL 2 (MS)
0/3	ETX	END OF TEXT (CC)	1/3	DC3	DEVICE CONTROL 3 (MS)
0/4	EOT	END OF TRANSMISSION (CC)	1/4	DC4	DEVICE CONTROL 4 (MS)
0/5	ENQ	ENQUIRY (CC)	1/5	NAK	NEGATIVE ACKNOWLEDGE (CC)
0/6	ACK	ACKNOWLEDGE (CC)	1/6	SYN	SYNCHRONOUS IDLE (CC)
0/7	BEL	BELL (MS)	1/7	ETB	END OF TRANSMISSION BLOCK (CC)
0/8	BS	BACKSPACE (FE)	1/8	CAN	CANCEL (MS)
0/9	HT	HORIZONTAL TABULATION (FE)	1/9	EM	END OF MEDIUM (MS)
0/10	LF	LINE FEED (FE)	1/10	SUB	SUBSTITUTE (MS)
0/11	VT	VERTICAL TABULATION (FE)	1/11	ESC	ESCAPE (MS)
0/12	FF	FORM FEED (FE)	1/12	FS	FILE SEPARATOR (IS)
0/13	CR	CARRIAGE RETURN (FE)	1/13	GS	GROUP SEPARATOR (IS)
0/14	SO	SHIFT OUT (MS)	1/14	RS	RECORD SEPARATOR (IS)
0/15	SI	SHIFT IN (MS)	1/15	US	UNIT SEPARATOR (IS)
			7/15	DEL	DELETE (MS)

*(CC) COMMUNICATION CONTROL; (FE) FORMAT EFFECTOR; (IS) INFORMATION SEPARATOR; AND (MC) MISCELLANEOUS COMMANDS

Figure 8.3: The ASCII Code and Control Character Mnemonics (Version ISO ANSI X3.4)

The ASCII code is probably the most widely used code today for data transmission between data terminal equipment. Nevertheless, problems arise when one adopts one ASCII code and then attempts to transmit to a receiver with a slightly different set. This is obviously prevalent when transmitting between DTEs around the world. Frustration arises, however, when the character sets are slightly different within the United States and between different vendors of terminal devices. The mismatched character sets are often referred to as "orphans and widows."

8.1.3 EBCDIC

The EBCDIC code (pronounced eb-si-dik)[4] is an 8-level code ($N = 8$) which provides 256 possible combinations ($S = 256$). Much of today's computational equipment operates on 8-bit processes. The genesis of this can be found in the microprocessors and random access memory (RAM) semiconductor architecture which appears to have adopted this 8-bit standard. A group of 8 bits is referred to as a byte or octet. It is quite commonplace to find a large number of 4-, 16-, and 32-bit oriented central processor unit (CPU) architectures, though some scientific CPUs still use a 60-bit word. The trend, however, is clearly toward multiples of the byte.

Figure 8.4 shows the EBCDIC character set. The bit positions are also numbered in this chart. The letter *t* is represented by:

B7	B6	B5	B4	B3	B2	B1	B0
1	1	0	0	0	1	0	1

Watch out for the bit position descriptors! The letter *t* is more commonly identified as 1010 0011 in which B0 is presented first and B7 last.

Also notice there is no parity bit for error detection. Error detection can, however, be accomplished in many more ways than the parity bit in ASCII. This will be expanded upon later in this chapter.

Many of the control characters in both the ASCII and EBCDIC codes will be clarified as we move through the text. The EBCDIC controls fall into the same four categories as the ASCII code: Format Effectors, Communication Control, Information Separators, and Miscellaneous Commands.

Obviously, EBCDIC is very accommodating. However, it is the most inefficient of the three mentioned codes based on bits per character transmitted.

8.1.4 General Remarks on Codes

Codes in data communications represent the binary means of communicating information between DTEs. Though two terminals agree to use a

HEXADECIMAL EQUIVALENT		00				01				10				11				← BITS 0 1
BITS	↓	00	01	10	11	00	01	10	11	00	01	10	11	00	01	10	11	← 2.3
4567		0	1	2	3	4	5	6	7	8	9	A	B	C	D	E	F	← HEXADECIMAL EQUIVALENT
0000	0	NUL	DLE			SP	&	-									0	
0001	1	SOH	SBA					/		a	j			A	J		1	
0010	2	STX	EUA		SYN					b	k	s		B	K	S	2	
0011	3	ETX	IC							c	l	t		C	L	T	3	
0100	4									d	m	u		D	M	U	4	
0101	5	PT	NL							e	n	v		E	N	V	5	
0110	6			ETB						f	o	w		F	O	W	6	
0111	7			ESC	EOT					g	p	x		G	P	X	7	
1000	8									h	q	y		H	Q	Y	8	
1001	9		EM							i	r	z		I	R	Z	9	
1010	A					¢	!	!										
1011	B						$	,	#									
1100	C		DUP		RA	<		%	@									
1101	D		SF	ENQ	NAK	(	)	-	'									
1110	E		FM			+	:	>	=									
1111	F		ITB		SUB	|	-	?	"								

Figure 8.4: The EBCDIC Code Character Set Used in the IBM 3270 Information Controller System

specific code such as ASCII, we must recognize that there are selected differences due to regional variations and equipment vendors.

Recall in section 8.0, Introduction and Overview, the statement was made that communication systems and digital processes are rapidly converging and becoming tightly interwoven. The various codes are a good case in point. Up until recently, code compatibility was left entirely to the end user. This user was responsible for implementing in his or her terminal the proper code to communicate with another remote terminal. The communication system attempted to be transparent—what goes in, comes out. There are indicators that the communication system can and will provide such services as code conversion—Baudot No. 2 in and ASCII ANSI X3.4 out. This is only an indicator of what is at hand, and more will be said as we move through the chapter.

8.2 CLOCKS, TIMING, AND SYNCHRONIZATION

There is a heartbeat associated with all digital equipment. This heartbeat, of course, is the internal clock of a digital system. All digital processes associated with a particular piece of equipment perform their respective functions on each pulse of the clock. Microprocessors, for example, operate on clock pulses whose period may vary from as little as 100 nanoseconds (or a clock that generates 10 million pulses per second) to as much as 1 microsecond (or 1 million pulses per second). It is amazing, isn't it, to realize that such a device as a microprocessor can perform up to 1 million operations each second?

8.2.1 Clocks

A digital wristwatch provides us with a perfect example of the clock or timing device. Notice how one can read hundreds of seconds (in the stopwatch mode), seconds, minutes, hours, days in the month, and months. The electronic clock provides us with a broad spectrum of time increments. What overall purpose would be served to have a wristwatch that would read seconds only?

Computational processes require or operate on each pulse of the clock while other operations are restricted to one operation every 10,000 pulses and 100,000 pulses and so on. The internal clock of a digital device can therefore provide different pulse rates to different operational functions, but all pulse rates are slaved together just as the hour reading is slaved to the minute and the minute to the second and so on.

8.2.2 Timing and Standards

There is a standard for the internal clock which is either a crystal or electronic oscillator. Every digital watch has one and so does the clock in the digital piece of equipment. Digital devices, because they do operate in binary, like to divide and multiply by two. If the standard to a microprocessor is a crystal oscillator that generates 1 million pulses per second, this internal clock, above all, provides 1 million pulses per second and then provides pulses at 500,000 per second, 250,000 per second, 125,000 per second, and so on. This same clock could provide 2 million, 4 million, and 8 million pulses per second as well.

These time pulses, as stated above, are used in the functional operations of the digital equipment. From the standpoint of data communications and data linking, each pulse from a clock source can produce one bit of information: one million pulses per second—one megabit per second (Mb/s or Mbps); one thousand pulses per second—one kilobit per second (kb/s or kbps); one hundred pulses per second—100 bits per second (b/s or bps); and so on.

There are several key digital rates in data communications. They include:

Multiples of 75 bps from 75 to 19,200 bps. These rates are commonly found in the current analog common carrier telephony systems;

56 kbps which is currently the lowest all-digital common carrier service;

1.544 Mbps which represents the digital service most commonly found in U.S. satellites.

This list is by no means complete, but it represents a good portion of the most popular rates used today.

8.2.3 Synchronization

One strives to maintain perfect time. With our digital wristwatches, we can contain our lost or gained time to within a minute each year. We will find ourselves resetting our watches every now and then to make up for those minor differences. In essence we are manually synchronizing our timepiece with the world around us, and we hope that those references are accurate.

With data transmission, synchronization between two terminals is a way of life and is accomplished electronically rather than manually. Synchronization prevents one system from over or under pacing another and causing system overloading. If the two system clocks are not exactly the same (and generally they are not), one unit can push binary information to the other faster than the other can operate on it. The result is lost information. If the two terminals operate with exactly the same time reference, then the transfer of information is one-for-one. Either terminal can serve as the reference, or an external reference provided by something else, such as the network or transmission carrier, can be used.

8.2.4 Synchronization Signaling

Timing pulses or signals sent from one source to another for purposes of establishing and maintaining synchronization can take on different forms. The obvious method discussed above is to transmit pulses. However, the most popular method of transmitting clocking data is to send alternating binary signals (10101010....). The receiver then uses these alternating signals to synchronize or slave its own clock and digital processes to this signal sequence.

There are two important synchronization processes performed on binary data: bit synchronization and character synchronization.

Bit Synchronization

Many times a terminal device can obtain synchronization or "sync" through the received binary information rather than a separate clocking signal. If both terminal devices have an established rate of operation, say 300 bps, then the receiver can slave its clock to the transmitter by electronically sensing when the signal state passes from a 0 to a 1 or a 1 to a 0. This is shown in Figure 8.5. These crossover pulses can be visualized as pulling the clock pulses into step even though the crossovers are somewhat sparse. Remember, both clocks are running very close together at the 300 bps rate. The crossover simply pulls the slave clock into sync. This is called bit synchronization because it performs sync on the bits.

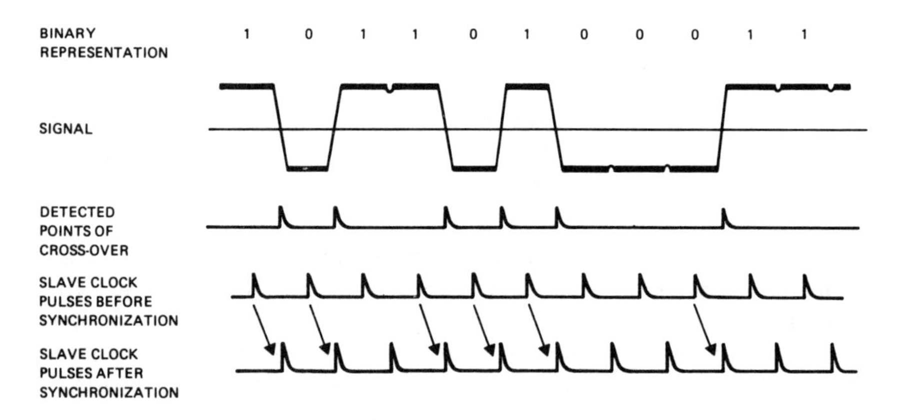

Figure 8.5: Representation of Bit Synchronization Process

Character Synchronization

Bit sync is necessary, but usually not sufficient, particularly if a long string of binary characters is transmitted which represents many characters like a sentence or paragraph of words. In these cases, we must also find the beginning of each binary coded character be it ASCII, EBCDIC, Baudot, or whatever. There are coded characters which are carefully designed to provide this character sync. In ASCII, it is the character SYN or 0010110. EBCDIC also has the SYN character, 0011 0010. There is also a character type sync sequence called a FLAG: 0111 1110. Much more will be said about these characters later. These coded sequences basically say, "Now that you have bit sync, here is the binary preamble you were expecting which establishes character blocking (sync)."

Think of binary synchronization as a micro process and character sync as a macro process. Without this type of synchronization, one must expect lost and garbled data.

8.2.5 Buffering Data

When two systems are communicating binary data and there is complete timing or clocking disparity between them (no forms of synchronization), buffering is necessary. A buffer is nothing more than binary storage with an electronic means of measuring when this storage is filled to capacity and can generate a signal to stop the binary filling process. Figure 8.6 shows a typical buffered process. Data is clocked in at 1 Mb/s until the memory module is filled. The electronic "fill" sensor sends a command to stop the data until the flow out at 56 kp/s empties memory. Another command is then sent to open the high volume flow again. Thus, the high volume input is turned on and off to accommodate the capacity of the memory system, while the data out is continuous but slow. Larger memory means that the input flow can stay on longer. In fact, if this memory board is very large, an entire file of data can be placed in memory before the slower output is opened.

A personal computer is a perfect example of this buffer application. The data bus structure can operate, for example, at 1 Mb/s. Yet data is moved to a printer, disk, or communication port at rates much less than the bus rate. One can purchase expanded memory boards called RAM (Random Access Memory) to accept entire files of printer, disk, or communication data (64k bytes and up). Now the computer CPU can be occupied with other matters while these buffer units move data at the lower, mis-

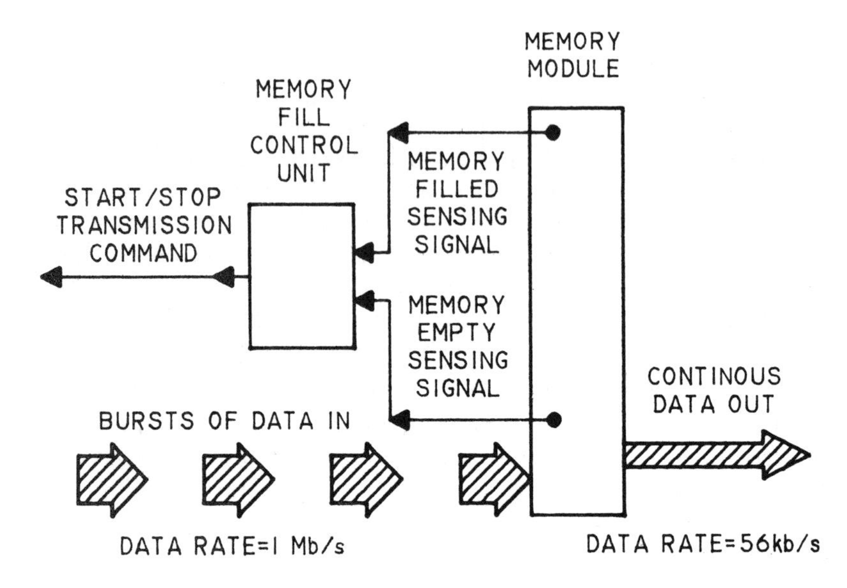

Figure 8.6: Buffering System which Compensates for Dissimilar Data Rate Flows

matched rates. The buffer is an effective means of isolating dissimilar clocking functions which arise in data transmission. This process is sometimes called "spooling."

8.3 PROTOCOLS

Data communication protocol is a formal set of conventions governing the relative timing and formatting of information exchanged between DTEs. Information cannot be sent to a terminal device over any kind of data link until the receiving device is prepared and ready. Protocols provide for this requirement and, needless to say, there are many kinds of them which vary from simple to complicated. One needs to have a variety of protocols available to match the true terminal and transmission abilities and requirements. Notice the transportation protocols that bikers observe versus those observed by semi-trucks. Small terminals like a TTY can and do operate under one set of protocols while major host processors operate under an entirely different set. What is good or efficient for one is not for the other.

There is a very structured architecture for protocols which will be defined later in this section. For the time being, however, the term protocol will be used in a very broad and less formal sense.

8.3.1 Transmission Link Modes

Data can be passed between two terminals in three modes: simplex, half duplex (HDX), and full duplex (FDX).

Some data terminals need only to receive data or transmit data. The airline departure and arrival display screens are examples of receive only terminals. A remote electronic rain gauge collecting and measuring precipitation and transmitting this data to a digital collection device is an example of a transmit only device. The data link only needs to accommodate data flow in one direction. This is called a simplex transmission mode.

A half duplex data transmission channel means that only one transmission can occur in one direction or the other at one time. Data can flow in or out but not both at the same time. As long as one terminal is transmitting data on a link, it is incapable of receiving anything until it ceases. Then the terminal must listen while the remote system sends data. Half duplex means that one can either send or receive data at one time but not both. A "push-to-talk" radio like a Citizen Band (CB) rig is an example of a half duplex transmission.

A data terminal device operating in a half duplex mode will receive data until the transmission is complete. There will be a slight delay before this same terminal can transmit information back on the same half duplex link to the remote station. In a sense, the data link experiences a momentary shock as the transmission direction is reversed and thus requires a period of time to settle down. This slight delay or halt is called "turnaround time" and typically requires 50 to 250 milliseconds (ms) on a commercial phone line.

Full duplex transmission means that a terminal can both receive and transmit information simultaneously. Thus, there is no data link turnaround time experienced. A human being has no problem operating in a full duplex mode listening and talking at the same time.

8.3.2 Data Transmission Modes

The coded binary data can transmit from one terminal device to another in one of two ways: asynchronous or synchronous.

Asynchronous Data Transmission

Recall that one can code binary data into Baudot, ASCII, or EBCDIC. In the asynchronous or "async" data transmission mode, each coded character or command is sent as a burst of binary data. There is no communication or link activity until a key on a keyboard at the terminal is depressed. These coded pulses, therefore, appear on the data link with each key stroke of the operator; otherwise the channel is quiet. This mode of operation is also called the START-STOP mode of transmission. The binary coded material typically varies in speeds from 75 bps (for Telex-type terminals) to 9600 bps.

There is a formatting phenomenon which occurs in the async mode that must be noted. The binary coded message receives a "start bit" and a "stop bit" or bits. Figure 8.7 illustrates this process using the ASCII code for the letter E. First note that a binary "1" is also called a MARK and that a binary "0" is also called a SPACE. The quiet data link in this case is in the MARK state or position. The second important point is to acknowledge the SPACE "start bit" which is then followed by the ASCII code for E. Third, note that an even parity bit is present which is appropriate for the ASCII code and, finally, the addition of the MARK "stop bit" followed by the quiet channel in the MARK position—ready for the next key stroke.

There are three stop bit conventions: the use of one stop bit, the use of one and one-half stop bits, or the use of two stop bits. This protocol convention, along with an agreement of the code (ASCII in this case), must be established before async communication between two data terminals can occur.

The start bit is used to establish bit synchronization. The receiving terminal accepts the start bit, strips it off, meters out seven bits of information, checks parity with the received parity bit, and is finally conditioned by the stop bit for the next character. The start-stop synchronization process enables characters to be transmitted at random since each character carries with it necessary synchronization information.

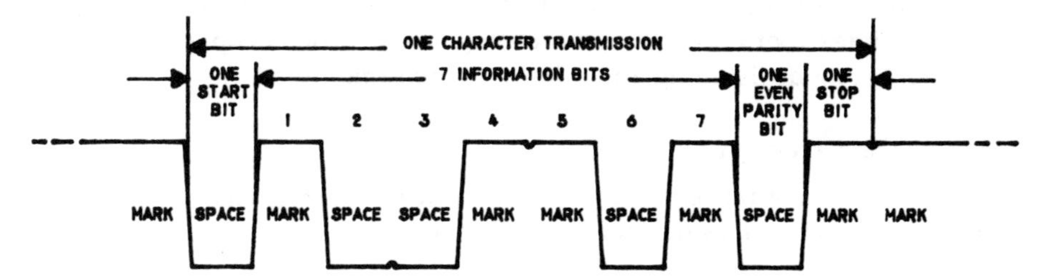

Figure 8.7: Bit Structure for the ASCII "Y" Asynchronous Transmission

Synchronous Data Transmission

Synchronous data transmission is a continuous stream of binary data with no intervals between the characters or commands. Once the transmission occurs, the transmitter and receiver must maintain synchronization for the duration of the lengthy message. Therefore, the receiver must be able to first establish bit synchronization and then character synchronization. Buffering obviously plays a key role in such a mode of operation. Figure 8.8 shows a graphic comparison of the async and sync processes.

8.3.3 Line Protocols

The asynchronous mode of data transmission was the first method for passing information. It is simple and relatively unencumbered with special formats and timing functions and, therefore, survives as a useful means of moving binary data. Yet, async transmission is inefficient in many respects because there is so much "dead time" between the coded transmissions, and there is up to 38 percent overhead: 5 bits of Baudot code, 1 start bit and 2 stop bits, or ⅝th information and ⅜th overhead.

The real break in complexity and the true development of data communication protocol begins with the synchronous mode of transmission. Though synchronous transmission is potentially more efficient, it also has the potential of becoming complicated. We will deal with three basic synchronous protocols: Binary Synchronous Communications (BSC or bisync), the High Level Data Link Control (HDLC), and the Synchronous Data Link Control (SDLC).

Binary Synchronous Communications (BSC)

The 1960's proved to be the dawn of data communications awareness to move data more efficiently and at far greater rates than the traditional async TTY process. IBM introduced the character oriented BSC protocol in 1966[5] which settled a developing chaos in high speed transmission where everyone was moving in many different directions. It soon provided an industry-wide de facto protocol, and BSC is still very popular and widely used today. There are probably 30 to 40 flavors of BSC, though we will discuss the more generic brand.

The BSC protocol uses a "handshaking" architecture between the DTEs. A simplified version of handshaking is shown on the left side of Figure 8.9, and the more formal data communication version is shown on the right side. There are three basic phases to this architecture: the establishment phase, the data transfer phase, and the termination phase.

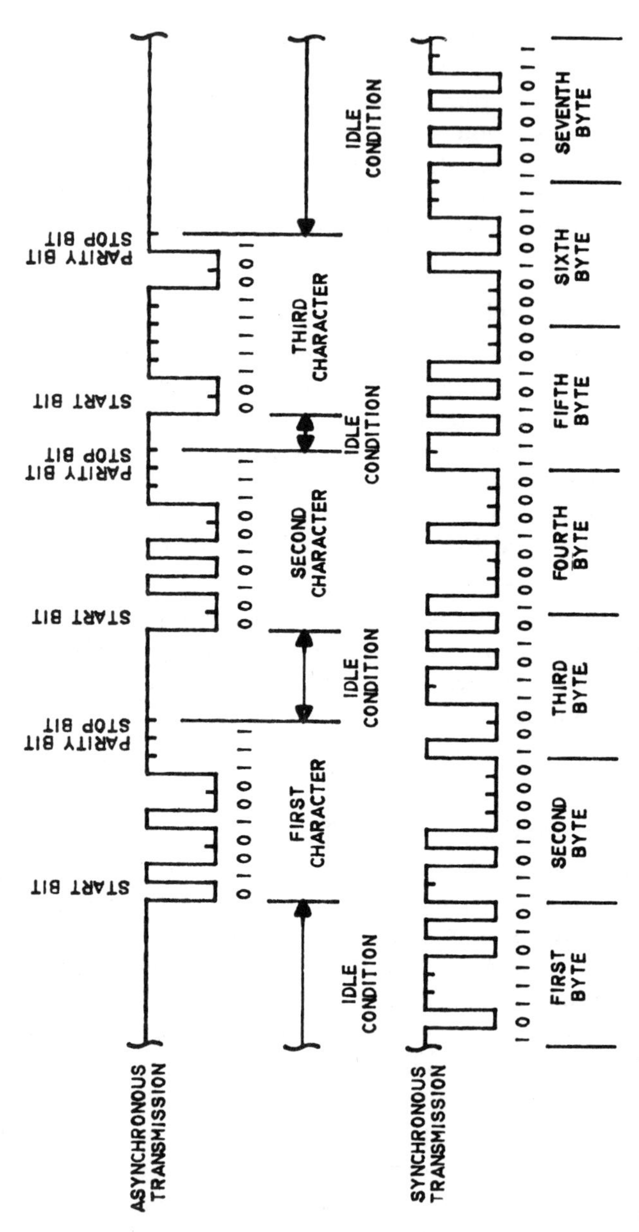

Figure 8.8: The Comparison of Asynchronous and Synchronous Transmission

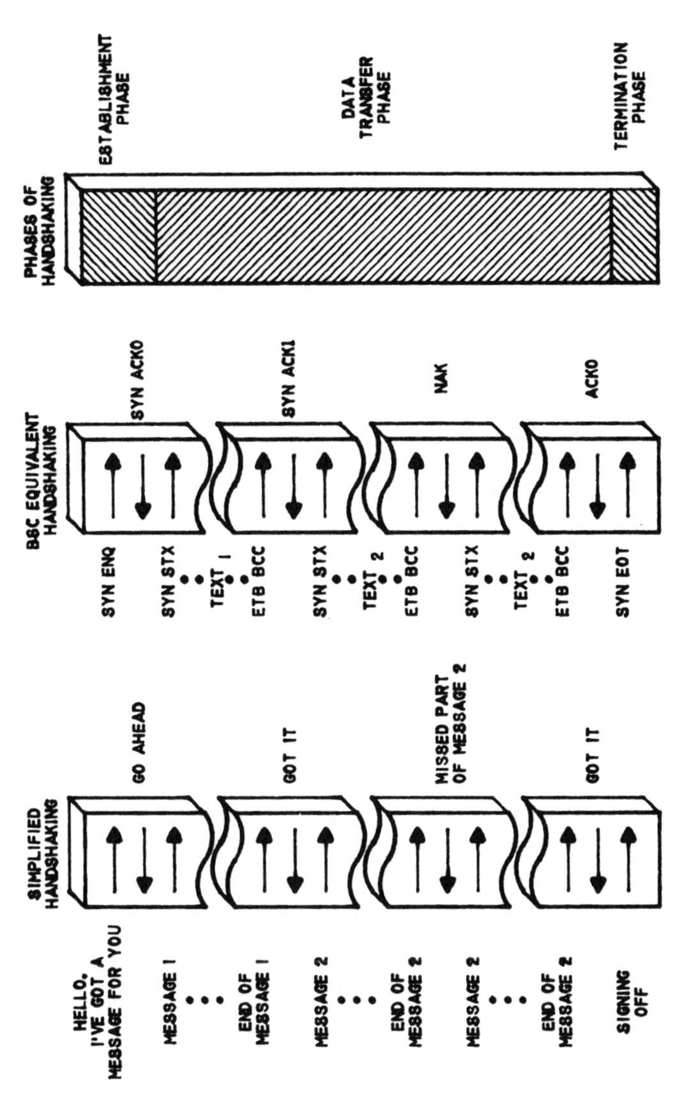

Figure 8.9: BSC Handshaking Process and Phases

In the establishment phase, the sending terminal generates the necessary "SYN" preamble so the receiver can establish both bit and character sync. In fact, this terminal can send, without restriction, as many SYN characters as necessary to keep the data channel alive or open until data is ready to send. Recall that SYN is one of the Communication Control mnemonics of the ASCII and EBCDIC codes. Typically one will see the SYN followed by ENQ in this phase which simply states, "Here is the necessary sync (bit and character) preamble (SYN), and I have a message; are you listening (ENQ)?" The receiver sends an acknowledgment "ACK" but not before it sends its own SYN. Remember, every time a transmission is started by either terminal, it must send the appropriate SYN preamble.

Now we enter the data transfer phase. Notice, there is a data communication shorthand used throughout. The BSC shorthand control characters are summarized in Figure 8.10. In this phase we send the required

CHARACTER	MEANING	TRANSLATION
ACK	AFFIRMATIVE ACKNOWLEDGEMENT	PREVIOUS TRANSMISSION BLOCK WAS ACCEPTED CORRECTLY, AND NOW READY FOR NEXT BLOCK. ALSO READY TO RECEIVE AFTER INITIALIZATION SEQUENCE.
DLE	DATA LINK ESCAPE	CONTROL CHARACTER TO IDENTIFY A TRANSPARENT MODE OF OPERATION.
ENQ	ENQUIRY	BID FOR A LINE CONNECTION.
EOT	END OF TRANSMISSION	TRANSMISSION IS CONCLUDED, RESET TO CONTROL MODE. ALSO AN ABORT SIGNAL WHERE THERE IS A MALFUNCTION.
ETB	END OF TRANSMISSION BLOCK	IDENTIFIES THE END OF A TRANSMITTED BLOCK WHICH STARTED WITH A STX OR SOH. FOLLOWED BY AN ERROR CHECK BLOCK.
ETX	END OF TEXT	IDENTIFIES THE END OF A BLOCK WHICH STARTED WITH A SOH OR STX. ALSO MARKS THE END OF SEQUENCE FOR ERROR DETECTION. FOLLOWED BY AN ERROR DETECTION BLOCK.
ITB	INTERMEDIATE TEXT BLOCK	CONTROL CHARACTERS USED TO TERMINATE AN INTERMEDIATE BLOCK OF CHARACTERS. FOLLOWED BY AN ERROR CHECK BLOCK.
NAK	NEGATIVE ACKNOWLEDGEMENT	PREVIOUS TRANSMISSION BLOCK IS UNACCEPTABLE AND RETRANSMISSION IS REQUIRED. ALSO A LINE BID NEGATIVE RESPONSE.
PAD		ADDED TO A TRANSMISSION TO PROVIDE ELECTRONIC IDLING.
SOH	START OF HEADER	IDENTIFIES THAT A HEADER FOLLOWS. HEADERS DATA CONTAINS ROUTING AND PRIORITY INFORMATION.
STX	START OF TEXT	IDENTIFIES THAT THE BODY OF TEXT FOLLOWS.
SYN	SYNCHRONIZATION CHARACTER	USED TO ESTABLISH AND MAINTAIN BIT AND CHARACTER SYNCHRONIZATION. ALSO USED AS A TIME FILLER TO KEEP THE LINE OPEN.

Figure 8.10: Simplified List of BSC Control Characters

SYN followed by STX (start of text). Next the sender pipelines a pure stream of data to the receiver, and finally, the sender signs off by transmitting an ETB (end of transmission block). When this is finished, there is an automatic BCC (block check character) transmitted which, like the parity bit, contains a mathematically derived character process to be used to check for errors in the newly transmitted data.

When the BCC is translated by the receiver and there appear to be no errors, the receiver sends a SYN ACK back—"received your data in good shape." If there is an apparent error as indicated by the BCC check, then the receiver sends a negative acknowledgment, NAK, back to the sender which asks for a retransmission. In this case, the receiver will repeat the transmission back.

The termination phase finds the sender with no other data to send and the receiver sending back a SYN ACK. Therefore, the sender closes by sending a SYN and EOT (end of transmission).

There is an embellishment on the ACK worth noting—the ACK0 and ACK1. By alternating between ACK0 and ACK1, it is possible to determine if an error was made in the data sent or in the SYN ACK sent back. Here is how it works. Figure 8.11 shows the normal sequence without errors. Notice the alternating ACK0 and ACK1. The sending terminal sends a message (text 2) and receives ACK0. Now it sends another message (text 3) and receives nothing; the sender sends an ENQ and waits. If the sender receives an ACK0, the message must have been lost. If the sender terminal receives an ACK1, then only the acknowledgment was lost. Think about it! It is useful to have this knowledge because sending an entire repeated file (text 3) is quite a bit more time consuming when only a mere SYN ACK1 was lost. This 1 and 0 embellishment can save a great deal of electronic energy.

Some data links have party lines; that is, there are several terminals connected to one host computer. Under these conditions, the host computer "polls" the remote terminals by routing addresses found in header information. Therefore, the SOH (start of header) command is placed in the front of the transmission sequence followed by STX and the text.

The PAD is sometimes inserted to add idle or delays in the transmission which provide time for the end terminal to prepare itself for the onslaught of data. The PAD is also used to provide a delay for line turnaround time in a half duplex mode of operation in which the reversing activity shocks the transmission media and settling time is required.

Very long messages are segmented into manageable blocks and transmitted as shown in Figure 8.12. Notice the SOH and header start up fol-

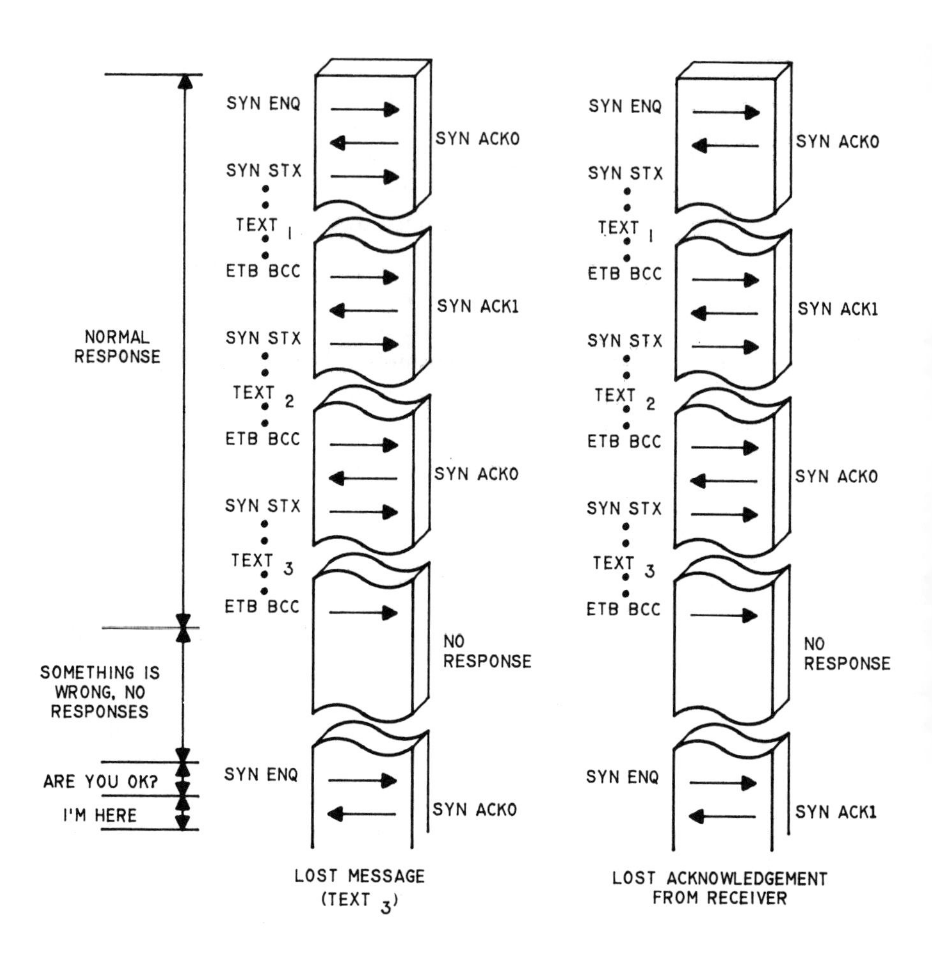

Figure 8.11: Example of a Problem in the Transmission of the Third Block of Text. Demonstration of the Use of Alternating ACK0/ACK1

lowed by an STX. Segmenting is accomplished by terminating with ITB (intermediate text block) and a BCC and a new STX, text, and an ITB/BCC; STX, text, and an ITB/BCC; and so on. An acknowledgment is not required until the end of the very long message is completed as noted by an ETX. Remember that the BCC is a block of error checking data, not a command.

SYN	SYN	SOH	HEADER	STX	TEXT ...	ITB	BCC	STX	TEXT ...	ITB	BCC	STX	TEXT ...	ETX	BCC

Figure 8.12: An Example of Long Message Transmissions in BSC Protocol

There is a problem in BSC when the transmitting terminal wants to send raw uncoded binary information (like a memory or register dump). The problem arises when a binary combination of real data appears to the receiver to be a communication command like ETX (0000 0011 in EBCDIC). One really wants the receiver to accept the binary data in the raw form, and not interpret it as a command. There is a provision for this called "transparency." The transparent mode of transmitting data is initiated by a DLE (data link escape) character. After this command, the receiver ignores all character recognition except another DLE character. When the receiver sees a DLE preamble in the raw data, it looks at the next full character which follows. If it is another DLE (DLE DLE) then it knows that the second DLE was in fact pure data and to throw away the first DLE binary sequence. If the DLE is followed by an ENQ, ETB, ETX, ITB, STX, or SYN, then the receiver reacts according to the data in 8.13, such as ending the transparent mode (DLE ETX).

Notice that once inside the transparent mode, there is still sufficient flexibility to handle large amounts of pure data just as there was in the non-transparent mode: DLE STX, text, DLE ITB, BCC; DLE STX, text, DLE ITB; and so on.

The BSC protocol, is referred to as character-oriented protocol because communication commands are part of the character coded messages in ASCII and EBCDIC. The receiver reacts to those character commands. BSC was originally designed for a half duplex channel type operation although full duplex channels accommodate it quite well.

The BSC protocol, as stated in the begining of this section, is still very much around although it is now being replaced by bit-oriented protocols like HDLC. The issue of information or text transparency always posed a difficult problem to BSC, a handshake half duplex type protocol.

High Level Data Link Control (HDLC)

HDLC[6] and HDLC-type protocols, such as IBM's Synchronous Data Link Control (SDLC),[7] approach transparency from an entirely different perspective. Information is transmitted in frames as shown in Figure 8.14. Each frame (sometimes called a block) begins and ends with a FLAG (0111 1110). The front flag is used to identify the start of the frame and establish bit synchronization. The closing flag is used to mark the end of the frame. Everything inside this frame is in multiples of 8. Unless otherwise specified the FLAG, ADDRESS, and CONTROL are 8 bits each. The FRAME CHECK SEQUENCE is 16 bits.

CHARACTERS	MEANINGS
DLE ENQ	RETURNS THE DATA LINK TO THE NORMAL MODE AND TELLS THE RECEIVER TO IGNORE THE BLOCK OF TEXT BEING TRANSMITTED.
DLE ETB	THIS TERMINATES THE TRANSPARENT BLOCK TRANSMISSION AND RETURNS THE DATA LINK TO THE NORMAL MODE. IT ALSO ASKS FOR AN ACKNOWLEDGEMENT.
DLE ETX	THIS TERMINATES THE TRANSPARENT TEXT TRANSMISSION AND RETURNS THE LINK TO THE NORMAL MODE. A REPLY IS ALSO ASKED.
DLE ITB	THIS TERMINATES THE INTERMEDIATE TRANSPARENT BLOCK AND RETURNS THE LINK TO THE NORMAL MODE, BUT DOES NOT REQUIRE AN ACKNOWLEDGEMENT. MUST BE FOLLOWED BY A BCC.
DLE STX	THIS STARTS A TRANSPARENT MODE FOR TEXT TRANSFER.
DLE SYN	THIS IS USED TO MAINTAIN SYNCHRONIZATION AND TIME IDLE CONDITIONS WHILE IN THE TRANSPARENT MODE.

Figure 8.13: The BSC Simplified Transparent Command Functions

The ADDRESS provides access from a host to multiple terminals on the data link. The CONTROL field provides the receiver with control type data which will be disccussed later. The INFORMATION field is entirely transparent and will accommodate any type of binary data. When the end FLAG is received, the terminal counts back 8 bits (through the FLAG) and 16 bits (through the FRAME CHECK SEQUENCE) and declares, "This is the end of the INFORMATION frame." The receiver then applies the 16-bit error detection code to the INFORMATION, CONTROL, and ADDRESS fields. It is simple indeed.

There is a provision called "zero bit insertion" which accommodates the unique case where a FLAG sequence of 0111 1110 appears in the raw INFORMATION field and should not be interpreted as an end FLAG but as data. In this situation, a 0 is inserted after the fifth 1 which is detected

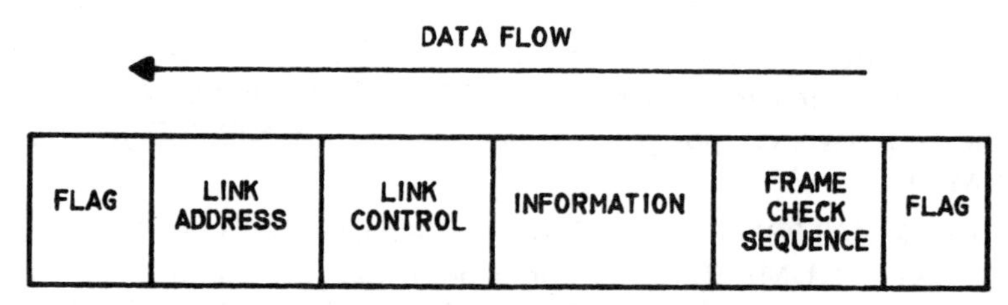

Figure 8.14: A Bit-Oriented (HDLC Type) Protocol

by the receiver and is promptly disregarded which restores the data sequence of 0111 1110. In other words, the receiver is designed to look for two bit patterns:

0111 1110 - legitimate FLAG, and

0111 1101 - potential zero bit insert.

If the system sees 0111 1101, it then looks at the next bit:

0111 1101 0 - toss out the second 0, or

0111 1101 1 - do nothing, pure data.

The CONTROL field requires further expansion as shown in Figure 8.15. There are three possible CONTROL fields as specified by a 0 in the

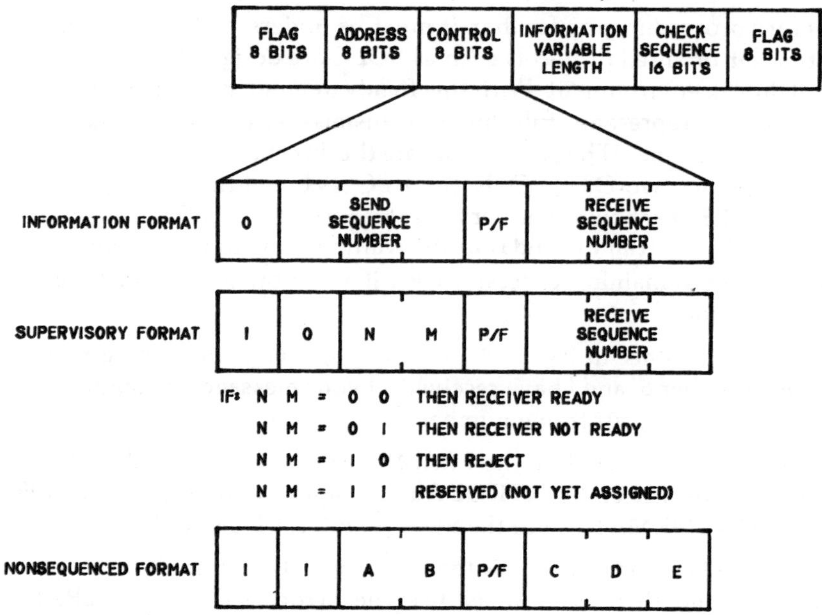

Figure 8.15: The HDLC Expanded Control Field

first bit position which defines an INFORMATION FORMAT, a 10 in the first two positions which defines a SUPERVISORY FORMAT, or an 11 in the first two bit positions which specifies a NONSEQUENCED FORMAT.

An INFORMATION FORMAT control field provides 3 bits for a SEND SEQUENCE NUMBER, 1 bit for a poll/final (P/F), and 3 bits for a RECEIVE SEQUENCE NUMBER. The SEND and RECEIVE SEQUENCE NUMBERS are used for error correction purposes. First, note that there are only 3 bits which can represent 8 numbers or 0 through 7. Therefore, each HDLC (or SDLC) frame is numbered in the SEND SEQUENCE slot as it is transmitted: frame 0, 1, 2, 3, 4, 5, 6, 7, 0, 1, 2, 3, 4, 5, 6, 7, 0, 1, 2 .... In other words, the transmitting terminal states that "I am sending you frame number 0, now frame number 1, now frame number 2, etc." This is all accomplished in the INFORMATION FORMAT, SEND SEQUENCE NUMBER mode of operation.

The RECEIVE SEQUENCE NUMBER is handled basically the same way. It states that "I have received your frame number 0, now frame number 1, now frame number 2, etc." Therefore, a data terminal can now say, "I am sending you my frame number 6, and I have received your frame number 1."

Before proceeding any further, it must be understood that HDLC and other bit-oriented protocols such as SDLC are designed to operate in a full duplex environment. With this firmly in mind, refer to Figure 8.16. The figure represents full duplex transmission of HDLC and SDLC numbered frames. The arrows indicate the direction of the transmission between the two DTEs. Only the RECEIVE and SEND SEQUENCE NUMBERS are shown. The very first frame states, "I, DTE_1, am sending you, DTE_2, my frame 0, and I am still waiting for your frame number 2." Follow the remaining process to see if you now get the drift of the transmissions.

The last sequence in Figure 8.16 states, "I, DTE_2, am sending you my frame number 6, and I have received all your messages through 4 and I am waiting for your frame number 5."

If an error is detected by the FRAME CHECK SEQUENCE code, the HDLC transmission CONTROL frame changes state from the INFORMATION FORMAT (IF) to the SUPERVISORY FORMAT (SF).

Figure 8.17 demonstrates how the two terminals communicate the error problem and then solve it. Notice the error was made in FRAME NUMBER 2 from DTE_1 to DTE_2. DTE_2 rejected this frame as having an

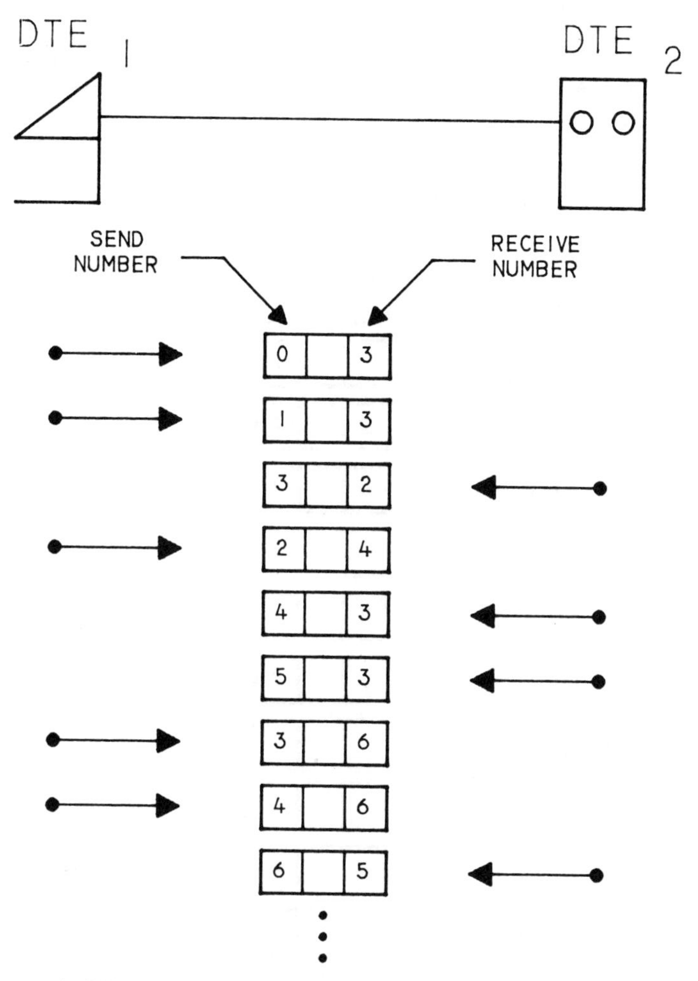

Figure 8.16: An HDLC Full Duplex Frame Transmission Flow with No Errors

error in it. The DTE_1 backed all the way down to FRAME NUMBER 2 and retransmitted it, plus it retransmitted frames 3 and 4. The final frame of this figure states that DTE_1 is sending frame 5 and waiting for DTE_2 frame number 6.

The NONSEQUENCED FORMAT of the CONTROL field will not be defined in detail here other than to say that this format is used to start or initiate a transmission between the end terminals.

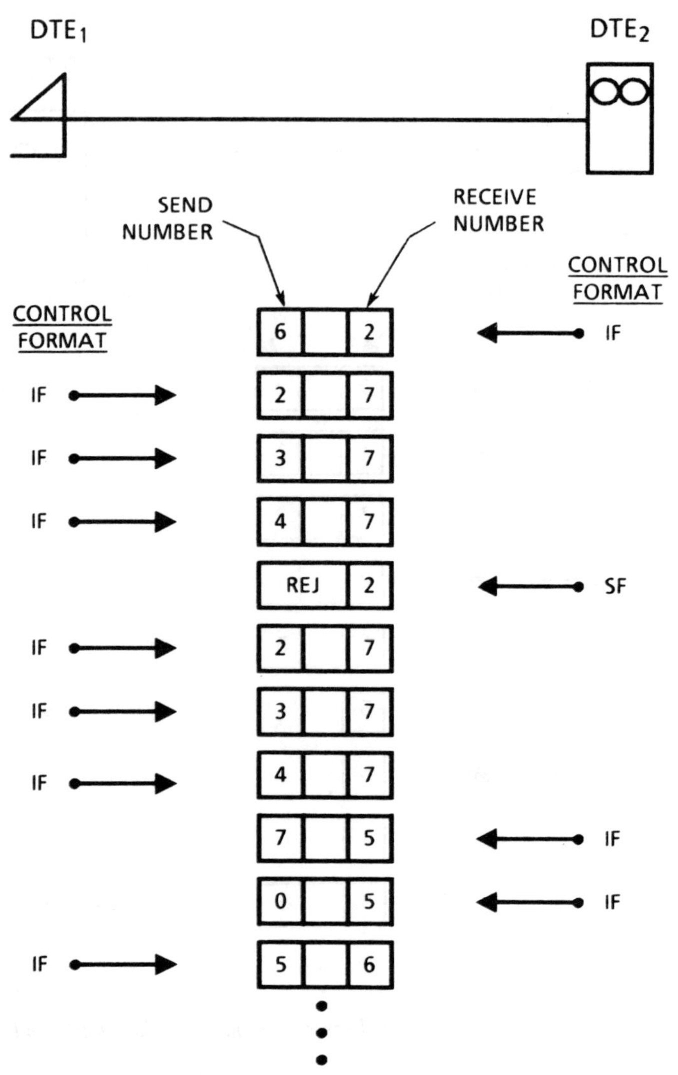

Figure 8.17: An HDLC Full Duplex Frame Transmission Flow with One Error in Transmission From DTE_1, to DTE_2

The HDLC protocol requires a primary station (like a host computer) and a secondary station (like a smart terminal). The POLE/FINAL or P/F position in the CONTROL field uses the primary/secondary relationship in the following fashion. When the primary station polls, P, a secondary, it basically requests a response from the secondary. The final, F, is sent by the secondary station with the final frame transmitted as a response to the poll command. If a final frame is not accepted by the primary or secondary, the primary initiates a time out. When there is no response after the time out, it again polls the secondary. In essence, the P/F bits are a very useful and simple means of communicating crucial frame status.

Other Line Protocols

The HDLC protocol is one of several bit-oriented protocols. Actually, IBM once again released the Synchronous Data Link Control (SDLC) bit-oriented protocol several years before HDLC, yet SDLC, for all practical purposes, is very close to HDLC in format. Another bit-oriented protocol is the Advanced Data Communication Control Procedure (ADCCP).

8.4 COMPUTER NETWORK ARCHITECTURES

It is very important to place the BSC and more specifically HDLC and SDLC type protocols in their proper perspective. Figure 8.18 shows how these protocols suit a single host computer with its communication control unit to a set of secondary type terminal devices. The communication controller is, in essence, a computer devoted entirely to terminal and communication housekeeping leaving the host computer available for heavier computational work such as managing data bases. The communication controller is extremely key in data communications and computer network architecture and requires proper attention.

8.4.1 The Role of the Communication Controller

The communication controller, like IBM's 3705 and 3725,[8] off-loads the communication housekeeping functions from the host. It is a mediator between the host and the communication lines. Thus it accepts and sends data to and from the host. It accepts and sends data to and from the data links or communication lines.

The controller establishes communications with the devices under its control by polling and addressing them[9] as well as dials and answers the communication lines to access the terminals.

The controller provides buffer services to facilitate data transmissions. It assembles incoming messages into units or blocks. It disassembles messages. It provides for retransmission if errors are detected. It generates and accepts the communication control characters and responds accordingly. The controller controls data traffic and flow by metering data accepted or transmitted to the terminals. And the controller keeps link performance records and performs link diagnostics. The communication controller in today's environment is foundational and extremely important.

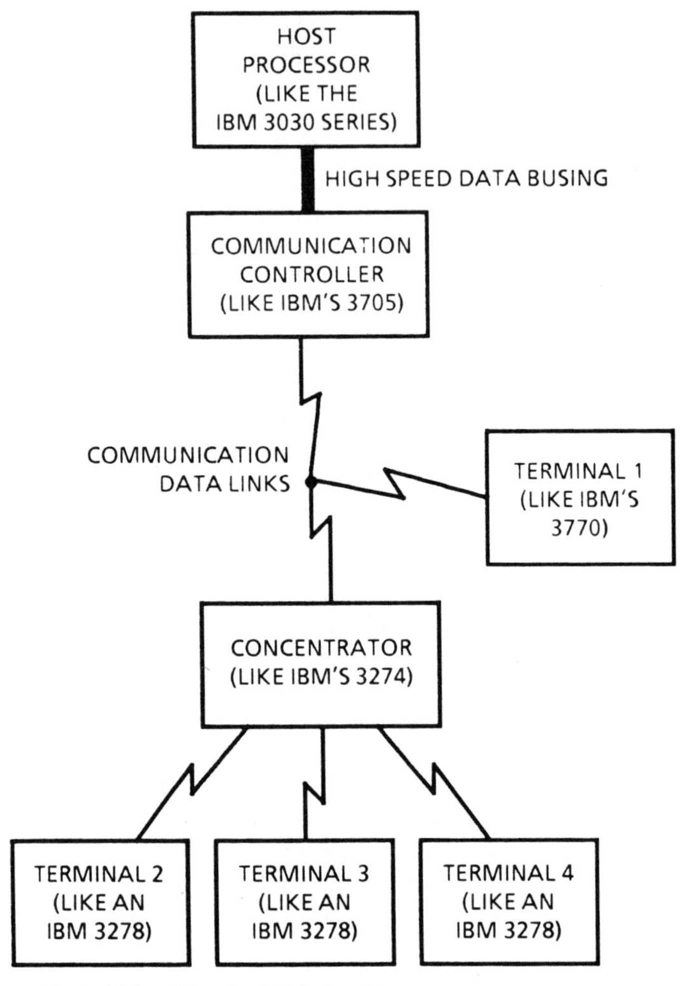

Figure 8.18: Typical Host/Terminal Relationship

8.4.2 How HDLC, SDLC, and BSC Fit into the System

The HDLC, SDLC, and to a far less extent BSC protocols are referred to as "data link control." Refer back to Figure 8.18. These protocols provide basic logical connections over the physical communication links shown in the figure. These data links are obviously between the communication controller and the family of terminal devices assigned and attached to it. These data link control protocols apply directly to the connections. Thus, the address and control functions embedded in the HDLC and SDLC type protocols access and control the appropriate terminals. In the past, the relationship has been the relatively dumb terminal interacting with the very smart host with its communication controller.

But today, smart computers are being reduced substantially in size while actually increasing in capability. Figure 8.19 shows a more typical modern-day configuration of hosts and terminals which present a dilemma that is solved only through specific computer network architectures.

Figure 8.19(a) begins to expose the limitations of the HDLC and SDLC data link control type protocols. First, terminals T_1 to T_4 only have the embedded protocol ability to communicate with the communication controller CC_1. The same may be said for terminals T_5 to T_8 into controller CC_2. Thus, there is no protocol convention yet established tc allow terminal T_1 to communicate directly to, for example, terminal T_8 especially if T_1 and T_8 have some intelligence. Today, this situation approaches the more realistic diagram shown in Figure 8.19(b). Under an expanded data link control protocol, one can move data to and from the domain of any one of the controllers. For example, an expanded protocol will accommodate controller CC_3 to communicate with terminals T_1, T_2, and T_3 and across to controller CC_2.

Now look at the situation where terminal T_1 needs to access an application program (AP) in HOST 1 which requires leaving DOMAIN 3, passing through DOMAIN 2, and accessing an AP in DOMAIN 1. Obviously, this must be accommodated.

In 1974, IBM released an architectural approach to address this very issue. They refer to it as their System Network Architecture (SNA).[10,11,12] SNA is still undergoing vernier tuning, but remains the very backbone of their entire product line. Others have followed IBM's lead in forming network architectures. However, SNA provides an excellent way of demonstrating the approach to networking.

8.4.3 System Network Architecture (SNA)

SNA[13,14] like other architectures is a set of specifications for the design and implementation of interactions among IBM communication products. SNA specified what class of product is responsible for specific functions, and how they interact with one another. SNA is also used to describe a set of products. Thus, users refer to "implementing SNA" into their system and organization. They have adopted SNA as their computer network architecture and will abide by those rules and regulations governing SNA processes.

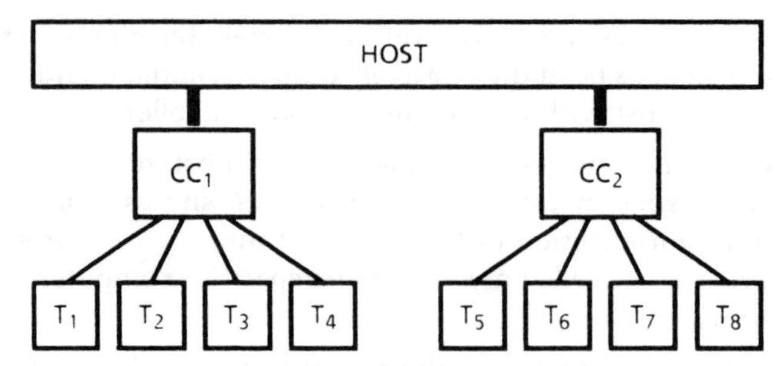

(a) EXPANDING COMPUTER CAPABILITY BY USING TWO COMMUNICATION CONTROLLERS INTO ONE HOST.

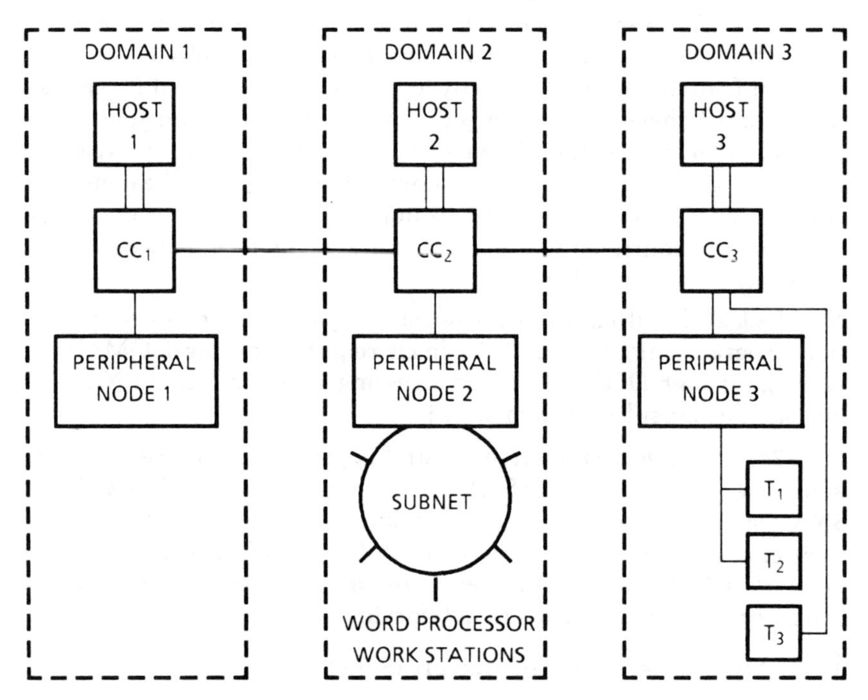

(b) TYPICAL MODERN DAY ASSEMBLAGE OF INTERCONNECTED COMPUTATIONAL COMPONENTS AND SYSTEMS

Figure 8.19: Hardware Configurations both Past and Present

The purpose of SNA and other network architectures is to provide services to users. In other words any terminal in Figure 8.19(b), if authorized and capable, can seek the services of any element of the entire system to use an AP (like an Information Management System), to pass electronic mail, or to generate graphics using the software resident in one particular host system.

Basic Components Of SNA

Consider a view of SNA or any of the network architectures from the standpoint shown in Figure 8.20. The key element is the "end user" which is relatively easy to conceptualize if the end user is you or your terminal—a good and accurate perception. However, you as an end user are probably connected to a computer running an AP. Thus the AP is also considered an end user. Now you can visualize you (end $user_1$) connected to an AP (end $user_2$).

A "logical unit" (LU) is the port into the SNA network for the end user. A "node" is a box or circle which contains network components. Nodes are connected to each other through "links." A communication controller can be classed as a node. A "physical unit" (PU) is the node resource manager.

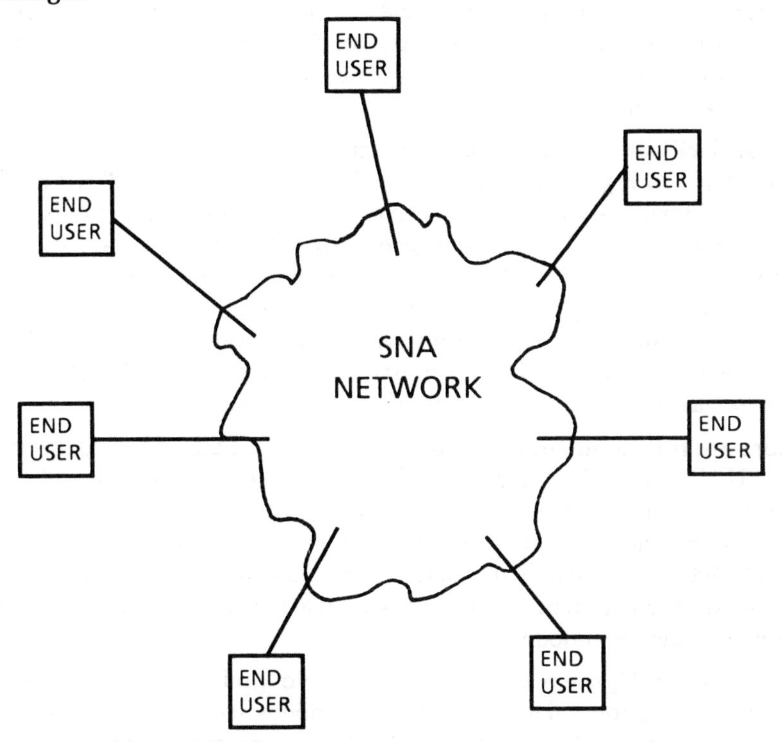

Figure 8.20: The End User and the SNA Network

There is a "System Services Control Point" (SSCP) which manages the entire SNA domain composed of several nodes of an SNA network. Refer back to Figure 8.19(b) and notice that there are three domains; therefore, there is going to be one SSCP in each domain of Figure 8.19(b). Think of the SSCP as the domain mastermind for providing the connecting services between end users.

In SNA, there are addresses called "Network Addressable Units" (NAUs). An NAU is an LU, a PU, or an SSCP. In other words, all LUs, PUs, and SSCPs represent an NAU. Each NAU has both a network name and a network address.

A "session" is a logical connection between NAUs. You as an end user connect yourself to an AP in a session. A NAU can establish a session with one or more other NAUs.

Finally, a "path control network" is the set of SNA network components responsible for routing data between NAUs. Refer back to Figure 8.19(b). Consider yourself as terminal T_1 working an AP in HOST 1. Everything between you and your AP is part of the path control network; your LU, your PU, your NAU, all the communication links, the communication controllers CC_1, CC_2, and CC_3, the AP's LU, the AP's PU, and the AP's NAU.

Visual clarification can be found in Figure 8.21. This figure represents DOMAIN 3 of Figure 8.19(b) which includes HOST 3, CC_3, PERIPHERAL 3, and terminals T_1, T_2, and T_3. Notice that HOST 3 is a stand alone "node" requiring a PU. Most host processors will have several APs; this one shows two. The DOMAIN SSCP is resident in the HOST 3 processor. Notice that CC_3 is a node and therefore requires a PU. But PERIPHERAL 3 which is also a node with a PU, actually supports terminals T_1 and T_2 each requiring an LU. Terminal T_3 is a stand alone type device and is therefore a node with a PU and of course an LU.

There are four basic sessions allowed: LU - LU, SSCP - LU, SSCP - PU, and PU - PU. The LU - LU session should be obvious: you as an end user operating a terminal connected to an AP. The SSCP - LU establishes LU - LU sessions and then monitors the session progress. The SSCP - PU is used to activate or deactivate various network components. The PU - PU session is used to download programs: activating, deactivating, and testing routes between nodes.

Also notice in Figure 8.21 that there are two kinds of nodes: "subarea nodes" and "peripheral nodes."[14] Subarea nodes are capable of receiving messages from any NAU in the total network and moving them toward any other NAU in the total network. Note that HOST 3 has an NAU and

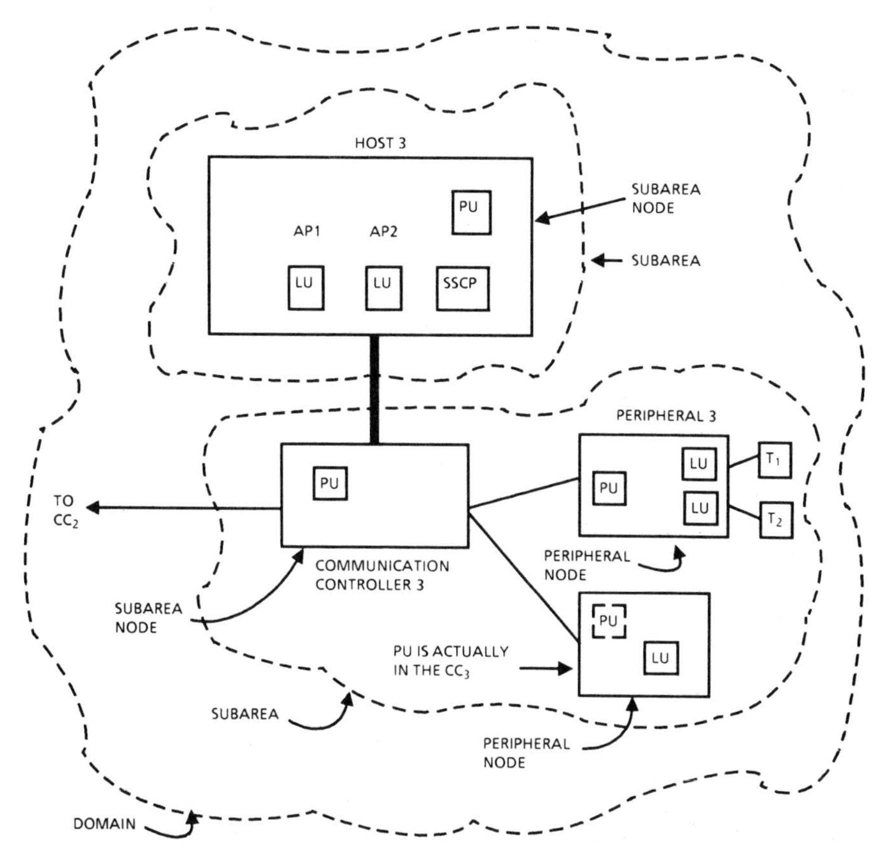

Figure 8.21: Expansion of the SNA Components

so does CC_3. The HOST 3 subarea node can therefore move information to the CC_3 subarea node because both have NAUs with network names and network addresses. A peripheral node communicates only with the subarea node to which it is attached. Refer to Figure 8.21.

The Effects of the SNA Architecture and Components on Data Communications

Data communications will clearly provide interconnections between the subarea nodes and between the domains of the SNA architecture. The two major elements of concern to the data communication planners are embedded in the NAUs and the path control network of SNA or other similar architectural components.

The NAU address contains two components: subarea address (like a telephone area code) and element address (a phone number within the area code). The NAUs use SSCP services to translate from a network name to network address each time a session is established. The network names are permanently assigned by the SNA administrator and never need to change. This is analogous to the telephone information operator who tells you the phone number if you provide the name of the person in that area code.

The SNA path control network is the common transmission network connecting the NAUs. The NAUs are naturally outside the SNA path control network. This network uses the network addresses to route message units between NAUs. Thus the SNA path control network is responsible for flow control (no blockage or overflow allowed), routing between nodes, error control, and translation of network addresses to local addresses for peripheral nodes. Figure 8.22 shows the interrelationships between the NAUs, the nodes, and the SNA path control progressing from the simple to the complex.

The top of this figure, part(a), shows end user to end user through SNA; part(b) shows the end users with their respective LUs which are connected to the path control network. Thus the end users with their LUs each represent a NAU as shown in part(c) of Figure 8.22.

An Emerging Model

The NAUs of this architecture form a layered hierarchy relationship to the SNA path control network. The originating NAU passes information down to path control which then moves this material across the node boundaries and then back up from the path control to the NAU as shown in Figure 8.22. There must be peer-to-peer protocols between these two NAUs so that what each receives and sends is interpretable by the other.

Take note of the emerging rules! Path control components in one node can only communicate to path control components in another node using their peer-to-peer protocols. Also path control and NAUs can only communicate vertically with one another. In summary, NAUs communicate as peers; path control components communicate as peers; all messages must pass vertically to and from the lowest layer; and all messages must pass through both layers.

The NAU is actually subdivided into four more layers. The highest layer is called the "NAU Services Manager," the next to the highest is called "Functional Management Data (FMD) Services" layer, the next is the "Data Flow Control" layer, and finally the "Transmission Control" layer.

The path control is subdivided into two layers: "Path Control" and "Data Link Control." This emerging SNA model is shown in Figure 8.23. The end user, as before, still enters the four-layer NAU at the top level—

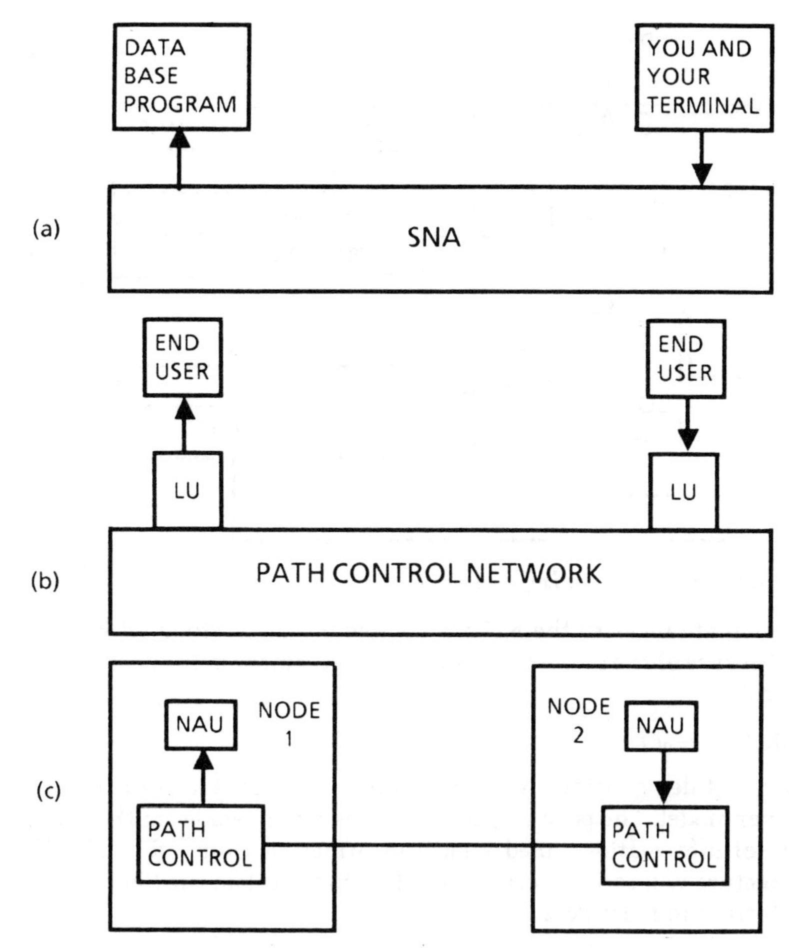

Figure 8.22: Progress Breakdown of SNA Component Relationships

NAU Services Manager. Remember, the user's message can only be passed vertically to the lowest Data Link Control layer before it is moved to another node.

Figure 8.23 shows the message flow between two nodes within the SNA layered models at each node. We will soon learn that each layer adds information to the basic AP message called a "Request/Respond Unit" (RU) as it moves down. The encumbered RU which is pure end user data now moves between the SNA layer 1 elements of nodes 1 and 2. The RU is then disencumbered as it moves vertically upward in the receiving node.

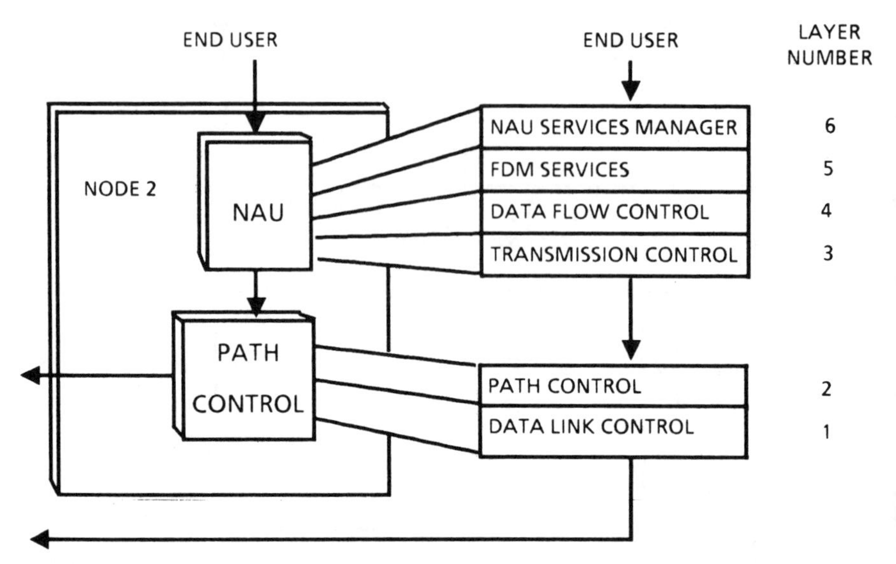

Figure 8.23: The SNA Six Layer Model

The encumbrances to the RU are completely understood within each node because of the pre-establishment or agreement in the peer-to-peer protocols.

The SNA Units of Data

Figure 8.24 demonstrates the SNA peer-to-peer protocols of the SNA six-layer model. The peer-to-peer protocols are embedded in this binary string of information called a Path Information Unit (PIU). The PIU architecture will soon emerge, but let us first explain the SNA layered architecture in more detail.

The NAU Service Manager layer exists for each NAU and provides both end user and network services. For a PU or an SSCP, this layer provides only network services. These services include such items as exception conditions (requests incorrectly entered, try again), initiates session-related components (FMD, data flow control, and transmission control), and so on.

The FMD Services layer operates in two different environments. For LU - LU sessions, the FMD Service is called "Session Presentation Services." In all other modes (SSCP - LU, SSCP - SSCP, and SSCP - PU), this session is called "Session Network Services."

The FMD in the LU - LU Session Presentation Services mode tracks items like carriage return, margin set, etc., and acts on those commands.

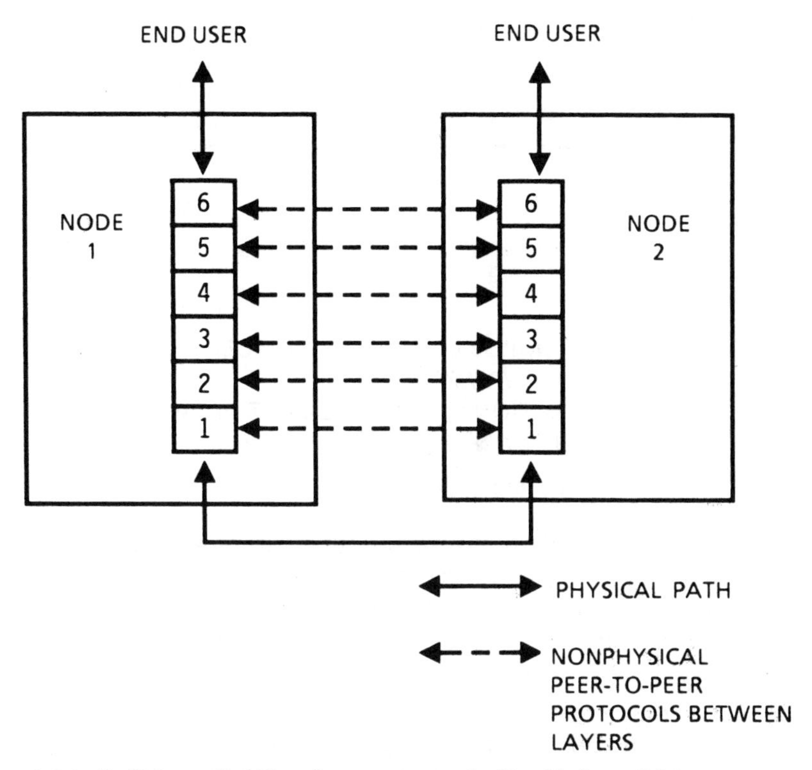

Figure 8.24: End User to End User Communication for Two Nodes in SNA

In other words, this mode handles data formatting for printers, displays, and graphics. It provides commands for component selection (turn on printer, go to card reader, use joystick commands, etc.). The FMD in the Session Network Services mode inserts commands to activate nodes, system start up and shut down, converts network names to network addresses, and synchronizes initiation processes between nodes.

The Data Flow Control layer establishes whether the mode of operation is full duplex, half duplex, and half duplex contention; provides chaining (disassembles and recovers very long messages as they are sent and received); and provides flow control of the RUs by controlling data flow and buffer capacity.

The Transmission Control layer performs such functions as session cryptography, session level pacing, and sequencing of information. This layer performs the actual construction of the PIU which is now shown in Figure 8.25.

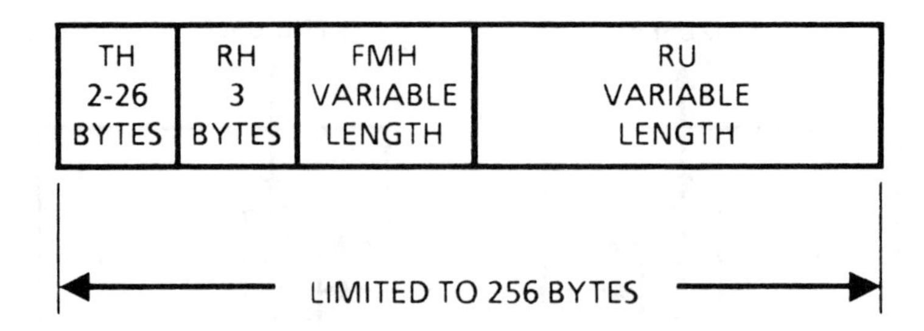

TH - TRANSMISSION CONTROL HEADER
RH - REQUEST/RESPONSE HEADER
FMH - FUNCTIONAL MANAGEMENT DATA HEADER
RU - REQUEST/ RESPONSE UNIT

Figure 8.25: An SNA Path Information Unit (PIU)

The path control layer performs the functions of message unit routing by selecting the virtual routing to the appropriate nodes involved. This layer performs nodal boundary functions, the most important of which is converting network addresses to local addresses.

The Data Link Control layer performs the very basic functions of establishing logical connections over the physical links, performs the physical transfer of data between the nodes, performs link-level error recovery, and serves channels and serial communication links. Sound familiar? It should. This is our old friend the HDLC or SDLC type protocol.

The Assembly Process of SNA Linkage

Figure 8.26 shows how the various layers contribute their respective peer-to-peer protocol entries on the raw end user information packet, the RU. Note the boundaries of the LU, the NAU, the path control network, and the node.

The Link Header (LH) is another name for our old friend the SDLC (or HDLC) protocol start FLAG, ADDRESS, and CONTROL fields. The Link Trailer (LT) is another name for the SDLC FRAME CHECK SEQUENCE and end FLAG fields. Figure 8.27 shows these equivalents.

Figure 8.27 also demonstrates the very essence of the SNA secret. Part(a) of this figure is the familiar SDLC frame. Part(b) is the retitled equivalent SDLC frame. Part(c) is merely a rescaling of the retitled SDLC frame. In all three SDLC frame replications (Parts a, b, and c of Figure 8.27), the INFORMATION field is now the center of our collective attention.

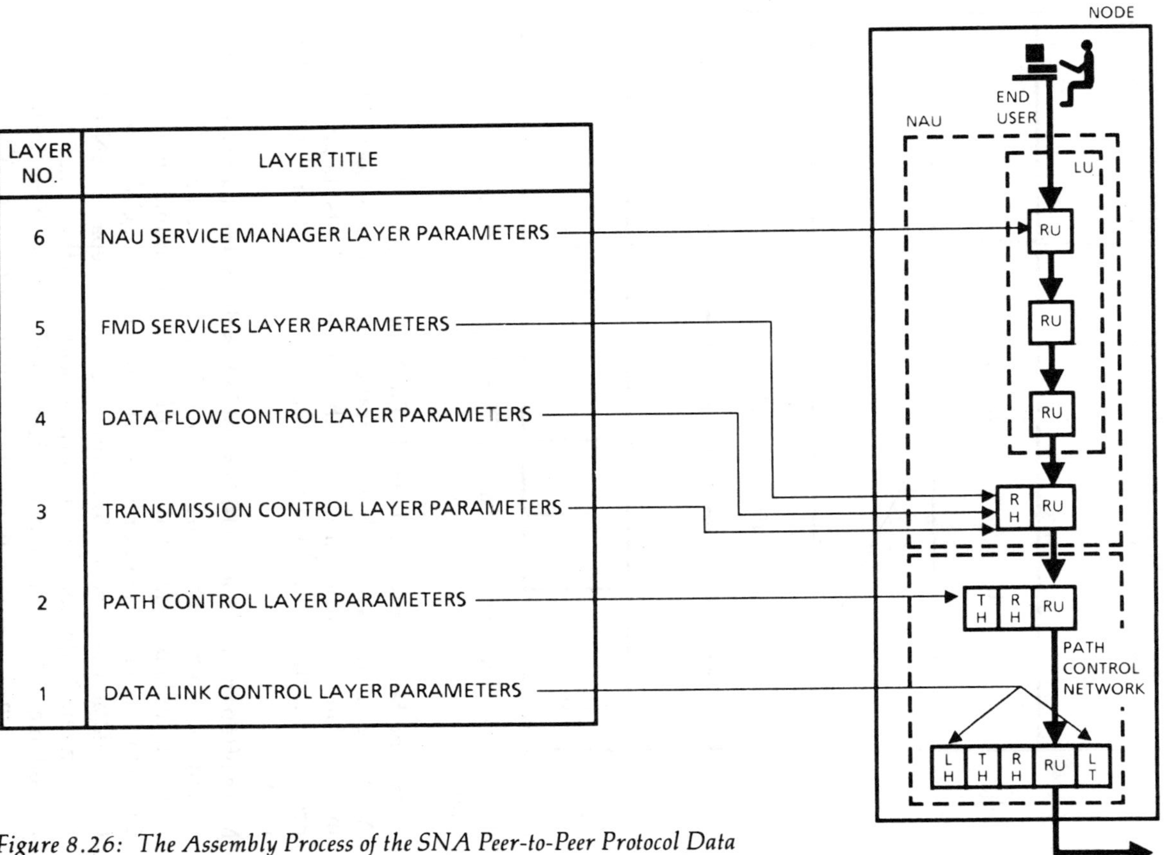

Figure 8.26: The Assembly Process of the SNA Peer-to-Peer Protocol Data

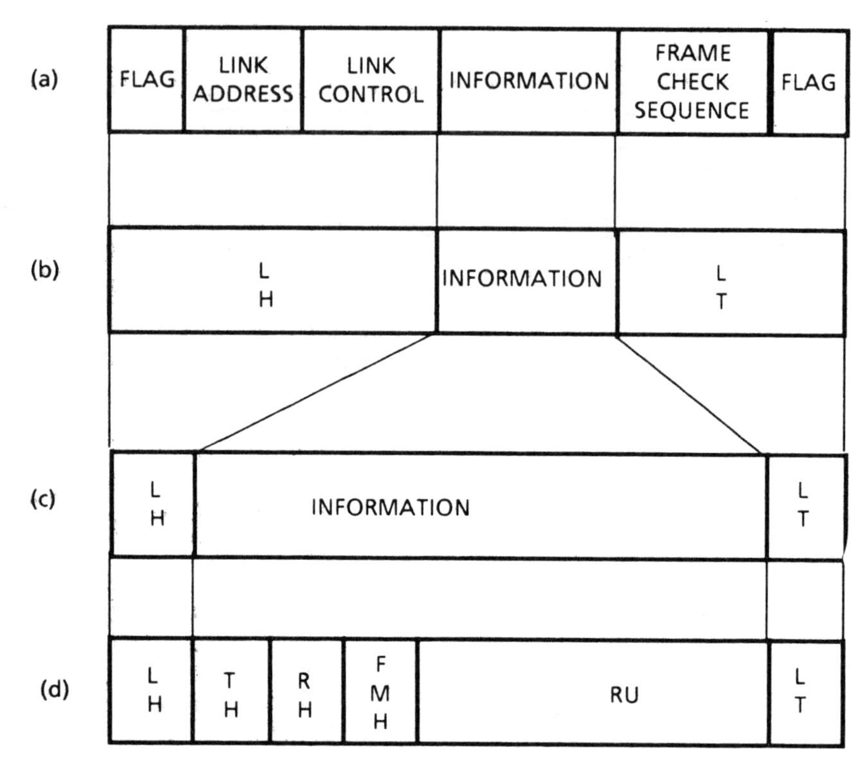

Figure 8.27: The Method by which the SNA Protocols are Buried in the SDLC Frame

Recall the fact that this INFORMATION field is entirely transparent. Anything inside is sent without alteration and anything goes. The SNA master plan capitalizes directly on this fact and plants its peer-to-peer protocols in this field along with the RU, as shown in Part(d) of the figure.

More often than not, an LU-LU session has the potential of connecting through a number of SNA nodes. For example, as an end user (LU_1), your terminal could be connected to an IBM 3705 communication controller CC_1 which is a node in your SNA domain in Boulder, Colorado. In turn, CC_1 could access a second 3705 node (CC_2) in a separate SNA domain, say Denver, Colorado, which in turn passes off to a third 3705 node (CC_3) located in an SNA domain in Los Angeles, California. This SNA domain also contains the host computer node which contains the AP of interest to you (LU_2).

Figure 8.28 is a representation of how the session data is acted upon as it

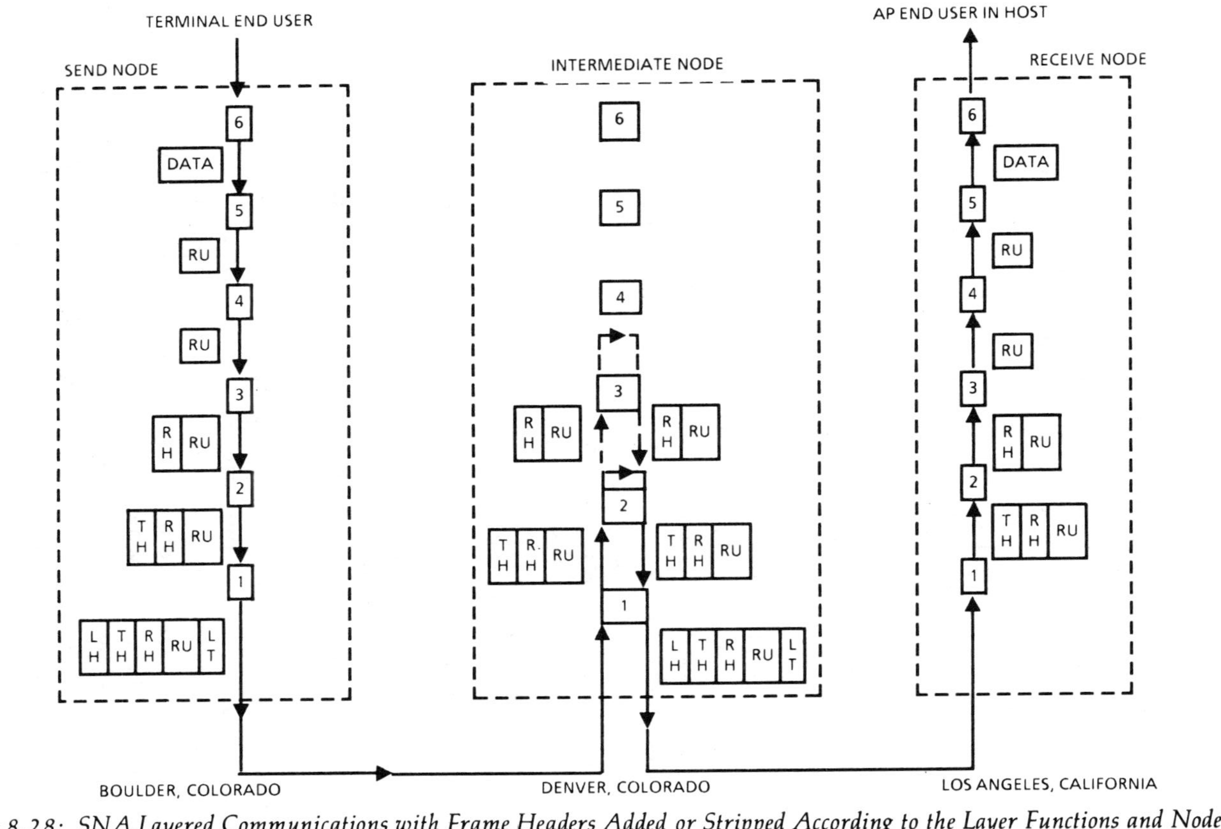

Figure 8.28: SNA Layered Communications with Frame Headers Added or Stripped According to the Layer Functions and Node Architecture

moves through the various SNA layers in the two end terminal nodes and the intermediate node CC_2. The SDLC data frame moves from the sender node into the intermediate node where layer 1 strips the FLAG, ADDRESS, and CONTROL (the LH) and the FRAME CHECK SEQUENCE and FLAG (the LT) which pertains only to the sender/intermediate node data link. At this point, the intermediate node architecture affects the next action. If this node which is the Denver communication controller CC_2 has more external domain connections other than to the Los Angeles domain, the transmission control header (TH) is stripped for the necessary location addressing (layer 3). A new TH header is developed and reassembled. If the Denver communication controller accesses Los Angeles only, then there is no need to penetrate layer 3. Only layers 1 and 2 are exercised.

Figure 8.29 summarizes the SNA units exchanged and the protocol hierarchy. To date, SNA has only formally defined 6 layers without a physical layer. This layer is beginning to appear more and more in the SNA total architecture. This physical layer is concerned with the transmission of raw bits over a communication channel. This layer thus addresses typical electrical issues such as what voltages constitute a binary 1 and 0; how long is one bit; is transmission simplex, half duplex, or full duplex; and how is the transmission established, maintained, and terminated. This layer also establishes the mechanical parameters of the interface such as plug type, size, number of pins, pin assignments, and position of the pins in the plug.

8.4.4 The Open Systems Interconnections (OSI) Reference Model

IBM is by no means unique in establishing a computer network architecture, although it clearly provides unprecedented leadership. Another computer network architecture includes Digital Equipment Corporation's DECNET.

Of far more significance, however, is the International Standard Organization (ISO) Technical Committee 97, Subcommittee SC 16 activities to establish a universal layered model called the Open Systems Interconnection (OSI) reference model.[15] The OSI is being developed as an international standard which will basically provide a solution to linking incompatible computers and heterogeneous networks. This model is shown alongside SNA in Figure 8.30. Clearly, OSI and SNA have very similar functional units or layers. An approximate correlation of the OSI reference model and IBM's SNA is also shown in this figure.

It should be noted that computer network architectures such as SNA or OSI are still undergoing change. This is appropriate because technology

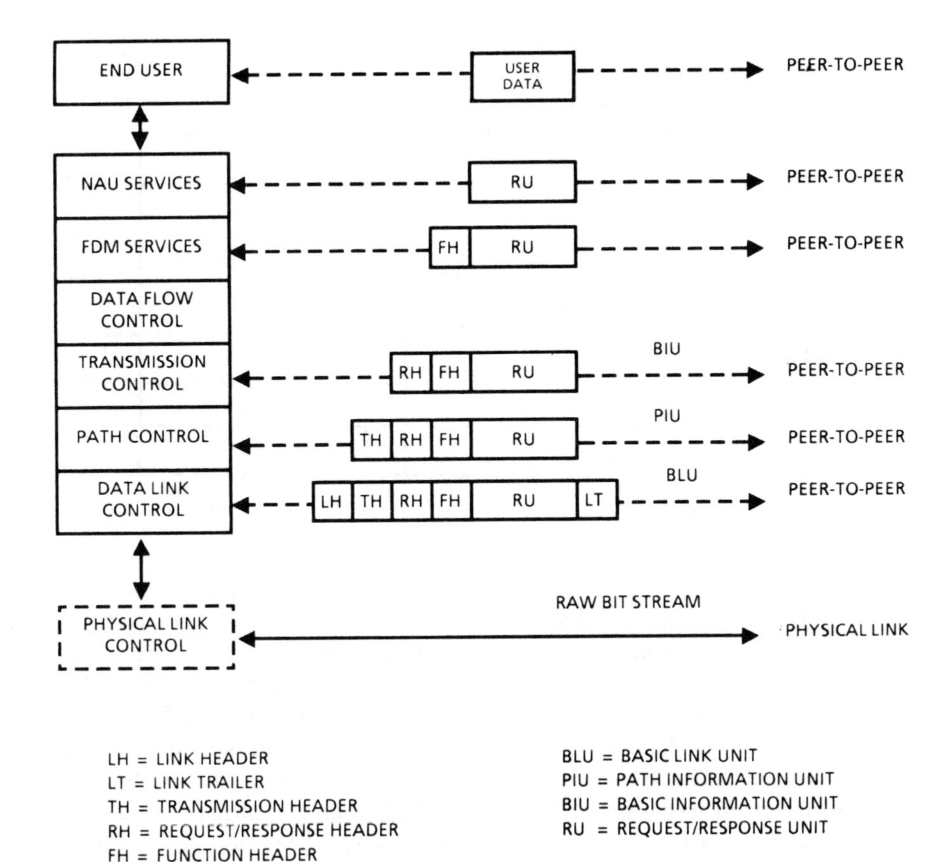

Figure 8.29: SNA Units of Exchange and Protocol Hierarchy Summary

is changing. Vendor attitudes are changing and therefore more emphasis is being placed on compatibilities and interoperability of equipment and software. For example, SNA 2 was announced in 1974 (SNA 1 was never released). Then in 1976 there was the announcement of SNA 3, and in 1979, SNA 4. Each announcement represented more flexibility and improved performance. These changes should not be confused with a complete restructure and overhaul of the computer network architecture, although the computer system operators may disagree.

8.4.5 Packet Switched Data Networks (PSDNs)

A packet switched data network (PSDN)[16] is intended to provide full duplex digital transport data services between two long distance, geographically separated data terminal equipment. The PSDN appears to be

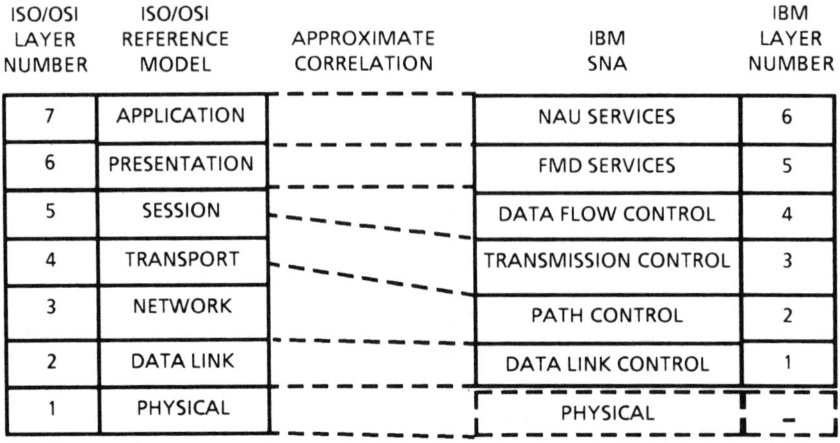

ISO/OSI LAYER NUMBER	ISO/OSI REFERENCE MODEL	APPROXIMATE CORRELATION	IBM SNA	IBM LAYER NUMBER
7	APPLICATION		NAU SERVICES	6
6	PRESENTATION		FMD SERVICES	5
5	SESSION		DATA FLOW CONTROL	4
4	TRANSPORT		TRANSMISSION CONTROL	3
3	NETWORK		PATH CONTROL	2
2	DATA LINK		DATA LINK CONTROL	1
1	PHYSICAL		PHYSICAL	–

Figure 8.30: The ISO Reference Model with the Approximate Correlation to the IBM SNA

a point-to-point connection between these devices. Typically, the PSDN provides the service of a long distance connection of a smaller terminal device such as a dumb terminal or a personal computer to a larger host processor which services several such devices at any one time. The PSDN is generally accessed through a local or Wide Area Telephone Service (WATS) dial-up to an Interface Message Processor (IMP). Once connected to the PSDN IMP_1, the terminal user states its destination and authorization access code. The PSDN IMP_1 then establishes the desired far-end computer connection through a second far-end IMP_2. Figure 8.31 shows this simplified connection. So far, this is identical to a normal long distance telephone dial-up connection where the IMPs appear to be local telephone end offices. IMPs, however, are basically communication controllers very much like the IBM 3705 or 3725 and can accommodate many terminals. In addition, the IMP can also accommodate one or more host processors. And the PSDN contains many IMPs as shown in Figure 8.32.

The IMPs are interconnected inside the PSDN network boundary. Thus, the IMP interconnections are shared by large numbers of diverse users. The PSDN thus provides virtual end terminal to end terminal communications while being able to share the long distance interconnect circuits.

This sharing process is accomplished in a very familiar way in which terminal data is divided up into bundles, typically 128 bytes long. Each bundle has various path and data link type headers placed on it by the originating IMP. Then this unit of data, called a packet, is placed inside

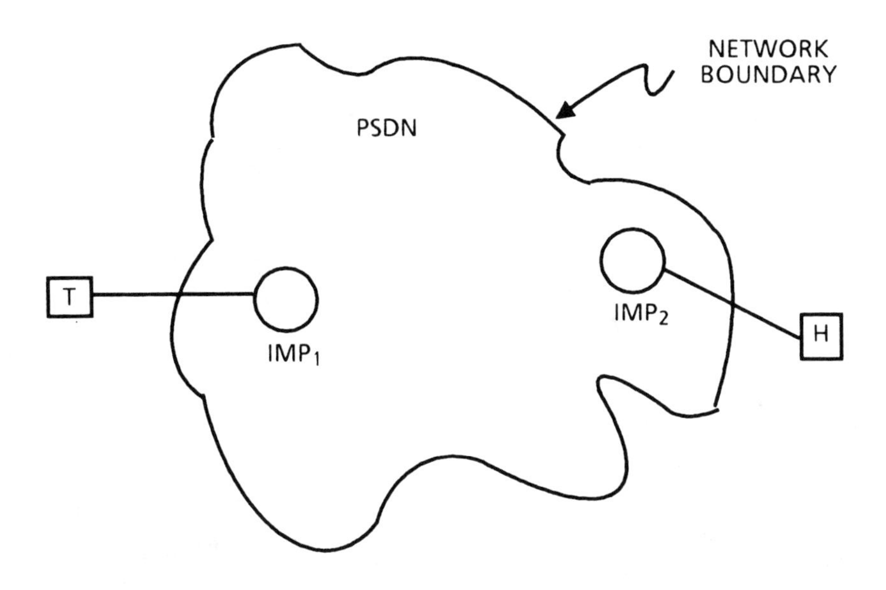

Figure 8.31: Simplified PSDN Access of a Terminal to a Host Processor

the HDLC INFORMATION field and transmitted out to another IMP. The originating IMP interleaves the various HDLC fields in which each HDLC field represents different end-to-end terminal equipment interconnects.

Sound familiar? It should because it basically represents the same functions as the SNA Physical, Data Link Control, and Path Control layers and the OSI Physical, Data Link, and Network layers. The access process and protocol to the PSDN is called X.25.[17,18]

Therefore, is X.25 compatible with the Physical, Data Link Control, and Path Control protocols of SNA or with the Physical, Data Link, and Network protocols of the OSI? Absolutely! If the SNA Path Control and Data Link Control are changed somewhat or an outboard protocol converter is used, it is compatible with SNA.[19] The X.25 is already accommodated in the corresponding OSI layers.

To expand on this slightly with SNA-when the PSDN X.25 is used between SNA LUs, only the Path Control Network layer of SNA is affected. When SNA data communications is required between an SNA DTE and a non-SNA DTE, then the outboard protocol converter is required.

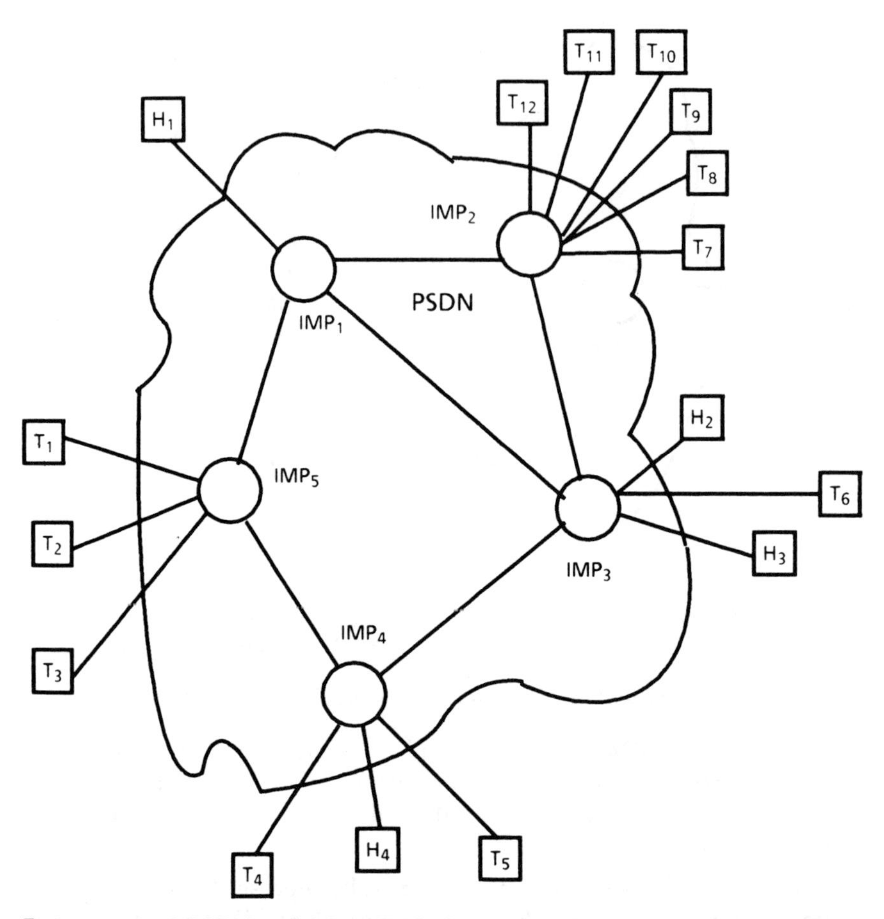

Figure 8.32: A PSDN with Many IMPs Supporting a Variety of Terminals and Host Processors

The X.25 is an evolving standard because it is closely tied to the total computer network architecture world. Basically, X.25 specifies the virtual setup and termination, error handling processes, PAD (packet assembler-disassembler) control, and a number of interactions of the data terminal equipment and the PSDN.

Figure 8.33 expands on the X.25 relationships to the DTE and the PSDN. An asynchronous data terminal, referred to as a START-STOP DTE, is defined by the X.28 standard. The PAD protocol standard is referred to as X.3. The X.29 is the protocol standard for a host communication controller which supports the X.3 PAD. Often, one will have to deal with X.25 in terms of X.28, X.3, and X.29.[18]

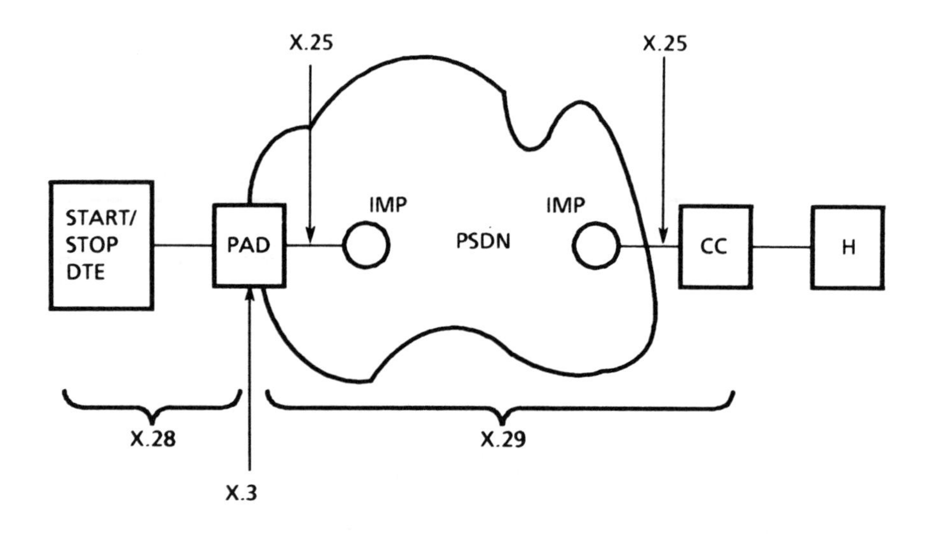

Figure 8.33: The Interrelationships of PSDN Standards

The relationship of the SNA PIU and the PSDN packet to the SDLC frame and the X.25 protocol is shown in Figure 8.34. Be aware of the fact that one or more PIUs can be sent as unqualified data packets which are concatenated by the "more data mark" or the "M-bit."

There are several key observations one should make in the PSDN and X.25. First, a PSDN is a means to an end, not an end in itself. This is because a PSDN only fulfills the first three bottom layers of either the OSI or SNA. Therefore, it meets a terminal-to-terminal need, not an end user-to-end user requirement. Figure 8.34 shows the other SNA layer controls inside the packet. In a simple terminal-to-host processor PSDN connection, these layer elements are often not included. This forces the user, for example, to place files in the host to be retrieved by local operators and then processed or delivered as a separate action. Without the proper higher level established architecture, there is always the potential for real bona-fide hassles at the host or back at the terminal if the requested service is complicated.

The second observation is best demonstrated by an example of a PSDN. Figure 8.35 shows an example of a six-IMP PSDN deployment in the western United States. A PSDN is placed into operation in the first place to share long distance interconnect services. The PSDN manager at-

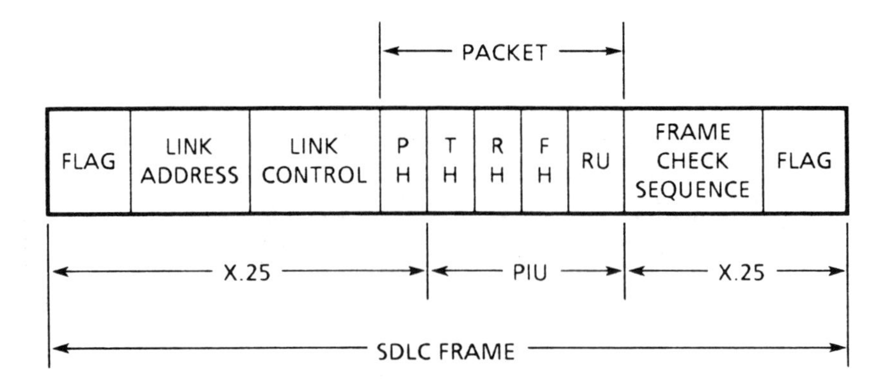

PH = PACKET HEADER
TH = TRANSMISSION CONTROL HEADER
RH = REQUEST/RESPONSE HEADER
FH = FUNCTIONAL MANAGEMENT DATA HEADER
RU = REQUEST/RESPONSE UNIT
PIU = PATH INFORMATION UNIT

Figure 8.34: Relationship of the SNA SDLC Frame to the PSDN Packet at X.25

tempts to maximize the traffic between any IMP, preferably filling the link capacity to 80% leaving 20% for a cushion. Since all traffic between subscribing terminals follows a "virtual" circuit, in one session between the San Francisco (SFO) IMP terminal and the Chicago (CHI) IMP host, the data may pass from SFO to Boise (BOI) to CHI one time and another time from SFO to Denver (DEN) to CHI another. The IMP software attempts to balance traffic loading in the network. If loading one data link appears to fall to 50% from 80% over a long period of time (say several months), the PSDN operator knows that this represents all loading on all links. This is because the IMP balances circuit loading. Therefore, the PSDN operator has the option of removing a link or two to raise the network loading back to 80%.

Let us say this was the case in Figure 8.36. A decision was made to reduce the number of links in the network from the top configuration to the bottom in order to raise the total network loading. All IMPs are still well covered and there are still large numbers of physical options available for virtual circuits between the IMPs. The PSDN has classically served the asynchronous START-STOP terminal users and smaller smart terminals into large host processors. Now that these PSDNs are loaded for efficiency, we can make a second observation.

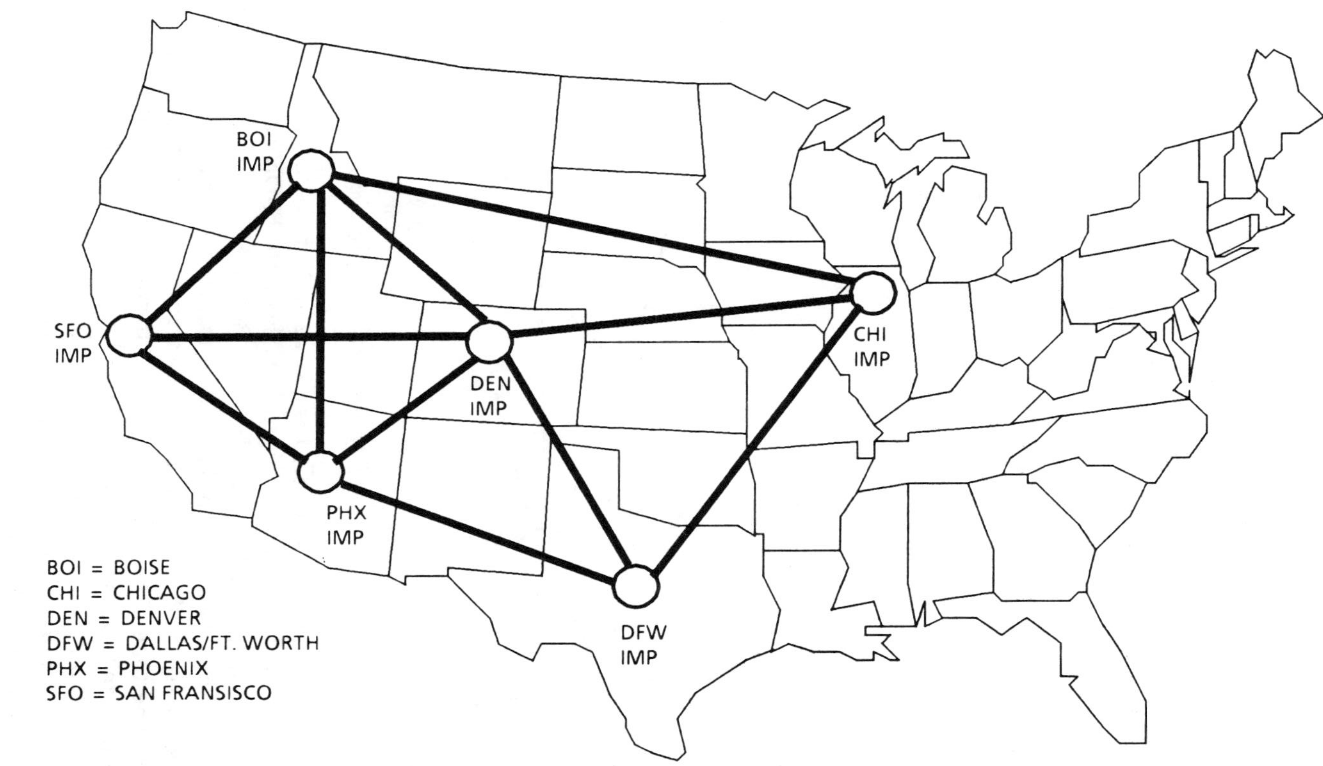

Figure 8.35: An Example of a PSDN Deployed in the Western United States

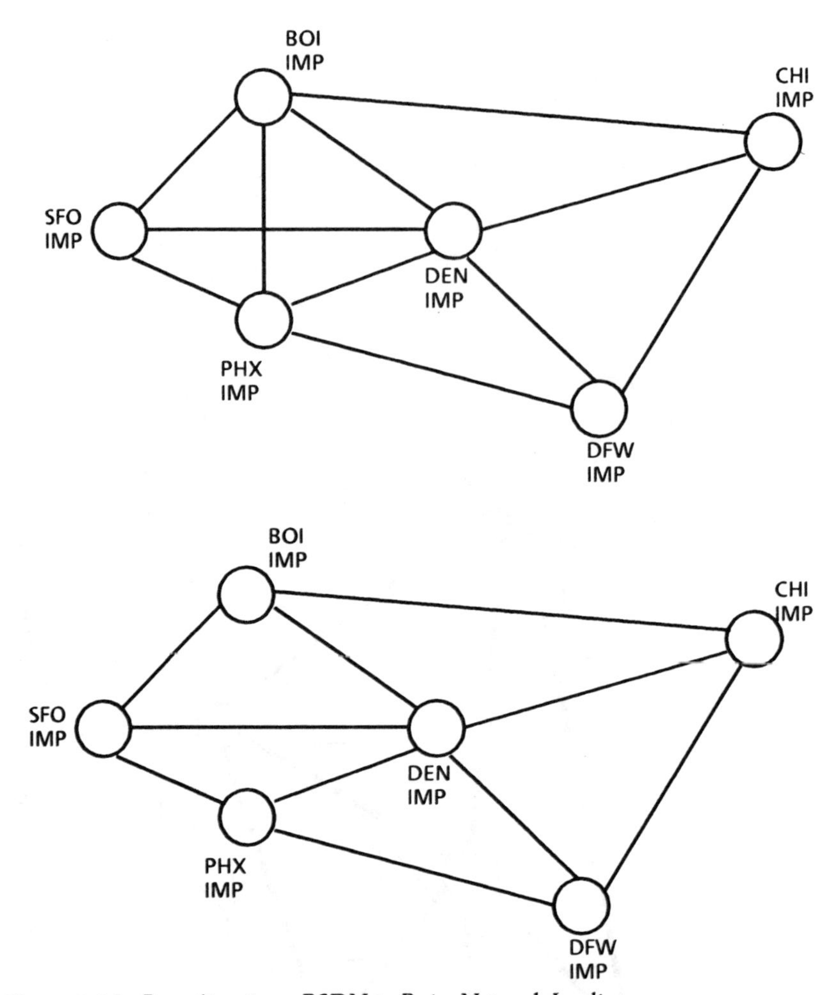

Figure 8.36: Reconfiguring a PSDN to Raise Network Loading

When one terminal, say in SFO, is a large host processor, and it dumps a large high-volume graphics file to a CHI host, the PSDN has the affinity to "drop data on the floor" in places like BOI, DEN, PHX, and DFW. The reason should be apparent. Large data transfers overload the IMPs and network especially if several large data transfers occur at approximately the same time. The PSDN has no front-end indication that it will be dumped on by a user. In other words, there is not yet a discriminator in the PSDN for sending terminal volume. In voice systems, calls are simply blocked or busied out. In data, the packet nature of the transmission makes it difficult to control.

8.5 DIGITAL MULTIPLEXERS AND CONCENTRATORS

An integral part of a data system is the digital multiplexer and concentrator. To a certain extent, the PSDN IMP is a form of multiplexer or concentrator. An example of their use was shown in Figure 8.18. Centralized networks often contain one or more levels of concentration in order to conserve central site resources or to reduce data link costs. The design problem is one of deciding on the size, quantity, and locations of these units. For small networks this can be done by inspection, and the design consists primarily of comparing costs between individual data links or multiplexer/concentrator lines.

8.5.1 Digital Multiplexers

Multiplexing (and concentrating) is a method for sharing a single communications channel among several users. There are basically three types: frequency division multiplexing (FDM), synchronous time division multiplexing (TDM), and statistical time division multiplexing (STATDM). In all three cases, the multiplexer/concentrator places the shared resources on a data link which has a fixed frequency bandwidth. If the output data link of a multiplexer or concentrator can support a 9.6 kbps synchronous data rate, then the sum of all input data rates will be equal to or less than this figure. Thus, a 9.6 kbps multiplexer could support two 4.8 kbps input lines, four 2.4 kbps input lines, eight 1.2 kbps input lines, and so on.

Frequency Division Multiplexer (FDM)

The 9.6 kbps output data rate requires a fixed amount of frequency bandwidth. The relationship of bandwidth to channel capacity[20] is defined as:

$$C = B \log_2\left(1 + \frac{S}{N}\right) \tag{8.2}$$

where

C = channel capacity in bps
B = bandwidth of channel
S = signal power
N = Gaussian noise power

This is referred to as the Hartley-Shannon Law and defines the absolute upper bound of data link capacity for a given bandwidth and signal-to-noise ratio. Typically, a 9.6 kbps data rate can be supported by a telephone type circuit with 4 kHz of bandwidth in which the link signal-to-noise ratio runs approximately 25-30 dB. The channel capacity, C, under such conditions is approximately 30,000 kbps. Therefore, 9.6 kbps is about one-third of the theoretical upper bound, which is reasonable under normal telephone system operating conditions.

The multiplexer technically slices the available bandwidth into isolated subfrequency channels to accommodate the input signals. Figure 8.37 shows this frequency division process, thus the term Frequency Division Multiplexing (FDM). In FDM each subchannel is isolated by a guardband.

Time Division Multiplexer (TDM)

TDM operates on a polled basis. The TDM unit contains small buffers to temporarily hold each of the terminal coded character inputs and outputs. Figure 8.38 shows a TDM configuration and the moving frames inside the data link. If any one of the four terminals has nothing to send, a blank code is relayed in any case. Blanks are very common if async terminals are used as inputs. Regardless of the terminal types, async or sync, the output of the TDM units is always synchronous.

Statistical Time Division Multiplexer (STATDM)

The STATDM takes advantage of the async terminal inefficiencies. Recall that async terminals only transmit data when a key is depressed on the terminal keyboard. The code contains a start bit, a character code, a parity bit, and at least one stop bit. The STATDM requires a buffer in case all the async terminal operators press a key on their keyboards simultaneously. Otherwise, the STATDM treats everyone's transmission on a "first-come, first-serve" basis. When terminal T_3 sends a character to the STATDM, the STATDM tags this character with an address code "3" and sends it on its way. The receiving STATDM knows from these tags to which connection to deliver the character. Figure 8.39 shows this process including the buffer in the STATDM. STATDMs are generally two to four times more efficient than a straight TDM process when async terminals are used.

The STATDM can generate an SDLC or HDLC frame for its data transmission. Figure 8.40 shows an example of how this frame is organized. There are many possible alternatives depending on the vendor selected.

Combining Multiplexer Processes

It is technically possible to combine multiplexer processes. If one had a 3 kHz, 9.6 kbps data channel available, it could be FDM split into two 4.8 kbps channels. One channel could then support a synchronous 4.8 kbps terminal while the other could support a STATDM process with several async terminals.

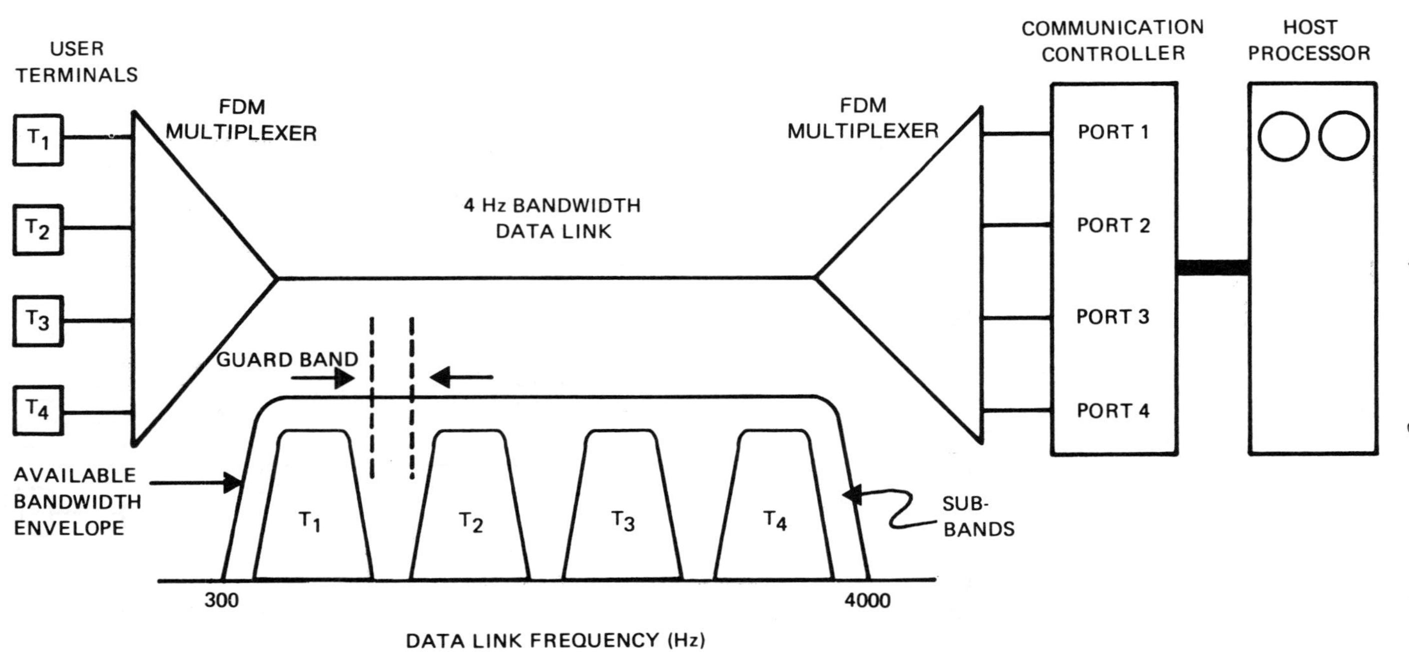

Figure 8.37: Frequency Division Multiplexing (FDM) a Data Link into Four Channels

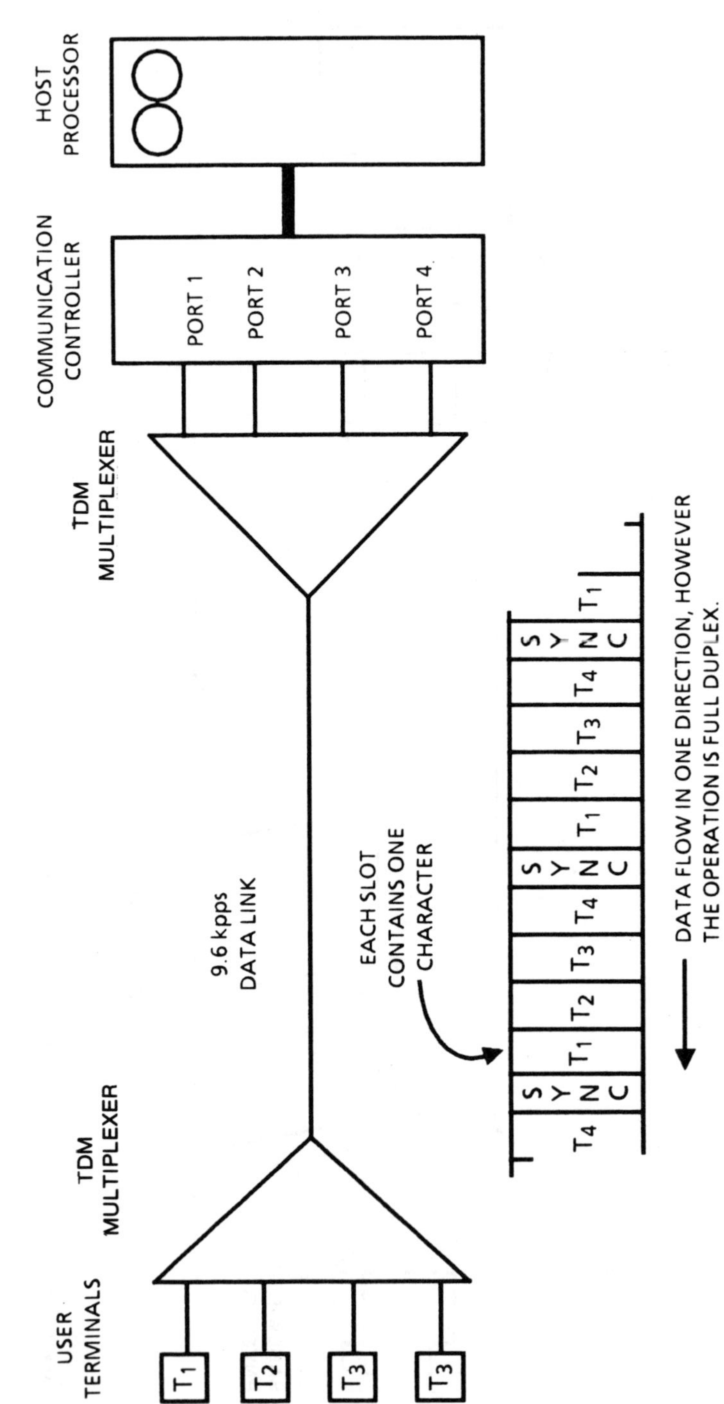

Figure 8.38: Time Division Multiplexing (TDM) a Data Link to Accommodate Four Terminals

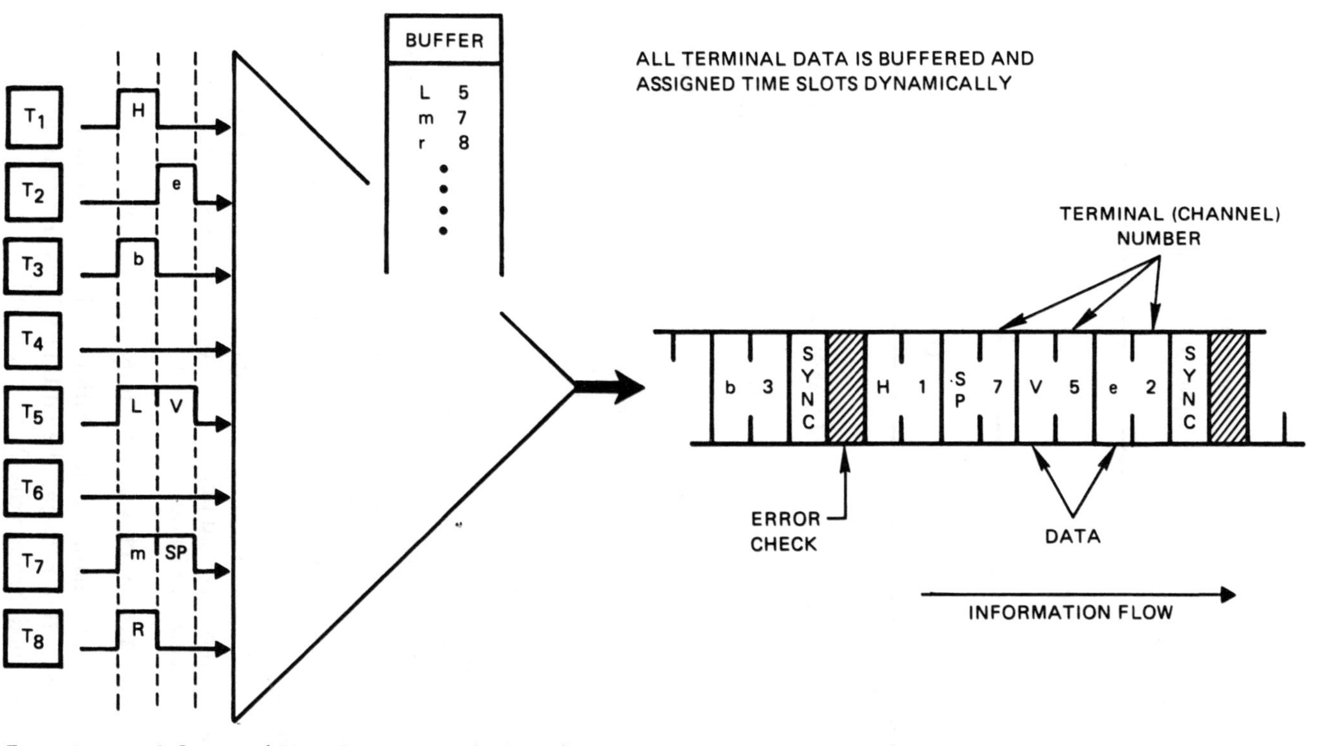

Figure 8.39: A Statistical Time Division Multiplexer (STATDM)

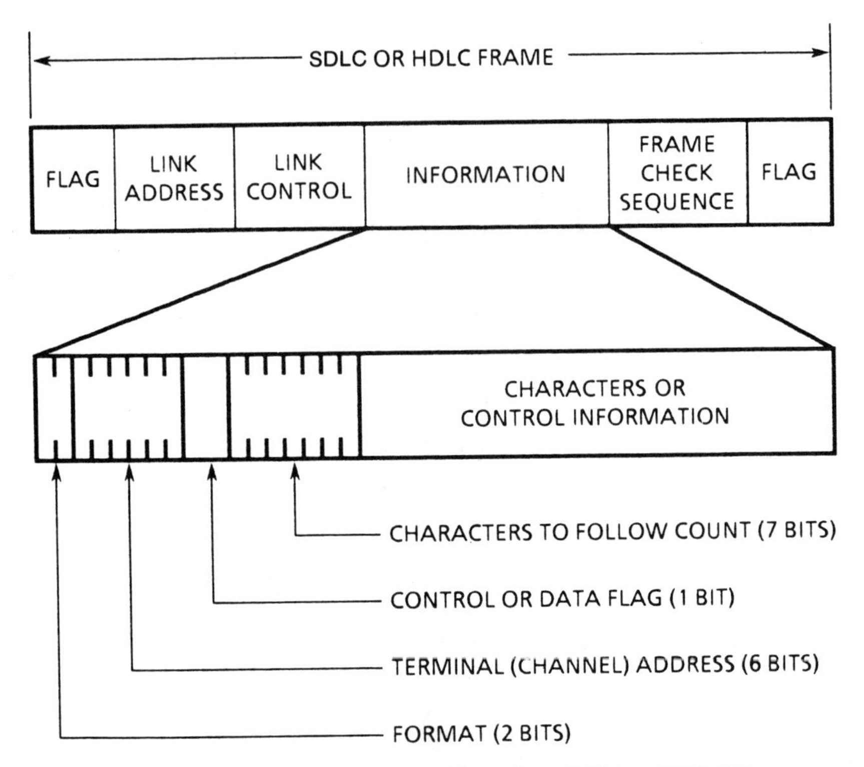

Figure 8.40: A STATDM Information Field Inside an SDLC or HDLC Frame

8.5.2 Digital Concentrators

Multiplexers are double ended; that is, they require multiplexer units at each end of the data link. Concentrators are single ended because the host communication controller hardware and software perform the channel breakout processes. Information processed by a concentrator is message interleaved rather than bit or byte interleaved, as it is normally done by the multiplexers. Also concentrators support some form of computer architecture like SNA.

As a result of these differences, the concentrator is considered to be part of the data terminal equipment while the multiplexer is considered part of the data communication equipment.

8.6 BASIC DATA TRANSMISSION

To this point, the data communication model has assumed a pure transparent connection between digital DTEs. This transparent connection was established through the DCEs. This model is redrawn in Figure 8.41.

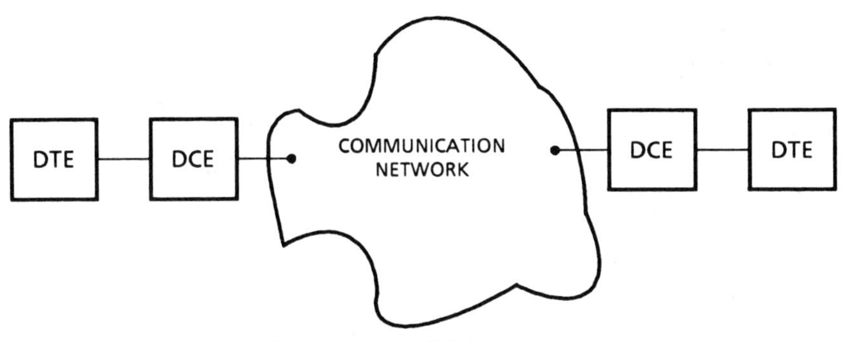

Figure 8.41: Basic Data Communication Model

We have established the parametric and protocol attributes of the DTE and now realize that the interconnection requirement will vary widely. Consider, for example, the variety of interconnection possibilities of Figure 8.42 in which simple terminals are connected to concentrators, simple terminals are connected to communication controllers, concentrators are connected to concentrators, concentrators are connected to a communication controller, communication controllers are connected to communication controllers, and communication controllers are connected to host processors. The data rates alone can vary from 75 bps to 15 Mbps.

Therefore, it is essential to understand the system architecture before attempting to arrange interconnection. If not properly understood, one could send a semi-truck down a four-wheel drive mountain trail. Assuming that one has some concept of the DTE-to-DTE architecture and requirements, one is ready to embark into transmission systems. These systems, like the DTEs have a wide variety of components and methods of operation. We have already been introduced to a DCE with the multiplexer.

8.6.1 The Interfaces between DTEs and DCEs

There are formal interfaces between all DTEs and DCEs. If one adopts the OSI model, many of these interface standards are spelled out in the physical layer of the model. Figure 8.43 lists many of these interface standards, which have been established and adopted by the International Telephone and Telegraph Consultative Committee (CCITT), the Electronic Industry Association (EIA), or by the American Telephone and Telegraph (AT&T) Bell System.

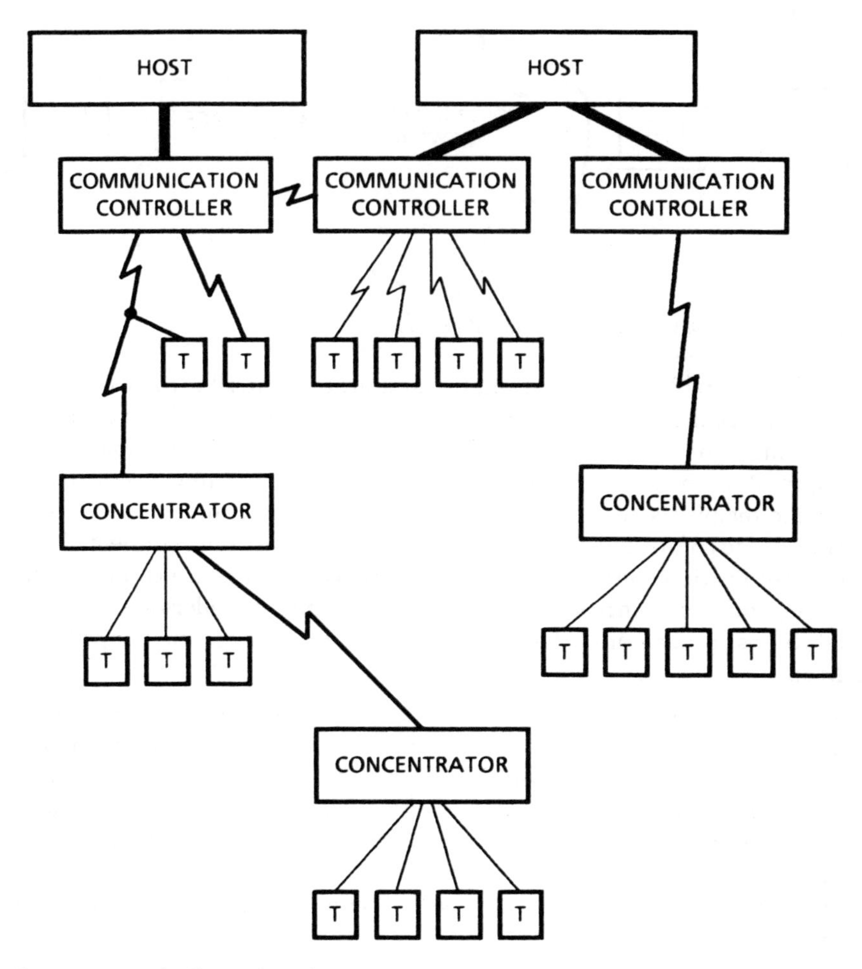

Figure 8.42: An Example of the Large Combinations of Interconnects, Each Requiring Different Interconnection Needs

These interface standards define combinations of the electrical and physical parameters to make a proper connection between the DTE and DCE. Electrical parameters include voltage levels, signalling requirements, timing, and grounding conditions. Physical parameters include the exact dimensions of the connecting plugs, number of pins and their exact position in the plugs, pin numbers, and color codes.

	EUROPEAN CCITT	NORTH AMERICAN EIA
LOW SPEED < 20 kbps	V.10 V.24,V.28	RS423 RS232C
HIGH SPEED > 20 kbps	V.35 V.11	RS422
MECHANICAL		RS449

Figure 8.43: The European and North American Digital Interface Standards between the DTEs and DCEs

The EIA RS232C Interface Standard

By far the most well established interface standard in the United States is the EIA RS232C. This standard[18] specifies a 25-pin connector. It makes lettered pin assignments for data, control, timing, and grounding. It specifies the exact dimensions of both the male and female plugs as well as states that the female plug is assigned to the DTE and the male plug to the DCE. It specifies that the cable length between the DTE and DCE should not exceed 50 feet. The RS232C specifies that data rates through this interface should not exceed 10 kbps and must operate in a bit serial mode of operation. It also specifies that it will support both sync and async operations.

INTER-CONNECTION SERVICE	DESCRIPTION	CCITT V.24 EQUIV.	RS-232 PIN NO	DATA		CONTROL		TIMING		GROUND
				FROM DCE	TO DCE	FROM DCE	TO DCE	FROM DCE	TO DCE	
AA	PROTECTIVE GROUND	101	1							●
AB	COMMON RETURN/SIGNAL GROUND	102	7							●
BA	TRANSMITTED DATA	103	2		●					
BB	RECEIVED DATA	104	3	●						
CA	REQUEST TO SEND (RTS)	105	4				●			
CB	CLEAR TO SEND(CTS)	106	5			●				
CC	DATA SEND READY	107	6			●				
CD	DATA TERMINAL READY (DTR)	108.2	20				●			
CE	RING INDICATOR	125	22			●				
CF	RECEIVED LINE SIGNAL DETECTOR	109	8			●				
CG	SIGNAL QUALITY DETECTOR	110	21			●				
CH	DTE DATA SIGNAL RATE SELECTOR	111	23				●			
CI	DCE DATA SIGNAL RATE SELECTOR	112	23			●				
DA	DTE TRANSMITTER SIGNAL ELEMENT TIMING	113	24						●	
DB	DCE TRANSMITTER SIGNAL ELEMENT TIMING	114	15					●		
DD	DCE RECEIVER SIGNAL ELEMENT TIMING	115	17					●		
SBA	SECONDARY TRANSMITTED DATA	118	14		●					
SBB	SECONDARY RECEIVED DATA	119	16	●						
SCA	SECONDARY REQUEST TO SEND	120	19				●			
SCB	SECONDARY CLEAR TO SEND	121	13			●				
SCF	SECONDARY RECEIVED LINE SIGNAL DETECTOR	122	12			●				
-	RESERVED FOR DATA SET TESTING	-	9							
-	RESERVED FOR DATA SET TESTING	-	10							
-	UNASSIGNED	-	11							
-	UNASSIGNED	-	18							
-	UNASSIGNED		25							

Figure 8.44: A Summary of the RS232 Interconnection Services

Figure 8.44 presents a summary of the RS232C interconnection services and compares them to the CCITT V.24 standard. This figure represents the array of available signals for interconnecting the DTE to a DCE. Please note, however, that they are available for use but do not necessarily need to be used. Figure 8.45 shows the RS232C (and equivalent V.24) interface signals and voltages.

Almost all of the pin-assigned signalling is available to alert the DCE that the DTE has information to send and allows the DCE to prepare itself for the interconnection. Thus, the dialogue between the DTE and DCE is, in effect, another handshaking sequence over and above any DTE protocols which will be embedded in the DTE-to-DTE information exchange.

		NEGATIVE VOLTAGE (-3 TO -25 VOLTS)	POSITIVE VOLTAGE (+3 TO +25 VOLTS)
DIGITAL CIRCUITS	BINARY STATE SIGNAL CONDITION	1 MARK	0 SPACE
CONTROL CIRCUITS		OFF	ON

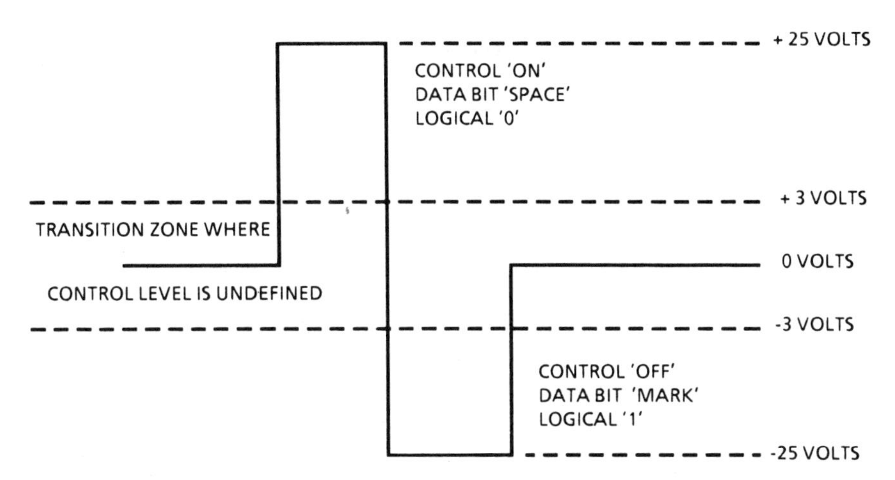

Figure 8.45: The RS232C (and the Equivalent V.24) Interface Signalling

There are excellent references available[21, 22, 23] which provide detailed layouts of various DTE-to-DCE handshaking processes. A simplified handshake between the DTE and DCE is provided in Figure 8.46.

Often the DCE must prepare itself for transmitting and receiving data. A user terminal DCE must contact a host DCE, wake it up, alert it to the fact that a data transmission link is required, pass test signals to the host DCE, and then return to the user terminal and indicate that everything is ready to go. This DCE task is often assumed by a communication modulator-demodulator or modem. Modems play a very essential role in data communications and will be discussed later in this chapter.

The Almost RS232C Interfaces

The EIA RS232C standard is probably the single most altered standard of the data communication world. It is a broadly accepted standard but quite old. It was intended to support DTE-to-modem interfaces. Today, how-

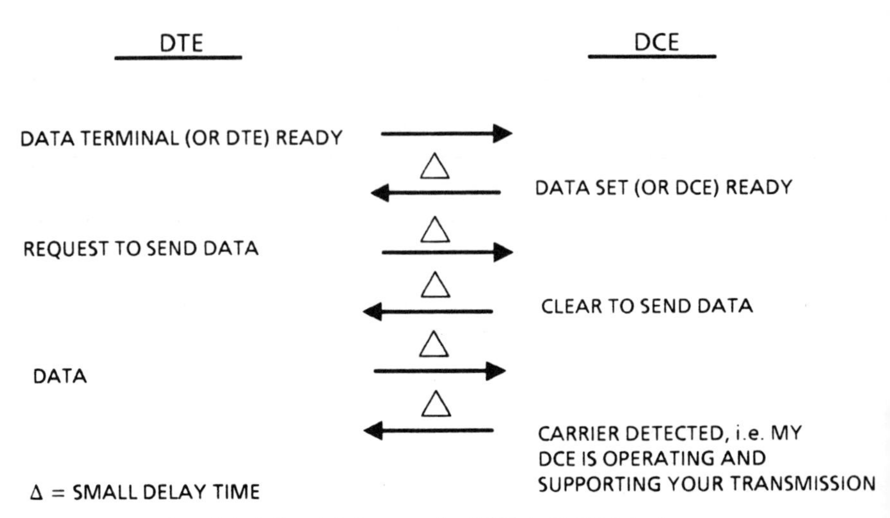

Figure 8.46: Simplified Handshake over the DTE-to-DCE Interface

ever, there are many more devices in the inventory of DCEs including the Network Interface Units (NIUs) of a Local Area Network (LAN).[24] LANs are discussed in detail in another chapter of this text. Other names for these devices include ADAPTERs or TIEs to the LAN. Many of the older modem-required functions have been abandoned in these devices leaving only a handful of active pins.

In some cases, a simple hardwire connection is made between the DCEs which requires absolutely minimum setup time compared to a telephone connection. Under a hardwired interconnect, the RS232C plugs provide very useful physical connection services. The terminal and host processor communication functions in the DTEs still retain the functions to connect to modems; therefore, the interconnection is wired to "fool" the DTE into thinking it sees the classic modem. When the DTE communication controller is connected in such a fashion, it would request a connection and receive it with lightening speed. I am sure it would say to itself, "Wow, that was fast!" A typical plug-to-plug connection might take the form displayed in Figure 8.47. It represents the bare essentials, indeed!

There are other RS232C improprieties which are relatively common. Possibly the next largest is exceeding the 10 kbps data rate, followed by exceeding the 50-foot length restriction between the DTE and DCE. Yet, each compromise appears to meet valid and legitimate technical, electronic, and physical needs.

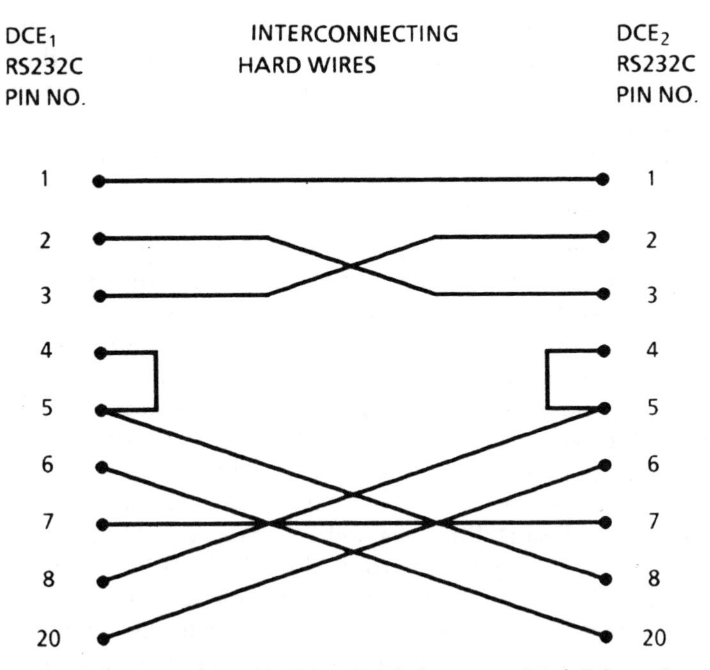

Figure 8.47: Typical Hardwired DCE-to-DCE Connection With RS232C

The EIA RS449 Interface Standard

In an attempt to meet and exceed the requirements which manifest themselves in the almost RS232C standards, the EIA produced a replacement. This replacement is referred to as the RS449[18] and calls for a larger 37-pin connector with an auxiliary 9-pin connector. RS449 calls for data rates up to 10 Mbps and DTE-to-DCE separations of up to 4000 feet. But RS449 is rapidly dissolving.

At the time RS449 was conceived and produced as a standard, there was no real perception of how rapidly computers were shrinking in size due to the major developments and breakthroughs in Very Large Scale Integration (VLSI). A rackmounted minicomputer can now sit on a table. In addition, disk storage devices have taken major leaps forward and they too are now occupying desk top space.

Why is this an issue with RS449? Because as these computers and peripherals are getting smaller and smaller, the interface plugs are getting larger and larger; and as most personal computer owners now realize, it is not the equipment volume anymore but all the interconnect wiring and plugs which inhibit clean, well layed-out systems.

In addition, it is one thing to declare 10 Mbps rates and 4000-feet length interfaces, and it is an entirely different issue to implement them with the appropriate electronic and wiring technology. Most agree that it is just as easy to implement these improvements on RS232C (or even smaller plug systems) than going to the larger wire bundles and plugs.

The EIA RS422 Interface Standard

Satellite transmission is an excellent means of moving digital data at rates of 1.544 Mbps or greater.[25] Such data rates are very popular in Computer Aided Design (CAD) graphic systems. And these data rates are also evident on the network side of TDM multiplexers as shown in Figure 8.48. The TDM multiplexer will support a wide variety of inputs on the channel side. Shown here are twenty-four 56 kbps channels. The multiplexer adds control and framing data which brings the total multiplexer output to 1.544 Mbps which is referred to as a basic T1 line rate. The input interface is specified as RS232C while the output is specified as RS422 (or V.11). RS422[18] defines a wire pair (balanced) interface which supports data rates from 10 kbps to 10 Mbps. Unlike RS232C, the RS422 only specifies the electrical (not mechanical) characteristics of the interface which include maximum data rates, voltage levels, minimum wire size, character impedances of the DTE and DCE, signal wave forms, test performance measurement specifications, and circuit protection.

The RS422 was developed in conjunction with RS423 where RS423 covered those standards associated with low speed unbalanced (single signal wire, grounded return) circuits. The original intent of these two standards established in the mid-70's was to support the development of RS449, where RS449 was to gradually replace RS232C. Basically, RS422 and RS423 define the electrical interface standards while RS449 specifies

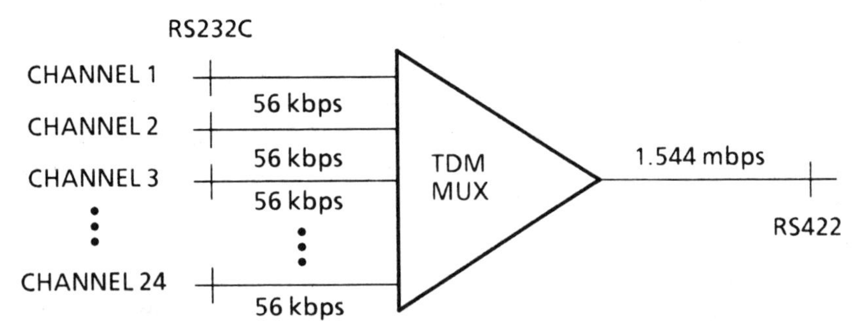

Figure 8.48: A TDM Multiplexer with RS232C Channel Interface Standard and RS232C Network Interface Standard

the mechanical and functional characteristics of the interface. Of the three standards, RS422 appears to be the survivor because it meets the high speed characteristics needed for a T1 type line rate.

Please note that RS422 does not specify the mechanical (connector) attributes. This is currently left to the carrier or equipment vendors.

Interface Developments

The interface between the DTE and DCE has been anything but static, and one must expect this trend to continue. Like the entire field of data communications, this interface must handle a very broad class of services and data rates. Up to this point in the text and for all practical purposes in industry at large, the DTE-DCE interface is copper wires and mechanical connectors whose characteristics are carefully articulated in various industrial and international standards.

There is evidence that fiber optics may play an increasing role in DTE-to-DCE interconnect. Fiber could potentially solve the bulky connector problems encountered in RS449 and even RS232C. Fiber certainly is less cumbersome than the RS232C ribbon wire or wire bundles. And fiber can easily handle the higher data rates and longer distances.

Will fiber come of age here in this world of microcomputers and mainframes? Fiber, by its very nature, causes difficulty within and between the currently defined OSI and SNA physical layers. And fiber is still operationally difficult to handle. Copper wire is something technicians handle with the twist of a screwdriver and a pinch of the pliers.

It is not known how long the jury will be out on fiber, but odds are it will serve a very practical role in the DTE-to-DCE interface.

8.6.2 The Digital-to-Analog DCE

Refer to our basic data communication model development in Figure 8.49. We began by deriving the simple-to-complex digital codes and protocols between the DTEs. Then we stated that the DTE-to-DTE connection was not necessarily a hardwire connection but a connection into a network through DCEs. Then we learned that the interface between a DTE and a DCE was well defined by industrial and internationally accepted standards. Now we are prepared to step directly up to the DCE and open this box and look in.

The DCE has classically been the well known communication modem which converts the DTE digital output into an analog signal that can be placed on conventional telephone lines. The modem thus modulates a signal and transmits it down the analog communication media. In turn,

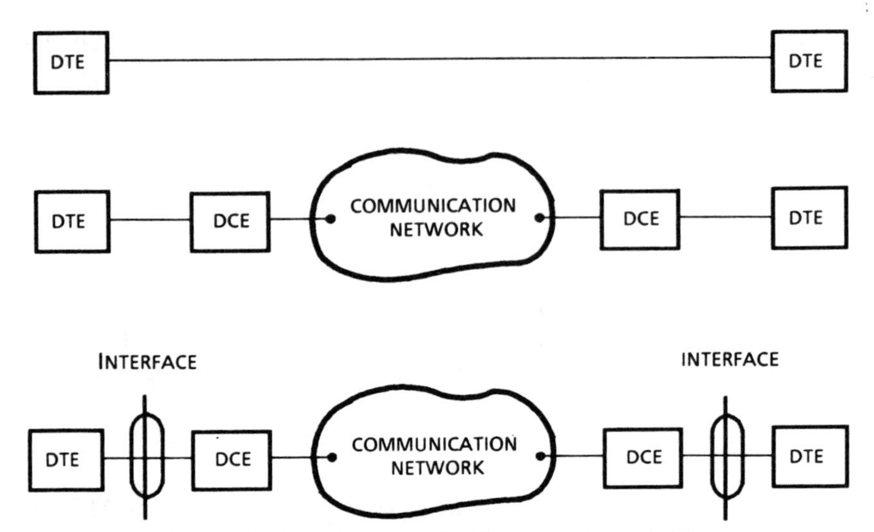

Figure 8.49: Progressing Detail of the Data Communication Model

the modem receives and demodualtes signals back into pure binary data. The modem thus modulates and demodulates signals.

Figure 8.50 graphically shows the modems inserted into our data communication model. Notice the binary signals going into the modem. The bandwidth of a telephone circuit is generally 3000 Hz as shown in this same figure. The modem must therefore condition the signals to fit within the 3000 Hz band. The telephone line introduces various line losses and signal distortion, noise, and interrupts which are typical of any analog system. The receiving modem must decode this mis-shaped signal and restore the original message.

Recall the Hartley-Shannon Law in equation (8.2) which defined the absolute upper bound of channel capacity for a communication link of a given bandwidth and operating at a given signal-to-noise ratio. Let us assume that our telephone line has the theoretical ability to handle up to 27,000 bps. Approaching one-third of this value (or 9.6 kbps) is considered extremely good. Modem costs generally reflect their relative efficiency: the more expensive modems can handle higher data rates than a cheaper unit as long as they both operate over identical telephone lines.

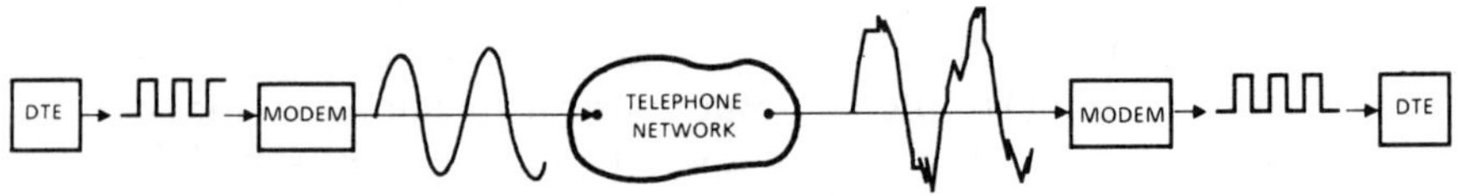

Figure 8.50: Signal Flow through the DCE Modems

Modem efficiency is accomplished in basically two ways: its modulation/demodulation approaches and the ability to perform automatic line equalization.

Modems employ one or a combination of three different signal modulation schemes which include amplitude modulation (AM), frequency shift key (FSK) modulation, and phase shift key (PSK) modulation. It appears that the most efficient modulation technique is combining PSK and AM. This technique is referred to as quadrature amplitude modulation.

Different modem signal frequency components encounter different amplitude attenuation and propagation delay times. These line properties do not affect voice communication performance but cause a great deal of problems for data transmission, particularly above 2400 bps. The modem can compensate for a portion of the problem by performing a line equalization function.

Figure 8.51 demonstrates the principle of modem line equalization. The telephone line normally exhibits attenuation distortion (loss) as a function of frequency in which the higher frequencies are attenuated more than the lower ones. Line equalization basically means that the modem's signal receiver raises the receiver gain in these higher frequency bands to compensate for the nonlinear losses.

Expensive modems conduct this process by transmitting signals between themselves and automatically make the appropriate adjustments. By this action, the modem-to-modem speed can be increased to support higher data rates.

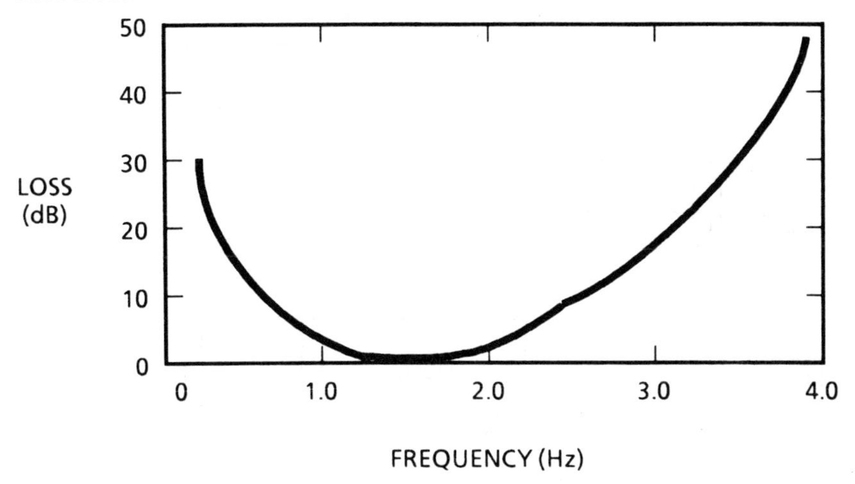

Figure 8.51: Attenuation Distortion Versus Frequency for a Voice-Grade Telephone Circuit

Figure 8.52 shows two different ways modems are coupled to the telephone lines. The first and oldest method is through acoustic coupling in which the connection to the telephone line is via the microphone and speaker of the handset. This handset is placed in a special modem cradle. The cradle is designed with special acoustic padding to prevent unwanted signals from entering the line or modem. The normal maximum data rate that is supported through acoustic coupled modems is 1200 bps.

In 1977 the Federal Communication Commission (FCC) implemented a certification program which allowed modem vendors to connect directly to the phone lines rather than through the acoustic coupling. The modems are registered in one of three ways: permissive, programmable, or fixed-loss loop. This physical interconnection is also shown in Figure 8.52.

The permissive connection is accomplished through the most common RJ11C telephone jack. The modem must limit the output line signal to a maximum of –9 dBm. The RJ455 represents the programmable data jack connection. The local telephone company sets the signal level through this device from 0 to –12 dBm which compensates for the distance of the telephone wire to the local central office. The fixed-loss loop allows modem transmission at a fixed –3 dBm, and the connection is made through the RJ415 data jack. The purpose of registration and different devices is to allow for maximum modem performance without damaging the public telephone system.

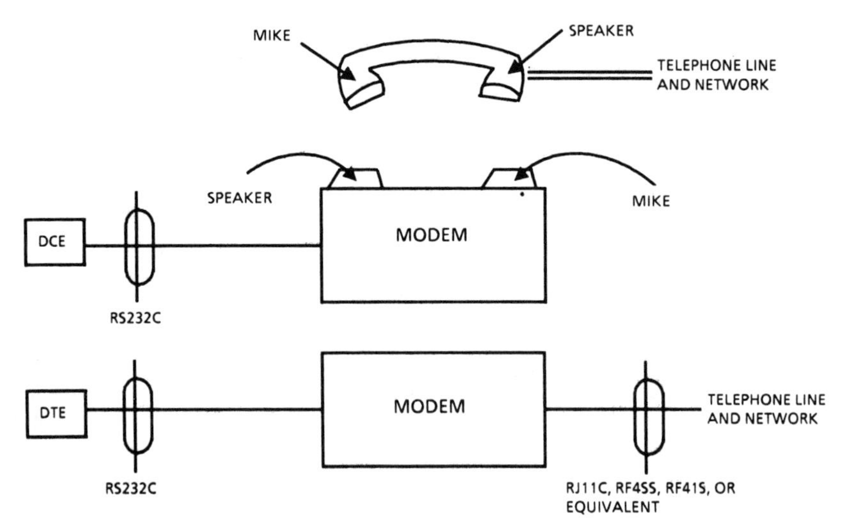

Figure 8.52: Two Line-Side Modem Interface Methods

High performance modems which use exotic modulation techniques and incorporate line adaptive equalization require hardwire interconnection to the telephone network as discussed above. The telephone line can be accessed either by dial-up or by a dedicated telephone wire which interconnects the two data facilities which house the DTEs. The best possible performance is, of course, dedicated private wire because the telephone company can have some control over the circuit quality since it is always the same circuit. When dial-up lines are used there is no guarantee that the same circuits are used. One must therefore match the performance of the modem to the performance of the telephone line.

Modems can also contain many other features including various self and line testing capabilities and unattended operation.

The purpose of many of the RS232C and RS449 pin functions is to accommodate the operation of the modem particularly in establishing a connection to the far end modem. This process is complicated and relatively time consuming. Therefore, the DTE must send a requesting command to the modem. With such a command, the modem then contacts the far end counterpart, sends it a carrier signal, receives a carrier, the far end notifies its DTE that a connection is in the making and gets prepared. Next the two modems perform the self-equalization process (if a part of the modem capability), and finally the two modems turn back to their respective DTEs and indicate that the data set (or modem) is ready. Some multiplexers, particularly those that operate in the FDM mode at an equivalent line side rate of 9.6 kbps or lower, incorporate the modem functions. It is, therefore, not necessary under these circumstances to require a modem on their output channel.

8.6.3 The Automatic Calling Unit (ACU) DCE

Terminal devices can operate in an unattended or automatic mode of operation. If they are hardwired together with a private dedicated telephone line, then the modems merely turn on and off according to the DTE commands. If dialing a telephone number is required, an Automatic Calling Unit (ACU) is needed. These units can be a stand alone piece of equipment as shown in Figure 8.53, or this function can be incorporated into the communication controller or a modem.

8.6.4 The Digital-to-Digital DCE

Historically, the DCE has been the modem (sometimes called a data set). However, modems typically operate at speeds of 9600 bps or less. The demand for greater data rates is on the increase. It is being driven by several factors including the structured communication architecture of

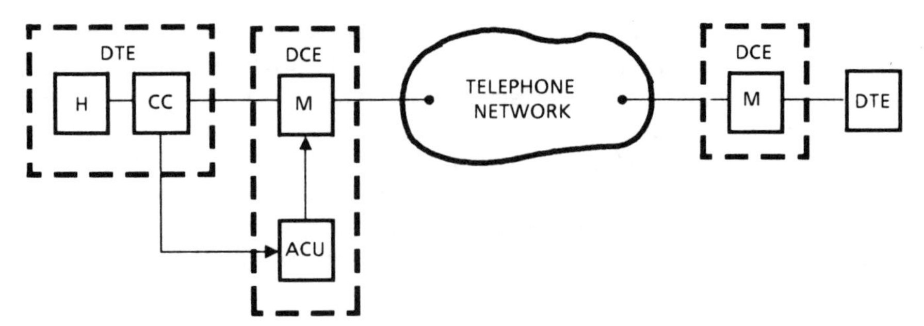

Figure 8.53: The ACU DCE in the Data Communication System

highly distributed systems. Graphics is also beginning to push the demand for higher data rates.

The Bell System has introduced a pure high speed digital service called DATAPHONE Digital Service (DDS) which provides 56 kbps of full duplex transmission. When this service is employed, a digital-to-digital DCE is used. Bell provides the user with a Data Service Unit (DSU) which performs decoding, timing recovery, synchronization, and generation and reconstitution of control signals. If the user furnishes a clock, then a Channel Service Unit (CSU) is substituted for the DSU.

There are several multiplexers which can be classed as digital-to-digital DCEs. Many of the multiplexers discussed in the previous section generate pure binary output. These units generally produce output rates of 56 kbps or 1.544 Mbps. They are basically digital-to-digital DCEs and feed a pure digital service provided by terrestrial or satellite carriers.

8.7 THE COMMUNICATION NETWORKS

A great deal has been said about telecommunication networks in this text. This discussion will only address those key data communication related issues.

8.7.1 The Public Voice-Grade Analog Network

Access to the public voice-grade analog network can either be dial-up or dedicated circuit. The dial-up or switched call has the advantage of being universally available and less expensive than leased dedicated circuits when usage is low. But a different circuit path is used with each dial-up call. Therefore, the circuit quality can vary from one call to the next. However, if a circuit fails, a call can be re-established. Switched networks are normally divided into several services. The three most common public voice-grade switched network groups include the common Direct

Distance Dialing (DDD), Wide Area Telephone Service (WATS), and the Other Common Carrier (OCC) dial-up services. The United States Government supports two additional broadly used switched networks including the Federal Telecommunication System (FTS) and AUTOVON (a Department of Defense system). All of these voice-grade switched systems reliably support data transmission up to 4800 bps.

A leased or dedicated private line has the advantage of being most cost effective if its usage is high. It has another advantage in that it is always available and a connection is virtually instantaneous between the two DCEs. And because it is a dedicated link, adjustments can be made to optimize its quality. These adjustments are referred to as line conditioning. While speeds of up to 4800 bps are possible without conditioning, generally at speeds above 1200 bps, some form of conditioning is required to overcome the effects of delay and attenuation distortions.

Conditioning on dedicated circuits is available in two forms: "C" and "D." "C" conditioning limits the amount of envelope delay distortion. The more compensation is made, the better the circuit performance. Therefore, one can order various levels of C conditioning from C1 to C5 of which C1, C2, and C4 are for commercial use while C3 and C5 are reserved for government lines.

"D" conditioning is for high performance circuits and limits C-notched noise and is normally used with 9.6 kbps modems. D conditioning is sometimes called High Performance Data Conditioning (HPDC)

There is naturally an added cost with each type and level of conditioning ordered for dedicated wire all of the way from a voice-grade dedicated unconditioned circuit to one that is both C and D conditioned. These dedicated data transmission lines are currently referred to as 3002 channels. One could therefore have a "Type 3002 with C4 conditioning."

By specifying and ordering Type 3002 non-conditioned and conditioned circuits, the telephone company guarantees its performance to a set of specifications corresponding to the type and level of conditioning.

Other voice-grade leased circuits are available from satellite vendors such as RCA, Western Union, and American Satellite. Also voice-grade leased circuits are available from other common carriers such as MCI, GTE SPRINT, and ITT.

8.7.2 The Public Sub-Voice Grade Networks

There are a variety of narrowband switched public sub-voice grade networks. These circuits include TWX which offers services to 150 bps and TELEX which offers a switched sub-voice grade capability at 50 bps.

There are a series of sub-voice grade leased services which support data rates up to 150 bps. These dedicated leased circuits include American Telephone and Telegraph (AT&T) 1000 lines plus those provided by Western Union, MCI, ITT, GTE SPRINT, and others.

8.7.3 The Public Broadband Networks

Broadband dedicated circuits are based on the T-carrier system. A T1 system[26] operates at 1.544 Mbps. The logic of this statement follows. The T1 is the result of 24 multiplexed analog voice channels. Each voice channel is sampled 8000 times per second and quantized into 128 possible discrete levels which are represented by 7 binary digits. Therefore, the resulting data stream is:

(8000 samples/sec) (7 bits/sample) = 56 kbps

There are 24 voice channels into one T1. And one bit is added to every 7-bit sample for timing and control. There is also one frame bit added to each group of 8-bit samples for the 24-voice channels. Therefore, there are:

(24 channels) (8 information bits/channel/frame) = 192 information bits/frame

and:

(192 information bits/frame + 1 control bit/frame) = 193 bits/frame

There are 8000 frames every second, therefore:

(8000 frames/sec) (193 bits/frame) = 1.544 Mbps.

Thus, broadband leased services come in data rates generally above 10 kbps, and specifically at 56 kbps or 1.544 Mbps.

8.7.4 Access to Packet Switched Networks

The packet switched digital network (PSDN) actually represents a hybrid of both dedicated and switched circuits. The PSDN, if you recall, is accessed through an Interface Message Processor (IMP) from the DCE. This is usually accomplished by a dial-up circuit, although dedicated connections to an IMP are possible especially if the DTE is a host processor which uses a communication controller with a large number of ports into the PSDN.

8.8 SUMMARY REMARKS

Figure 8.54 attempts to summarize the data communication model for this chapter. The approach used in this presentation is a bit unorthodox in that most approaches begin with the public network and move out toward the DTEs.

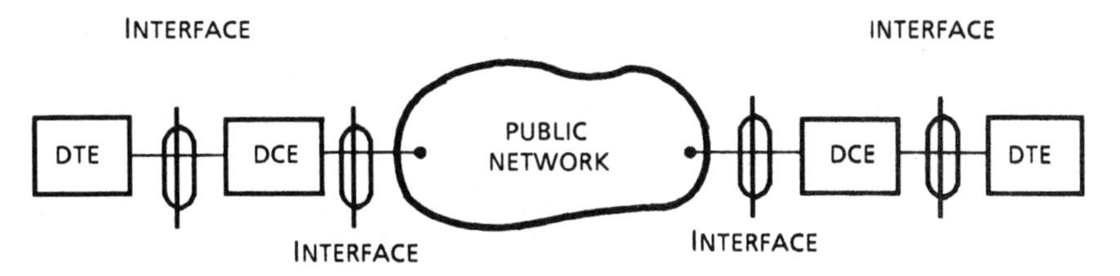

Figure 8.54: The Total Data Communication Model

Our textbook deals heavily with these networks: their components, attributes, and policies governing their operations. This gave rise to the approach used here by first assuming that the DTE-to-DTE link was initially best served as transparent or a solid connection. Then the DCE was exposed as a necessary part of the interconnection. And finally the ultimate reliance on the public networks themselves was discussed. This placed more emphasis on the DTE and DCE than is normally given. Hopefully, the overlap of this approach with the other chapters of the text gives the reader a good overview of computer networking and data communication proportion.

REFERENCES

1. Housley, Trevor, *Data Communications and Teleprocessing Systems,* Prentice-Hall, Inc., Englewood Cliffs, N.J., 1979, pp. 16-18.

2. *Reference Data for Radio Engineers,* Fifth Edition, Howard W. Sams & Co., Inc., A subsidiary of International Telephone and Telegraph Corporation, New York, N.Y., 1974, pp. 30-38, 39.

3. McNamara, John E., *Technical Aspects of Data Communications,* Digital Press, Digital Equipment Corporation, Bedford, MA, 1977.

4. Doll, Dixon R., *Data Communications, Facilities, Networks, and Systems Design,* A Wiley-Interscience Publication, John Wiley & Sons, New York, N.Y., 1978, p. 241.

5. *IBM Systems Reference Library,* "General Information-Binary Synchronous Communications," File No. TP-09, Order No. GA27-3004-2, IBM Systems Development Division, Publications Center, Research Triangle Park, NC, 1970.

6. Tanenbaum, Andrew S., *Computer Networks,* Prentice-Hall, Inc., Englewood Cliffs, N.J., 1981, pp. 167-172.

7. *IBM Systems Reference Library,* "IBM Synchronous Data Link Control General Information," File No. GENL-09, Order No. GA27-3093-2, IBM System Communications Division, Publications Department, Research Triangle Park, NC, 1979.

8. *IBM Systems Reference Library,* "Advanced Function for Communications System Summary," File No. GENL-09, Order No. EA27-3099-1, IBM Systems Communications Division, Publication Center, Research Triangle Park, NC, 1975.

9. *IBM Systems Reference Library,* "IBM Data Communication Device Summary," File No. S370-09, Order No. GA27-3185-1, IBM Corporation, Research Triangle Park, NC, 1980.

10. Gray, James P., and Charles R. Blair, "IBM's Systems Network Architecture, *Datamation Magazine,* April, 1975, pp. 51-56.

11. *IBM Systems Reference Library,* "Systems Network Architecture, General Information," File No. S370-09, Order No. GA-3102-0, IBM Systems Development Division, Research Triangle Park, NC, 1975.

12. *IBM Systems Reference Library,* "Systems Network Architecture Introduction," File No. S370-09, Order No. GA27-3116-0, IBM Systems Communications Division, Research Triangle Park, NC, 1976.

13. Cypser, R. J., *Communications Architecture for Distributed Systems,* Addison-Wesley Publishing Co., Reading, MA, 1978.

14. *IBM Systems Reference Library,* "Systems Network Architecture Format and Protocol Reference Manual: Architectural Logic," Order No. SC30-3112, IBM Corporation, Triangle Research Park, NC, 1980.

15. Folts, Harold C., "Coming of Age: A Long-Awaited Standard for Heterogeneous Nets," *Data Communications Magazine,* January, 1981, pp. 63-73.

16. Tanenbaum, *op. cit.,* pp. 7-9, 115-119.

17. Drukarch, C. Z., et. al., *X.25: The Universal Packet Network Interface,* Proceedings of the Fifth International Conference on Computer Communications, North-Holland Publishing Company, Amsterdam, Holland and New York, N.Y., October, 1980.

18. Folts, Harold C., *McGraw-Hill's Compilation of Data Communications Standards,* Edition II, McGraw-Hill Publications Co., New York, N.Y., 1982.

19. *IBM Systems Reference Library,* "X.25 NCP Packet Switch Interface General Information," Order No. GC30-3080, IBM Corporation, Triangle Research Park, NC, 1980.

20. Shannon, Claude, E. and Warren Weaver, *The Mathematical Theory of Communication,* The University of Illinois Press, Urbana, IL, 1964, pp. 100-107.

21. Nichols, E. A., J. C. Nichols, and K. R. Musson, *Data Communications for Microprocessors with Practical Applications and Experiments,* McGraw-Hill Book Co., USA, 1982.

22. *HP Training Manual,* "Data Communication Testing," Manual Part No. 5952-4973, HP Delcon Division, Mt. View, CA, 1980.

23. *HP Training Manual,* "Guidebook to: Data Communications," Parts No. 5955-1715, General Systems Division, Santa Clara, CA, 1977.

24. Franta, W. R. and I. Chlamtac, *Local Networks,* Lexington Books, D. C. Heath and Company, Lexington, MA, 1982.

25. Martin, James, *Communication Satellite Systems,* Prentice-Hall, Inc., Englewood Cliffs, NJ, 1978, pp. 204-207.

26. Bellamy, John, *Digital Telephony,* A Wiley-Interscience Publication, John Wiley & Sons, New York, N.Y., 1982, pp. 50-53.

chapter 9

AN INTRODUCTION TO THE LOCAL AREA NETWORK CONCEPT

John E. Hershey
The BDM Corporation
4800 Riverbend Road
Boulder, Colorado 80301

William J. Pomper
8032 Fox Ridge Court
Boulder, Colorado 80301

9.0 INTRODUCTION

It is doubtful that anyone is entirely sure what a local area network (LAN) is. It is truly a case in which "I'll know it if I see it" applies with a vengeance. Some LANs, or LAN predecessors, have been neither "local" or true networks. Yet, all LANs seem to share some visceral attributes.

Rather than trying to define a LAN in terms of what it is, it may be instructive to attempt a definition or at least a connotation, in terms of what it does. There are many ways to do this; the way we have chosen is through a discussion of a famous LAN predecessor — the Additive Link On-Line Hawaiian Access (ALOHA) system. The study of this particular system is valuable for three reasons. First, it is historically important. It is a unique system that did the job for which it was developed, and in the course of its construction and operation, spurred both research and technology. Second, the reasons for the ALOHA development lie in economics and this is as valid as any other motivator for LANs today. Third, a variant of the ALOHA operating protocol is an important channel or "medium" sharing scheme today.

The history of ALOHA started in the fall of 1968 at the University of Hawaii. There was a need to link computers to computers and consoles to computers. What made the problem particularly difficult was the geography. As described by Abramson (1970), the sites that needed to be linked together included locations near Honolulu, Hilo, and locations on the islands of Oahu, Kauai, and Maui along with other locations within 200 miles of Honolulu. See the map in Figure 9.1 for a pictorial layout.

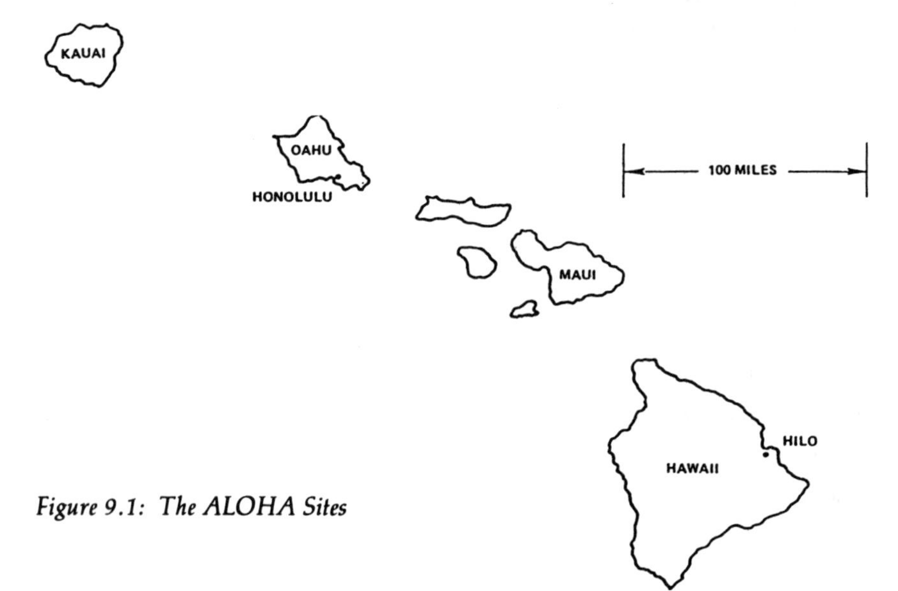

Figure 9.1: The ALOHA Sites

Telephone circuits could have been used for each link, but the options available using telephone were more or less limited to either dedicated lines or dial-up service every time a block of information needed to be sent. That both of these options were deemed unacceptable is readily apparent once we examine the nature of the traffic to be passed.

The ALOHA traffic consisted of short data blocks — often just a single line of alphanumeric characters entered by an operator at a terminal for the purpose of remote job entry. The communications are thus seen to be sporadic bursts with random and perhaps long inter-block generation intervals. The communication flow is also highly asymmetric; little, if any, traffic flows in the other direction, i.e., from computer to console site. Telephone circuits were clearly not a good match for this type of traffic. Dedicated lines were deemed too costly and wasteful; dial-up service for every block, while also very expensive, was too slow.

The ALOHA system was devised to provide a cost-efficient channel for transmitting the sporadically generated information described above. What was done involved the allocation of two 100 kHz wide UHF channels. One channel served as the incoming channel from all outstations to a computer that served as an interface message processor for the main computer. The other channel served as a channel from the interface message processor to all outstations.

The traffic from the outstations to the main computer consisted of packets of 704 bits allocated as follows: 80 eight-bit characters, 32 identification and control bits, and 32 parity bits. The transmission rate selected was 24,000 bits per second. The time for transmission of a packet was 35 milliseconds; 29 milliseconds were required for the 704 bits of the packet, and the additional 6 milliseconds were used for synchronizing the receiver. If the incoming packet were received correctly, i.e., if the parity checks were passed, then a confirmation message was sent to the originating station. A packet would not be accepted and no confirmation message sent if the parity checks were incorrect. This could happen via one of two mechanisms. Either channel noise could cause bit inversions or deletions, or two packets from two different outstations could overlap or "collide." If this happened and the outstations received no confirmation, they would both repeat their packets until an acknowledgment was forthcoming. To prevent the outstations from continually overlapping their packets, the retransmission epochs were randomly chosen by both outstations.

This type of transmission protocol, wherein each information source (outstation) transmits whenever it wishes, regardless of the state of the channels' usage, is called "pure contention." It is very inexpensive to implement as no channel sensing apparatus, which would detect another user's presence, is required. Neither is any sophisticated timing or synchronization system needed. This, the simplest variant of the ALOHA Protocols, is very effective for low channel utilization scenarios. (Channel utilization is a measure of the rate at which a channel is actually carrying traffic referenced to its theoretical maximum rate.) The ALOHA protocol is important because it is the ancestor of the current "collision detection" protocols in use today. But more important, at this stage, the ALOHA system was a unique way of sharing a medium — a UHF channel — among many users without the benefit of strict isochrony, i.e., there was no central time reference, rather the traffic generators operated totally independently of each other.

The most important thing to note about ALOHA, as we have said, is that it provides for the sharing of a common medium. But ALOHA's geometry was certainly not "local." After all, it extended over hundreds of miles. What, then, is "local"? What should the term denote? The term is fuzzy and hence any demarcation will also be fuzzy. Cotton (1982) has, however, offered what we believe is a useful way of viewing the term. He suggests that "local" implies system serving users in the same building or on the same campus. Building on Cotton's order of magnitude breakdown for distances and diagramming it we have the picture shown in Figure 9.2.

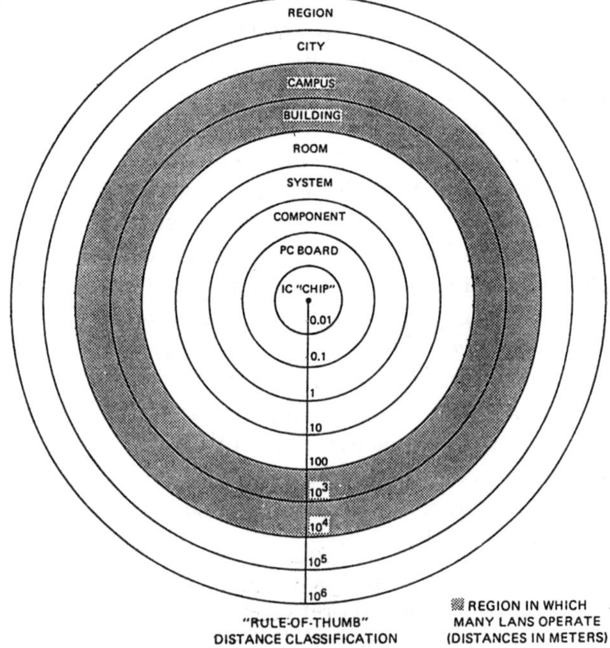

Figure 9.2: "Rule-of-Thumb" Distance Classification

The UHF channel of ALOHA was, in a sense, a bus. Clark et al. (1978) advise that although a LAN medium and a computer bus structure seem very close at times, we should be inclined to view them as separate functions on a philosophical ground. The reason is that a LAN, in general, connects autonomous nodes together rather than items that depend upon a central function. Philosophy aside, it is still very useful to think of many LAN mediums as a bus. The bus structure is quite prevalent for many LANs today. A bus structure is diagrammed in Figure 9.3 along with its cousin "the unrooted tree." (More will be said on network topologies in a later section.)

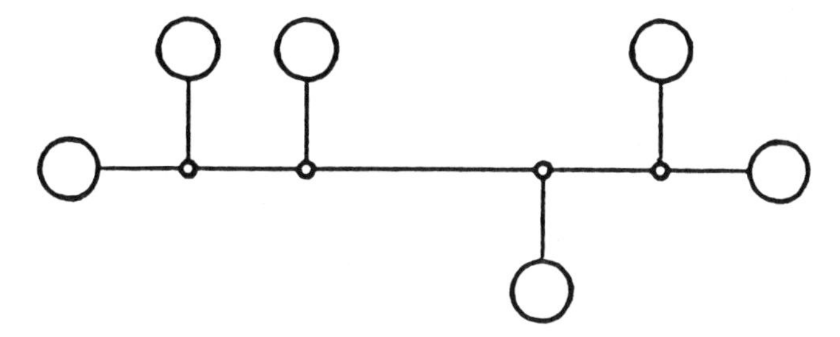

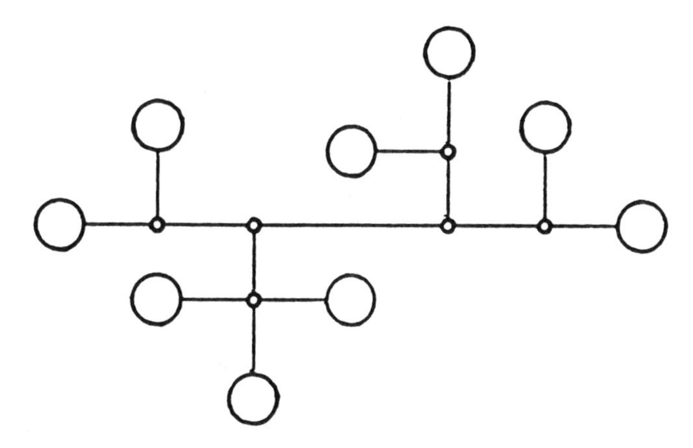

Figure 9.3: Bus and Related Topologies

9.1 LOCAL NETWORK ARCHITECTURES AND ACCESS CONTROL

Local networks are generally divided into two broad and ill defined categories. These are BASEBAND local networks and BROADBAND local networks. Baseband and broadband are terms which relate to whether or not modulation techniques are employed on the transmission medium. A LAN defined as a baseband network implies that the data, when transmitted over the medium, is placed directly on the medium in digital form. That is, the data is represented by shifts or changes in a DC voltage level.

Broadband implies that some form of modulation is used when transmitting data over the medium. Broadband LANs use techniques developed by the cable TV industry. In many cases, broadband LANs use multiple carriers (frequency division multiplex or FDM) to put several independent channels on to the transmission medium. An example of a baseband LAN is the ETHERNET. Examples of broadband LANs are WANGNET and MITRENET.

9.2 NETWORK TOPOLOGIES

There are a number of basic network topologies commonly used in LANs. By network topology, we mean how nodes on the network are interconnected, and this can affect a number of key items. The topology of the network affects its reliability, the complexity of the routing scheme required at each node, and, in many cases, the type of access control mechanism to be employed.

9.2.1 The STAR Network

A generic star network is illustrated in Figure 9.4a. In the star network configuration, all nodes are interconnected through a single central node or switch. All communications between nodes take place through this central node. A Private Branch Exchange (PBX) is a good example of a central controller in a star type of network, and indeed the newer, all digital, PBXs are increasingly being used for local networking in addition to voice applications.

Notice that in the star configuration, the routing decisions at nodes other than the central node are quite simple. All traffic is sent to the central node. At the central node the routing decision is simply a mapping of a message address to an outgoing link. The routing logic is fairly simple since there are no multiple routes between the central node and the destination nodes.

The star configuration requires that resources of sufficient capacity, so that the central node is capable of handling connections between many nodes, be concentrated at one central location. This can make the central node complex. For this reason, the reliability of the star network is questionable. Since the central node is crucial to continued network operation, a failure here can bring down the entire network. Of course, there are the obvious advantages to having one centralized control point in a network for operations control and monitoring as well as maintenance.

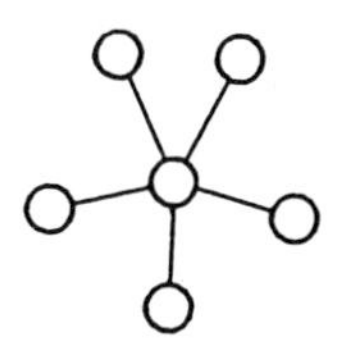

Figure 9.4a:
The Star Topology

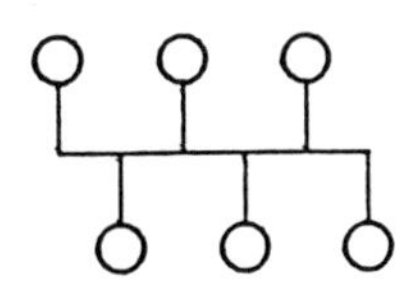

Figure 9.4b:
The Bus Topology

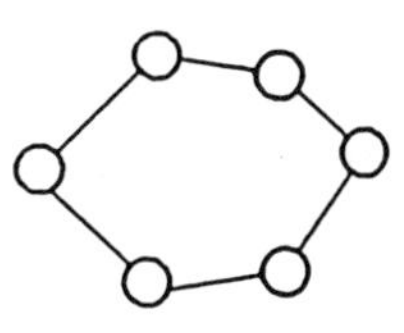

Figure 9.4c:
The Ring Topology

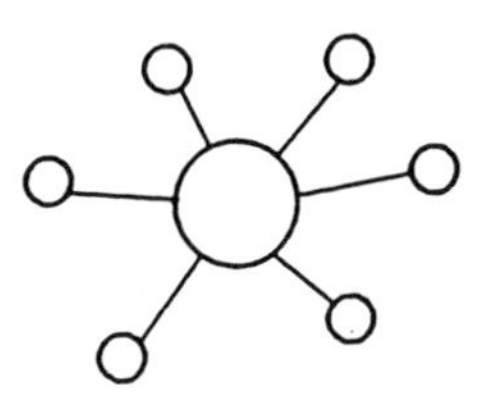

Figure 9.4d:
A Variant of Ring Topology

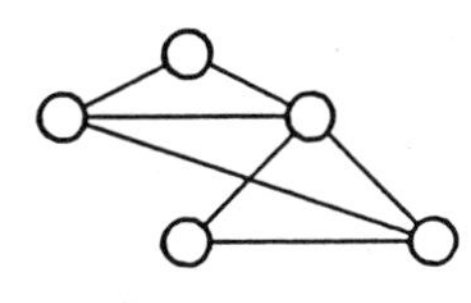

Figure 9.4e:
The Mesh Topology

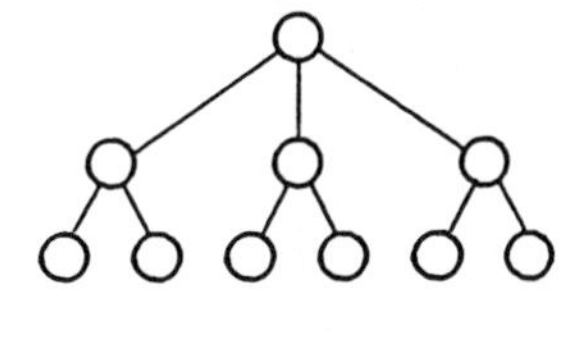

Figure 9.4f:
The Tree Topology

9.2.2 BUS Networks

The bus topology is shown in Figure 9.4b. It is probably the most commonly used topology in LANs. As shown in the figure, each node is interconnected via a single common transmission medium. Data transmitted by a node propagates in both directions on the bus to the other network nodes. Notice that no routing decisions are required at the originating node. However, receiving nodes must screen incoming traffic for an address, unless they wish to receive all traffic passed over the bus.

Bus networks are generally resistant to the failure of individual nodes, since traffic does not pass through the nodes as it propagates over the bus. However, it should be pointed out that there are node failure mechanisms which can cause the entire network to fail. These mechanisms are primarily related to the fact that access to the bus must be controlled and certain types of node failures can cause the failed node to jam the other nodes' transmissions. Another advantage of the bus architecture is that it tends to minimize the amount of cabling required to interconnect dispersed nodes (when compared with the star configuration, for example).

Since nodes on the bus share a common transmission medium, this implies that some form of access control or multiple access scheme is required in order to prevent nodes from jamming or interfering with each other's transmissions. In broadband networks the access control can be some form of frequency division multiple access (FDMA), time division multiple access (TDMA), or a combination of the two. Baseband networks cannot use FDMA techniques, and thus baseband LANs must use some form of TDMA.

9.2.3 RING Networks

A network using ring topology is shown in Figure 9.4c. This type of architecture is closely related to the bus architecture previously discussed. The network nodes are simply connected in a loop or ring configuration. In Figure 9.4c, nodes are in the transmission path, i.e., messages passing between non-adjacent nodes must pass through, or be relayed by intermediate nodes. This is a common architecture for LANs. Traffic generally flows in one direction around the ring. A somewhat different physical implementation of a ring is shown in Figure 9.4d. Lines from each node are brought to a central location called a relay or wire center. Here the nodes are interconnected in a ring configuration through use of relays or switches. Use of a relay center allows easy bypassing of a failed node plus it simplifies the task of adding new nodes to the ring.

The routing logic required at each node is simple. The originating node just sends the messages to the next node on the ring. Other nodes simply pass the message around the ring in the same direction until it reaches the destination node. The destination node recognizes its address and removes the message from the ring.

Ring networks are generally designed so that a node which fails can be easily bypassed. The access control structure for this type of network is generally distributed so that, unlike the star network with its centralized control structure, failure of a node in the ring does not necessarily cause failure of the entire network.

9.2.4 Other Topologies

Two other network topologies of some interest are the mesh or unconstrained topology (Figure 9.4e) and the tree topology of Figure 9.4f. The tree topology is very ordered and allows for a well defined routing algorithm at each node. Note that there is only one route between each pair of nodes.

Perhaps the best example of the mesh network is the public switched telephone network. Such networks generally evolve because of geographic constraints on node locations and a desire to optimize the utilization of the circuits interconnecting the network nodes. Such unconstrained networks generally require relatively complex routing decisions at each node as compared to the networks with more constrained architectures. Both topologies described above are rarely used in LANs.

9.3 ACCESS METHODS

The important network topologies used for LANs, i.e., the previously discussed star, bus, and ring architectures each present different problems in the control of user access to the medium. In the star network, if the central node has sufficient capacity, an outlying node can transmit at any time without interfering with another node's transmission. Such is the case because each node is independently connected to the central node. Alternately, for example, a node may be required to ask the central node for permission to transmit prior to starting a transmission. Another alternative is for the central site to poll each of the other nodes in turn.

Most LANs in use today are based on either the bus or ring architecture. In the bus and ring configurations typically there is no designated intelligent central controller. The control function is usually distributed among the network nodes. In addition, since the nodes share a common transmission medium, it is possible for nodes to interfere with each others' transmissions. Various distributed access control schemes are used in LANs.

Certain control strategies are used more often for bus types of LANs while others are frequently used for ring networks. These strategies are described below. While certain strategies are favored for certain network architectures, in general, each of the access control methods described below could conceivably be used on either type of network.

9.3.1 BUS Access Control

The "pure contention" scheme described in the introduction to this chapter is one type of a distributed control scheme applicable to bus oriented LANs. However, while simple and relatively inexpensive to implement, pure contention strategies have certain drawbacks particularly as the utilization of the medium increases. Since nodes can transmit at any time, transmissions by two or more stations can overlap or "collide" causing loss or corruption of messages. In order to minimize the time periods in which transmissions could overlap, the slotted ALOHA protocol was developed. This protocol required that stations start a transmission only at the beginning of well defined, fixed length, time slots rather than at any time. Transmission duration is slightly shorter than the length of a time slot. By constraining transmissions to begin only at these fixed times, the amount of time wasted per collision is reduced from a maximum of two transmission times to one transmission time (or the duration of one time slot). The reason for this reduction is that since stations can only start a transmission at the beginning of a time slot, collisions can only occur near the beginning. If such a collision does occur, only a single time slot is affected. This is not the case for the pure contention mode of operation. For example, in the pure contention mode, a station could be nearly finished with its transmission when a second station decides to begin a transmission. Now the collision occurs near the end of the first station's transmission and at the beginning of the second station's transmission, and both transmissions are corrupted. Thus, up to two transmission periods are wasted.

Unfortunately, the slotted ALOHA technique requires that precise timing information be known at each station. Thus, each station must include a certain amount of hardware to synchronize its transmissions.

Both of the ALOHA protocols mentioned above are not very efficient in terms of channel utilization. In an effort to minimize the number of collisions and thus to increase the utilization of the medium, the so-called channel sensing schemes were developed. What is meant by channel sensing? In simple terms, channel sensing means that a station listens to the channel prior to transmitting to determine if another station is using it. If the channel is free, the station sends its data packet. If the channel is in use, the station wishing to transmit keeps testing the channel until it detects that the channel is free, and at that point it sends its data packet. This method of operation has become known as Carrier Sense Multiple Access (CSMA). It is also often referred to as listen-before-transmit. (We will examine this access control mode again in our discussion of ETHERNET.) Channel sensing schemes such as CSMA are applicable when the propagation delay over the medium is short when compared to the transmission time durations. This is the case for LANs, i.e., the time it takes to transmit a packet of information is generally much longer than the maximum propagation delay between stations or nodes on the LAN. In certain LANs where high data rates are employed, the minimum allowable packet size must be constrained to ensure that packet transmission time is greater than twice the maximum propagation delay between stations or nodes. The reason for this requirement is explained under the discussion of ETHERNET.

The CSMA approach is widely used in bus oriented LANs. While this approach tends to minimize collisions, it does not eliminate them. For example, consider the case in which two stations become ready to transmit while another station is transmitting. Both will sense that the channel is busy and wait for the channel to clear. When this occurs, they will detect that the channel is free at approximately the same time and will begin simultaneous transmissions, hence, a collision occurs. This type of CSMA operation, where a node or station with data ready to send transmits as soon as the channel is free, is often referred to as "1-persistent CSMA."

One of two methods is usually employed to detect collisions in CSMA LANs. The first method requires that a node sending a packet must wait a certain amount of time for an acknowledgment from the intended receiver that the packet was received correctly. If an acknowledgment is not received within a certain period, a collision or other error condition is assumed to have occurred, and the packet is retransmitted. The second method, known as listen-while-transmit, requires that a node sending a packet listen to the channel while it is transmitting. Essentially, the node receives its own transmission and, in effect, compares what it receives to what it is transmitting. If the two do not agree, a collision is assumed to have occurred. Upon detection of a collision, the node will wait a random amount of time and retransmit the packet. This type of CSMA operation (listen-while-transmit) is called CSMA/CD (Carrier Sense Multiple Access with Collision Detection).

CSMA and CSMA/CD are by far the most often used distributed access control schemes employed with bus oriented LANs. However, it is certainly possible to use other modes of access control with these types of LANs.

9.3.2 RING Access Control

Conceptually, there are many access control schemes which could be employed on LANs using the ring architecture illustrated in Figure 9.4c. However, only some of these access control schemes are of real interest: token passing, message slots, and register insertion.

Recall the basic characteristics of ring networks. Data packets or messages are passed around the ring from node to node in one direction only. Nodes are active, i.e., messages pass through each node on their way around the ring. Therefore, LANs which use the ring architecture include provisions for easily bypassing failed nodes. Only a limited number of messages can be "in transit" around the ring at a given time.

Most of the popular access control schemes for ring architectures use the idea that "somehow" permission to enter messages is passed around the ring from node to node. It is the "somehow" we'll now explore.

Many LANs which use the ring architecture use the concept of token-passing for access control. These networks are often referred to as "token-rings." The idea here is that a special bit pattern called a control token is created and sent around this ring from node to node. Receipt of the token gives a node permission to transmit a message if it so desires. A node which receives the token either pa•ses the token on to the next node, or it holds the token, sends a message, then sends the token on to the next node. Note that the token passing concept is simply a special form of polling.

An alternative to token passing is the use of message slots. Message slots are somewhat analogous to empty envelopes, or empty packets and are circulated around the ring from node to node. Slots are usually of a fixed size in terms of number of bits. A node with a message waiting to be sent waits for an empty slot to pass by. When an empty slot is detected, the node marks it "full," then fills the slot with the message and sends it on around the ring. The message may be removed by either the recipient, or by the oiginator after the message has been circulated completely around the ring.

A third alternative for ring access control is called register insertion. The technique is common at each node and involves the insertion of a shift register into the ring at times, and it is removed at other times. The technique operates as follows. A node having a message to transmit loads the message into a shift register. The node then monitors the traffic on the ring. When the node senses that the channel is idle, or that a point separating two contiguous messages has been reached, it switches the shift register into the ring. The message data is then shifted out of the register onto the ring. Any data received is, by the same mechanism, shifted into the shift register. At this point the register has become an integral part of the ring. It cannot be removed from the ring until it contains no useful information. Also, since it is now a part of the ring, no further messages can be by sent the node until the register is removed from the ring. One method used to decide when to remove the shift register from the ring is to wait until the message has completed a circuit around the ring and returned to the sender. The shift register again fills with the message. Once the register is full, it (and the message) are removed from the network. Another technique for register removal is to wait until the network has been idle for a sufficient time so the register empties (i.e., contains no messages or other useful data), and switch out the register.

Of the three access control schemes described above, token passing has become the dominant strategy for ring access control as far as LANs are concerned. It also is the one which is most likely to become a standard for ring networks. Token passing is also used in some LANs which employ the bus architecture. Again, it is important to be aware that any of the control strategies mentioned for the bus architecture could be used in ring networks.

9.4 EXTENDING LOCAL AREA NETWORKS

LANs allow the sharing of information and resources within a limited geographic area. Many LAN users need to share either information or resources over much larger areas than can be obtained with current LAN technology. For example, a corporation may have offices or divisions in different cities and may need to communicate information among these offices. If these offices are large, each may have its own LAN, and the corporation might want to interconnect these LANs. There are advantages to interconnecting these LANs as opposed to interconnecting individual terminals or computers at the different locations. The reason is, if one desires to interconnect one group of terminals or computers, say M devices total, to another group of devices, say N devices, and one wants each device in the first group to be able to communicate with each of the devices in the second group, one rapidly runs into the well known MxN problem if separate line are used to interconnect the devices. It takes a total of MxN lines to interconnect the two groups of devices. This is illustrated in Figure 9.5a. Group A has two devices (M=2). Group B has three devices (N=3), and the total number of lines required is MxN=6. As the number of devices grows, the number of lines rapidly becomes quite large as does the associated cost of the lines and their supporting hardware and software.

The advantage of interconnecting two LANs, which might support two separate groups of devices, is obvious in Figure 9.5b. Only one connecting line between the two LANs is required to allow all of the devices in each group to intercommunicate.

Two terms relating to interconnecting LANs have come into vogue recently. These are "bridge" and "gateway." These terms loosely describe the complexity of the interface required to interconnect two LANs. Bridges are simple interfaces usually used to interconnect LANs which have the same architecture and use similar access protocols as shown in Figure 9.6a. Use of simple bridges also generally implies that the two LANs are in close proximity to each other. Figure 9.6b shows a bridge used to connect two LANs which use bus architecture.

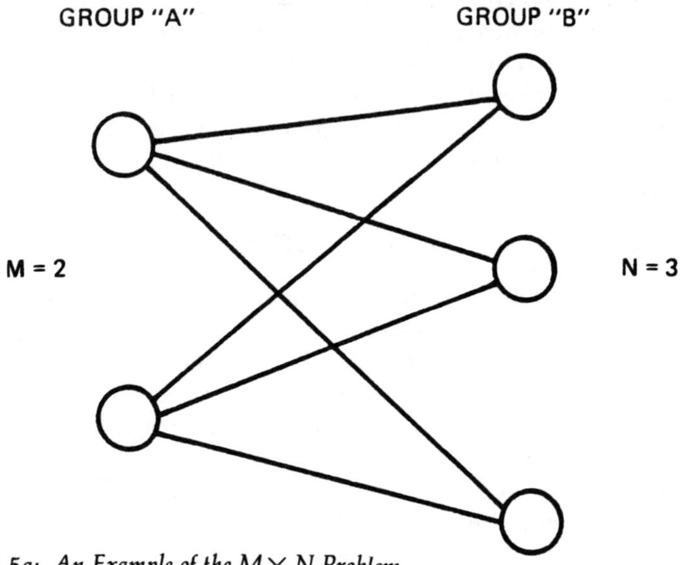

Figure 9.5a: An Example of the M × N Problem

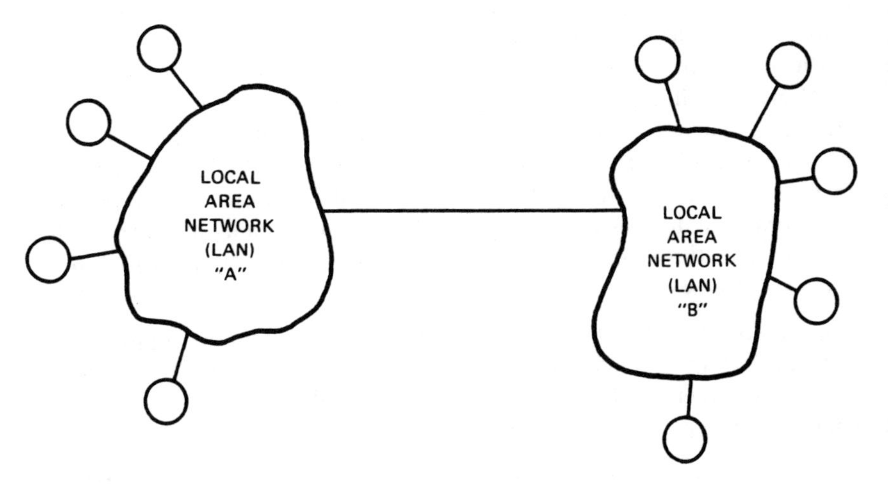

Figure 9.5b: Interconnecting LANs Allows Full Intercommunication between Two Groups of Devices

The function of the bridge is straightforward. It provides a bidirectional interface between the two LANs. Bridges normally perform certain message screening functions. In other words, not all messages sent on one network are repeated on the other LAN. Only those messages sent on one LAN which are addressed to nodes on the other network are allowed to pass between networks via the bridge. Bridges do not perform any protocol conversion. Messages are simply repeated on the second network using the appropriate access scheme. Notice that because bridges can perform a message screening function they can prevent communication between certain nodes on the different LANs. This feature allows bridges to be used in security applications wherein one wishes to prevent communication between certain entities.

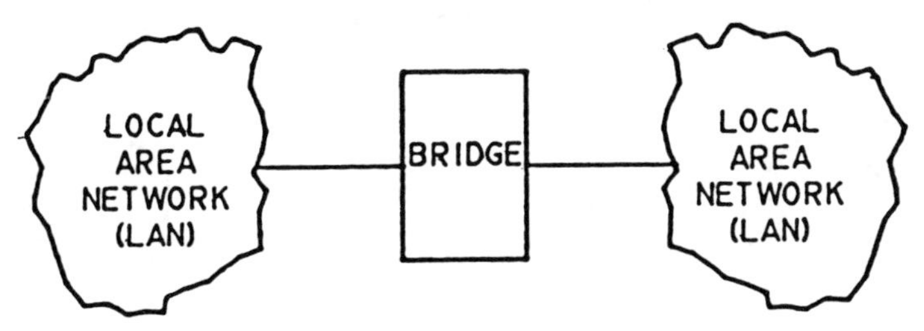

Figure 9.6a: Two Similar LANs Connected by Bridge

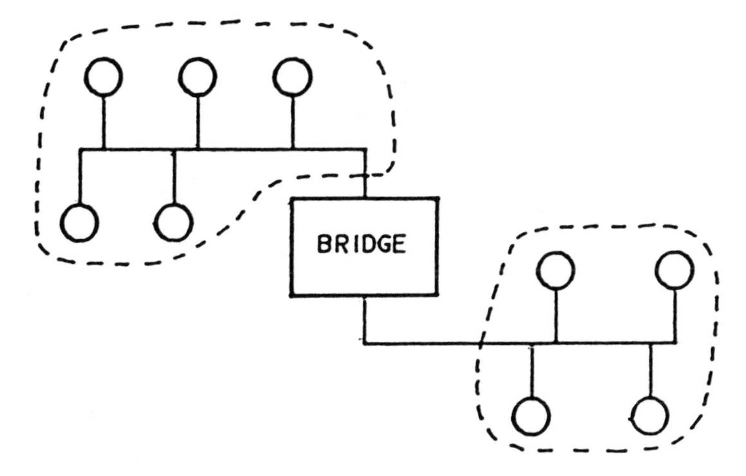

Figure 9.6b: An Example of Two Bus-Oriented LANs Connected by a Bridge

The term gateway implies a much more sophisticated interface. Gateways are required when interfacing dissimilar LANs. Another example of a situation which requires a gateway is given in Figure 9.7. Here we have several LANs, similar or dissimilar, with a large geographic separation, interconnected by an X.25 public data network (PDN). The gateways illustrated must perform a much more complex set of tasks than the bridge previously discussed. For example, the gateways must provide for protocol conversion between the LAN protocols and the X.25 network protocol and vice versa. They must also provide for operation with the various access control schemes used by the LANs as well as the PDN. Other examples of functions which may be required at the gateways include data rate conversion, address mapping from one network to the other, and packet or message size conversions between the different networks.

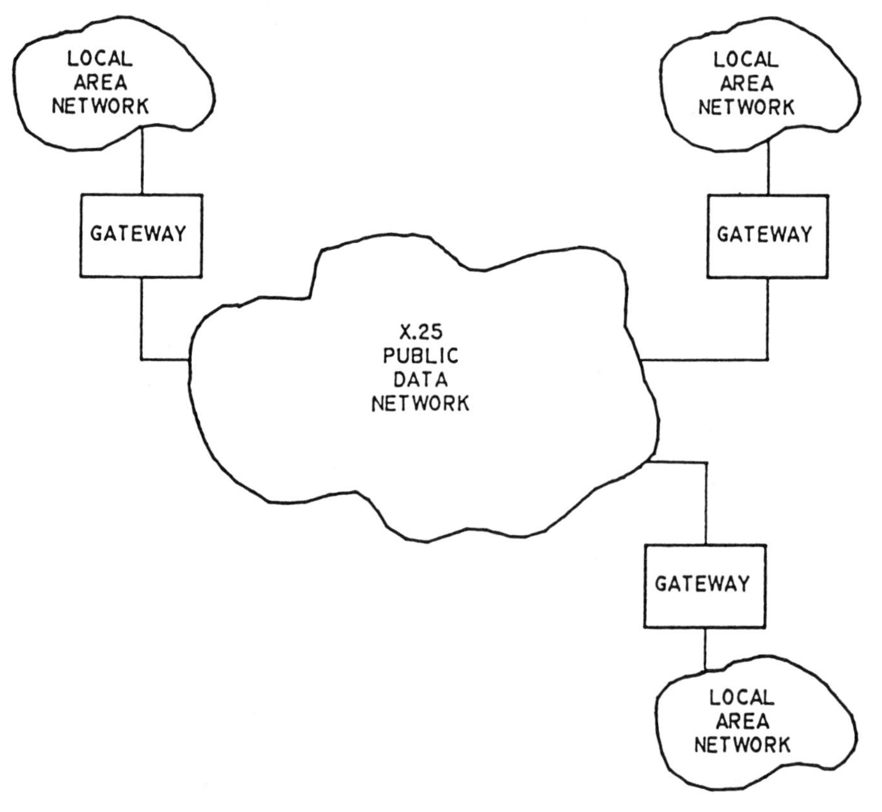

Figure 9.7: An Example of Internetworking Using a Public Data Network and Gateways

9.5 ETHERNET

Perhaps the most important LAN is the Ethernet because it is widely used and its behavior has been thoroughly studied and documented. It is an evolution of the ALOHA concept and yet, in another sense, it is the dual of the ALOHA. This will become clear as we examine the medium architecture and traffic flow control.

The Ethernet concept (see Metcalfe and Boggs, 1976) is a "controlled contention" scheme. It is centered about the LAN medium, or "ether," which is usually taken to be a coaxial cable capable of supporting up to 10 megabits a second over a kilometer or so. The geometry of the nodes on the cable is either an unrooted tree or a bus. The Ethernet is a packet system; there is a field for the address and port of origination. There is also a checksum for error detection purposes.

A node can originate a packet transmission only if it does not sense a transmission in progress. The node looks for presence of "carrier" or bit alternations indicating a packet is in transit, hence the term "carrier sense" or CS. Because the Ethernet allows many nodes to share the same medium, it is a "multiple access" or MA system — thus the partial acronym CSMA. But the Ethernet nodes continue to monitor the medium even when they are transmitting a packet to see if another packet "collides" with the one they are sending. If so, the transmitting node will abort the transmission. This continual monitoring for "collision detection" or CD completes the acronym for the Ethernet system: CSMA/CD. A natural question would be, "If the node does not begin transmission until the medium is clear, how could a collision take place?" The question is a good one, and the answer will lead us to one of the most important design rules of an Ethernet system.

Consider the Ethernet shown in Figure 9.8. The distances between cable joinings are given in meters. Assume that the cable propagates signals at 2/3 the speed of light or approximately $2 \cdot 10^8$ meters per second. Let us assume that Node A wishes to send a packet. Node A senses the medium, finds it vacant, and starts to send its packet. Suppose that Node E also wishes to send a packet. Node E will not learn of A's packet until 3 microseconds after A begins to transmit, and Node A will not learn of the collision until Node E's packet arrives at Node A, which could be as much as 6 microseconds after Node A starts its transmission. If Node A were to send a very short packet, say less than 6 microseconds, Node A might never learn of a collision, as Node A would cease monitoring the medium after it had sent its packet. To ensure that a node will always know if a collision occurs, a fundamental tenet of an Ethernet is that a packet has a *minimum* length equal to twice the propagation time between the two most distant nodes. For our example in Figure 9.8, we see that Nodes C and G compose the node pair exhibiting the greatest distance, 1000 meters, and hence the minimum packet length for the Ethernet of Figure 9.8 is 10 microseconds.

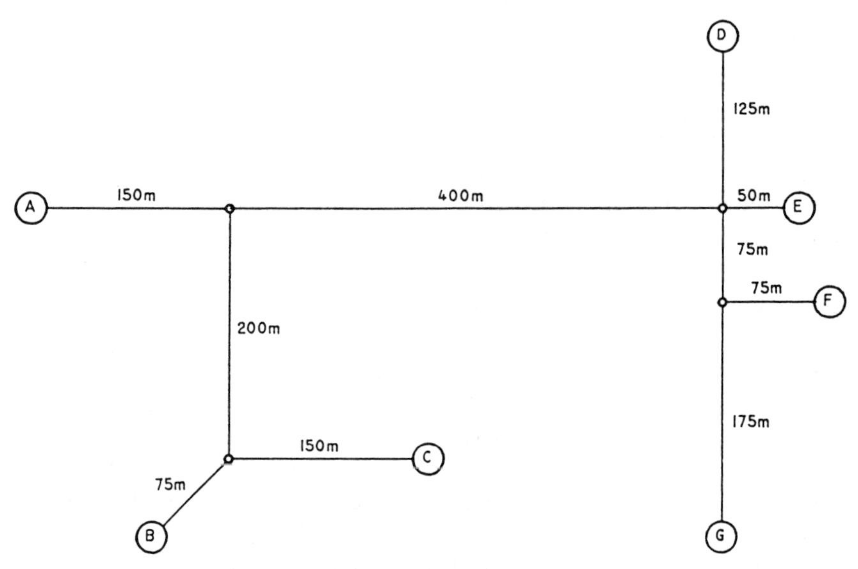

Figure 9.8: Unrooted Tree For Ethernet Example

Unlike ALOHA's pure contention, Ethernet has a definite traffic flow control algorithm. A node, if it detects a collision, will not retransmit immediately but will backoff its retransmission epoch according to the output of a random number generator. The amount of backoff, which translates indirectly to congestion delay, is greatly affected by the number of times the node must try to seize the ether without collision. Even if a packet is sent without collision, it may still arrive corrupted because of noise. The receiving node's *hardware* examines the packet's checksum. If the checksum is incorrect, the packet is discarded without being passed to the node's *software*. The Ethernet LAN thus provides datagram packet service; higher levels in the node's protocol must take care of accounting, sequencing, response, and so on.

There is no master clock for an Ethernet, yet in a sense, there is a type of synchronous timing because time "slots" are established. In a time slot, there may be no traffic, a collision, or the start of collisionless transmission. The transmission time is determined by the packet length. A new time slot is determined by the end of a packet transmission.

The Ethernet can be a very efficient LAN even under high load providing the designer judiciously sets the system parameters. As in most cases of network architecture, it is folly to zero in on only one parameter to the exclusion of all others. For example, it should be obvious that the efficiency of a heavily loaded Ethernet can be forced arbitrarily close to unity by letting the packet size grow without bound. Remember that all we required is a *minimum* packet size. The maximum allowable packet size must be chosen, not solely for net efficiency, but with user traffic and user response considerations clearly in mind and properly weighted.

9.6 THE MEDIUM

The medium is, of course, that part of a LAN through which the messages pass from sender node to receiver node. The most familiar and common medium is copper, usually in the form of a coaxial cable and sometimes in the form of a twisted wire pair. The physics of electrical conduction are heavily dependent on the skin resistance of a conductor (see Gallawa, 1977; Hull et al. 1983). The skin resistance causes some frequencies to be propagated with less delay than others. How this affects us becomes readily apparent when we consider that a rectangular pulse can be shown by Fourier decomposition to have a spectrum whose magnitude falls off slowly with increasing frequency. Thus, a short pulse will exhibit spectral energy over a relatively wide frequency band. Because the spectral components will each propagate at slightly different velocities, what starts out as a rectangular pulse on a coaxial cable becomes distorted after some distance. The distortion takes the form of pulse broadening, i.e., the pulse width becomes greater, and the pulse height becomes irregular as the energy becomes spectrally diffused over a greater time period than the pulse's original width. If we send many pulses close together, i.e., operate at a high bit rate, then the pulses will start to overlap and interfere with each other, and we will have intersymbol distortion and a consequent reduction in signal-to-noise ratio with a concommitant increase in errors.

For extremely high bit rate LANs, then, it is no wonder that much interest is being directed to optical fibers. Optical fibers have a number of ancillary attributes which also commend their use for many unusual LAN environments. Some of these additional attributes are:

Increased difficulty of passive "wiretapping"

Immunity to external radio frequency interference and, especially, for military users, resistance against electromagnetic pulse (EMP) threats

Immunity to ground loop problems

Smaller and lighter materials.

A very significant amount of progress has been made in fiber optics over the last decade, and it is now a viable candidate for the LAN medium. Perhaps it is a good idea to review the basic concepts of fiber optics before proceeding with a discussion of its role in modern LANs. First of all, a fiber is merely a waveguide for light. The light is contained by virtue of Snell's law which causes total internal reflection and hence confinement of the photons within the medium. The optical fiber consists of a "core" of material surrounded by "cladding." The difference in the refractive indices between the core and the cladding invokes Snell's law and confines the light to travelling within the core. Following Hull et al. (1983), we shall quickly review the two most important dichotomies of fiber technology: single mode versus multimode and step index versus graded index.

A beam of light through a fiber may enter at a variety of angles determined by the size of the numerical aperture. The particular path followed by the beam through the fiber is the "mode" of the transmission. As you can visualize, a mode whose path involves many reflections travels a greater distance in the core than a mode whose path is "launched" into the fiber at an angle nearly parallel to the fiber. If a transmission involves a set of modes, then we will have pulse distortion somewhat similar to the case just considered for coaxial cables. The answer to avoiding distortion can be pursued in one of two ways. Either we can restrict our fiber optic system to operating in a single mode or we can choose a graded index fiber instead of one with a step index. A step index fiber has a constant index of refractivity. The velocity of light is determined by this index and thus light paths or modes that traverse more core material (by virtue of more reflections) will take longer to travel the fiber. In a graded index fiber, the index of refractivity is carefully varied as a function of the distance from the fiber core's center. The variation in refractivity is engineered so that all modes will propagate at the same rate. A graded index fiber is, in a sense, an example of spatial deconvolution of the pulse distortion function.

The use of optical fibers in LANs has been growing rapidly over the past five years. The fiber is not coaxial cable, of course, and so a companion technology has grown up to support optical LANs. This technology is the technology of optical power division — the sharing of the optical power among the various ports. Perhaps the most popular structure for doing this is through a device known as a "star coupler." (See Hudson and Thiel, 1974; Rawson and Metcalfe, 1978; and Rhodes, 1982, for further elaboration.) First of all, a star coupler is more than a device; it is also a network architecture. The star coupler is the optical power sharing device that sits at the network's "hub" or center as shown in Figure 9.9. The star coupler may be either reflective or transmissive. Generic schematics for these two types of couplers are shown in Figures 9.10a and 9.10b. In a passive, reflective star coupler, the light entering on a fiber is reflected by a mirrored surface and spread over all the fibers in the bundle entering the coupler. The light that falls on the fiber cores is carried to the nodes; the light that illuminates the cladding or falls interstitially between the fibers is wasted. Each node then receives back a portion of its transmitted light. Generally, each node communicates through a beam splitter when using a passive reflective star coupler; this allows each node to use a single fiber as a bidirectional medium. With a passive transmissive star coupler, the light entering on a fiber is spread over all of the exiting fibers. Generally, each node has two fibers — one fiber is used for transmitting to the coupler, and the other fiber is dedicated to receiving from the coupler.

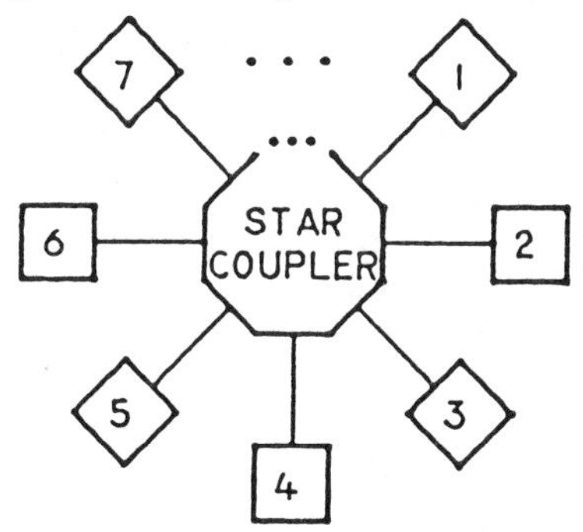

Figure 9.9: A Star-Coupled Optical Fiber LAN

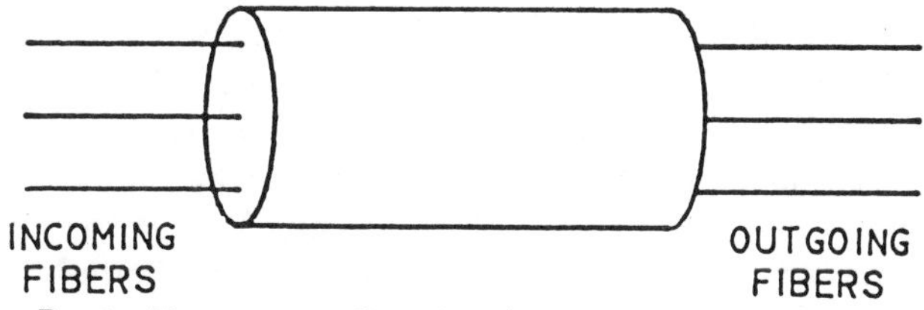

Figure 9.10a Passive Transmissive Star Coupler

Figure 9.10b: Passive Reflective Star Coupler

Fiber LANs, as might be expected, have different optimal operating regions than their metallic medium relatives. The reason for this stems from the much greater channel capacity afforded by the optical fiber.

Two differences are especially noteworthy regarding CSMA systems. The first of these is packet length. To achieve a high network efficiency when using a fiber optic LAN, it is necessary that the packets be relatively long, perhaps on the order of a few thousand bits. The second item concerns what is known as the "training sequence." In order for a receiving node to synchronize itself to a burst of data, it must first recover or "phase" to the clock or bit transition intervals of the incoming data. Recall the extra 6 milliseconds used in the ALOHA packet transmission. This extra time was used for this purpose. The same requirement pertains to Ethernet and other CSMA architectures. An Ethernet packet is thus preceded by a training sequence or preamble. For Ethernets operating at 10 Megabits per second or below, a typical rate for a coaxial based CSMA system, the length of the training sequence is insignificant compared to the packet length — typically the training sequence will take a small fraction of the time it takes to send the packet. In an optical LAN operating at, say, 100 Megabits per second, this is not so. The overhead for the training sequence may be as much as twenty percent.

9.7 SECURITY ISSUES FOR LANs

Public interest and awareness of communications security has never been greater than in the past decade. The interest was probably ignited by the reluctant belief that there was, indeed, a threat to personal privacy posed by a wide range of snoops, and our recent political history served only to accentuate the feelings of uneasiness. These trepidations, coupled with the increasingly dazzling display and accessibility of sophisticated electronic equipment, crystallized the fear that, in the eighties, the electronic interloper had both means and motive to exploit several opportunities to practice his art. Also, as has been pointed out countless times, our traffic was becoming more and more digital and thus more and more amenable to computerized analysis.

Cryptography, an art which has reluctantly become a science, encompasses the techniques for rendering messages unreadable to all but intended, authorized, recipients, was seized upon as a shelter, albeit a partial shelter, in which to seek safe haven from the snoops. Traditionally, cryptography had been the reserved purview of the black chambers of government; the secret organs which served to protect the officially designated secrets of a country. With the computer privacy legislation of the early seventies, it became clear that some sort of public encryption standard was required. Following a fantastic set of machinations, see Hershey (1983), the National Bureau of Standards promulgated such an official standard in 1977 (FIPS PUB 46). Entitled the Data Encryption Standard, or DES, the standard specified an algorithm which, if properly implemented in accordance with other official government directives would provide sufficient protection against realistic threats for an adequate time period. It is almost certain that the DES will play a large role in LAN security. But before we discuss cryptography further, it would be best to review communications security as an integrated discipline.

Communications security (COMSEC) involves four distinct disciplines:

a. Cryptographic Security

b. Physical Security

c. Transmission Security

d. Emanations Security

Each of these disciplines constitutes an entire professional field in its own right, but no area will do the whole job in general. COMSEC is not a plan or a piece of equipment, rather, it is a condition; the condition of freedom from communications exploitation. To achieve this condition, all four of the disciplines listed above must be considered and must interact in a synergistic manner.

There is, quite frankly, often a reluctance to proceed through a formal COMSEC cycle. Viewed from the operational side, the manager who has to get something done and done on time, which may or may not be a realistic demand schedule, often views COMSEC related activities in a hostile manner. The reasons are twofold. First, COMSEC related measures often "get in the way" and can delay an activity. Although responsible for both the project's product *and* its security, the manager pushes the former responsibility to the fore. This is probably because of the second reason which is that the COMSEC discipline does not have a product. Rather, it deals in denial of product. Americans are product conscious and have extreme difficulty in dealing with a product-less activity.

Let us return now to the four disciplines enumerated at the start of the section and comment on each of them:

Cryptographic Security: This is the protection that derives from the use of a cryptographic system, cryptosystem for short. A cryptosystem transforms the plaintext, the classified electrical data, into unintelligible data which is termed ciphertext. This transformation is called "enciphering" or "encryption." There are two important attributes to the enciphering process. First, the process must be reversible, i.e., there must be an inverse process, called "deciphering" or "decryption," in order that the authorized recipient can recover the plaintext.

Second, the enciphering transformation rests on a "cryptographic algorithm" which is usually known to the opposition. The security of the encipherment resides in a protected piece of information called the "cryptographic keying variable." The cryptographic keying variable must be held by both sender and receiver in order for communication to be possible. The security of the communications depends upon the security of the keying variable. The period of time that a particular keying variable is used is termed the cryptoperiod.

Physical Security: It is imperative that the systems which protect communications be absolutely inviolable to physical penetration. The opposition's mission would be relatively trivial if they could somehow enter the "chokepoint" of sensitive information flow. It is essential, then, that appropriate physical safeguards be applied and maintained to defeat physical tapping of plaintext communication lines, plaintext handling equipment, and electronic alteration (sabotage) of any communications security equipment. Of paramount importance is the protection of cryptographic material, i.e., the cryptographic variables.

Transmission Security: Once a plaintext has been completely encrypted into ciphertext, there may still be concerns regarding its transmission. These concerns usually fall into two classes. First, it may be necessary to disguise the fact that a particular site is transmitting information to another specific site, or, in extreme cases, even transmitting at all. Second, it may be necessary to conceal the volume, or times that a particular site is transmitting data. These concerns are the concerns of transmission security.

Emanations Security: Emanations security comprises two problems: classified and unclassified signal emanations associated with plaintext processing equipment. If a machine processing classified information radiates the information, or signals directly correlatable to the classified information, there is a possibility that the emanations can be monitored and exploited. This is similar to the intercept and exploitation of unprotected plaintext communications.

All of the four disciplines just cited could bear on LAN security; we will look briefly at all of them with the exception of emanations security. But before we can do this, we must make an attempt at characterizing the threat to our LAN. What or who is the opposition or interloper, what are they capable of doing, and what are they *willing* to do? The first threat breakdown is traditionally a dichotomy between passive and active attacks. A passive attack is confined to examination and analysis of traffic in the LAN medium. An active attack comprises those measures that the opposition may do to alter, delete, replay, or otherwise insinuate messages flowing within the LAN medium. The study of this latter class, the active attacks, is much more involved than the passive class of attacks and for purposes of this introductory chapter, we will consider only passive attacks.

Passive attacks have their own dichotomy. May the opposition gain access to all messages in the LAN medium, or are they limited to a port which, depending on the port's interface hardware or software, may or may not access all messages flowing in the LAN's medium? An example of this restriction is the Hyperbus LAN which uses four individually selectable high speed busses as a medium to help avoid congestion. A port on the Hyperbus can be hardware configured so that it cannot access a subset of the four busses. A port so configured cannot receive *or* send messages on the subset so proscribed.

The encryption of messages sent on the LAN medium is an extremely involved issue. We first of all have to decide what we wish to protect. Are we concerned if the passive monitor can discern who is sending how much traffic to whom, or are we concerned with simply protecting the information sent? The former case is far more complicated than the latter for if the address and destination fields are encrypted, then each node will have to possess a cryptographic module and cryptographic key and process each message header to determine if the message is for the node. (See Shirey, 1982).

If it is desired only to encrypt the message itself, then a variety of cryptographic schemes may be used. The previously mentioned DES may be employed as the basis of a number of cryptographic processes or "modes." These modes are specified in a publication from the National Bureau of Standards entitled Federal Information Processing Standard Publication No. 81. A worthwhile companion document that addresses synchronization issues is by Pomper (1982).

The DES cryptosystems are examples of what are termed "classical" cryptographic systems. A classical cryptographic system is a system that requires its subscribers to securely transport cryptographic variables ("keys," etc.) prior to communication. This implies the existence and use of a channel separate from the LAN medium. Indeed, the keys are often physically transported by trusted courier.

There is, however, another solution, and this is provided by what has come to be known as a "public key" cryptographic system. A public key cryptosystem (PKC) does not require keys to be separately distributed. A PKC depends on asymmetric complex or "trapdoor" functions — functions wherein the forward mapping, i.e., the computation of $f(x)$ given x is easy, but the recovery of x given $f(x)$ is hard. A number of such functions and related schemes have been found and can be used as a secure cryptovariable transfer channel. The study of PKCs is an exciting contemporary field, and there is much literature on the subject, their implementation and analysis. There are many software packages available for implementing PKCs and even some hardware. The reader is referred to the classic paper by Diffie and Hellman (1977) to help understand these interesting and potentially useful mathematical artifices.

9.8 MEASUREMENT OF LAN PERFORMANCE

The National Bureau of Standards (NBS) is, of course, the leading US authority on metrology, the science of measurement. Their work has gone so far as to include a measurements program on a large network in excess of 250 nodes designed by the NBS Institute for Computer Sciences and Technology and installed on the NBS campus at Gaithersburg, Maryland. An excellent article by Amer et al. (1983) describes the results of the network measurement center. We will draw on this article and relate the important concepts that we believe will be useful to other LAN architects.

The NBS LAN (NBSnet) is a CSMA/CD architecture, but its protocol is not exactly Ethernet's. NBS was eager to study a host of variables and their interactions. To create conditions that did not occur with sufficient frequency, the NBS devised a measurement system that had both a passive mode, i.e., performed only monitoring of net traffic, and an active mode, it could insert traffic and study the network's behavior.

Some of the parameters studied by NBS included basic statistics, such as the data packet size distribution, throughput utilization, distribution of packet interarrival times, collision counts, and so on. In addition, NBSnet researchers are interested in studying some complex "forefront" LAN issues such as network fairness, the equitable division of LAN bandwidth or access, and network stability which is the condition in which network utilization never decreases under ever increasing traffic load. One of life's great surprises, experienced by one of the authors, was learning that throughput for a complex communications system is not necessarily a nondecreasing function of offered load. (Indeed, for some telephone switches, throughput can be drastically reduced by a relatively modest increase in offered load above a critical value.)

The NBS study had some interesting fallout which should serve as an important advisory to LAN architects and implementers. What the study disclosed were both hardware and software errors. Software, as we know is prone to latent errors. Proving program correctness is a field in its infancy. It is not surprising, then, that there was a software error in the user board protocol software. What is noteworthy is that analysis of the measurement program's output led to its quick discovery even though its effect on the network's operation was insignificant *but would not have remained so* under higher network loads. In Amer et al.'s words:

> "NBSnet operations staff was notified of the problem and subsequently programmed a correction into all new user-board software. It is interesting to note that except for insignificant (and unnoticed) degradation of network service, the . . .[problem]. . .was invisible to both network users and the operations staff. Without a measurement center to summarize or report traffic and performance characteristics, the problem could have gone undetected. As the load grew, this problem eventually would have significantly degraded performance."

The message is clear. Users, implementers, and architects of new LANs or LAN protocols are well served by real-world experimentation with and statistical measurement of the behavior of their new systems.

9.9. LANs AND PBXs

As mentioned previously, the star network has significant implications for local area networks. This is so because the medium is often already in place. The voice signals carried in a PBX have little energy above 4 kilohertz. The copper medium is, however, capable of carrying digital signals well above 4 kilohertz without appreciable degradation or attenuation. It should be possible, then, to piggyback a local area network on an already extant PBX installation.

One possibility that is very exciting and will probably develop into widespread realization is the use of Data Over Voice (DOV). By modulating a baseband LAN signal with a carrier, we can shift the spectrum well above the voice frequencies. This high frequency data signal can now coexist or "ride along with" the analog voice signal to the PBX switch. At the PBX switch site, the high frequency data signal can be stripped off and put onto all the other lines that enter the PBX switch. A high pass filter would prevent low frequency, analog voice crosstalk, Thus, the PBX would stay as it was and perform its intended function of switching voice circuits while the high frequency, data signal would function as a LAN medium with a star architecture. Access to this LAN would be via pure contention, CSMA/CD, token-passing, or almost any other access scheme.

It is too early to tell what the implications of this and similar concepts will be. The potential seems to exist for very serious competition to installation of a dedicated baseband LAN for many situations.

9.10 STANDARDS

The emergence of LAN technology and the proliferation of LANs over the past few years has lead to the realization, by both manufacturers and users alike, of the need for LAN standards. Standardization benefits manufacturers because (hopefully) they can produce a great quantity of similar components instead of small numbers of different specialized components. Thus, the cost to produce "standard" components should be less. Users benefit because they can easily specify a standard LAN. In addition, standard equipment should be available from a number of vendors. Thus the user is not necessarily locked into a particular vendor just because the user happened to have bought a LAN from that particular vendor. This competition among suppliers should also mean lower costs to the user.

The Institute of Electrical and Electronic Engineers (IEEE) 802 Standards Committee is the primary organization developing LAN standards in the United States. This committee is developing standards for various LAN media and access schemes.

The main areas of effort are in standards for baseband and broadband bus networks using CSMA/CD access, baseband and broadband bus networks using token passing and for baseband ring networks using token passing.

The 802 Committee is also working on standards for higher level (logical link) protocols as well as those areas related to internetworking such as addressing, gateways and network management. In addition, this committee has also been chartered to develop a standard for metropolitan area networks which are essentially extended LANs, i.e., networks which extend to about 50 kilometers.

As Myers (1982) points out, however, "An IEEE standards committee is not a dictatorial body that can arbitrarily impose a standard on the world." But there is widespread fervent hope that order will proceed out of the committee's labors and, indeed, thirteen large and diverse companies have recently endorsed the IEEE P802.3 draft standard which addresses CSMA/CD (Office Administration and Automation, 1983).

REFERENCES

1. Abramson, Norman, "The ALOHA System — Another Alternative for Computer Communications," Proceedings of the 1970 Fall Joint Computer Conference, pp. 281-285.
2. Amer, Paul, Robert Rosenthal and Robert Toense, "Measuring a Local Network's Performance," *Data Communications*, April, 1983, pp. 173-182.
3. Clark, David, Kenneth Pogram and David Reed, "An Introduction to Local Area Networks," *Proceedings of the IEEE*, Vol. 66, No. 11, November, 1978, pp. 1497-1517.
4. Cotton, Ira, "Local Area Networks — An Overview," in *The Local Area Networking Directory*, (Second Edition), Phillips Publishing, Inc., 1982.
5. Diffie, Whitfield and Martin Hellman, "New Directions in Cryptography," *IEEE Transactions on Information Theory*, Vol. IT-22, No. 6, 1976, pp. 644-654.
6. Gallawa, Robert, "Conventional and Optical Transmission Lines for Digital Systems in a Noisy Environment," *OT Report* 77-114 (National Technical Information Service, Springfield, Virginia, 22161, Accession No PB 265799), 1977.
7. Hershey, John, "The Data Encryption Standard," *Telecommunications*, September, 1983, pp. 77+.
8. Hull, Joseph, A.G. Hanson and L.R. Bloom, "Alternative Transmission Media for Third-Generation Interface Standards, *National Telecommunications and Information Administrative Report*. No. 83-121, 1983.
9. Hudson, Marshall and Frank Thiel, "The Star Coupler: A Unique Interconnection Component for Multimode Optical Waveguide Communications Systems," *Applied Optics*, Vol. 13, No. 11, November, 1974 pp. 2540-2545.
10. Metcalfe, Robert and David Boggs, "Ethernet: Distributed Packet Switching for Local Computers Networks," *Communications of the ACM*, July, Vol. 19, No. 7, 1976, pp. 395-404.
11. Myers, Ware, "Toward a Local Network Standard, *IEEE Micro*, August, 1982, pp. 28-45.
12. Nelson, J., "802: Progress Report," *Datamation*, Vol. 29, November 9, September, 1983, pp. 136-152.

13. Office Adminstration and Automation February, 1983, p. 11.

14. Pomper, William, "The DES Modes of Operation and their Synchronization," Proceedings of the International Telemetering Conference, Vo., XVIII, 1982, pp. 837-851.

15. Rawson, Eric and Robert Metcalfe, "Fibernet: Multimode Optical Fibers for Local Computer Networks," *IEEE Transactions on Communications,* Vol. COM-26, No. 7, July, 1978, pp. 983-990.

16. Rhodes, N. Lee, "Interaction of Network Design and Fiber-Optic-Component Design in a Local Area Network," Proceedings of Globecom, 1982, pp. E-1-1-E.1.5.

17. Shirey, Robert, "Security in Local Area Networks," Proceedings of Computer Networking Symposium, December, 1982, pp. 28-34.

chapter 10

TELECOMMUNICATIONS SYSTEMS

S.W. Maley
Department of Electrical Engineering
University of Colorado
Boulder, Colorado

10.0 INTRODUCTION

Telecommunications, in its many forms, is in a state of rapid expansion throughout the world. The capabilities of existing systems and the character of current expansion are governed by several factors: the influence of governmental regulatory agencies, the economics of the telecommunications industry, and the technology of telecommunications. This chapter is concerned specifically with the influence of technology on the current and future trends in expansion of telecommunication throughout the world.

Following a general discussion of current trends in the evolution of telecommunications, foundations are laid for a more detailed study of its technological aspects first by a reasonably broad study of communication signals and the manner in which they represent information then by a study of some aspects of information theory which provides a means of numerical measure of information in signals, of information handling capabilities of telecommunication systems, and of coding procedures for matching a signal to a system.
Telecommunication systems are then categorized and the various categories are analyzed with the objective of providing insight into current and future trends in system design.

10.1 TRENDS IN TELECOMMUNICATIONS

Many types of telecommunication systems are in use throughout the world today and the variety is becoming steadily greater. It may be said that communication capability in almost all types of systems is rapidly expanding. The underlying reasons are discussed in detail in other chapters of this volume. Very briefly one could cite the improved standard of living in many parts of the world. Further, there is a strong desire to improve relations and cooperation among people on all levels, personal as well as the various levels of government. The potential contribution of telecommunications toward these ends is complex, but certainly improved communication will be beneficial over an extended period of time.

In recent times the demand for telecommunications capability has, on a worldwide basis, always exceeded the supply, and despite the rapid expansion there seems no possibility of satisfying that demand in the foreseeable future. The reasons are simple. Telecommunications systems are characterized by long lives and by high initial capital requirements. Indeed, at the present time, new telecommunications capability is consuming a substantial proportion of the investment capital available throughout the world. The rate of expansion of telecommunications capability is thus limited only by the availability of resources that can be currently devoted to new systems.

Examples of increasing demand for telecommunications are commonplace. Consider nonbusiness telephone usage. The number of telephones in homes is increasing rapidly in the lesser developed countries. In the more highly developed countries, the rate of increase is much less; however, in such countries, the proportion of telephone calls that are made to distant cities is increasing; that is, the average distance over which calls are made is increasing. This greater proportion of toll-calls requires significant expansion in intercity communication links. It may therefore be said that expansion is occurring throughout the world to accomodate non-business telephone demand although the character of the expansion takes different forms in different countries.

The other major category of communications will be called business communications although it will include government, military, and educational communications as well as all other types that cannot be considered personal. This is a very broad category of communications, and most aspects of it are increasing rapidly. Business, in the broad sense used here, is changing in character because it is being conducted over greater distances than in the past. Even small businesses now are frequently finding it necessary to communicate with other businesses in distant cities. This increase in business communications is taking several forms, voice, data, and facsimile. Perhaps the most dramatic increase is in data communications which may be described as communication between computers. This has occurred as a consequence of the ongoing electronics revolution which has so drastically reduced costs of electronic data processing systems. These systems reduce costs of record keeping and information handling to such a degree that businesses cannot remain competitive without them. Their use requires communication links between the computers or terminals at the various locations at which the business operates. This linking of computer terminals, called data communications, is one of the most rapidly growing areas of telecommunications.

The increasing use of telecommunications seems destined to continue at least into the near future but there is another significant factor that may influence future trends in telecommunications expansion. That factor relates to the cost of energy. Modern business involves a great amount of travel with the consequent consumption of large amounts of energy to permit face-to-face discussion. The impending shortage of low cost energy has dictated that alternatives be found for at least part of such travel. Some travel could be avoided by using the telephone in place of face-to-face conversation; undoubtedly this has occurred, to some extent, during the past several years while fuel costs and thus travel expenses have increased. It is apparent that the telephone does not provide adequate communication capability to substitute for all face-to-face communication, but a significant improvement, such as a versatile video telephone, could reduce the required amount of travel significantly. A video signal requires much greater communication capacity than a voice signal, so the cost of video telephone service is significantly greater. Despite this fact, video telephone communciation offers the possibility of great savings in cost as well as energy and time. Such possibilities are emphasized here because of the role they may play in future expansion trends for telecommunication systems.

The foregoing discussion makes it clear that telecommunication systems are being expanded as rapidly as is possible with the resources available. A question may be asked as to the role played by research and development. This is easily answered; technological progress determines the amount and quality of telecommunications capability that can achieved in return for the investments of capital that can be made. The potential technological benefits over the course of the next few years appear to be highly significant; their nature will be the subject of the remaining sections of this chapter.

In conclusion it may be said that a high rate of expansion is occurring throughout the world in most types of telecommunications. The rate of expansion is limited, not by demand, but rather by the worldwide resources available. The intensifying shortage and consequent higher cost of energy appears destined to influence trends toward more telecommunications and less travel. Technological development will greatly increase the telecommunication capability achievable for the available investment capital. This latter point will be examined in detail in the following sections of this chapter.

10.2 COMMUNICATION SIGNALS

10.2.1 Representation of Information by Signals

The function of a telecommunication system is the transfer of information from one location to another. The information transferred is in the form of so-called communication signals. Such signals need to be introduced before embarking upon any further discussion of systems. There are various kinds of signals representing various kinds of information sources. For example a voice signal, which often takes the form of an electrical voltage varying in time, is obtained from an actual voice (by means of a microphone) and it represents the voice in the sense that it can be converted by means of a loudspeaker into an understandable replica of the voice from which it was obtained. Similarly a video signal is obtained from a picture, and it can be converted back into a replica of that picture. A digital or data signal, on the other hand, represents a sequence of symbols, usually alphabetic or numeric characters. It is usually obtained from a computer or from a device such as an electric typewriter (more specifically, a typewriter having an electrical output). In the latter case, depression of one of the keys not only prints an impression on the paper but it also generates an electrical signal having a graph of amplitude versus time which is characteristic of the particular key that was depressed. Each key has a different graph. The electrical signal then is the communication signal representing the symbol labeling the key that produced it. Just as in the case of voice and video signals, a digital communication signal can be converted back into the symbol that it represents (by means of a typewriter with electrical control capability or by means of a teletype console or other similar device). In the case of a digital signal generated by a computer, the same sort of data signal is generated, but it occurs as a consequence of control signals in the computer rather than a keystroke as in the case of the electric typewriter.

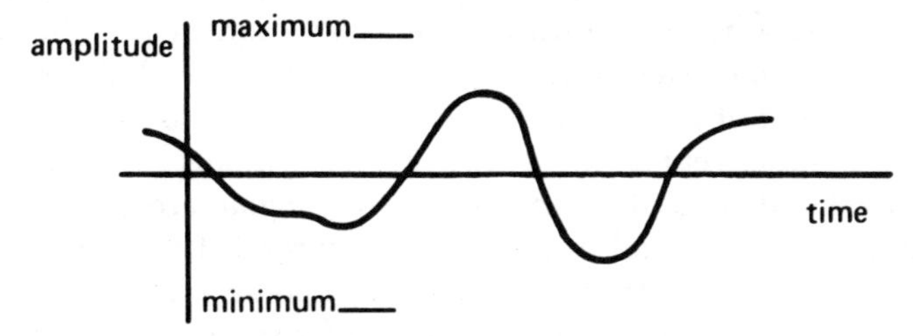

Figure 10.1 Typical Analog Signal

Analog Signals

The signals discussed above are of two distinct types. Voice and video signals are of the first type and a typical signal is sketched in Figure 10.1. The amplitude varies with time in a manner determined by the information source that generated the signal. The amplitude can take any value between two extreme values labeled maximum and minimum amplitude which are determined by the device that converts the information into the signal. A signal that may take any amplitude within a range of values, is called an analog signal. It is apparent that within a specified length of time there can be an infinite number of different analog signals.

Digital Signals

The other type of signal is the digital (or data) signal; a typical example is sketched in Figure 10.2. In this case the signal represents a sequence of decimal digits. Each digit is represented by a pulse having an amplitude equal to the digit it represents. The digits represented by the various pulses are given in parentheses below the pulses. This is only one of the many possible ways to represent decimal digits by pulses. In another version, the pulses could be widened to completely fill in the interpulse spaces. In another version both positive and negative pulses could be used. In still another version, pulses of only two possible amplitudes could be used but with four such pulses for each decimal digit. An example of this type of signal is sketched in Figure 10.3. The first four pulses have amplitudes 1,0,1, & 0 and, as indicated by the decimal numbers in parentheses, represent the digit 5. The second block of four pulses represents the digit 7. The rule relating the decimal digit to the block of four binary (two-level) pulses is, in this case, based upon conversion of the decimal digit to a four digit binary number, and the representation of that four digit binary number by four binary pulses, least significant digit first. The major difference between analog signals and the data signals sketched in Figures 10.2 & 10.3 is the fact that each of the pulses in the data signals can have only a finite number of amplitudes, ten in Figure 10.2 and two in Figure 10.3. This is in contrast to the analog signal which can have an infinite number of amplitudes. This limitation to a finite number of possible pulse amplitudes is a consequence of the fact that the signal represents a sequence of symbols selected from a finite set (or alphabet), and it is the reason such signals are often referred to as digital signals (digital in this usage is intended to imply a finite symbol set rather than a strictly numerical symbol set). The finiteness of the pulse set in the case of digital signals has far reaching consequences in the design of communication systems to handle such signals. This will be discussed further in later sections of this chapter.

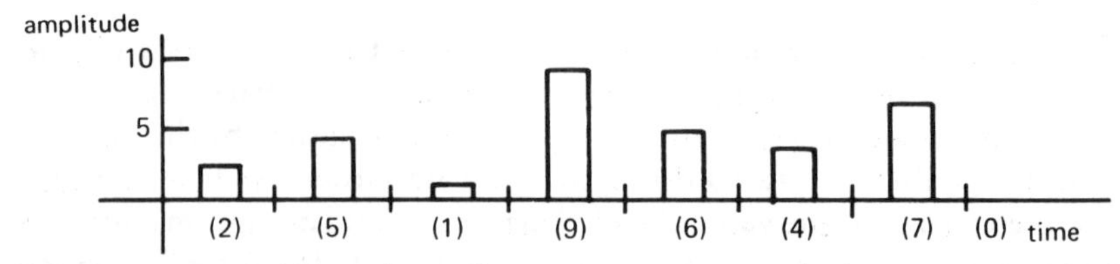

Figure 10.2 Typical Digital Signal

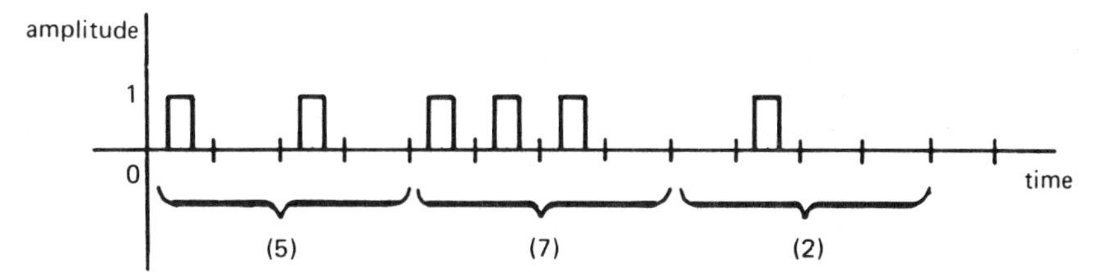

Figure 10.3 Typical Binary Digital Signal

10.2.2 Frequency Domain Representation of Signals

One of the characteristics of both signals discussed above is the range of amplitudes they may assume. In the case of the signal sketched in Figure 10.1, the range extends from the value marked minimum amplitude to the value marked maximum amplitude. In the case of the signal of Figure 10.2, the range is discrete rather than continuous and extends from 0 to 9 in steps of one. Similarly the signal of Figure 10.3 has a discrete range of zero and one. Another and much more important physical characteristic of all signals is the bandwidth. In order to explain this characteristic, refer to the signals of Figure 10.4. There are graphs of three different signals. The shape of each is called sinusoidal because it is a mathematical sine function of time, but each has a different frequency. The frequency, by definition, is the number of times the cycle (a cycle consists of one complete positive and one complete negative excursion of the amplitude) occurs in one second. Only a sinusoidal wave can be assigned a precise frequency because the definition requires the shape of the graph to be that of a sine function. It follows that signals such as those sketched in Figures 10.1, 10.2, and 10.3 cannot be assigned a precise frequency; however there is an alternative way to characterize these signals in terms of the concept of frequency. First, note that any signal can be shown to be equal to the sum of a number of sinusoidal signals, different signals in the sum having different frequencies, different amplitudes, and different phases. (A change in phase corresponds to shifting the signal to the right or left on the time scale.) The number of sinusoidal signals in the sum may be finite or infinite depending upon the signal that is to be expressed as the sum. Furthermore there is only one way that the sinusoidal signals in the sum can be selected. These facts can be verified experimentally or theoretically, but no attempt at such verification will be made here. The matter is more fully explored in the appendix. The representation of a signal in terms of the amplitudes, frequencies, and phases of its sinusoidal components is called a frequency domain representation. Another frequency domain description expresses a signal in terms of the power and frequency of the various sinusoidal components; it is called the power spectrum and will be discussed further. The sinusoidal signals in the sum all have different frequencies, and the range of values of frequencies (that is the difference between the highest and lowest frequencies) is called the bandwidth of the signal.

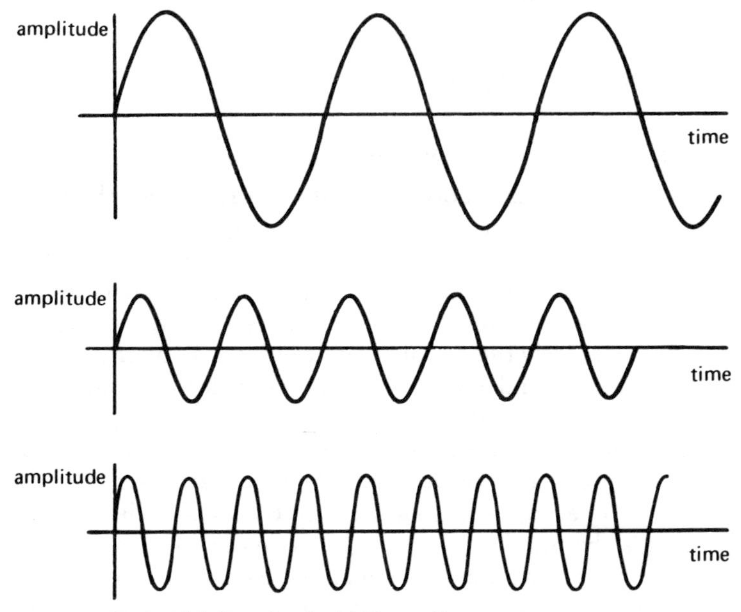

Figure 10.4 Sinusoidal Signals of 3 Different Frequencies

It is an extremely important characteristic of a signal because the cost of transmission of a signal is fundamentally dependent upon its bandwidth. Generally speaking higher bandwidth signals cost more to transmit. Frequency and thus bandwidth are measured in cycles (of the sine wave) per second. This unit of measurement is often called the hertz (one hertz is one cycle per second). Often frequencies and bandwidth are so large that the kilohertz, megahertz, and gigahertz are more convenient (these are 10^3 hertz, 10^6 hertz and 10^9 hertz, respectively).

10.2.3 Sampled Signals

Now that the concept of bandwidth has been introduced, a method for changing an analog signal to a digital signal can be discussed. Suppose the signal of Figure 10.1 is passed through a gate which passes the signal for only a short period of time at periodic intervals, the resulting signal, called a sampled signal, is sketched in Figure 10.5. A theoretical analysis shows that if the number of samples per second is at least two times the highest frequency of the signal being sampled (the signal of Figure 10.1 in this case) then the original signal can be exactly recovered from the sampled signal by passing it through a low-pass filter.[3] (The output of the filter may differ from the original signal in amplitude but that can be corrected by an amplifier or an attenuator). This is true regardless of the width of the samples. There are several reasons for sampling; one is the fact that the sampled signal has unused time intervals between pulses. These intervals can be used for some other purposes. For example the pulses of another sampled signal could be inserted. Then the transmission of the single composite signal effectively accomplishes the transmission of the two analog signals which were sampled and interleaved. At the receiving end of the transmission system, the samples for the two analog signals are separated and passed through individual filters to recover both of the original signals. This combining of two signals into one in such a manner that they can be recovered is called multiplexing;[3] and this particular technique is called time-division multiplexing; another method will be discussed later. In this example, only two signals were multiplexed, but in principle any number can be multiplexed. In practical systems oftentimes thousands of signals are multiplexed.

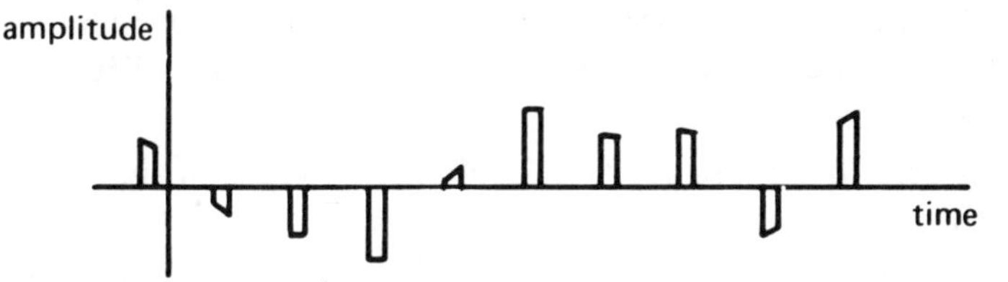

Figure 10.5 Sampled Signal Obtained from the Signal of Figure 10.1

10.2.4 Digital Representation of Analog Signals

Sampling a signal is also used as a first step in the conversion of an analog signal to a digital signal. It may be noted that the tops of the pulses in Figure 10.5 take the shape of the signal being sampled. If the tops were all flattened at their average value, the signal recovered from the resulting distorted, sampled signal by passage through a filter would not be exactly the same as the original signal, but if the width of the samples is narrow compared to the sample spacing, the difference between the original signal and the recovered signal (which could be termed the error) can be made sufficiently small that the information carried by the original signal can be extracted from the approximate replica of it obtained by passing it through a filter. More specifically the error approaches zero as the width of the sampling pulses approaches zero. Thus the error caused by the flattening of the top of the pulses can be made arbitrarily small.

But why flatten the pulse tops if it causes errors, however small? The reason is that the flattening makes all pulses have the same shape. The advantage of this is that each pulse can be completely specified by its amplitude; that is by a number. The number describing the pulse amplitude must be rounded off; consequently it is not an exact description of the pulse amplitude. The error can be made arbitrarily small by choice of the number of digits used in the number describing each sample amplitude. If the pulse amplitudes are changed to the values given by the rounded off numbers, the resulting signal is called a quantized version of the original signal. Each pulse in a quantized signal has one of only a finite number of amplitudes and is thus a digital signal. Therefore, if an analog signal is band limited (has finite bandwidth) it can be converted to a digital signal with arbitrarily small error; and, of course, the digital signal can be converted back to a close approximation to the analog signal. It can be concluded that an analog signal can, if desired, be converted to a digital signal and then transmitted over a digital communication system. The converse problem, that of converting a digital signal to analog form is trivial because actually no conversion is needed; an analog communication system can handle a digital signal without any change if the frequencies present in the digital signal can be handled by the analog channel. If not, then the frequencies can be shifted by some sort of modulation.

It may be said that there are two fundamental types of signals, analog and digital. An analog signal can, if desired, be converted to a digital signal for transmission and then converted back to analog form. Conversely, a digital signal can, if desired, be transmitted without modification, by an analog communication system. The relative advantages and disadvantages of the two types of signals and systems will be discussed in detail later.

10.2.5 Modulation

An analog signal obtained from an information source can be converted into any of a great variety of different but equivalent analog signals, each of course, carries exactly the same information. A similar statement can be made concerning digital signals. The process used to convert one form of signal into another is called modulation, and the inverse of the process is called demodulation (since no information is lost in the conversion to a different form, the inverse of the conversion, which recovers the original form is always possible). The objective in changing a signal to a different but equivalent form (modulation) is the matching of the signal to a communication system. More specifically a communication system is designed to handle signals of a specific form; and, if a signal does not have the proper form, its form must be changed by a suitably chosen modulation process. It can be said that modulation is a reversible process used to match a signal to a communication system.

Pulse Modulation

One class of modulation processes consists of methods of changing the form of pulses in a pulse type of signal, and another class consists of methods of changing the band of frequencies that makes up the signal. The class concerned with the changing of pulse shapes will be discussed first.

Pulse Amplitude Modulation (PAM)

A diagram of several pulse type signals and the processes for conversion between types is shown in Figure 10.6. An analog signal is shown in the diagram to permit inclusion of the process of conversion from an analog signal to a digital signal by sampling.[3] The result is a pulsed analog signal which has been discussed. If the tops of pulses are flattened to their average value, the result is a pulse amplitude modulated (PAM) signal.[3] This is still an analog signal since the pulse heights can be any value, and it is distorted due to the flattening of the tops of the pulses. As mentioned above, this signal can be converted to an approximation of the original analog signal. The error in the approximation approaches zero as the pulse width in the sampling process approaches zero.

Pulse Duration Modulation (PDM)

An analog PAM signal can be converted to other forms of pulse signals. One of these forms is called a pulse duration modulated (PDM) signal or a pulse width modulated (PWM) signal.[3] (The terminology PDM will be used here.) It is obtained from the PAM signal by replacing each PAM pulse with a PDM pulse having a duration which is a function of the amplitude of the PAM pulse. Of course the PDM signal can be converted back to the PAM signal with no distortion or loss of information. All PDM pulses have the same amplitude. The relationship between the PAM pulse amplitude and the PDM pulse duration is under the control of the system designer. A possible relationship is given in the graph sketched near the PDM signal in Figure 10.6. There are, of course, some constraints on the relationship; for example, if the slope of the graph in Figure 10.6 were too small, the pulses could overlap. The control over the relationship between PAM pulse amplitude and PDM pulse width possessed by the system designer has important consequences. To facilitate a description of those consequences, assume the relationship between PAM pulse amplitude and PDM pulse duration is linear. (It is not neces-

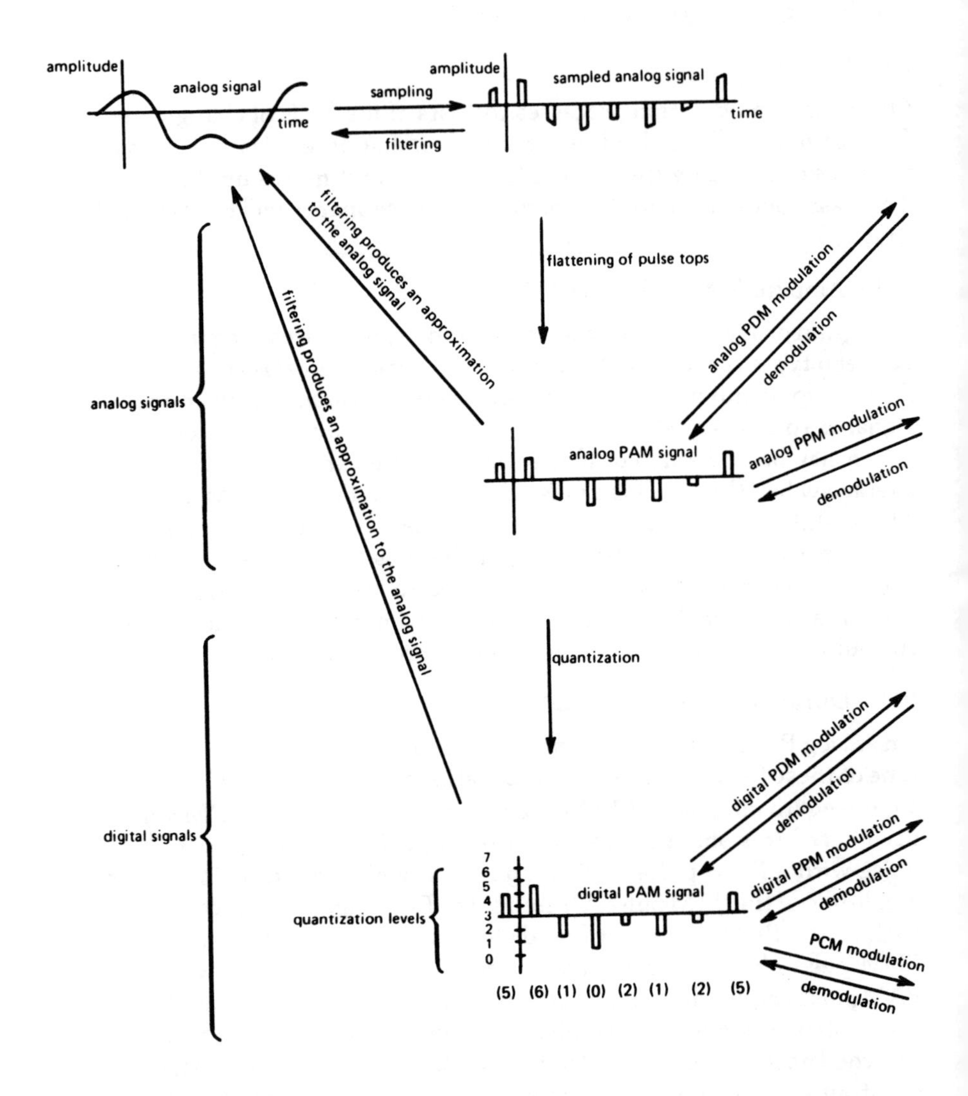

Figure 10.6 Various Forms Taken by Pulse Type Communication Signals and Processes for Conversion between Forms

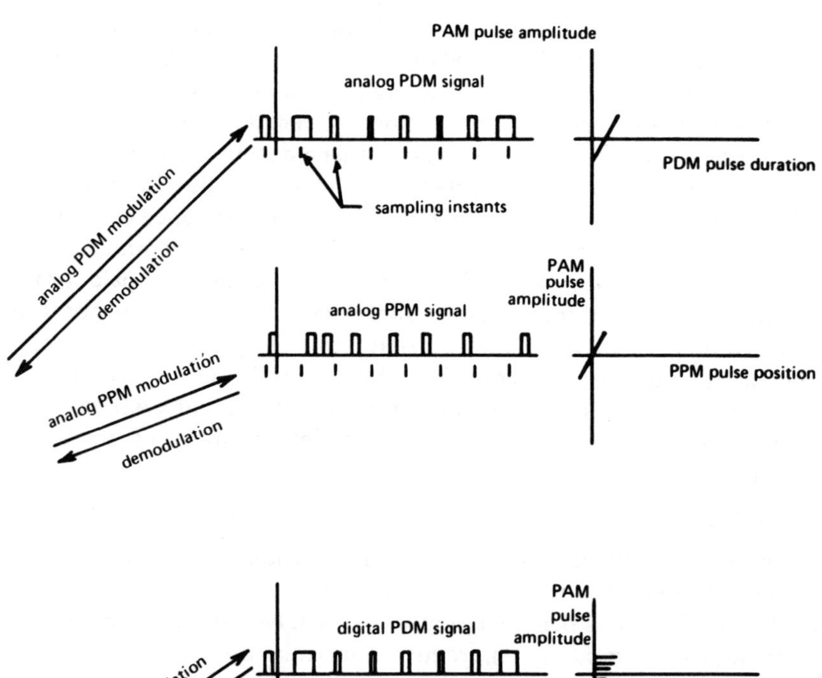

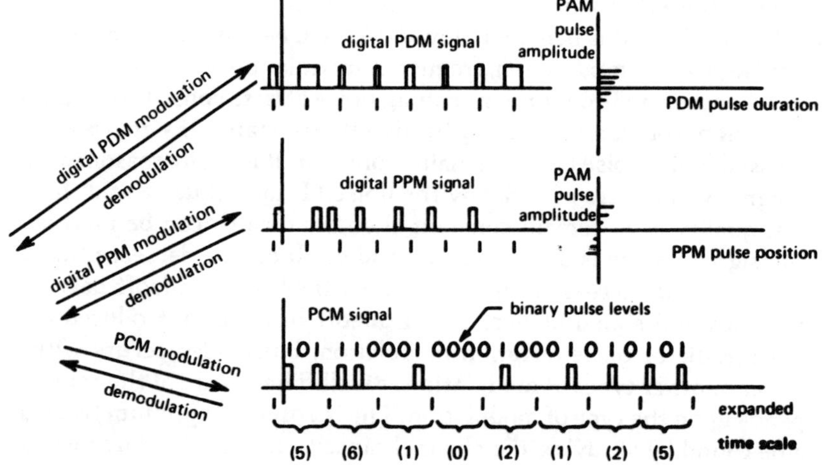

Figure 10.6 (Continued)

sarily linear but such an assumption simplifies the reasoning relation to the following ideas and yet does not diminish their generality.) A decrease in the slope of the linear relation implies a greater range of values of pulse duration which in turn implies a greater bandwidth. However, a greater range of values of pulse duration also results in greater tolerance to degradation of the signal by noise. Thus a designer using PDM can make the signal more or less tolerant of noise by adjusting the bandwidth, the greater the bandwidth the greater the resistance to degradation by noise. Since greater bandwidth requires greater communication system capacity and consequent higher cost, a price must be paid for the aforementioned greater resistance to degradation by noise. The question as to the degree of resistance to noise the designer should employ depends upon its cost in terms of bandwidth (and thus also in terms of investment capital); this in turn is different for differnt systems. Each system design therefore requires a complete analysis of costs and benefits.

Bandwidth-Signal Power Tradeoff

It has just been argued that a telecommunication system designer can diminish the effects of noise on signals at the expense of increased bandwidth if PDM is being employed. (As will be seen, the same is true for pulse position modulation, frequency modulation, and phase modulation.) Another method for diminishing noise effects consists of increasing the signal power but keeping bandwidth constant. (If signal power is increased while noise power remains constant, it is easily reasoned that the signal will be less affected by the noise.) It may thus be said that if signal quality is to remain constant then bandwidth can be increased while signal power is decreased or bandwidth can be decreased while signal power is increased. In other words there is a tradeoff between bandwidth and signal power. This tradeoff has been introduced as a characteristic of pulse duration modulation, but is also occurs with a number of other types of modulation as well. The nature of the tradeoff depends upon the type of modulation. The theoretical optimum tradeoff can be found by studying the channel capacity theorem of information theory. This will be discussed in section 10.3.

Pulse Position Modulation (PPM)

Returning now to the various forms of pulse signals as diagrammed in Figure 10.6, another alternative form obtainable from the analog PAM signal is the pulse position modulated (PPM) signal.[3] In this signal there is one pulse for each of the PAM pulses and all pulses are of the same amplitude and width. However their position with respect to the PAM pulses they represent is shifted right or left on the time scale an amount dependent upon the amplitude of the PAM pulse they represent (of course the PPM signal must lag behind the PAM signal by an interval sufficiently long to permit the necessary signal processing). The inverse of the conversion process, of course, recovers the PAM signal exactly. The relationship between the pulse position and the pulse amplitude may have any form; a linear relation is sketched to the right of the PPM signal in Figure 10.6. The communication system designer has control over this relationship subject to certain constraints just as in the case of PDM. If the slope of the line in the linear relationship of Figure 10.6 is decreased (it must not be decreased to an extent causing pulses to overlap), the bandwidth is increased, and at the same time the tolerance of the signal to degradation by noise is increased. Thus with PPM the designer also has the capability of tradeoff between signal power and bandwidth just as was true with PDM.

Digital Pulse Modulation (PCM)

The analog PAM signal can be quantized to convert it to a digital PAM signal. In the quantization process a finite number of discrete amplitude levels are selected and each analog PAM pulse is changed in amplitude to the nearest discrete level.[3] Then each pulse can be represented by a number having a pre-specified number of digits (the number of digits is the same for every pulse and is determined by the number of quantization levels and the number system used.) The discrete levels used in the quantization process may be equally spaced or non-equally spaced. Sometimes closer spacing is used for small amplitudes than for large to improve the quality of low level signals: this is a form of companding (section 10.2.6). The changing of the amplitude of a PAM pulse to one of the discrete amplitude levels causes an irreversible loss of information, and the original amplitude cannot be recovered by any sort of inverse operation. The changes are called quantization noise. However if the discrete amplitude levels are spaced sufficiently close, the loss of information can be made negligibly small, and the analog signal obtained from the digital PAM signal by filtering will differ negligibly from the original analog signal.

The digital PAM signal obtained from the analog PAM signal can be completely described by a sequence of multidigit numbers, each number representing the amplitude of one pulse. In a sense the sequence of numbers may be described as a code for the signal; for this reason the signal is sometimes called a pulse code modulated (PCM) signal. A PCM signal can take a variety of forms, in fact a different form for every number system that can be used to express the pulse amplitudes. An example may clarify this point. Suppose that the quantization in Figure 10.6 utilizes 10 levels, then each pulse can be described by one decimal digit, and the signal could be coded into a sequence of single decimal digits. Now suppose each decimal digit is written as a four digit binary number. Then a digital PAM (or PCM) signal having pulses of either of two possible amplitudes (binary pulses) can be generated to correspond to this binary description of the signal. Such a signal is sketched in Figure 10.6; successive blocks of four pulses represent decimal digits. Binary PCM signals are widely used. It may be noted that less distortion of the signal would occur if the quantization had been into 16 discrete levels rather than 10. The 16 levels could still be converted into 4 binary pulses; this is the procedure that is customarily used. More generally, if binary pulses are to be used, the number of discrete amplitude levels is selected to be a power of 2; then the number of binary pulses needed to represent each PAM pulse is the logarithm (base 2) of the number of levels.

The digital PAM signal can be converted to a digital PDM or to a digital PPM signal in the same manner that the analog PAM signal was converted to the analog PDM or to the analog PPM signal. In the case of a digital PDM signal, the pulse duration can be any of a discrete set of values, one value for each of the possible amplitudes of the digital PAM pulse it represents. Similarly a discrete PPM signal has a pulse shifted by any of a discrete set of values, one for each possible amplitude of the digital PAM pulse it represents.

The various forms of pulse signals shown in Figure 10.6 have been discussed. Each of these signal forms is used in practical systems. The characteristics of different communication systems dictate the use of different forms of signals; the choice must be made by the system designer. The presentation of pulse signals in Figure 10.6 is by no means complete; there are many others; these are simply the more common.

Modulation Terminology

Some discussion is needed concerning the terminology of the conversion between signal forms. The conversion from PAM signals to PDM signals is usually called pulse duration modulation or pulse width modulation with the adjectives analog or digital added, if necessary, to clarify the meaning. The conversion from PDM to PAM is called pulse duration demodulation (or pulse width demodulation). Similarly pulse position modulation and demodulation are used for the conversion between the PAM and PPM signals. The conversion from one PCM (or digital PAM) signal to another is called pulse code modulation (or digital PAM). The term modulation fundamentally means a reversible change in the form of the signal and demodulation refers to the reverse change which exactly (in principle) recovers the original signal. The conversion from an analog signal to a sampled analog signal is a reversible conversion, but it is usually not called modulation, but rather sampling. The conversion from the sampled analog signal to the analog PAM signal does not have a standardized name and often is considered as part of a conversion from an analog signal directly to an analog PAM signal and is then referred to as pulse amplitude modulation.

The discussion of the signals of Figure 10.6 may have implied that the various pulse signals are all alternative forms for analog signals; this is not true. In many systems the information source (a computer, for example) generates digital signals which can then be converted to any of the various alternative digital signal forms.

Modulation for Frequency Shifting

The other class of modulation processes involves changing the band of frequencies present in the signal. As explained above, any signal can be expressed as the sum of a number of sinusoidal signals each of different frequency, amplitude, and phase. A graphical representation of the signal consists of a graph of the power (or the square of the amplitude) of the sinusoidal wave at the various frequencies. Examples are shown in Figure 10.7. In Figure 10.7a it is assumed the signal is a sum of a finite number of sinusoidal signals; the height of the vertical lines indicates the power of the individual sinusoidal signals. In Figure 10.7b it is assumed that the signal is a sum (or integral) of an infinite number of sinusoidal signals, one at every frequency, for which the graph is non-zero. In either case, the important characteristics of the signal are the upper, f_2, and lower f_1 frequencies present in the signal and the difference between them, which (as mentioned above) is called the bandwidth (meaning the width on the frequency axis of the band of frequencies in the signal). Such graphs were introduced in section 10.2.2 and are called frequency domain descriptions of signals. More specifically those in Figure 10.7 are called power spectra. These are incomplete descriptions because they do not give the phase of the various sinusoidal components of the signal. Therefore, a power spectrum is not a description of a single signal but rather of an entire class of signals. The sinusoidal signals of power and frequency specified by the power spectrum may have any phase relationship among themselves. Since there is an infinity of ways the phases of the various sinusoidal signals can be chosen, there is an infinity of signals, all of which have exactly the same power spectrum. The incompleteness of the power spectrum as a description of signals may, at first, seem to be a disadvantage; but communication systems are not designed to handle precisely defined, prespecified signals but rather to handle broadly defined classes of signals. Furthermore these classes of signals generally have frequencies in the same band; therefore the power spectrum is an ideal description of signals for use in the study of communication systems.

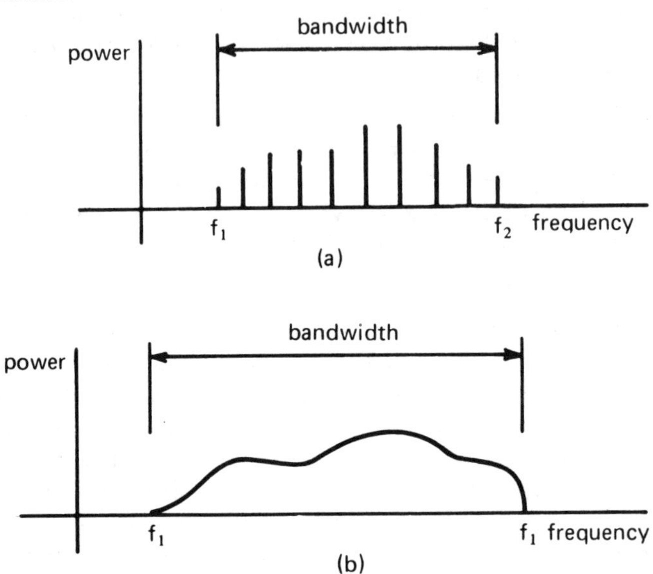

Figure 10.7 Frequency Domain Representations of Signals

Another reason for the use of the power spectrum is that communication links (the transmission facilities in a communication system) are customarily defined in terms of the band of frequencies that they will transmit. In the overall design of a communication system, the characteristics of the signal to be transmitted over the communication link must be of a nature that can be handled by the link, and since the link is usually defined in terms of the power spectrum that can be transmitted over it, it is clear that the most useful type of signal description is the power spectrum.

The problem of matching a signal to a link may be described by referencing Figure 10.8. The signal has the power spectrum shown in part (a) of Figure 10.8 (the signal to be shifted in frequency is often called the baseband signal), and the channel has the capability of transmitting signals in the frequency band from f_2 to f_3. The matching of power levels is trivial since they can be changed using amplifiers or attenuators. Nothing more will be said of this aspect of the matching problem.

Practical and economical communication links often are such that f_2 and f_3 are both much larger than the highest frequency, f, present in the signal. In order to match the signal to the link its frequency band must be shifted up into the region between f_2 and f_3. This can be done by any of several modulation processes. The definition of modulation in this case is the same as in the case of pulse modulation; it is a reversible transformation of the signal. The inverse transformation which recovers the original signal is called demodulation. There are three fundamental types of modulation whereby frequency bands may shifted; they are amplitude modulation (AM), frequency modulation (FM), and phase modulation (PM).

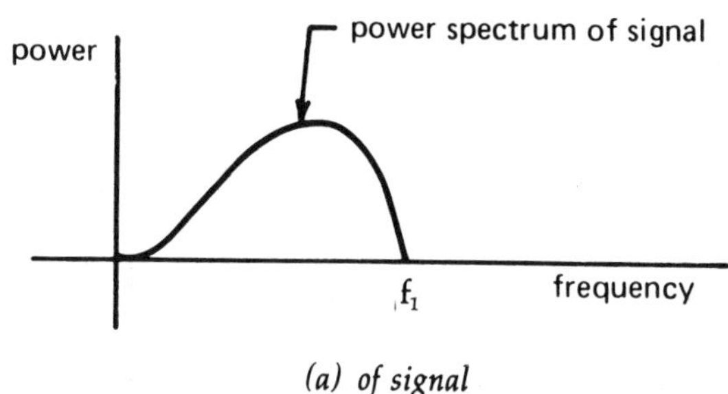

(a) of signal

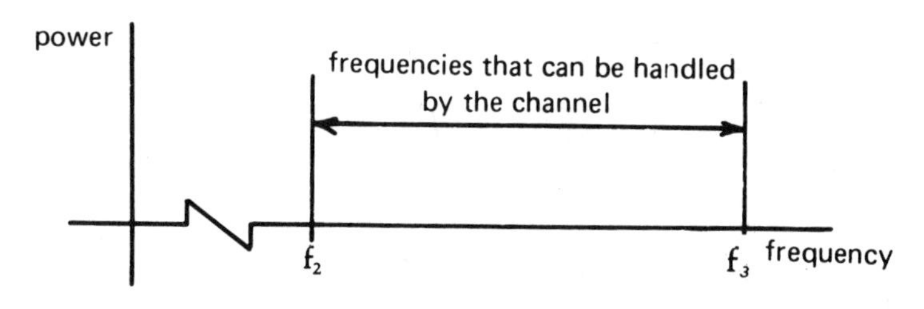

(b) of the Communication Link

Figure 10.8 Power Spectrum

Ampltiude Modulation (AM)

There are three common types of amplitude modulation, double sideband, double sideband-suppressed carrier, and single sideband;[3] they are described in terms of how they shift frequencies in Figure 10.9. Double sideband ampltiude modulation shifts the power spectrum and its image with respect to the zero-frequency line up to the region between f_2 and f_3 (it is assumed that $f_3 - f_2 = 2f_1$ in this discussion). In addition a so called carrier is added to the signal at the center frequency, f_c. The presence of the carrier requires much more power thus making transmission less efficient than double sideband-suppressed carrier. However its presence permits use of less complex (and thus less expensive) receivers. The two

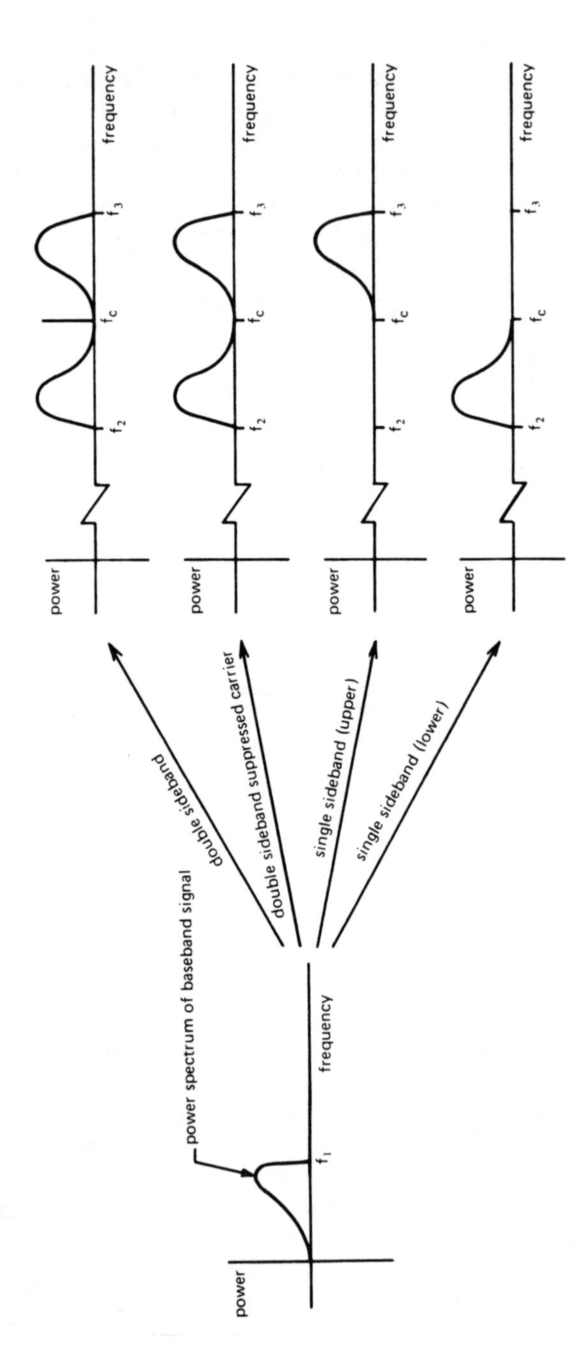

Figure 10.9 Frequency Shifting Characteristics of Various Types of Amplitude Modulation

halves of the double sideband amplitude modulated signals are called upper (right hand) and lower (left hand) sidebands. It is possible to shift only the upper sideband or only the lower sideband. This is called single sideband amplitude modulation. There are a variety of amplitude modulation techniques intermediate between double sideband and single sideband; these shift parts of both sidebands and are called vestigial sideband (VSB) amplitude modulation. The bandwidth of amplitude modulated signals ranges from f_1 (the bandwidth of the modulating signal) for single sideband to $2f_1$ for double sideband with intermediate values for vestigial sideband. Single sideband AM is widely used because its narrow bandwidth permits more radio links to operate in the same area within a specified range of frequencies (to prevent interference, radio links operate on non-overlapping frequency bands.) This conservation of the spectrum is becoming increasingly important because of the steadily increasing use of radio systems. The disadvantage of single sideband is the complexity of equipment compared to that needed for double sideband. Vestigial sideband has bandwidth and equipment complexity characteristics intermediate between those of double sideband and single sideband; it is used in commercial television broadcasting.

Frequency Modulation (FM)

Frequency modulation[3] can be described in terms of the diagrams in Figure 10.10. Here again the band of frequencies making up the baseband signals is shifted up into the region between f_2 and f_3. (In this discussion it is assumed that $f_3 - f_2$ is greater than $2f_1$.) This process, however, does not preserve any of the characteristics of the shape of the power spectrum of the baseband signal. Furthermore, the bandwidth, $f_3 - f_2$, of the modulated signal is usually somewhat greater than for amplitude modulation; it is approximately given by $2mf_1$ where m is the modulation index. It is equal to or greater than one and is often in the vicinity of five.

Thus a typical FM signal has several times the bandwidth of the equivalent AM signal. In view of the electromagnetic spectrum crowding problem mentioned above, it may be asked why FM is ever used. The answer lines in the fact that FM, like some of the pulse modulation techniques mentioned previously permits the system designer to control the bandwidth (by choice of the modulation index, m) and simultaneously the vulnerability of the system to degradation by noise. Just as discussed before, if the quality of the signal is to be kept constant (in terms of noise), the designer can increase the bandwidth and decrease the transmitter power or alternatively decrease the bandwidth and increase the transmitter power. Because of this tradeoff it is possible to transmit very high quality signals using very low transmitter power and correspondingly high bandwidth. For this reason FM is used in many applications in which transmitter power is limited. These include such diverse uses as transmission of biological information from birds in flight to recorders on the ground and transmission from earth satellites and other space vehicles to the earth.

FM is also used in radio telemetry systems because signal fading, which is common on radio links, primarily affects signal amplitude, not the frequency (which carries the data). Therefore the calibration problems are greatly simplified.

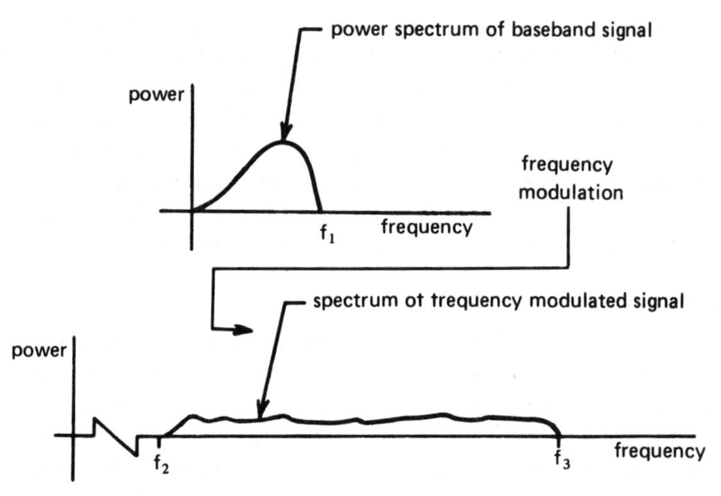

Figure 10.10 Frequency Shifting Characteristics of Frequency Modulation

Phase Modulation (PM)

Phase modulation has characteristics which are very similar to those of FM, and it has the same bandwidth-transmitter power tradeoff.[3] For this reason no further discussion of phase modulation is needed.

Digital Modulation Terminology

AM, FM, and PM take particularly simple forms in cases of digital baseband signals. Because of the simplicity of the processes, they are easily described by simple phrases which are often used as names for the modulation processes. Those phrases (names) are so widely used that they need to be introduced at this time. In the case of digital baseband signals, AM is called amplitude shift keying (ASK), FM is called frequency shift keying (FSK), and PM is called phase shift keying (PSK). A device, used at a digital terminal, that performs such modulation on the outgoing signal and the corresponding demodulation on the incoming signal is commonly called a modem (modulator-demodulator).

Multiplexing

The types of modulation that involve the shifting of frequency bands to facilitate the matching of a signal to a channel also provide the means for a second method of multiplexing, frequency division multiplexing. The first method, time division multiplexing (TDM) was introduced in the discussion of sampling and pulse amplitude modulation.

Frequency Division Multiplexing (FDM)

Multiplexing in general means the combining of a number of signals into one, called the multiplexed signal, in such a way that the separate signals can be recovered from the multiplexed signal.[3] A multiplexing system is sketched in block diagram form in Figure 10.11a. Frequency division multiplexing is described in terms of power spectra in Figure 10.11b. Each of the individual input signals has its band of frequencies shifted to a different band. These are then added together to produce the multiplexed signal. The demultiplexing process consists of extracting the various bands of frequencies from the multiplexed signal by means of filters. The outputs of the various filters are then frequency shifted, using the same modulation process as used for multiplexing, to the proper band so as to exactly recover the individual signals that were originally multiplexed. The modulation process (FDM) used for the frequency shifting can be AM, FM, or PM; often single sideband AM is used (as is done here) because the narrow bandwidth permits more signals to be multiplexed into the frequency band available for use. It is apparent from the diagram that the multiplexed signal has a bandwidth equal to the sum of the bandwidths of the signals multiplexed.

Time Division Multiplexing (TDM)

Time division multiplexing, which was discussed previously, is described diagramatically in Figure 10.11c. The interleaving and the separation processes are illustrated. This shows multiplexing for a sampled signal, but obviously any sort of pulse signal can be time division multiplexed. It can be shown that just as is true for FDM, the minimum possible bandwidth of the multiplexed signal is the sum of the bandwidths of the signals originally sampled. Multiplexing permits many individual, relatively narrow bandwidth signals to be transmitted by a single wide bandwidth communication link. The only justification for multiplexing results from the fact that in many cases a single wide band link is less expensive than many narrow band links.

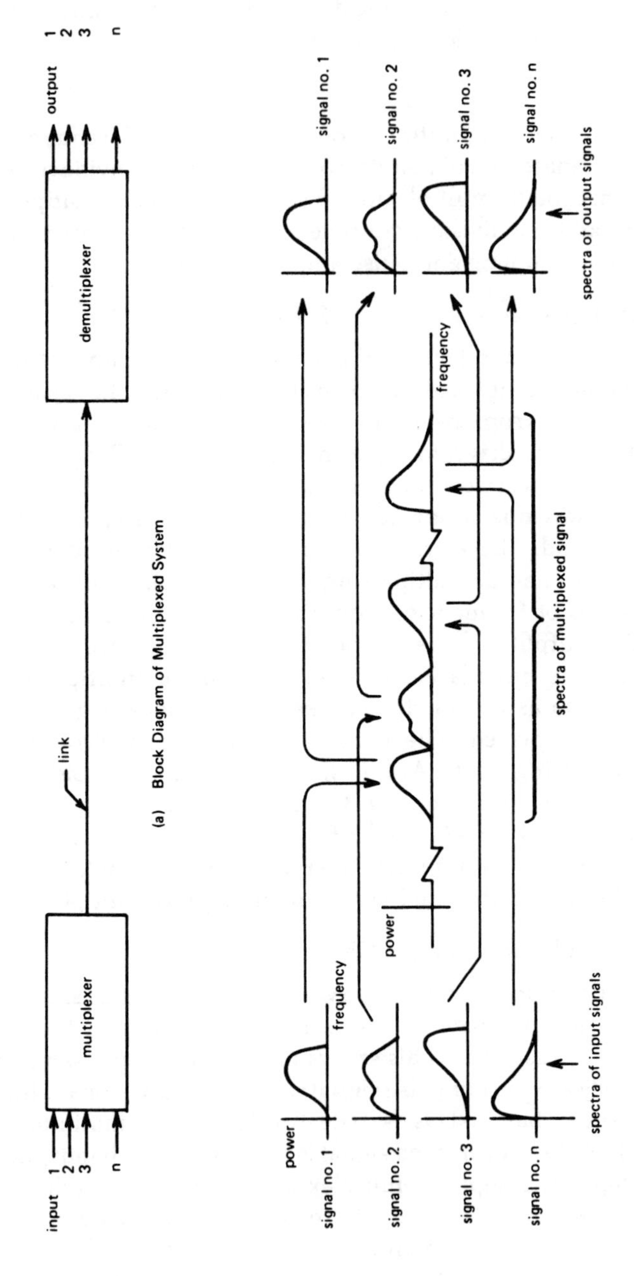

(a) Block Diagram of Multiplexed System

(b) Frequency Division Multiplexing Described in Terms of Power Spectra

Figure 10.11 Multiplexing Techniques

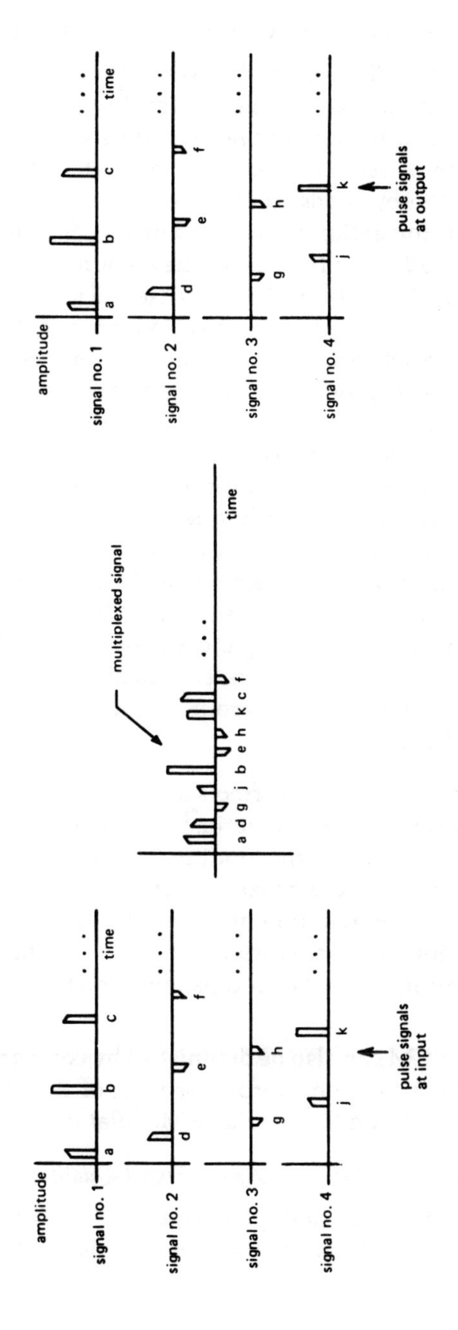

(c) Time Division Multiplexing Described in Terms of Sequences of Pulse Type Signals

Figure 10.11 (Continued)

10.2.6 Effects of Noise on a Communication Signal, Compandors

Little has been said in this discussion of the effects of noise upon communication signals. Every device that generates, transmits, modulates, amplifies, multiplexes, etc., contributes noise to the signal. It may be said that the contamination of signals by noise is inevitable. The noise distorts the signal and thereby tends to obscure its information content. The degree to which information is obscured or lost depends upon the amount of noise power added in relation to the amount of signal power. It also depends upon the type of signal and its use. All signals have some degree of noise tolerance; that is the information can be extracted from the signal despite the presence of noise. The noise tolerance is relatively low for video signals; in other words the noise power must be very small compared with signal power to produce a high quality picture. In the case of audio signals, speech is intelligible even with substantial amounts of noise power, although music requires lower noise power levels for pleasant listening. Digital signals are highly tolerant of noise; indeed it is possible to transmit digital signals with low error rates in the presence of noise power of levels nearly of the magnitude of the signal power. As noise power levels increase, communication is still possible but with lower quality. Video signals produce pictures somewhat distorted or with less detail, audio signals are still understandable but with greater difficulty due to the noise in the background, and digital signals have more errors in the symbols.

Since a constant average power level affects low level signals more than high level signals, analog signals are sometimes modified so as to increase the level of the low level signals at the input to a communication system and then restore the levels to their proper values at the output of the system. This reduces the degradation of the signal due to noise in the system. The compression of the amplitudes at the input and the subsequent expansion at the output is called companding and the device used is called a compandor.

Quantization noise in PCM can also be diminished by companding. This can be accomplished by using non-uniform spacing of the quantization levels. (see previous section on Digital Pulse Modulation).

Signal-to-Noise Ratio (SNR), A Measure of Analog Signal Quality

In the analysis of telecommunication systems, a quantitative measure of signal quality is needed. In the case of analog signals, that measure is customarily taken to be the signal-to-noise ratio which is the signal power divided by the noise power. This measure will be abbreviated SNR. It is a number which is usually in the range of about 2 up to 100,000 or so; but in some cases it is outside these limits. Because of the large numbers that SNR often takes, it is customary to express them in decibels; to do so the logarithm (base 10) is taken and the result is multiplied by 10. The range 2 to 100,000 then becomes 3 dB to 50 dB (decibel is abbreviated dB).

Error Probabilities in Digital Signals. A Measure of Digital Signal Quality

In the case of digital signals, their quality can be measured in terms of the SNR (signal-to-noise ratio) just as in the case of analog signals, but it is more meaningful to use a different measure, namely the probability of error, BER (meaning bit error rate), of a symbol. This measure usually lies in the range 10^{-2} to 10^{-8} meaning error rates ranging from 1 digit in 100 to 1 in 100,000,000. Since SNR and BER are both measures of the quality of digital signals, it may be expected that they are related. This is true, and the relationship depends upon the type of modulation used. The relationships are somewhat involved but approximations (good for signal-to-noise ratios of about 10 or higher) can be written concisely for ASK (amplitude shift keying), FSK (frequency shift keying) and PSK (phase shift keying). The approximations for binary signals are

$$BER \cong \frac{1}{2} e^{-\alpha(SNR)} \quad \text{for non-coherent systems} \tag{10.1}$$

and

$$BER \cong \frac{\frac{1}{2} e^{-\alpha(SNR)}}{\sqrt{\pi \alpha (SNR)}} \quad \text{for coherent systems} \tag{10.2}$$

The paramter α is to be chosen 1/4 for ASK, 1/2 for FSK, and 1 for PSK. Different expressions are needed for coherent and non-coherent systems. The distinction between such systems is discussed below.

Signal Quality Using Coherent Demodulation

Coherent systems are those which a reference signal is transmitted from the modulator to the demodulator to be used in the demodulation process. When this is done it is possible to reduce the noise power in the output of the demodulator to a greater extent than is possible for a noncoherent system (a system without such a reference signal). Thus a coherent system gives a higher quality signal (a signal of higher SNR). However, there is a price to be paid since coherent systems are more complex and costly (due to the necessity of transmitting the reference signal to the receiving end of the system).

A study of the above relations between BER and SNR shows that for a specified SNR a coherent PSK system gives the lowest BER followed by non-coherent PSK, coherent FSK, non-coherent FSK, coherent ASK and non-coherent ASK in that order. The best choice of a system depends upon costs and other factors. All six of the systems are in wide use.

10.2.7 Error Control in Digital Systems

It is apparent that occasional errors in digital signals are inevitable because of the ever present noise that contaminates the signal. In some systems those occasional errors are so serious that substantial effort is justified in reducing their number. They can be reduced by reducing noise levels in the system, but there is a limit to the improvement that can be achieved by this method for reasonable cost. However there is another method which is widely used; it involves the coding of the information in such a manner that errors can be detected and corrected. The codes used for this purpose are called error detecting and error correcting codes.(7) The concepts underlying such codes will be briefly discussed. Error detecting codes will be considered first.

Error Detecting Codes

The simplest code for error detection makes use of a so called single parity digit. To illustrate its use, assume the digital signal is coded into successive blocks of 5 binary digits. (This assumption does not limit the applicability of the results of this discussion because any digital signal can be recorded into blocks of any length and of any set of symbols.) Now assume several of the successive blocks are as shown in Figure 10.12. One additional binary digit is added to each block making the total block length 6 digits. The added digit is called a parity digit, and it is chosen so as to make the total number of 1's among the 6 digits an even number. (An alternative to the code described here could be based on selection of a parity digit to make the total number of 1's an odd number.) At the receiving end of the communication system the number of 1's in each block is counted; if it is an odd number, the block is in error and a signal is sent to the transmitting end of the system causing the block to be transmitted again. Then a parity check (a count of the number of 1's) is made again at the receiver. If the parity check is correct (an even number of 1's) the block is accepted as correct; if not, a signal is sent causing another retransmission. This procedure is then repeated until the parity check on the received block is correct. (In practice, more than one re-transmission is rare.) It is apparent that the use of the single parity digit will not detect all possible errors. It will detect all possibilities of a single error but it will not detect any double errors. (If two errors occur in a block, the number of 1's will still be an even number). More generally it may be said that the use of a single parity digit in each block permits detection of an odd number of errors but it will not detect an even number of errors. This code, even though it will not detect all possible errors is useful because, in many telecommunication systems the great majority of errors occur not more than one to a block; so in such a system the use of a single parity digit would detect them. In systems in which the probability of two errors per block is substantial, the single parity digit is not particularly successful in reducing the error rate; but there are other codes which make use of more than one parity digit per block (each chosen to have even parity on a selected set of digits within the block) which are capable of detecting double errors. Generally it may be said that an error detecting code can be devised to detect any type of error pattern; so codes are in essence, designed to fulfill the needs of a particular communication system. The example above illustrates that the use of an error correcting code can reduce the error rate, but only at a price, the price being the reduction of the rate at which the information is transmitted by the channel. Without parity digits, all of the digits transmitted carry information, but with parity digits, only 5/6 of the digits carry information; so the information rate is reduced to 5/6 of the value possible without parity digits. There would be an additional reduction in the information rate due to the necessity of retransmitting a block when an error occurs. The system designer must decide whether the advantages gained by reduction of the error rate are worth the cost in terms of the reduced information rate. The reduction in the information rate in the example above is deceptively large. Often the length of the block to which the parity digit is added is somewhat larger, which results in a smaller reduction in the information rate.

Error Correcting Codes

The other type of code used for error control is called an error correcting code.[7] Such codes will be illustrated by a simple example. Consider information grouped into blocks of 4 binary digits. Let 3 parity digits be added to each block for a total block length of 7 digits. Let the information digits be labeled I_1, I_2, I_3, and I_4 and let the parity digits be labeled P_1, P_2 and P_3. Each parity digit will be selected for even parity (an even number of 1's) among itself and a selected subset of the information digits; the rules for determining each parity digit are given in the encoding-decoding table in Figure 10.13. The top row of the table is labeled P_1. P_1 is selected so there will be an even number of 1's among I_1, I_2, I_4 and P_1. P_2 is selected so there will be an even number of 1's among I_1, I_3, I_4 and P_2; and finally P_3 is selected so there will be an even number of 1's among I_2, I_3, I_4 and P_3. An example may make this clear; assume the information digits are $I_1 = 1$, $I_2 = 0$, $I_3 = 0$ and $I_4 = 1$. P_1 is selected so there will be an even number of 1's among I_1, I_2, I_4 and P_1 that is among 1, 0, 1 and P_1; so P_1 must be 0. Similarly the rules dictate that $P_2 = 0$ and $P_3 = 1$; so the code word, complete, with its parity digits is, 1001001. This is transmitted instead of the block 1001, of information digits only. It is evident that the "cost" of the use of this code is the reduction of the information rate to 4/7 of that which would be possible without the addition of parity digits.

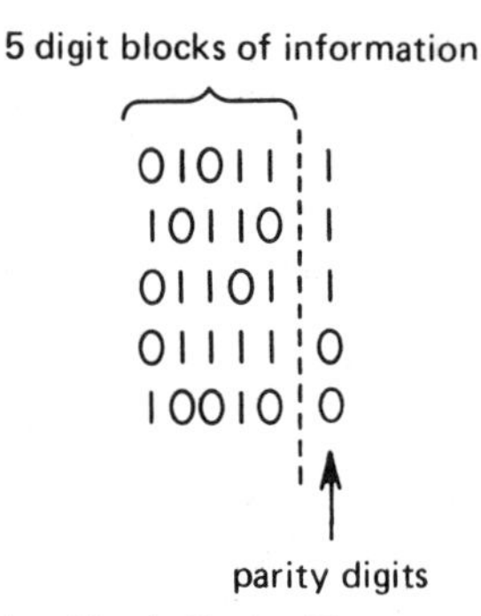

Figure 10.12 A Code Utilizing Single Parity Digits

	I_1	I_2	I_3	I_4	P_1	P_2	P_3
P_1	✓	✓		✓	✓		
P_2	✓		✓	✓		✓	
P_3		✓	✓	✓			✓

Figure 10.13 Encoding-Decoding Table For a Simple Error Correcting Code

To illustrate the use of this code assume the above codeword, 1001001 is transmitted over the communication system. At the receiving end of the system the parity checks indicated by the encoding-decoding table are made. More specifically the number of 1's among I_1, I_2, I_4 and P_1 is determined. This is called parity check number 1, and if the number of 1's is even the parity check is said to be correct, otherwise it is said to have failed. Similarly, parity checks no. 2 and 3 are made according to the definitions given by the 2nd and 3rd rows of the encoding-decoding table. Suppose that an error occurs and the block is received as 1011001. At the receiver the first parity check is correct but the second and third both fail. Then it is assumed that only one error has occurred and reference to the encoding-decoding table indicates the location of the error. (For systems for which the probability of double errors is appreciable this code is unsuitable but other codes are available for use.) If the error was in I_1 the table shows that I_1 enters into parity checks 1 and 2 but not 3; so the first two parity checks would fail, and the third would be correct. This is not the result of the parity checks so the error cannot be in I_1. Each of the other possible positions is considered, and the only position where an error could cause the first parity check to be correct and the other two to fail is I_3; therefore the error must be in digit I_3. I_3 is then changed, producing 1001001 which was the block transmitted. This procedure is called error correction, and it gives the code its name, an error-correcting code.

Efficiency of Codes

The procedure for making the parity checks and performing the corrections is automatically done by computer type circuits in the receiver. In this example the "cost" of the use of the error correcting code is the reduction in the information rate by the factor 4/7; alternatively the efficiency of the code may be said to be 4/7. It may generally be said that error detection and error correction are accomplished by the placement of parity digits among the information carrying digits or, in other words, by the incorporation of redundancy into the message. (The parity digits are redundant because they carry no message-information.) This redundancy provides protection against errors by permitting error detection or correction in much the same manner that redundancy in natural languages makes them resistant to errors. The redundancy in error control codes or in natural languages reduces communication efficiency by necessitating longer sequences of symbols to carry the information codes. Generally the greater the probability of error, the greater the redundancy, the less the efficiency, and thus the greater the cost of implementation. Error control codes take a great variety of forms. Many have longer block length than those discussed; there are also codes that don't have a block structure, but in general codes can be devised to overcome errors with any sort of characteristics.

10.2.8 Coding for Information Security

Codes as discussed above are used strictly for error control purposes, that is for protection of digital messages from distortion and thus errors due to noise. Another important use for codes is in security of information, or, in other words coding for secrecy.[5] This need, at one time, existed primarily in military systems, but now it exists in all systems that transmit information, protected by privacy laws, concerning all citizens. Techniques for coding and decoding digital information for secrecy purposes have been standardized and equipment is widely available.

10.3 QUANTITATIVE CHARACTERIZATION OF TELECOMMUNICATION SIGNALS AND SYSTEMS

Telecommunicaton systems exist to transmit information from one place to another. In order to describe and discuss such systems quantitatively, a method of measuring information must be devised. It is also necessary to devise a method of measuring the capacity of a telecommunication system to transmit information. These needs are met by modern information theory, the primary results of which will be discussed below.

10.3.1 Information Theory

Information theory will be discussed in terms of digital systems, although the ideas can be extended in a straightforward manner to analog systems. Digital information takes the form of a sequence of symbols, each symbol being selected from a set of finite size. Assume a digital information source uses an alphabet consisting of symbols A,B,C,D, and E; then a message may take the form ACDBAABDECBE, another may be BCEADCA. The information source can be studied statistically to determine probabilities for every possible message. Strong arguments can be made for defining a measure of information that assigns higher information content to messages of lower probability. One may reason that a message of low probability is more informative because it is more unexpected and in that sense, contains more information than one that is expected. (The information is greater when the surprise of receiving the message is greater.) A great variety of other arguments supports the basic idea that lower probability should correspond to greater information. This is the most fundamental concept in information theory. To implement it, assume the probabilities of occurrence of the various symbols are $p(A)$, $p(B)$, $p(C)$, $p(D)$ and $p(E)$, and that the probabilities of occurrence are independent of previous symbols in the sequence. The arguments can be modified for the case of non-independence yielding the same concepts but with greater mathematical complexity. One possibility for the definition of information content is that of making it equal to the reciprocal of the probability. If this definition is used, the information content in symbol A would be $1/p(A)$ and that in symbol B would be $1/p(B)$. The two symbol message AB would have probability $p(A)\,p(B)$ and information content $1/(p(A)p(B))$ which is $(1/p(A))(1/p(B))$, the product of the information content of A and that of B. The information content of a sequence of symbols would be the product of the information in the various symbols. Such a definition of information could be used, but it would be awkward. An additive measure of information would be much more convenient than a multiplicative measure. A simple modification of the definition produces an additive measure of information; the information content, I, of a message will be defined as the logarithm of the reciprocal of the probability, p, of the message;[2] or

$$I = \log \frac{1}{p} \tag{10.3}$$

Thus the information content of the symbol A is $\log 1/p(A)$ and that of B is $\log 1/p(B)$. Now the information content of the message AB is $\log 1/p(A)\,p(B)$ which is $\log 1/p(A) + \log 1/p(B)$ or the sum of the information in A and B. Generalizing, it may be said that the information in a sequence of symbols is the sum of the information in the various symbols in the sequence. The logarithm in the formula for information may be for any base that is convenient, but almost all analytical investigations are conducted using the base 2, in which case the unit of information is called the bit.

10.3.2 Entropy

Considering again a sequence of the symbols A, B, C, D, and E, the information content of the message would depend upon the number of the various symbols present in the message. If the message is reasonably long then the numbers of the various symbols can be closely estimated from their probabilities, and it is possible to calculate the average information content per symbol. A straightforward averaging procedure produces the result.

$$H = p(A)\log\frac{1}{p(A)} + p(B)\log\frac{1}{p(B)}$$

$$+\, p(C)\log\frac{1}{p(c)} + p(D)\log\frac{1}{p(D)} + p(E)\log\frac{1}{p(E)} \tag{10.4}$$

The average is called the entropy because of the close conceptual correspondence with thermodynamic entropy and has the units of bits per symbol. It is easily shown that to maximize the entropy (which will then maximize the efficiency of utilization of a communication link) the probabilities of the various symbols should all be equal. Let k be the number of symbols, then maximum entropy H_{max} occurs if the probabilities of all symbols are equal, in other words $p = 1/k$ for all symbols. Then the above formula gives $H_{max} = \log k$. Sometimes a recoding of information into a new set of symbols is done to increase the entropy and thus the efficiency of communication.

Methods of measuring the information I, in a message (of any number of symbols) and for the average information per symbol, have been introduced. One more convenient quantity is needed to quantitatively describe messages, that is the average information rate, r. If the message source produces n symbols per second and the entropy is H then the information rate (average) is

$$r = nH \tag{10.5}$$

and it is measured in bits per second.

10.3.3 Channel Capacity

The second major use of information theory concepts, that of characterizing the information handling capabilities of a channel will now be discussed. The discussion will be phrased in terms of a system handling a digital PAM signal but the result can be generalized to any system.

The amount of information in a pulse depends upon the number of possible levels, K, the pulse can take. Each level can represent a different symbol, so the maximum information represented by a pulse is the maximum information per symbol for a set of K symbols; this was previously found to be $\log K$. The maximum rate at which information can be transmitted over a channel is called the channel capacity, C, and the expression for it is

$$C = N \log K \tag{10.6}$$

where N is the number of pulses per second the channel can handle.

Another and more useful expression for C is obtained by expressing K in terms of the signal-to-noise ratio, SNR, at the receiving end of the channel. The relationship is[2]

$$K = \sqrt{1 + SNR} \tag{10.7}$$

where SNR is the signal-to-noise ratio.

N is then expressed in terms of the minimum bandwidth, B, required to transmit N pulses per second. The relation is

$$N = 2B \tag{10.8}$$

Using these relations, C can be expressed as

$$C = B \log (1 + SNR) \tag{10.9}$$

The logarithm is usually taken to base 2 in which case C has units of bits per second.

10.3.4 Fundamental Theorem of Information Theory

Now that methods of measurement of information in signals and methods of measurement of information handling capability of channels have been introduced, it is possible to state what is sometimes called the fundamental theorem of information theory.[2] It is convenient to present the theorem in the diagram below. A key point concerning the fundamental theorem is that if the information rate, r, does not exceed the channel capacity, C, the information in the signal can be transmitted by the channel. The theorem does not say the signal can be transmitted, but rather that the information in the signal can be transmitted. The signal can be transmitted only if it is matched to the channel; the matching conditions are $n \leq N$ and $k \leq K$. If these are not both satisfied, the signal is not matched to the channel and cannot be transmitted. A signal that is not matched to a channel (assuming $r \leq C$) can always be recorded into a signal that does satisfy the matching conditions and can therefore be transmitted. The reader is referred to other sources for a discussion of recoding techniques.

Signal Description

pulses per second = n
number of levels = k

entropy $= H = \sum_{i=1}^{k} P_i \log \frac{1}{P_i}$

average information rate = $r = nH$

Channel Description

pulses per second = $N = 2B$
number of levels = $K = \sqrt{1 + SNR}$
channel capacity = $C = N \log K$
or $C = B \log (1 + SNR)$

Fundamental Theorem of Information Theory

If $r \leq C$, then the information in the signal can be transmitted by the channel with arbitrarily small probability of error.

Matching the Signal to the Channel

If $r \leq C$, the channel will transmit the information but not necessarily the signal. If

$$n \leq N \text{ and } k \leq K$$

then the channel will transmit the signal. (Some modulation may be needed for spectrum shifting.) If these conditions are not both satisfied, it is always possible to recode into a signal which does satisfy both conditions.

10.4 TYPES OF TELECOMMUNICATION SYSTEMS

Telecommunication systems can be broadly categorized in three ways; point-to-point vs. broadcast systems, cable vs. radio systems, and analog vs. digital systems. These categorizations are discussed further in the following sections.

10.4.1 Point-to-Point versus Broadcast Systems

Perhaps the most fundamental classification of telecommunication systems is into switched, point-to-point systems on the one hand and into broadcast systems on the other. A switched point-to-point system provides communication between any two (or sometimes among a small number larger than two) of a group of terminals. This is accomplished by a communication link from each terminal to a switching center. The switching center (or centers) are capable of connecting the communication links of any two terminals. The most familiar example of a switched point-to-point system is the worldwide telephone system in which the many switching centers in many different locations are connected together to effectively function as one large (worldwide) switching center, which is capable of connecting together any two of the terminals connected to any of the individual switching centers. The communication links between terminals (telephones) and switching centers are customarily cables strung on poles or buried underground, but in some cases they are radio links. Similarly, the communication links between switching centers can be cables or radio links. The switched point-to-point type of communication system basically provides for two-way communication between two persons or between two machines (computer or computer terminals) or between a person and a machine. Sometimes the communication involves several persons or machines; and in some cases such a system has very limited one-to-many or broadcast capability.

A broadcast type of system, on the other hand, provides for one way communication from one terminal to many, (although some broadcast type systems such as cable television can be made to have very limited point-to-point capability.) A familiar example is a television broadcasting station. The uses, operation, regulation, technology, and financing of broadcasting type of communication systems are so different from those of switched point-to-point that analysis of them requires separate study. Broadcast systems will not be treated, in detail, in this chapter.

10.4.2 Cable versus Radio Systems

A second type of categorization of communication systems is based upon the type of communication link used. The two types may be referred to as the cable type and the radio type. Each of the two types has unique characteristics which strongly influence the choice of which to use in various applications. A cable system requires use of a cable or waveguide, preventing application in any communication system where terminals are located in moving vehicles. Such a communication system requires use of a radio type of link. (Isolated exceptions such as guided wave communication with vehicles on railroad tracks occur infrequently.)

Radio type systems radiate signals that propagate through the troposphere. The signals can be received by any radio receiver located in regions having sufficient signal power. For this reason two or more radio systems operating in the same region will interfere with each other if their signals are in the same frequency band. Interference can be avoided by operating each system in a different frequency band, an excellent solution to the interference problem unless there is a demand for more systems than there are frequency bands available. This is a problem, called the spectrum crowding problem, that is becoming common in the more heavily populated regions of the world. A solution is the use of cable communication links for new systems in those regions. Cable systems do not interfere with each other or with radio systems. (This is not a completely accurate statement, but the interference is very small and can be made even smaller if necessary.) For this reason much of the future expansion of communication systems will involve cable links.

10.4.3 Analog versus Digital Systems

The third and last major category of communication system involves analog communication links as opposed to digital communication links. This classification of systems is not nearly as clear-cut as those previously discussed because an analog link can handle digital signals by the simple expedient of inserting a modem at each end of the link, and a digital link can handle analog signals by inserting samplers and filters. However the distinction between the two types of links is important in understanding the current trends in communication system design. Most of the existing communication capacity utilize analog links, but there are strong arguments favoring digital links for new designs. Digital links are more efficiently adaptable to transmission of data than are analog links. Since demand for data communication is expanding rapidly, and since the cost of digital communication equipment has steadily compared to analog equipment (this is partly a consequence of the drastically reduced costs of computer type circuitry made possible by the development of integrated circuits) digital communication links are likely to be widely used in future communication systems.

10.5 Communication Systems

Communication systems consist of many components. This analysis will be concerned with three major components, cable communication links, radio communication links, and switching systems and their interconnection into networks.

10.5.1 Cable Communication Links

The most widely used communication links may be broadly categorized as cable links. They may be twisted wire pairs, coaxial cables, optical fibers, or any of a variety of waveguides. They may be used in any application for which it is possible to run a cable between the communicating terminals. The most obvious communication systems that cannot use cables are those involving moving vehicles. Radio links are necessary in such systems. There are also cases in which it would be possible, in principle, to run a cable; but, it is not practical because of right-of-way problems.

A significant advantage of cable communication links is that they do not radiate and therefore do not require a frequency allocation. This not only eliminates the problem of obtaining a frequency allocation, but, more importantly, it permits an unlimited number of communication links to operate in the same region without interference. This could not be done with radio links because of the limited radio spectrum.

Cable Communication Links with Repeaters

Cable communication links attenuate signals and are thus limited in length to a few kilometers unless amplifiers or repeaters are employed. The use of repeaters in cable links permits long distance communication. The relationship among parameters for such systems is

$$SNR = \frac{P}{n\,[NB\,]e^{2\alpha D/n}} \qquad (10.10)$$

where SNR is the predetection signal to noise ratio at the receiver; it is a measure of the quality of the communication signal (section 10.2.6), and n is the number of repeater sections in the system. (If the distance between repeaters is d, the total length, D, of the system is $D = nd$.) P is the signal power at the output of the transmitter or a repeater (assumed to be the same). α is the cable attenuation constant in nepers per unit length. B is the bandwidth of the signal and N is the spectral density of the noise contributed by one repeater section. Thus NB is the total noise contributed by one repeater section.

The distance between repeaters is $d = D/n$; it is determined primarily by the attenuation characteristic, α, of the cable and secondarily by signal power levels, noise power levels, and the number of repeater sections. Repeater spacing varies from a few hundred feet for wide-band twisted-pair links to 30 km or more for low loss waveguide such as optical fiber. The total power required for communications is nP which is given by $nP = (SNR)n^1 [NB]e^{2\alpha D/n}$ · B and SNR are dictated by the use to be made of the system. The choice of equipment, that is cable and repeaters, determines α and N; these determine repeater spacing, d, within moderate limits. A specific value for d is selected; since, $d = D/n$, the expression for total power is

$$nP = (SNR)\, n^2\, [\,\mathrm{NB}\,]\, e^{2\alpha d} \qquad (10.11)$$

Since d is fixed, the total length of the system is proportional to n; and, thus, the total power needed to communicate is proportional to the square of the total distance. This is also true for a radio relay system; so it may be considered a general law of communication.

Many of the existing long distance cable links are coaxial cable systems used by the common carriers. For heavy traffic routes, a bundle of many coaxial cables handles a large amount of communication traffic with repeaters spaced at intervals of a few km.

Dispersion

In the previous section the problem of repeater spacing in a long distance, high capacity cable communication link was considered. It was found that repeaters must be inserted in the cable at regular intervals to restore the power level of the signal. Another important phenomenon that must be considered in cable systems is distortion due to dispersion. Dispersion occurs in all communication links, but it is more severe in cables than in radio links. It is the variation of velocity of propagation of signals with frequency. It distorts the signals in a manner easily apparent by viewing the graph of a signal. A pulse signal, for example, after passing through a dispersive link, becomes spread out in time and takes a broadened shape much different than at the input to the link. Each pulse in a sequence of pulses may become so broadened that it overlaps adjacent pulses in the sequence and the detection of the information in the pulse sequence becomes difficult. The distortion may become so extreme that the individual pulses are obliterated.

The nature of dispersion is such that the distortion becomes progressively worse as the signal travels along the link. Sufficiently short communication links produce tolerable distortion due to dispersion; but longer links particularly cable links requiring repeaters, often produce enough distortion to require correction.

The velocity of propagation of waves can be considered, as a first approximation, to be a linear function of frequency. Since the spreading of pulses is due to the different velocities of the various frequencies in the signal, the amount of broadening of the pulse is proportional to the bandwidth, B. It is also proportional to the length of time taken by the signal to traverse the link which in turn is proportional to the length, L, of the link. The amount of broadening permissible without causing pulse overlap is proportional to the interval between one pulse and the next; this in turn is inversely proportional to bandwidth, B. Thus, to avoid excessive pulse overlap

$$\frac{C}{B} \leq BL \tag{10.12}$$

where C is a constant dependent upon the amount of overlap permissible. Thus, the maximum permissible values of B and L that could be used without excessive overlap of pulses are related by

$$B^2 L = \text{constant} \tag{10.13}$$

In pulse systems, the pulse repetition frequency, PRF, is often specified rather than bandwidth. Since PRF is proportional to B, the above relation can be written

$$(PRF)^2 L = \text{constant}$$

In deriving this result, it was assumed that as the PRF increases, the width of the pulse decreases and consequently its bandwidth increases proportionally. If the shape of the pulse remains unchanged, as PRF increases, then the bandwidth of a single pulse is constant, and the amount of broadening is proportional to just L rather than the product BL. Then the relationship is

$$BL = \text{constant} \tag{10.14}$$

or

$$(PRF)\, L = \text{constant}$$

Distortion due to dispersion is predictable; and therefore it can be completely eliminated by signal processing. In other words the distorted signal can be restored to its original, undistorted form by use of a device usually called an equalizer. Although the distortion, in principle, can be completely overcome regardless of its severity, the signal processing required is simpler and more practical if the degree of distortion is limited. Therefore, in a long link, it is customary to insert equalizers at uniform intervals. In a typical cable link the equalizers are inserted at intervals that are several times the interval between the repeaters that are used to restore signal power.

In some cable systems the dispersion is so great that the equalizers need to be spaced at intervals less than that required between repeaters determined strictly for power restoration. Since equalizers and repeaters are colocated or actually combined into one unit, the repeater spacing in such systems is said to be dispersion limited rather than power limited as it is called otherwise.

In a sufficiently long analog, cable link there is no alternative to the use of equalizers. The case of a digital cable link is somewhat different as is discussed in the next section.

Regenerative Repeaters

A digital communication link long enough to require repeaters may use the type of repeater known as a regenerative repeater. A regenerative repeater is a type of repeater that generates a new pulsed signal rather than amplifying the one received. The input circuitry detects the pulsed signal at the input and produces an output in the form of a sequence of symbols containing the information in the signal. The sequence of symbols is then used as the input to a pulse generator which generates a distortion free signal carrying the same information as the input signal.

Pulsed signals become distorted due to dispersion. If the distortion is severe, the detector may make errors in the sequence of symbols extracted from the input signal. To avoid such errors, it may be necessary to use an equalizer on the input signal. The amount of equalization needed is less than for an analog system in which almost all distortion is eliminated. In this case it is only necessary to remove enough distortion to permit the determination of the symbol corresponding to each pulse. This could be called partial equalization and can be accomplished with simpler and less expensive equipment than full equalization in an analog system. If the regenerative repeaters are spaced sufficiently close, the distortion, due to dispersion, can be limited so that detection is possible without equalization. Thus, in digital links repeaters can be of the relatively simple and inexpensive regenerative type spaced at intervals requiring no equalizers. This is a significant cost advantage of digital links as compared with analog links because equalizers are complex and expensive.

Noise in a digital link causes distortion of pulses which, in turn, causes the detector in a regenerative repeater to interpret some of the pulses incorrectly and therefore to make errors in the symbol sequence. Unlike distortion due to dispersion, distortion due to noise cannot be corrected. It may be said that noise in a link with regenerative repeaters takes the form of errors in the symbol sequence. This noise, of course, accumulates from one repeater section to the next as is true in analog links.

The relative ease of overcoming dispersion effects in a digital link compared with an analog link strongly favors digital signals in systems having links with high dispersion.

Optical Fiber Communication Links

Optical fiber links which have been in development during the past ten years or so are increasingly being used for new systems. They are competitive with coaxial cable on a cost basis at this time and offer the potential of significantly lower costs in the future. The channel capacity of a single fiber is of the same magnitude as a coaxial cable; but in this case too, it could increase significantly in the future. Optical fibers are characterized by substantial dispersion and for that reason are generally used for analog signals only if the system is relatively short. In long optical fiber links the signal is usually digital to simplify the problem of overcoming dispersion effects. Fibers are used to some extent in cable television systems to transmit analog video signals over short distances and in telephone systems to transmit digital signals over longer distances such as between switching centers.

Cable communication links take many forms and are used in many applications. The majority of such links are incorporated into switched networks which are discussed in the next section.

Communication Networks

The most common communication network, the telephone system, uses twisted wire pairs to link individual telephones to a central switching facility which can provide a communication link between two telephones in the network or between a telephone in the network and a terminal in another network. The most common of these systems handles analog signals (section 10.2), and are primarily intended for voice communication. They can, however, handle data by use of a modem (section 10.2.5), at each terminal. Most of these modems use FSK or PSK modulation and are noncoherent; some of the higher capacity modems are coherent.

The switching facility in these networks may be the central office of the local telephone company or it may be a private branch exchange (PBX) on the customer's premises. A PBX may be privately owned or alternatively part of the local telephone system. Most switching systems in use are of the space division type, but a substantial proportion of the newer installations use time division switching (section 10.4.2). A time division switching system for analog signals must convert the signals to PAM signals (section 10.2.5) before switching and then perform the reverse transformation after switching.

Multiplexed Signals

Communication links between switching facilities may be in the form of twisted wire pairs, each pair carrying one signal, or they make use of multiplexing schemes which combine many signals into one. This one signal is then transmitted over a single communication link, which could be a single wire pair if the number of multiplexed signals is not too great; or it could be a coaxial cable, an optical fiber, or a microwave radio system either terrestrial or satellite. Most multiplexing systems in use today use frequency division multiplex (FDM) (section 10.2.5), although time division multiplexing (TDM) (section 10.2.5), is frequently favored in newer systems.

Multiplexing has been described here in terms of use between switching centers, but it is also used between a switching center and a group of remote terminals. A multiplexer centrally located among the terminals permits use of a single communication link to the switching center rather than a link for each terminal as would be required otherwise.

The justification for multiplexing is purely economic. It is only justified if the total cost of the multiplexing and demultiplexing equipment and of the single wide band channel is less than the cost of the many narrowband channels that would be needed if multiplexing were not employed.

Digital Networks

Some of the newer telephone systems handle digital signals only. Data can be handled directly by such systems, but voice signals which are analog in nature (section 10.2), must be converted to digital signals. The conversion is done by a coding operation that involves sampling of the analog voice signal followed by digital coding of the samples (section 10.2.4). The resulting digital signal is transmitted and switched as necessary. Then at the receiving terminal it must be converted back to its original form by a decoding process which is the inverse of the coding process. The coder, processing the outgoing signal at the terminal, and the decoder, processing the incoming signal at the same terminal, are often packaged together into a component identified by the acronym CODEC.

Switching facilities used for digital signals are of the time division type (section 10.5.3) and may be part of the public telephone system or may be privately owned and located on private property. Communication links between switching centers or from a switching center to a remote group of terminals takes the same form as for analog signals, but multiplexing, if it is used, is time division multiplexing (TDM) (section 10.2.5).

Private Networks

It has been common throughout most of the history of the telephone industry for organizations (business, government, educations, etc.) to have private telephone systems complete with switching centers permitting internal communication within a building or group of closely spaced buildings, without using the public telephone network. Such systems allow access to the public network from any of the terminals but, in many cases, handle more internal communication than external.

Recently these private systems have increased in scope to include facilities of an organization in different cities throughout a country or even throughout the world. Such a private system is commonly called a private network. The communication links between cities in a private network are usually leased from a common carrier; since, due to right-of-way problems, it is rarely practical for an organization to own and operate its long distance links. Railway companies are a noteworthy exception. Leased links between distant cities now are often satellite links.

The cost of long distance links in private networks can be lowered by the use of TASI equipment (sec. 10.5.3). TASI equipment has long been in use on the more expensive links of public telephone systems (such as submarine cable links); but it has only recently become available for use on private networks.

Local Area Networks (LAN's)

The networks discussed above have been designed and used primarily for voice communication. They are adaptable to other types of signals (data, video) but usually are well suited only to voice communication. The expanding need for data communication has led to a variety of new networks particularly suited to data. The smaller such networks, intended for use in a single building or in several closely spaced buildings, are generally privately owned and are called local area networks (or LAN's). They are used to link computers, computer terminals, printers, digital storage units, and other such devices. They will play a key role in the so-called office of the future wherein electronic handling of information supplants the handling of information recorded on paper.

LAN's can be rather loosely categorized as either of two types. One type is the switched network in which each terminal is connected directly to a switching center. This type of network is the same as described above for voice communication and is supplied by some of the same manufacturers that supply switched voice networks. Some of the more recently designed systems make a switched data system available along with a switched voice system. The other type of LAN makes use of a common communication link (usually a coaxial cable) to which all terminals are connected. The transmission capability of the cable is shared by all terminals on a time-division basis (as in time division multiplexing, section 10.2.5), or on a frequency division basis (as in frequency division multiplexing section 10.2.5), or both. The more sophisticated systems can handle voice and video as well as data.

Widespread use of LAN's dictates a degree of standardization among the manufacturers of terminal equipment to permit a variety of terminals to be connected to a single system. Efforts towards standardization of interfaces have been under way for some time with limited success.

Integrated Services Digital Networks (ISDN's)

Large scale networks in the past have been largely analog in nature. These have been extended and modified to handle digital signals as the need arises. As digital components and systems became more available and more attractive economically, parts of the older networks were converted to the digital type and new additions have been of the digital type. The economic as well as operational advantages of digital networks have become such a dominant driving force in telecommunications that planners are now attempting to establish standards for new, completely digital networks for handling all types of telecommunications. Such systems are referred to as integrated services digital networks (or ISDN's) and are expected to make extensive use of satellite communication links and of optical fiber links.

10.5.2 Radio Communication Links.

The majority of communications is handled by cable networks as described above. Another class of communication systems that serves to complement networks is single link communication systems. Many of these use radio communication links and are of a somewhat more specialized nature than the cable networks discussed above. Examples include mobile radio communications, maritime radio communications, and aircraft communications. Others include point-to-point radio which takes a variety of forms. Although this section is primarily concerned with what may be called "stand-alone" radio communication links, it should be noted that radio communication links often serve as links in networks which consist primarily of cable.

Radio Spectrum Crowding

The most obvious distinction between cable communication links and radio links is that the former needs a cable or waveguide between the communicating terminals. Another equally significant difference is that the radio link needs a spectrum allocation, or an authorization to use a specified band of frequencies. Such an authorization precludes other users from using frequencies in that band for any radio communication link close enough to interfere. Since the usable radio spectrum is finite, this limits the amount of radio communication in any region. Such limitation, due to spectrum crowding, exists in a number of regions today.

Spread Spectrum Systems

There are types of radio communication links that permit two or more radio links to operate on the same frequency band in the same region. This is accomplished by a signal transformation procedure that greatly expands the bandwidth. This characteristic gives the signal its name, spread spectrum. The spectrum is spread by a time-varying procedure which can take an infinite variety of forms. The time-varying procedure used by the transmitter must also be used by the receiver to detect and decode the signal. Therefore many signals, each using a different procedure, may use the same frequency band in the same region. Although many spread spectrum systems may share the same frequency band, that band is so large that this is not generally looked upon as a method to overcome the spectrum crowding problem, but it may someday contribute to lessening the problem. Spread spectrum systems at this time, are primarily of military interest.

Types of Radio Communication Links

There are low frequency, high power, long distance radio systems which are used for navigation to disseminate standard signals. They also have military uses. There are high frequency (HF) systems that use ionospheric communication links. These are used for amateur radio communications; and for military communication. They once provided transoceanic links for telephone networks, but they have now been supplanted by satellite links and by submarine cable.

Radio communication in the very high frequency (VHF) and ultra high frequency (UHF) bands serves a variety of users. These communication links are restricted to line-of-sight communication paths so they are limited to distances of 200 km or less (one exception, troposcatter links, is discussed later). They require only modest amounts of power and small antennas and are widely used for mobile and aircraft communication.

Cellular Radio Telephone Systems

One particular use for VHF and UHF radio links is for mobile telephones in which case the system is incorporated into a telephone network.

Mobile telephone service has existed for several decades but has not expanded to fill the need for such service because of limitations on the radio spectrum available. This problem can be overcome by dividing a radio-telephone service area into small cells and using low power transmitters in the cells. The low power permits the reuse of frequency bands in non-adjacent cells. In a system of many cells, a single frequency band can be used many times. Such a system permits very significant expansion of radio telephone service. The system is complex because it is necessary to change frequency bands when a vehicle crosses the boundary from one cell to the next. This change is handled automatically. These so-called cellular radio-telephone systems have been under study, development, and test for about 20 years; it appears they will go into wide scale use in the near future.

Tropospheric Scatter Radio Links

Troposcatter (tropospheric scatter) communication links achieve communication using the scattering of electromagnetic waves by the natural irregularities of the troposphere. This permits communication beyond line-of-sight distances; however, high power levels and large antennas are needed, making such systems impractical for most commercial uses. Their unique properties are useful in military communications.

Radio Communication Links with Repeaters

The line-of-sight limitation on radio communication links operating at VHF and higher frequencies requires the use of repeaters for long distance communication. A long distance terrestrial link has repeaters, usually on towers, spaced typically 30 - 70 km apart. Such systems generally use microwave frequencies because the bandwidths are greater and the antennas and other equipment are less costly than for VHF or UHF frequencies.

The relationship among the significant parameters of a microwave relay system is

$$SNR = \frac{P}{n\,[NB]\,\dfrac{\left(4\pi \dfrac{D}{n\lambda}\right)^{2}}{G_t G_r L}} \tag{10.15}$$

SNR is the predetection signal-to-noise ratio at the receiver; it is a measure of the quality of the communication signal (section 10.2.6). n is the number of repeater sections in the system (if the distance between repeaters is d, the total length, D, of the system is $D = nd$). P is the signal power at the output of the transmitter or a repeater (assumed to be the same). G_t and G_r are the gains of the transmitter and receiver antennas respectively. L is an efficiency factor for the transmitting and receiving transmission systems. λ is the wavelength. B is the bandwidth of the signal and N is the spectral density of the noise contributed by one repeater section. Thus NB is the total noise contributed by one repeater section.

The distance between repeaters is D/n and is governed by the line-of-sight requirement. It can be increased by increasing the height of the towers. The economics of tower construction thus determines D/n and therefore n. It may be assumed that SNR and B are dictated by the system's use. λ is determined by the frequency assignment for the system. The designer then must select appropriate equipment so that S_t, N, G_t, G_r and L satisfy the above relation.

It is of interest to note that the total signal power, nP, needed to operate a radio relay system is given by

$$nP = (SNR)[NB]\frac{\left(4\pi\frac{D}{\lambda}\right)^2}{G_tG_rL} \tag{10.16}$$

Two significant observations can be made. First, the total power required to communicate is proportional to the square of the total distance, D. Second, the total power required is independent of the number of repeater sections. This is in contrast to cable systems with repeaters in which total power is strongly dependent upon the number of repeater sections. Consequently repeater spacing in radio relay systems is not governed by power considerations. Most radio relay systems operate at VHF and higher frequencies (50 MHz and higher) at which signals propagate in a straight line from transmitter to receiver. This so-called line-of-sight, signal path requirement is a primary consideration in the spacing of repeaters. Due to earth curvature the repeaters must be placed on tall towers (or hills or buildings). Taller towers permit greater repeater spacing and fewer repeaters, but they are more expensive. Repeater spacing is thus strongly dependent upon the economics of tower construction.

Another factor in the determination of repeater spacing relates to meteorological conditions. Heavy precipitation along the signal path results in severe reduction of signal power level (called fading.) Systems are customarily designed to have excessive signal power levels under normal operating conditions so that during moderate fading (30-40 dB) conditions communication is sustained (although with diminished quality). The excessive signal power under normal conditions is called the fade margin. It is chosen to be the maximum signal fade under moderate to severe fading conditions. Extreme fading conditions will disrupt communication despite the fade margin. Some geographic regions are subject to extreme fading conditions more often than others. To reduce the probability of communication loss in such regions, it is common to reduce the repeater spacing which reduces the signal path length and lowers the probability of a deep fade. In such regions fading statistics, as well as the line-of-sight requirement, are important criteria for determination of repeater spacing.

Satellite Radio Links

Note that a satellite communication system has one repeater (and thus two repeater sections). The above relation thus applies, in general terms, with $n = 2$.

Because of the high capacity and great distance possible with satellite systems, they are becoming widely used throughout the world as high capacity links in large networks. The more recent satellite systems serve many users and have various forms of switching capability to handle varying traffic levels for the users. Two of the methods for sharing capacity among users are known as Frequency Division, Multiple Access (FDMA) and Time Division, Multiple Access (TDMA) and will be discussed in this chapter and in the one on Communication Satellites.

10.5.3 Switching Systems

A broad interpretation of the term switching system is assumed in this section. Conventional space division and time division switches are discussed as well as some of the newer systems that are used to accomplish selective communication among terminals of a network.

Space Division Switching Systems

Presently, telephone switching centers are predominantly of the space division type. These switches establish electrical contact between the appropriate terminals for the duration of the communication. Such switches are suitable for analog signals as well as pulse signals. The complexity and cost of space division switches are approximately proportional to the square of the number of communication terminals handled by the switch. This places such switching systems at a disadvantage compared with the newer types of time division switching systems with the result that time division switching systems are being increasingly used in newer systems.

Time Division Switching Systems

A signal in the form of a train of pulses requires an electrical connection through the switching system only during each pulse. This permits switching equipment to be shared by many terminals with potential cost savings. The cost of time division switches is approximately proportional to the number of communication terminals it handles. (This is in contrast to space division switches discussed in the previous section.) The components of time division switches are of the same sort as used in computers. Such components have been highly developed and refined in recent years, and their cost has been going down. These characteristics have given time division switches a significant advantage over space division switches for many applications. Their increasing popularity seems destined to continue for the forseeable future.

Time division switching systems were described above as being suitable only for pulsed signals. This is not a serious limitation because analog signals can be sampled and converted to pulsed signals (section 10.2.3). Some switching systems in fact convert all input signals from analog to digital by sampling; then the switching is done by a time division switch, and finally the signals are all converted back to analog at the output. It has been found that such systems can be less expensive than space division switches despite the extra equipment needed for two conversions between analog and digital signals. Digital signals, of course, are in a form to be directly handled by time division switches. Thus in a digital network, time division switches are the most logical choice.

Message Switching

A class of networks has evolved which may be called message switched networks. These are intended for one-way communication of written (or, in some cases, voice) messages or of data which need not be communicated in real time; such services include the so-called electronic mailbox. The network consists of communication links connecting network nodes at which receiving, storage, and transmitting equipment are located. A message and the name of its intended destination are sent to a node where they are temporarily stored. The control system then selects a route, through the network, from node to node, depending upon the destination and upon network traffic conditions, for the transmission of the message. The transmission from each node to the next is accomplished when the communication link becomes available. The message is then stored until it is possible to transmit it to the next node. The selection of a route through the network may be done before the message leaves the first node; or it may be done by making a decision at each node for the transmission to the next node. These networks are sometimes called store and forward communication systems. Message switched networks are usually incorporated into other, larger, general purpose networks.

Packet Switched Networks

A message switched network as discussed in the previous section can handle a great variety of message lengths. The length of time required for a message to be delivered to its final destination depends upon its length, its destination, and network traffic conditions. Such a network is only practical if speed of delivery is a secondary concern.

For some purposes, such as data transmission in a computer system, the delay in delivery of a message to its destination may need to be limited. A message switched system can be adapted to such use by suitable control procedures. One of the modifications is a restriction on the length of the message to be handled. At first this seems an unworkable sort of restriction; but it simply means the message must be divided into short blocks, usually called packets, each of which is transmitted separately. Different packets of a single message may be routed differently through the network; and, the sequence of arrival of packets at the destination may be different than the sequence at the source. The packets are reordered, if necessary, and the message is reassembled by the control system at the node nearest the destination. Some packet switched networks offer message delivery fast enough to be used for some services that normally require real-time communication.

TASI Systems

In an effort to improve the efficiency of utilization of high cost voice channels (such as submarine cable channels), Bell Laboratories developed a so-called Time Assignment Speech Interpolation (TASI) system. This system utilized detectors to determine the begining and end of each burst of speech in a voice signal. It assigned a channel to transmit that burst of speech; then, at the end of the burst the channel was reassigned to handle a burst of speech in another voice signal. The switching of a voice signal to a channel, during a burst of speech, must, of course, be done at both ends of the communication link, and it must be done for the signals in both directions. This requires a highly sophisticated switching device at each end of the communication link and a control system to coordinate their switching. Such equipment is used at the two ends of a bank of channels. It permits a bank of say, N, channels to handle more than N voice communications because, during each pause in a voice signal, the channel it had been using handles a burst of speech in another voice signal. At the end of the pause, when another burst of speech starts, a channel (not necessarily the one transmitting the previous burst) is assigned to transmit it.

For a two-way telephone conversation in a standard telephone system, each of the one-way channels will carry a voice signal usually less than 50% of the time. It follows that TASI equipment has the potential of approximately doubling the efficiency of utilization of those channels by handling twice the number of voice signals as there are channels. In a system utilizing TASI, any time the number of voice signals being handled is greater than the number of channels, there is the possibility, due to the random nature of voice signals, that all of the channels are in use when one is needed for a burst of speech. A channel cannot be assigned until one becomes available due to a pause in the speech in one of the channels. In this case, part or maybe all of the burst will be lost. The TASI system is not perfect; but for a sufficiently large group of channels it can carry twice as many voice signals as there are channels without objectionable degradation due to loss of short intervals of speech.

Until recently, the complexity and cost of TASI systems limited their use to only the most costly channels in the telephone system. Cost reductions in electronic components and circuitry have recently permitted use of TASI systems for a broader class of networks. In particular TASI is now used in some private networks.

FDMA Demand Assignment

Operation of wide band communication systems to accomodate a number of users can be accomplished by assigning different frequency bands to the various users. This is referred to as Frequency Division, Multiple Access or FDMA. This is similar to frequency division multiplexing. It works well if each user continuously utilizes a fixed bandwidth; but if users need bandwidths only part time or if they have need for different bandwidths at different times, some sort of switching is needed. Such a system is called a Frequency Division, Multiple Access, Demand Assignment System. This situation arises in satellite communication systems and is discussed in another chapter.

TDMA Demand Assignment

The use of a wide band channel by many users can be done on a time sharing basis rather than by frequency division as discussed in the previous section. This is similar to Time Division Multiplexing, previously discussed. When this is done it is called a Time Division, Multiple Access or TDMA system. In this case too, a user's needs may vary with time. This can be handled by a switching scheme which adjusts the time slots allocated to the various users according to demand. This is then called a Time Division, Multiple Access, Demand Assignment system. Such systems are used with satellite communication links and are discussed further in another chapter.

Addressable Signals

Another approach to the problem of switching a message to the proper communication terminal makes use of an address added to the message which is then broadcast to all of the terminals on the system. The terminals, each of which has a unique address, have circuitry which decodes the address of all messages broadcast and accepts only those having the proper address. Such addressable systems are widely used.

Addressable Signals in CATV Systems

Cable television systems use addressable signals to access terminals of individual subscribers to activate or deactivate signal decoders. This is also done by subscription television (STV) broadcasting stations.

Paging Services

Paging services use addressable radio signals to communicate with portable receivers carried by individuals. Most of these systems simply alert the individual that communication is desired; the actual communication is then achieved using the telephone system. Paging services were originally confined to a city or region, but are being expanded to provide service on a nationwide basis and eventually, perhaps, on an even wider basis. This is done by using a network of transmitters and by distributing signals to the transmitters by a private network. The actual broadcast of the paging signals is often done by commercial broadcasting companies providing the service under contract to the paging company.

Addressable Signals in LAN's

Addressable signals are used in some types of Local Area Networks (LAN's). The address of the terminal to which the message is directed is added as a prefix to the message which is then transmitted over the common cable and appears at the input to each terminal on the network. Only the terminal (or terminals) to which it is addressed actually accepts the message.

Addressable Signals in Spread Spectrum Radio Systems

Spread spectrum radio systems use a common frequency band for many communication links. A transmitted signal can be received by many receivers, but it can be detected and decoded only by those using the same time-varying spectrum spreading procedure used by the transmitter. Thus, in a sense, the transmitted signal may be said to be addressed to only certain receivers. This type of system is also of interest for signals that must be kept confidential (section 10.2.8).

10.6 COMMUNICATION TRAFFIC

Communication networks are capable of handling varying amounts of communication traffic up to levels approaching their capacity. As that level is approached, delays occur in the provision of service. The amount of degradation of service depends upon traffic levels in relation to the amount of equipment. The relationship between quality of service, level of communication traffic, and amount of equipment (number of trunks, size of switches, etc.) is the subject of traffic theory (also known as queueing theory or the theory of stochastic service systems). Familiarity with these relationships is needed in the planning or expansion of communication networks. Some of the more fundamental relations of traffic theory are briefly discussed below.

10.6.1 Blocking Formulas

In a switched network, an attempt to initiate communication between two terminals may fail because the switching facilities or the trunks that are needed are already in use. The attempt to communicate is said to have been blocked. There are formulas giving the probability of blocking in terms of traffic volume and amount of equipment for several types of networks. Three of the most widely used formulas will be discussed. The discussion will be phrased in terms of a number, say N, of trunks handling telephone calls from a much larger number of telephones; but the formulas are applicable to a variety of related communication problems. The amount of traffic is defined to be 0.01 times the total number of seconds of use of all the telephones in a one hour interval. In other words, during a one hour interval, the number of seconds of use of each telephone is determined. Then these numbers are all added, and the sum is divided by 100. The unit of traffic is the CCS which is taken from the phrase, hundreds of call-seconds per hour. Traffic is sometimes described in terms of the number, n, of calls per hour and the average holding time, h, of a call. Then the number of CCS is $nh/100$. Another unit of traffic is the Erlang. Let T be the number of Erlangs of traffic; T can be calculated by finding the number of CCS of traffic and dividing the result by 36. Thus $T = CCS/36$.

Poisson Blocking Formula

The Posisson blocking formula takes its name from the Poisson probability distribution on which it is based. It gives the approximate probability, $P_p(N,T)$, of blocking for N trunks carrying T Erlangs of traffic. It does nor matter whether the blocked calls are diverted from the system, as for Erlang B, or they are delayed, as for Erlang C. It is given by

$$P_P(N, T) = 1 - \sum_{i=0}^{N-1} \frac{T^i e^{-T}}{i!}$$

Although this is not an exact result, it is widely used because it is simple and because it gives answers that are in many cases, close to those given by more complicated blocking formulas. Computations of blocking probabilities based on this formula are usually done with the aid of charts or tables or by use of a computer. Charts or tables are available in most books on traffic theory.

Erlang B Blocking Formula

In the early years of the 20th century, A.K. Erlang derived two blocking formulas that bear his name. The Erlang B formula is applicable to systems in which blocked calls are diverted to, and handled by, alternate facilities. The probability, $P_B(N,T)$, of blocking, or, in other words, the probability that diversion is necessary for a system with N trunks handling traffic of T Erlangs (this is the total traffic handled by the N trunks and the alternate facilities) is given by[1]

$$P_b(N, T) = \frac{\dfrac{T^N e^{-T}}{N!}}{\displaystyle\sum_{i=0}^{N} \frac{T^i e^{-T}}{i!}}$$

Charts and tables of this formula are given in traffic theory books. The Erlang B blocking formula is applicable to a variety of practical problems.

Erlang C Blocking Formula

The Erlang C blocking formula is applicable to a group of N trunks handling T Erlangs of traffic assuming that a blocked call is simply delayed until one of the N trunks becomes available. The probability, $P_c(N,T)$, of blocking, or, in other words, the probability that there will be a delay, is

$$P_c(N,T) = \frac{\dfrac{T^N e^{-T}}{N!}\left(\dfrac{N}{N-T}\right)}{\sum_{i=0}^{N} \dfrac{T^i e^{-T}}{i!} + \dfrac{T^N e^{-T}}{N!}\left(\dfrac{T}{N-T}\right)}$$

Charts and tables of this formula are given in books on traffic theory. The Erlang C blocking formula is applicable to a wide variety of communication systems.

10.6.2 Delay Formulas

In a system in which blocked calls are delayed, the probability of blocking is given by the Erlang C blocking formula as discussed in the previous section. The quality of communication in such a system is dependent upon the probability that a call will be delayed; but it is also dependent upon the length of delay. The length of delay is a random variable which is characterized by a probability distribution function. Delay formulas relate the same three measures as do blocking formulas, that is, quality of service, the level of traffic, and amount of equipment. The relationship is, however, somewhat more involved than in the case of blocking formulas because the measure of quality is a function instead of a single number. Another complication results from the fact that a procedure must be specified for selecting one of the calls that is being delayed for service by the next trunk that becomes available. The procedure may be, for example, selection of the call that has been delayed longest or it could be selected randomly. Each procedure leads to a different delay formula. Formulas for these two cases are given in the following sections.

Delay Distribution for Order of Arrival Service

In order of arrival service, the call waiting the longest gets the next available trunk. Let d be the delay in seconds; then, for the delayed calls only, the probability distribution function, $P_a(d, N, T)$, or more specifically, the probability that the delay will be d seconds or longer, is[1]

$$P_a(d, N, T) = \exp\left[-(N-T)\frac{d}{h}\right]$$

where N is the number of trunks, T is the traffic in Erlangs and h is the average holding time in seconds. If the distribution function for delay is desired for all calls rather than just those delays, the above result is simply multiplied by the probability of blocking as given by the Erlang C formula for $P_c(N,T)$. The average delay, $\bar{d}$, for delayed calls only, is

$$\bar{d} = \frac{h}{N-T}$$

The average delay for all calls would be this multiplied by $P_c(N,T)$ as previously mentioned. Charts for determination of $P_a(d,N,T)$ are found in books on traffic theory.

Delay Distribution for Random Service

In random order of service, the next available trunk is assigned to a call, selected at random, from all calls that are being delayed. The probability distribution function, $P_r(d,N,T)$, or, in other words, the probability that the delay will be d seconds or longer is the solution of the partial differential equation[1]

$$\frac{\partial^2 P_r(d,N,T)}{\partial d \partial T} + \frac{1}{T} \frac{\partial P_r(d,N,T)}{\partial d} - \left(1 - \frac{N}{h} - \frac{T}{h}\right) \frac{\partial^2 P_r(d,N,T)}{\partial T} + \frac{1}{h} P_r(d,N,T) = 0$$

subject to the boundary conditions

$$P_r(d,N,O) = e^{-\frac{N}{h}d}$$

and

$$P_r(O,N,T) = \frac{N}{N-T}$$

where N is the number of trunks, T is the traffic in Erlangs and h is the average holding time in seconds.

An explicit expression for $P_r(d,N,T)$ in a convenient form for presentation here does not seem possible; but charts and tables for evaluation of $P_r(d,N,T)$ are given in books on traffic theory. Furthermore many engineering computer systems have programs for computing $P_r(d,N,T)$ in their software libraries.

10.7 CONCLUSIONS

This chapter has been concerned with the technological aspects of the evaluation of telecommunication systems. A foundation for the study was laid first by a review of the various forms assumed by communication signals and of the methods (modulation and multiplexing) for conversion among them. Next a review of information theory was undertaken. It provided a means for the measurement of the information in a signal and for the information handling capacity of a communication channel as well as providing guidelines for recoding a signal into a form adapted to a channel.

For purposes of studying trends in telecommunication system design, systems were categorized in three different ways: point-to-point versus broadcast systems, cable versus radio systems, and finally analog versus digital systems.

Point-to-point switched telecommunication systems (telephone and related types of systems) are being expanded worldwide as rapidly as available investment capital will permit. Demand seems destined to exceed supply indefinitely. Such communication systems can make use of cable communication links or radio links. Although it is possible, in theory, to operate many closely spaced unguided wave links in the same frequency band, it is presently impractical; therefore different frequency bands must be used for different systems. The crowding of the frequency spectrum therefore provides a strong argument for the use of cable systems, (many of which can operate on the same frequency band) for expansion of capacity of systems already having substantial capacity. Cable systems, in contrast to microwave and other radio systems, suffer from substantial dispersion-related signal distortion. Such distortion can be completely corrected. The correction can be accomplished by less complex and less costly equipment in a digital system. This fact, in addition to the fact that digital signal handling equipment of all kinds has recently become very attractive economically compared to equipment for analog systems, strongly suggests that digital cable (or waveguide) systems will play a prominent role in the future expansion of point-to-point telecommunication systems. A significant added incentive for the use of digital systems is the versatility of such systems with regard to the variety of signals that can be handled. Analog systems, of course, are similarly versatile but with significantly less ease of adaptability.

Broadcast types of communication systems are predominantly of the radio type (radio, television, etc.) Expansion of this type of telecommunications system is occuring at a rapid rate in some parts of the world but at a slow rate in most developed countries. There are no apparent new technological developments of sufficient importance to significantly influence the current trends in the expansion of such systems. The other type of broadcast system, as exemplified by cable television (CTV or CATV) systems, is presently in a state of rapid expansion in highly developed countries. Such systems seem destined to play a greatly expanded role in future society. Their importance lies in their ability to offer a much greater variety of programs than is available from over-the-air broadcasts. Cable television systems have the potential for providing point-to-point, dedicated, communication service as an adjunct to the principle intended use of such systems. For example, some systems are providing data service, under contract to commercial customers.

REFERENCES

1. Beckmann, Petr, *Elementary Queuing Theory and Telephone Traffic,* Lee's abc of the Telephone, Training Manuals, Geneva, IL, 1977.
2. Beckmann, Petr, *Probability in Communication Engineering,* Harcourt, Brace & World, 1967.
3. Carlson, Bruce A., *Communcation Systems,* McGraw-Hill, Inc. 1975.
4. Martin, James, *Telecommunications and the Computer,* Englewood Cliffs, NJ, Prentice-Hall, Inc. 1976.
5. Martin, James, *Security Accuracy and Privacy in Computer Systems,* Englewood Cliffs, NJ., Prentice-Hall, Inc. 1973.
6. Panter, Philip F., *Modulation, Noise, and Spectral Analysis,* McGraw-Hill, Inc. 1965.
7. Peterson, W.W. and E.J. Welden, Jr. *Error-Correcting Codes,* The M.I.T. Press, 1972.

chapter 11

THE TRANSMISSION OF ELECTROMAGNETIC SIGNALS

Warren L. Flock
Department of Electrical Engineering
University of Colorado
Boulder, Colorado

11.1 ELECTROMAGNETIC WAVES AND THE ELECTROMAGNETIC SPECTRUM

11.1.1 Electromagnetic Fields and Waves

The practice of telecommunications is dependent almost exclusively on the use of electromagnetic waves. In some applications, the waves travel or propagate through the earth's atmosphere, as from one microwave repeater station to another or from an earth station to a satellite. In other situations the waves are confined to the interior of, or otherwise guided by, cables or waveguides. We discuss first some basic characteristics of electromagnetic waves and then their applications to telecommunications.

The simplest type of electromagnetic wave is a plane wave of infinite extent propagating through a vacuum. A plane wave travels everywhere in the same direction instead of spreading out over a range of directions, as do the ripples generated when a stone is thrown into a pond. Actual waves are not of infinite extent and are often spherical or of some form other than planar near their source. At large distances from sources, however, waves tend to be good approximations to plane waves, just as a portion of the earth's surface may appear flat, or planar, even though it is actually a portion of a spherical surface.

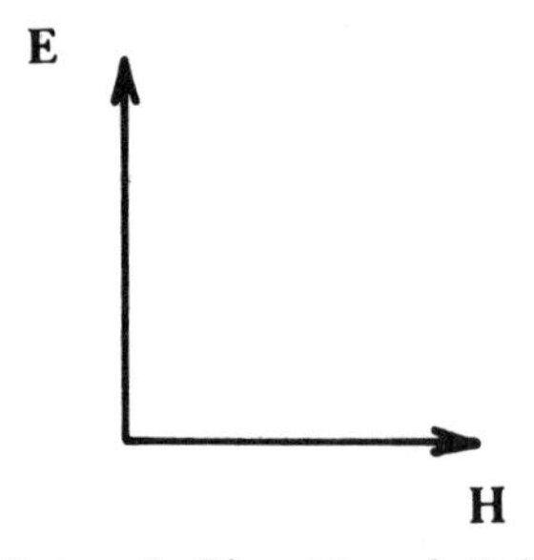

Figure 11.1: The **E** *and* **H** *Vectors of a Plane Linearly Polarized Electromagnetic Wave.*

Plane electromagnetic waves are characterized by an electric field intensity, **E**, having units of volts/meter (V/m) and by a magnetic field intensity, **H**, having units of amperes/meter (A/m) and directed perpendicular to **E** as in Figure 11.1 (**E** and **H** are vector quantities that have directions as well as magnitudes. Another example of a vector quantity is force. Temperature has a magnitude but not a direction and is a scalar quantity.) An electromagnetic wave with **E** in the vertical direction is called a vertically polarized wave; if **E** is in the horizontal direction the wave is called a horizontally polarized wave.

The quantity, electric field intensity, can be defined by Equation (11.1) which states that a charged particle having an electric charge of q coulombs experiences a force **f** that is equal to $q\mathbf{E}$ and is therefore in the same direction as **E**

$$\mathbf{f} = q\mathbf{E} \qquad \text{N} \qquad (11.1)$$

If a force described by Equation (11.1) exists in a volume of space, an electric field of magnitude E exists in that volume. The unit of force in the SI-MKS system is the newton (N), which is the force required to accelerate a mass of 1 kilogram (kg) at a rate of 1 meter per second^2 (1 m/s^2). An electric field imparts motion to charged particles, and charged particles in motion constitute an electric current. In good conductors such as metals, one or a small number of electrons of each atom is relatively free and can move easily from one atom to the next; when an electromagnetic wave impinges on a metallic structure such as an antenna, an electron current flows. A charged particle moving with velocity **v** in a magnetic field experiences a force as indicated by Equation (11.2)

$$f = q\,(\mathbf{v} \times \mathbf{B}) \qquad \mathrm{N} \tag{11.2}$$

where **B** is magnetic flux density in webers/m^2 and is related to the **H** of Figure 11.1 in isotropic media by $\mathbf{B} = \mu\mathbf{H}$. For the case of nonmagnetic materials $\mu = \mu_0$ which has the value of $4\pi \times 10^{-7}$ henry/m. The force **f** involves the vector product $\mathbf{v} \times \mathbf{B}$, and is perpendicular to both **v** and **B**. The force therefore imparts circular motion to the charged particle. The angular velocity of rotation ω_B of the charged particle in its circular orbit is given by $|\omega_B| = |qB/m|$, and the radius r of the circular motion is given by mv/Bq, where m is the mass of the charged particle. Equation (11.2) and the relation $\mathbf{B} = \mu\mathbf{H}$ define magnetic-field intensity in the same way that Equation (11.1) defines electric field intensity.

However, the force on a charged particle due to the electric field of an electromagnetic wave is greater than the force due to the magnetic field by the ratio c/v, where c the velocity of light is 3×10^8 m/s and v is particle velocity. The force due to the magnetic field of a wave itself is therefore usually of little importance, but the forces on charged particles due to the earth's magnetic field play an important role in ionospheric propagation.

In 1865 James Clerk Maxwell summarized all the then accumulated knowledge about electricity and magnetism and added exceedingly important insights of his own. The equations which he used for this purpose have become known as Maxwell's equations. They are usually considered to be four in number although some assert that two are contained in the other two and are unnecessary. We must forego a mathematical analysis, or even a complete statement of the equations or explanation of their notation, and instead refer the reader not yet familiar with electromagnetic fields to textbooks on the subject, such as those by Johnk;[1] Jordan and Balmain;[2]; and Ramo, Whinnery, and Van Duzer.[3] Two of Maxewell's equations in the differential form applicable to "free space" are given below, however, as Equations (11.3) and (11.4).

$$\nabla \times \mathbf{E} = \frac{-\partial \mathbf{B}}{\partial t} \tag{11.3}$$

$$\nabla \times \mathbf{H} = \frac{\partial \mathbf{D}}{\partial t} \tag{11.4}$$

The quantities **E**, **B**, and **H**, have already been defined. The symbol ∇ stands for the vector operator del, which involves spacial derivatives. Partial derivations with respect to time t appear on the right sides of the equations. The quantity **D**, electric flux density, has units of coulombs/meter^2 (C/m^2) and is related to **E** for isotropic media by

$$\mathbf{D} = \epsilon \mathbf{E} \tag{11.5}$$

where*

$$\epsilon = \epsilon_o K \tag{11.6}$$

with $\epsilon_o = 8.854 \times 10^{-12}$ farad/meter and K being the relative dielectric constant. K is a nondimensional quantity having values of 1 for a vacuum, close to 1 in the lower atmosphere, 2 to 10 for common dielectric materials, near 81 for water, and even higher values for certain materials such as titanium dioxide.

For present purposes, suffice it to say that the equations show that an electric field is generated by a magnetic field that varies with time, and a magnetic field is generated by an electric field that varies with time. Thus time-varying electric and magnetic fields are directly related, and it is not possible to have one without the other. A combination of time-varying electric and magnetic fields, furthermore, results in an electromagnetic wave; manipulation of Equations (11.3) and (11.4) allows deriving the wave equations for and determining the properties of electromagnetic waves. Some basic characteristics of such waves will now be mentioned. One characteristic is that, for a wave in homogeneous, isotropic, lossless media, traveling in the +z direction, as indicated by + superscripts

$$\frac{E^+}{H^+} = \eta \quad \Omega \tag{11.7}$$

where η is the characteristic impedance of the medium and has units of ohms (Ω). η is determined by the electric and magnetic properties of the medium. For ordinary dielectric materials for which $\mu = \mu_0$

*K is also denoted by ϵ_r, the relative permittivity.

$$\eta = \sqrt{\frac{\mu_o}{\epsilon_o K}} \quad \Omega \tag{11.8}$$

When $K = 1$, as is approximately true in the troposphere, $\eta = 377\ \Omega$. The velocity of propagation of an electromagnetic wave under the same conditions is given by

$$v = \sqrt{\frac{1}{\mu_o \epsilon_o K}} \quad \text{m/s} \tag{11.9}$$

When $K = 1$, $v = 1/\sqrt{\mu_0 \epsilon_0} = c$, where c has the value of about 3×10^8 m/s and is commonly referred to as the speed of light. A characteristic of a wave of any kind is the relation between wavelength, frequency, and velocity, namely

$$\lambda f = v \tag{11.10}$$

where λ is wavelength in m, f is frequency in Hz (cycles per second), and v is velocity in m/s. Finally for a wave in a lossless medium propagating in the +z direction

$$E = E_o e^{-j\beta z} \tag{11.11}$$

where e is the base of natural logarithms, β, the phase constant, is equal to $2\pi/\lambda$, and j is the square root of minus 1. This expression is based upon the assumption of a sinusoidally varying electric field intensity, and the factor $e^{-j\beta z}$ shows that the phase of the sinusoidal variations lags with increasing distance z as shown in Figure 11.2.

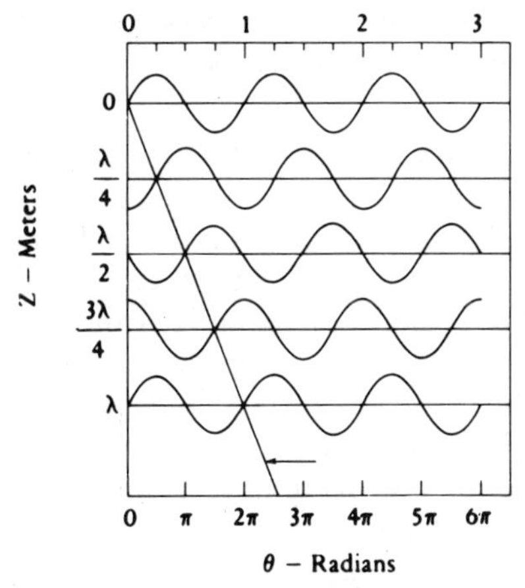

Figure 11.2: Phase lag θ as a function of distance z, where $\theta = \beta z = 2\pi z/\lambda$. The individual sinusoidal curves, for 5 different values of z, show the instantaneous values of electric field intensity E as a function of time for a wave having a frequency of 1 MHz.

Radiation

Maxwell's equations for **E** and **H** in free space predict the possibility of electromagnetic waves, but how are such waves originally initiated or launched? A partial answer, applying to the case of an electric circuit, involves the following considerations.

1. When an electric current flows, a magnetic field is generated in the surrounding region, and if the current varies with time the magnetic field likewise varies with time.
2. A varying magnetic field generates a varying electric field.
3. The combination of varying electric and magnetic fields constitutes, and propagates as, an electromagnetic wave.

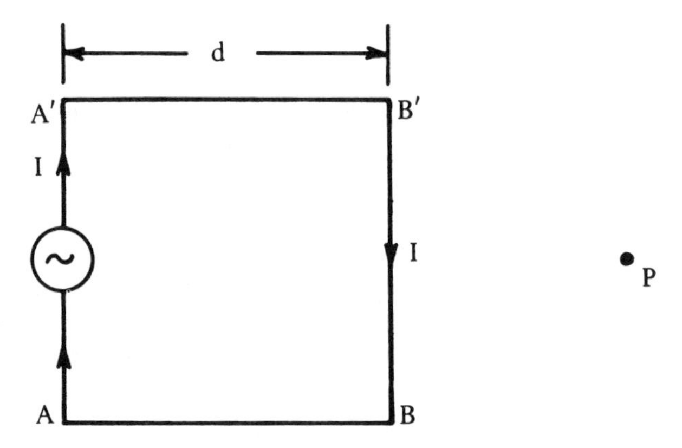

Figure 11.3: A simple low frequency circuit. Waves are radiated from the two arms AA' and BB' to the point P which is to be taken as being at a rather large distance from the circuit.

In the case of a low-frequency circuit as shown in Figure 11.3, however, the currents in the two arms AA′ and BB′ are in opposite directions in space. Thus if one considers that a wave is radiated from the arm AA′ to P, and that another wave is radiated from BB′ to P, the fields of the two waves will tend to be of opposite polarity and to cancel. The exact degree of cancellation will be determined by the factor $2\pi d/\lambda$ which is the phase lag of the wave from AA′ in propagating the distance d to BB′. At low frequencies for which $d \ll \lambda$, this phase lag is negligible, and the waves from AA′ and BB′ cancel out almost completely. If the frequency is raised and λ is correspondingly reduced, however, a condition may be reached for which $d = \lambda/2$, and the contributions from AA′ and BB′ add in phase. Under such a condition the radiation process in the direction of P is efficient. When the frequency is raised sufficiently, however, the situation is complicated because the current will no longer flow in the same direction with respect to the source throughout the circuit. The subject of radiation from a simple circuit will not be pursued here further, but certain practical antenna structures are discussed in section 11.2.3.

11.1.2 The Electromagnetic Spectrum as a Natural Resource

Electromagnetic waves exist over a tremendously wide range of frequencies which extend from nearly zero as a lower limit to as high as 10^{23} or more. Included in this range are radio waves and light. The term "spectrum" refers to a class or group of similar entities arranged in the order in which they possess a certain characteristic. One can refer to a group of people whose political opinions range from extremely conservative to extremely radical as representing a wide political spectrum. The term electromagnetic spectrum is used here to refer to electromagnetic radiation or waves of all possible frequencies arranged or displayed as a function of frequency. The availability of electromagnetic radiation is beneficial to man in various ways, and this radiation, or the electromagnetic spectrum, can be considered as a major natural resource. The earth receives its energy from the sun in the form of electromagnetic radiation, and electromagnetic radiation is essential to all life on the earth. The electromagnetic spectrum is also used for communications and various other purposes, either in free space or in transmission lines or waveguides.

The electromagnetic spectrum and the environment are closely related in various ways, in addition to the fact that the spectrum is essential to life on the earth. Mankind lives in an environment permeated by electromagnetic radiation of both natural and man-made origin, and it is important to understand as well as possible just what effects the radiation has on people and biological matter. The designer and user of telecommunication and other electronic equipment, on the other hand, must consider the electromagnetic environment in which the equipment must operate. He must, for example, consider both the natural and man-made radiation in the frequency range of the equipment. Also he must take care to minimize interference caused by his equipment. The earth's atmosphere is part of our environment and affects our ability to use the electromagnetic spectrum. Electromagnetic waves that are incident upon the earth's atmosphere from the outside, as from the sun, tend to be sufficiently attenuated or reflected that they have negligible intensity at the earth's surface, except in two wavelength bands that are referred to as the optical and radio windows (about 0.4 to 0.8 μm and 1 cm to 10 m, respectively). Wavelengths falling within the windows are affected also in some degree by the atmosphere, but usually not as drastically as those without. Other ways in which the environment and the electromagnetic spectrum are related involve the poles or towers and wires that may be employed to utilize the electromagnetic spectrum for telecommunications. The early years of telecommunications were characterized by very large numbers of unsightly telephone and telegraph wires, but these were reduced in number by development of underground cables and carrier systems. Further developments in communications involving more extensive utilization of broadband telecommunications services — cable television, closed circuit television, video teleconferencing, digital data transmission, etc., — may also have effects on the environment. Travel for some business and educational purposes can be increasingly reduced by planning more on the use of versatile broadband telecommunication facilities.

Some of the characteristics of the radio spectrum as a natural resource, as specified by the Joint Technical Advisory Committee (JTAC) of the Institute of Electrical and Electronic Engineers (IEEE) and the Electronic Industries Association[4, 5] are:

1. The radio spectrum is utilized but not consumed. Present usage does not interfere with future usage or cause any degree of deterioration of the resource. A portion of the spectrum that is assigned or otherwise occupied for one use at a given place and time, however, may not be available for other uses in the same area and at the same time.
2. The resource has dimensions of space, time, and frequency and all three dimensions are interrelated.
3. The spectrum is an international resource.
4. The resource is wasted when assigned to tasks that can be done more easily in other ways or when it is not correctly applied to a task.
5. The spectrum is subject to pollution by man-made radio noise.

It has been argued in certain quarters that the radio resource should be placed on the open market to be bought and sold as any other commodity. A counter argument is that telecommunications has the potential for providing social benefits which are not proportional to market value.

The above listing of characteristics mentions space and time as well as the spectrum itself. Hinchman[6] has pursued this view further and has pointed out that efficient use of the spectrum involves time, wave polarization, radiated power, antenna directivities, and terminal locations, as well as frequency. This reasoning led him to propose the term "electrospace" for the radio resource, electrospace being a quantity that can be represented by an eight-dimensional matrix involving frequency, time, polarization, power, direction of propagation, and three spatial dimensions. Another variation is the treatment presented in the JTAC report, Radio Spectrum Utilization in Space.[7] In this case, frequency, orbit, and number of earthward beams per satellite are regarded as the orthogonal axes pertinent to satellite communications. It is suggested here that efficient use of the spectrum requires, among other factors, suitable atmospheric characteristics (or a suitable medium or waveguide structure of some kind). Thus the concept of electrospace can be enlarged to encompass the subject of atmospheric effects on electromagnetic waves.

Utilization of the spectrum and electrospace requires governmental regulation, licensing, and international cooperation so users can operate without interfering with other parties and without being interfered with. Obviously chaos would result without regulation and licensing procedures. The regulation of telecommunications is treated in chapters 1 through 4 of this volume.

From the practical viewpoint, utilization of the electromagnetic spectrum involves communication systems, and these must have adequate signal-to-noise ratios for successful operation. Signals are generally man-made, but there are practical limits to the signal intensities which can be provided. Consequently, noise may be the factor which limits the use of the electromagnetic spectrum in a particular application. Noise is generated both in the receivers of communication systems and externally to the receivers. The external noise may be of natural origin or man-made. In this respect, what is noise to one party may be the desired signal to another party and vice versa. Man-made signals that are not in accordance with regulations and good practice or are otherwise unwanted may be considered to be a form of pollution.

11.2 TRANSMISSION LINES, WAVEGUIDES, AND ANTENNAS

11.2.1 Transmission Lines

Electromagnetic waves utilized for telecommunications travel from one location to another in transmission lines or waveguides or through the atmosphere. Even when most of the path from one location to another is through the atmosphere, transmission lines or waveguides almost invariably form the connecting links between the transmitter and the transmitting antenna and the receiver and the receiving antenna. The terms, transmission line and waveguide, are sometimes used in such a way as to be indistinguishable, but more commonly, and in this chapter, a transmission line is a two-or-more-conductor line such as a coaxial line, parallel-wire line, or microstrip line. The waveguides of interest in this chapter, however, are hollow metallic guides or dielectric guides. Coaxial lines are an extremely important form of transmission line that have the advantage that the electric and magnetic fields are confined entirely to the space between the inner and outer concentric conductors, assuming the use of sufficiently high quality construction and connectors. Coaxial lines can operate from dc (direct current or zero frequency) up to frequencies in the GHz range, but attenuation increases with frequency and ultimately becomes a limiting factor.

The type of wave normally utilized in coaxial lines is designated as a TEM wave. This notation indicates that both the electric field (**E**) and the magnetic field (**M**) are confined to the transverse plane (**T**) in the lossless case. The transverse plane is perpendicular to the length or axis of the line. The configurations of the electric and magnetic fields are illustrated in Figure 11.4. Lines representing the direction of the electric field are radial, extending from one conductor to the other, as in Figure 11.4a, but alternating in polarity with time and distance. Lines representing magnetic-field intensity have the form of closed circles as in Figure 11.4b. The **E** and **H** fields are everywhere perpendicular to each other, and their magnitudes vary inversely with radial distance r from the center of the line.

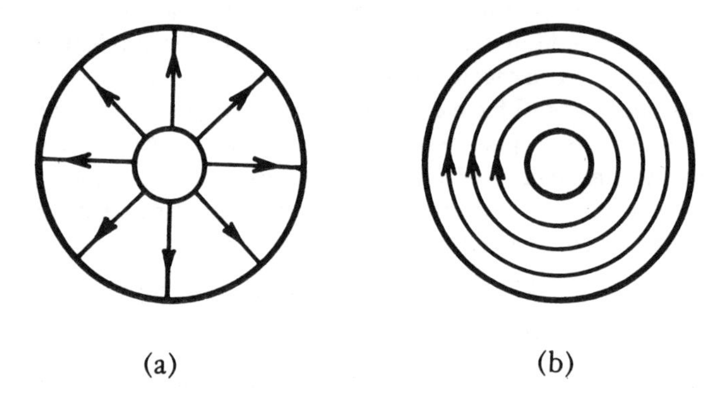

Figure 11.4: (a) Lines representing electric field intensity **E** *in a coaxial line of inner radius a and outer radius b. (b) Lines representing magnetic field intensity* **H** *in the same coaxial line as in (a).*

In coaxial lines, however, reference is commonly made to the voltage V between the inner and outer conductors, and the current I flowing in the conductors, rather than to **E** and **H**. The relations between the magnitudes of these quantities are

$$E = \frac{V}{r\ln(b/a)}, \quad V = Er\ln(b/a) \tag{11.12 a\&b}$$

where the value of **E** is that at the distance r from the center of the line, a is the radius of the inner conductor, b is the radius of the outer conductor and $a \leq r \ll b$. Also

$$H = \frac{I}{2\pi r}, \quad I = H2\pi r \tag{11.13 a\&b}$$

The ratio of V to I for a positively traveling wave is

$$\frac{V^+}{I^+} = \frac{E^+}{H^+}\,\frac{\ln(b/a)}{2\pi} = \frac{\eta\ln(b/a)}{2\pi} = Z_o \tag{11.14}$$

η is the characteristic impedance of the medium between the inner and outer conductors, and Z_o is the characteristic impedance of the line. An alternative approach to the analysis of two-conductor transmission lines that makes use of equivalent distributed circuits shows that Z_o is also given by

$$Z_o = \sqrt{\frac{L}{C}} \tag{11.15}$$

where L is inductance per unit length of line and C is capacitance per unit length or line. Coaxial lines that are available commercially have standard values of Z_o such as 50 ohms, 75 ohms, etc.

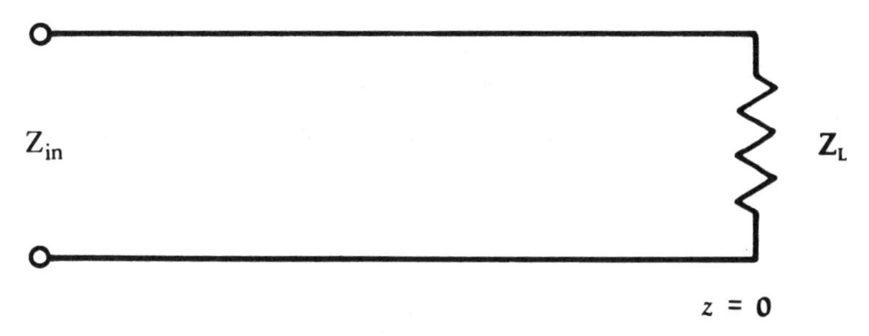

Figure 11.5: Transmission line terminated by a load impedance Z_L at $z = 0$.

The general solution of the wave equations which can be derived from Equations (11.3) and (11.4) describes waves that travel in both directions in a medium or along a transmission line. In most practical applications, a source is applied at one end of a line and travels in the assumed positive direction. Reflection at the end of the line or at discontinuities along the length, however, may cause a negatively traveling wave to exist as well as the original positively traveling wave. Reflection in transmission lines or of electromagnetic waves in general is similar qualitatively to the reflection of sound waves or water waves from obstacles in their path. But what determines the volatage of the negatively traveling wave in a transmission line? The answer can be found by writing equations for the voltage and current at the end of a transmission line, as shown in Figure 11.5.

The total voltage on the line, $V(z)$ (meaning the voltage V expressed as a function of the coordinate z), is the sum of the voltages of positively and negatively traveling waves as indicated by

$$V(z) = V_m^+ e^{-j\beta z} \quad V_m e^{j\beta z} \tag{11.16}$$

for the lossless case. The superscripts identify the peak voltages of the positively and negatively traveling waves. A similar equation applies to current, $I(z)$, but whereas $I^+_m = V^+_m / Z_o$ it develops that $I^-_m = -V^-_m / Z_o$ so that

$$I(z) = \frac{V_m^+}{Z_o} e^{-j\beta z} - \frac{V_m^-}{Z_o} e^{j\beta z} \tag{11.17}$$

If these equations are applied at $z = 0$ where $e^{\pm j\beta z} = 1$ and $V(z)/I(z) = Z_L$, the load impedance, it is possible to solve for the ratio, V^-_m / V^+_m. Identifying this ratio as a reflection coefficient, Γ_L,

$$\Gamma_L = \frac{Z_L - Z_o}{Z_L + Z_o} \tag{11.18}$$

This relation shows that if $Z_L = Z_0$, $\Gamma_L = 0$, and there is no reflected wave. Conversely if $Z_L \neq Z_0$, Γ_L has a finite value. Γ_L will be real and either positive or negative if Z_L and Z_o are real. If instead of taking the ratio of the voltage of the negatively traveling wave to the voltage of the positively traveling wave at $z = 0$, we take the ratio for any arbitrary value of z we have

$$\frac{V_m^- e^{j\beta z}}{V_m^+ e^{-j\beta z}} = \Gamma(z) = \Gamma_L e^{2j\beta z} = \frac{Z(z) - Z_o}{Z(z) + Z_o} \tag{11.19}$$

$$V(z) = V_m^+ e^{-j\beta z} [1 + \Gamma(z)] \tag{11.20}$$

$$I(z) = \frac{V_m^+}{Z_o} e^{-j\beta z} [1 - \Gamma(z)] \tag{11.21}$$

The term $1 + \Gamma(z)$ can be represented by a phasor diagram as in Figure 11.6.

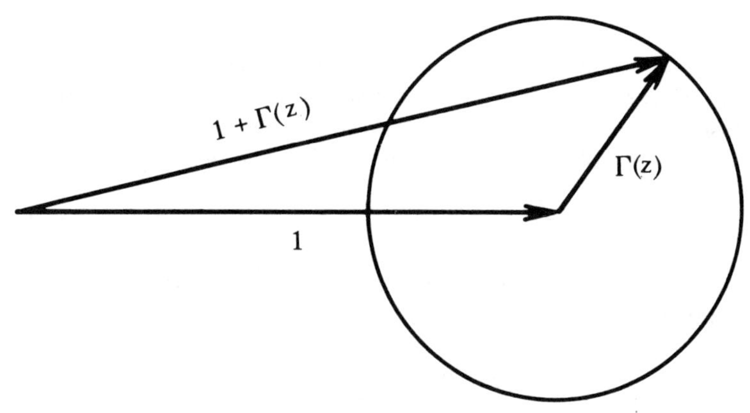

Figure 11.6: The circle shows the locus of $1 + \Gamma(z)$.

The horizontal phasor represents the unit phasor [the 1 of $1 + \Gamma(z)$], and the circle represents the locus of $1 + \Gamma(z)$ for a particular magnitude $|\Gamma(z)|$. The diagram also shows $\Gamma(z)$ and $1 + \Gamma(z)$ for one illustrative condition. Consideration of the diagram shows that the voltage $V(z)$ varies from a maximum value proportional to $1 + |\Gamma(z)|$ to a minimum value proportional to $1 - |\Gamma(z)|$. The ratio of maximum to minimum voltage along the line is called the standing-wave ratio (SWR) and is given by

$$SWR = \frac{1 + |\Gamma(z)|}{1 - |\Gamma(z)|} \tag{11.22}$$

In most applications in telecommunications it is desirable to make $Z_L = Z_0$ in order to avoid having a reflected wave, and therefore a standing wave, on the line. As any obstacle, imperfection, or departure from a perfectly matched condition results in a small reflected wave, it is usually necessary to tolerate a SWR value greater than unity, such as 1.1. In certain applications, such as the use of a short-circuited stub (short auxiliary section of line) to produce an impedance match in the main line, a high SWR is deliberately produced.

Transmission line calculations for determining the value of Γ_L, SWR, and input impedance Z_{in} given the load impedance Z_L, or calculating Z_L from knowledge of the SWR and position of a voltage minimum, etc., are facilitated by the use of a graphical technique employing the well-known Smith chart. Reflection coefficient $\Gamma(z)$ plots directly on the chart in polar coordinates, and corresponding normalized impedance values $z(z)$ [where $z(z) = Z(z)/Z_0 = r(z) \pm j\,x(z)$] can be determined from sets of super-imposed circles representing constant values of r and x. Coaxial lines can be used for long distance transmission, including transmission across an entire continent or ocean when repeater stations are suitably placed. Coaxial lines, however, are subject to losses which increase with frequency. If the losses are not too great, the attenuation coefficient α which describes the decrease of voltage in accordance with $e^{-\alpha z}$ is given by

$$\alpha = \frac{r}{2}\sqrt{\frac{C}{L}} + \frac{g}{2}\sqrt{\frac{L}{C}} \qquad \text{Np/m} \qquad (11.23)$$

The quantity r is series resistance per unit length of line, and g is shunt conductance per unit length. r is given by

$$r = \frac{1}{2\pi}\left(\frac{1}{a} + \frac{1}{b}\right)\sqrt{\frac{\omega\mu}{2\sigma}} = \left(\frac{a}{2\delta} + \frac{b}{2\delta}\right) r_{dc} \qquad (11.24)$$

where a and b are the inner and outer radii of the coaxial line, $\omega = 2\pi f$ where f is frequency, σ is conductivity of the inner and outer conductors of the line, and δ is skin depth, $1/\sqrt{\pi f \mu \sigma}$. The form of the expression involving δ is of interest because it indicates that the resistance is increased above the dc value by the factors shown.

An important point to notice is the increase of attenuation with frequency. This condition makes it necessary to convert to the use of waveguides at higher frequencies in order to keep attenuation at a reasonable level.

The familiar television lead-in line is an example of another type of transmission line, the two-wire line. The characteristic impedance of this type of line, with air between the wires, is given by

$$Z_o = 120 \ln \left(\frac{2s}{d}\right) \qquad (11.25)$$

where d is the diameter of the wire and s is the center-to-center spacing. Values of Z_0 for two-wire lines are commonly near 300 ohms. The electric and magnetic fields are not confined as in a coaxial line, and radiation from two-wire lines restricts their use to lower frequencies than can be utilized in the case of coaxial lines.

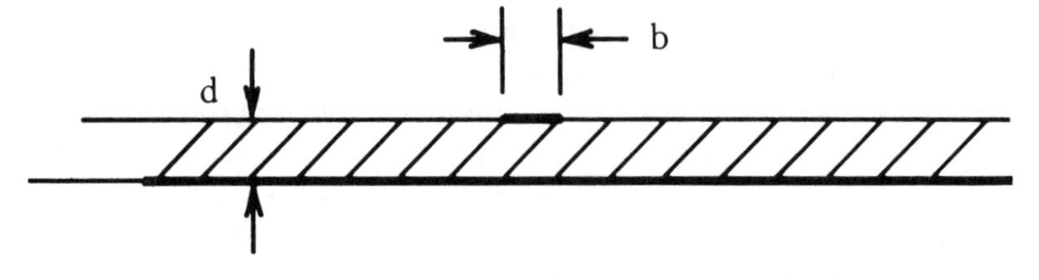

Figure 11.7: Microstrip Line

Microstrip lines have come into prominence for certain applications of microminiaturization. A microstrip line is constructed as suggested in Figure 11.7, which represents a metallic base overlaid with a thin dielectric layer of thickness d.

A transmission line is formed by a metal strip of width b on the top surface of the dielectric layer. The characteristic impedance of such a line is given approximately by

$$Z_o = \sqrt{\frac{\mu_o}{\epsilon_o K}}\ \frac{d}{b} \tag{11.26}$$

where K is the relative dielectric constant of the dielectric.

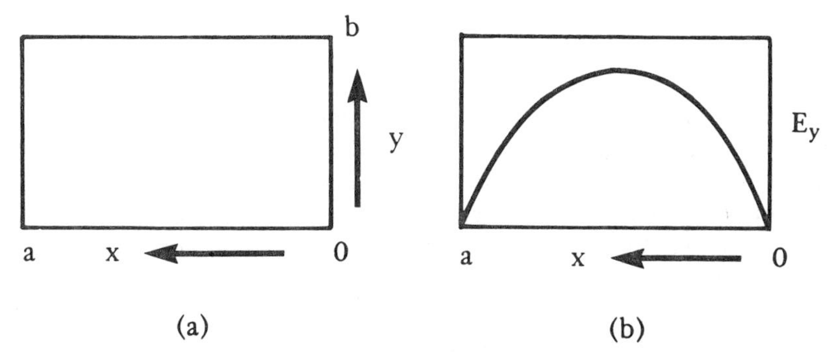

Figure 11.8: (a) Geometry of rectangular waveguide. (b) Plot of E_y versus x for TE_{10} mode.

11.2.2 Waveguides

Hollow metallic waveguides are practical for microwaves and millimeter waves, and dielectric fibers are coming into prominence for use at optical frequencies. Metallic waveguides may be rectangular, circular, or elliptical in cross section. Various configurations or modes of electric and magnetic fields can exist in waveguides. The modes are of the TE or TM variety, the designation TE indicating that the electric field has components in the transverse plane only and the designation TM indicating that the magnetic field has components in the transverse plane only. The direction along the length of the guide is normally taken as the z direction and the transverse plane is then the x–y plane. A TEM wave, the type found in coaxial lines, cannot propagate in a hollow metallic waveguide.

For discussing rectangular waveguide, consider Figure 11.8a which shows a guide with its wide dimension a in the x directions and with a smaller dimension b in the y direction. The widely used TE_{10} mode has a y component of electric field intensity E_y and x and z components of magnetic field intensity H_x and H_z as described by Equations (11.27) through (11.29).

$$E_y = E_{ym} \sin\left(\frac{\pi x}{a}\right) \tag{11.27}$$

$$Hx = \frac{-E_{ym}}{Z_{TE}} \sin\left(\frac{\pi x}{a}\right) \tag{11.28}$$

$$H_z = \frac{jE_{ym}}{\eta}\left(\frac{\lambda_o}{2a}\right)\cos\left(\frac{\pi x}{a}\right) \tag{11.29}$$

$$Z_{TE} = \eta\sqrt{1 - \left(\frac{f_c}{f}\right)^2}$$

and is the characteristic impedance for a TE mode. η is the characteristic impedance of the medium filling the guide, and λ_o is free-space wavelength. f_c is cutoff frequency, and f is operating frequency. A plot of the magnitude of E_y as a function of x is shown in Figure 11.8b, where the x and y axes are drawn so that the z direction is into the paper. E_y must be zero at $x = 0$ and $x = a$ because it is tangent to the highly conducting walls at these positions. The subscripts, 1 and 0 in this case, refer to the cycles of variation of the electric and magnetic fields in the x and y directions, respectively.

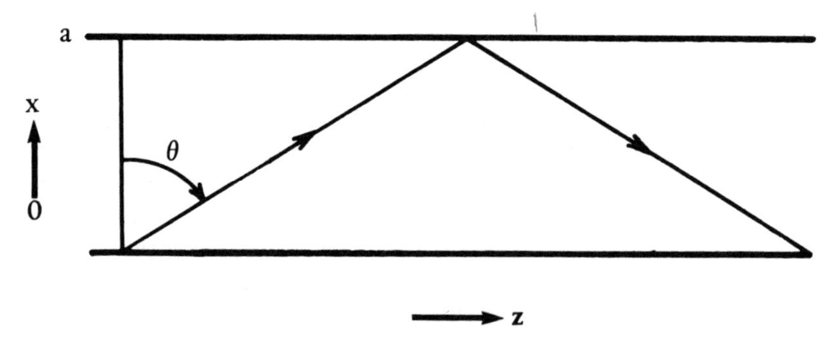

Figure 11.9: Representation of a wave that is reflected back and forth from one side of a guide to the other.

A valid picture of propagation in a waveguide is presented in Figure 11.9 where the diagonal lines and arrows depict a wave that is reflected back and forth from one side of the guide to the other at an angle θ from the perpendicular to the walls. Under this condition the wavelength in the z direction λ_z is related to the unlimited-medium wavelength λ_0 by

$$\lambda_z = \frac{\lambda_0}{\sin\theta} \tag{11.30}$$

whereas

$$\lambda_x = \frac{\lambda_0}{\cos\theta} \tag{11.31}$$

A wavelength is the distance between two surfaces of constant phase that differ in phase by 2π radians or 360°. Normally when speaking of wavelength one refers to λ_0 but one can refer to wavelength in any direction. In order to satisfy the requirement that the electric field intensity E_y be zero at both $x = 0$ and $x = a$, referring to the TE_{10} mode,

$$a = \frac{\lambda_x}{2} = \frac{\lambda_0}{2\cos\theta} \tag{11.32}$$

and hence

$$\cos\theta = \frac{\lambda_0}{2a} \tag{11.33}$$

At a rather high frequency, or small wavelength λ_0, $\cos\theta$ will be rather small, corresponding to a rather large angle θ. If the wavelength is increased, however, a limiting value of λ_0 is reached corresponding to $\cos\theta = 1$ and $\theta = 0,°$ which refers to a wave bouncing back and forth in the x direction but making no progress in the z direction. The particular wavelength satisfying this relation plays an important role in waveguide theory and is designated as λ_c, standing for cutoff wavelength. It can be seen that $\lambda_c = 2a$ and that for any angle,

$$\cos\theta = \frac{\lambda_0}{\lambda_c} \tag{11.34}$$

Referring now to Equation (11.30) for λ_z, and making use of $\sin\theta = \sqrt{1 - \cos^2\theta}$

$$\lambda_z = \frac{\lambda_0}{\sqrt{1 - \left(\frac{\lambda_0}{\lambda_c}\right)^2}} \tag{11.35}$$

and since $\lambda_z f = v_z$

$$v_z = \frac{v_0}{\sqrt{1 - \left(\frac{\lambda_0}{\lambda_c}\right)^2}} \tag{11.36}$$

Alternatively as $\lambda_c f_c = \lambda_o f$,

$$v_z = \frac{v_o}{\sqrt{1 - \left(\frac{f_c}{f}\right)^2}} \tag{11.37}$$

Note that λ_z, also commonly designated as λ_g for wavelength in the guide, is greater than the unlimited-medium wavelength (which is the "free-space" wavelength if the guide is filled with air). Similarly the velocity v_z is greater than v_0 the unlimited medium velocity (c or 3×10^8 m/s if the guide is filled with air). v_z is a phase velocity, however, and not the velocity with which information or energy is transmitted. It is important to note that only frequencies greater than f_c, the cutoff frequency, corresponding to wavelengths less than λ_c, can propagate in a guide. A principal reason why the TE_{10} mode in rectangular guide is so widely used is that it has the lowest cutoff frequency of any mode. Thus, for a range of frequencies above its cutoff frequency, the TE_{10} mode is the only mode that can propagate, and the undesirable possibility of having more than one mode propagate at the same time is avoided. Equations (11.35) through (11.37), however, apply to any TE or TM mode.

Waveguides having elliptical and circular cross sections are utilized in addition to rectangular waveguide. Elliptical guide is difficult to analyze quantitatively but supports a mode that is similar to and compatible with the TE_{10} mode in rectangular guide. Commercially available elliptical guide has the advantage of flexibility with respect to rigid rectangular guide. Elliptical guide can be bent and twisted and for this reason does not have to be cut to precisely accurate lengths in order to join sections of guide together.

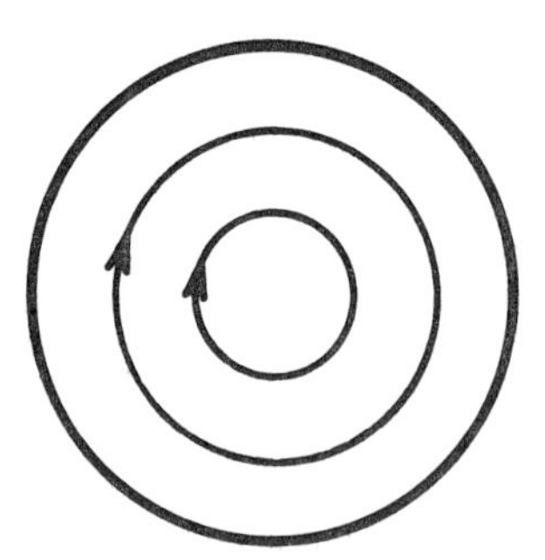

Figure 11.10: The E_ϕ field of the TE_{01} mode in circular waveguide is represented by the inner circles and arrows.

Associated with each mode in a metallic waveguide is a pattern of surface current which flows on the inner surfaces of the walls. The currents are integral features of the modes, and the modes cannot exist without the currents. The waveguide walls are highly conducting but not perfectly conducting, and the flow of current results in the attenuation of waves propagating in the guide. The attenuation of microwave frequencies is less than that for coaxial lines operating at the same frequency, but for most modes it increases with frequency above a range of frequencies for which attenuation is near minimum. A particular mode in circular metallic waveguide, however, has long been of interest because attenuation essentially decreases indefinitely with frequency for this mode, the TE_{01} mode of circular guide. The subscripts for circular guide refer to the cycles of variation of the fields in the circumferential and radial directions. The TE_{01} mode has only a ϕ component of electric field intensity (E_ϕ) as in Figure 11.10. The mode also has radial and longitudinal components H_r and H_z of magnetic field intensity. The electric and magnetic fields are described by

$$E_\phi = j\eta \frac{f}{f_c} BJ_o(k_c r) \tag{11.38}$$

$$H_r = -E_\phi / Z_{TE} \tag{11.39}$$

$$H_z = BJ_o(k_c r) \tag{11.40}$$

where $J_0(k_c r)$ is the Bessel function of zero order. $J'_0(k_c r)$ is the derivative of $J_0(k_c r)$ and goes to zero at $r = a$, where a is the radius of the guide. For a given power transmitted in the guide, or for a given E_ϕ, Equation (11.38) shows that as the frequency f increases, the coefficient B decreases. As B decreases, H_z decreases, and the surface current flowing in the wall in the ϕ direction also decreases, as the surface current density is numerically equal in magnitude to H_z. Thus the attenuation decreases with increasing frequency. The problem with this mode is that it does not have the lowest cutoff frequency of modes in circular waveguide. The modes having a lower cutoff frequency all have an H_ϕ field, and associated with H_ϕ there is a current flowing in the z direction along the length of the guide. But this current is eliminated by constructing the inner wall of a tightly wound helix of fine insulated wire.[7] The Bell System has advertised that its circular waveguide, which is 2 inches in diameter, can operate at frequencies to 100 GHz and carry 230,000 conversations. Application of the guide has seemed close at hand, but in recent years optical fibers have become highly promising, and it appears that millimeter-wave systems utilizing the TE_{01} mode in circular guide will be overshadowed by optical-fiber systems.

Optical Fiber Communication

The invention of the laser spurred interest in the use of light for the transmission of information, and the production of optical fibers having losses under 20 dB/km in 1970 at the Corning Glass Works caused heightened interest in the use of such fibers. By 1974 a loss as low as 2 dB/km had been achieved in the laboratory and prospects for the application of fibers appeared to be highly favorable.[8] Laboratory attenuations of a fraction of a dB/km had been achieved by 1978, and fibers were available commercially from several sources, including Corning Glass Works, which sold fibers having an attenuation ≤ 10 dB/km.

Optical fibers for communications consist of a center glass core having an index of refraction of about 1.5, surrounded by cladding (concentric and cylindrical) having a slightly lower index of refraction. The need for cladding arises from the fact that electromagnetic fields are not confined entirely to the interior of a single fiber but occur as evanescent fields outside the fiber.[3] Thus anything in contact with a single fiber would disturb its transmission characteristics. When cladding is employed, however, the fields are negligible at the outer surface of the cladding.

Although circularly symmetric TE and TM modes, like those in hollow metallic waveguides, can occur in fibers, the modes of most importance have three E and three H field components and are designated as EH_{Lm} and HE_{Lm} modes. All exhibit a cutoff frequency except the HE_{11}, which is called the dominant mode and is the mode used in single-mode operation. Analysis shows[8] that the number of modes that can be transmitted in a fiber, N, is given by

$$N \simeq v^2/2 \simeq (ka/\lambda)\, n_1^2\, \Delta \tag{11.41}$$

where $v = (2\pi a/\lambda)\,(n_1^2 - n_2^2)^{1/2}$, a is the radius of the core, λ is the wavelength, n_1 is the index of refraction of the core, n_2 is the index of refraction of the cladding, and Δ is defined by $n_2 = n_1(1 - \Delta)$. In order to achieve single-mode operation a must be small, a value of 5 μm being suitable. For multimode operation, a is typically near 60 μm. For both single and multimode operation the outer diameters of fibers are usually in the 75 to 100 μm range. If the values of n_1 and n_2 are constant as a function of radius in the multimode case, significant dispersion can occur in fibers, with the result that when a single pulse is transmitted, for example, a smeared out version is received because the different modes have different velocities. If n_1 is made to vary appropriately with radius, however, the multimode dispersion is considerably reduced. The term, graded index, is applied to the variation of n_1 with radius. Waveguide dispersion (variation of velocity within a mode) and material dispersion (variation of index of refraction of glass with frequency) also contribute to total dispersion in both single-mode and multimode fibers. Whether it is practical to reduce multimode dispersion sufficiently to achieve total dispersion as low for multimode fibers as for single-mode fibers is not clear.

11.2.3 Antennas

Antennas are of many types [9, 10] and only some of the varieties used most commonly in telecommunications will be mentioned. At microwave frequencies, paraboloidal-reflector and horn antennas are widely utilized, and at lower frequencies dipole and other thin-wire antennas are commonly employed. Directivity, gain, and antenna pattern are the basic antenna parameters that are considered here. Directivity is a measure of the degree to which an antenna concentrates radiation in a given direction instead of radiating it uniformly in all directions. Thus, by definition, directivity D is given by

$$D = \frac{4\pi}{\Omega_A} \tag{11.42}$$

where Ω_A is the solid angle of the antenna and 4π is the solid angle in steradians or radians^2 that surrounds a point in space. The solid angle Ω subtended by an area A at a distance r from a point of observation is $A\perp/r^2$ where $A\perp$ is the projection of the area on a plane perpendicular to the line of sight. Ω_A is given approximately by

$$\Omega_A = \frac{4}{3}\,\theta_{HP}\,\phi_{HP} \tag{11.43}$$

where θ_{HP} and ϕ_{HP} are half-power antenna beamwidths in two orthogonal directions, corresponding to the θ and ϕ coordinates of a spherical coordinate system. Gain G is related to directivity D by $G = \kappa_0 D$ where κ_0 is ohmic efficiency. In the theoretical case of a lossless antenna, gain and directivity have the same value but gain is less than directivity for an actual antenna.

In this case of paraboloidal-reflector antennas, an antenna feed is placed at the focus of a paraboloid with the result that a surface of constant phase is generated over the aperture which lies in the plane AA′ of Figure 11.11. For this type of antenna having an aperture area A:

$$D = \frac{4\pi A'_{eff}}{\lambda^2} \quad (11.44)$$

$$G = \frac{4\pi A_{eff}}{\lambda^2} \quad (11.45)$$

where A'_{eff} is the effective area for directivity, A_{eff} is the effective area for gain, and $A > A'_{eff} > A_{eff}$. The antenna feed blocks part of the aperture, and that is one of the reasons that $A > A'_{eff}$. Also the illumination of the reflector by the antenna feed usually is tapered intentionally so that the illumination is less intense near the edge of the reflector. Tapering minimizes antenna side lobes and minimizes the danger of spillover of radiation but results also in a decrease in effective area. (Spillover refers to radiation from the antenna feed that misses the reflector.) A_{eff} is less than A'_{eff} because of ohmic losses. Values of A_{eff} generally vary between about 0.5 A and 0.7 A. A value of 0.54 A is commonly used for conventional antennas.

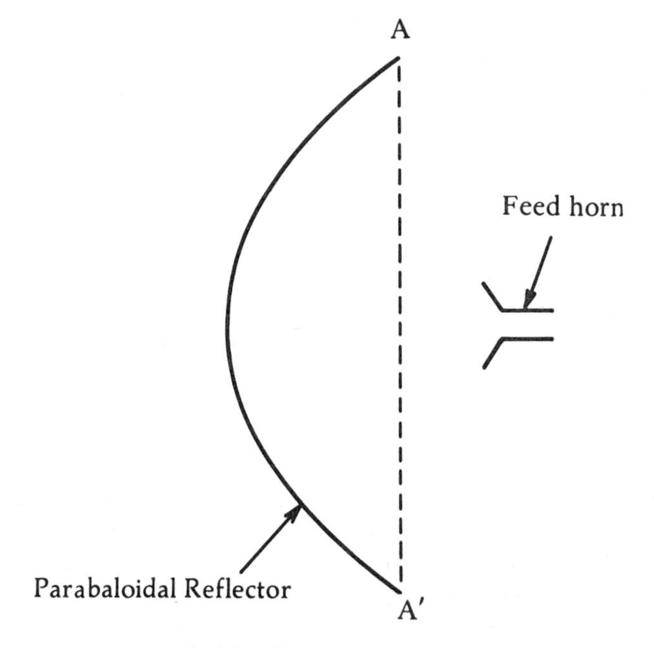

Figure 11.11: Paraboloidal-reflector Antenna

Other forms of microwave antennas are the Cassegrain and horn antennas, illustrated in Figures 11.12 and 11.13. The Cassegrain antenna uses a paraboloidal main reflector and a hyperboloidal subreflector. The waveguide feed approaches the main reflector from the rear, with the resultant advantage that a preamplifier can be placed there. Much the same advantange applies to the horn antenna.

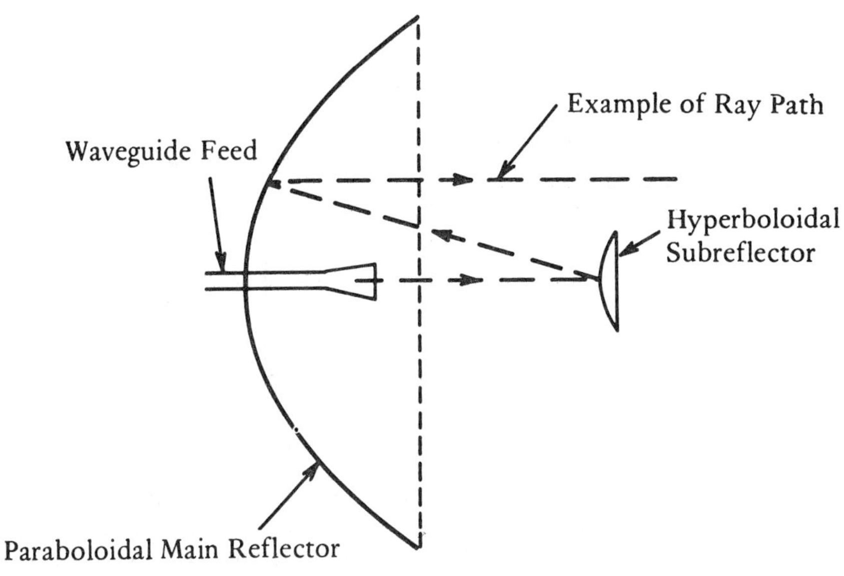

Figure 11.12: Cassegrain Antenna

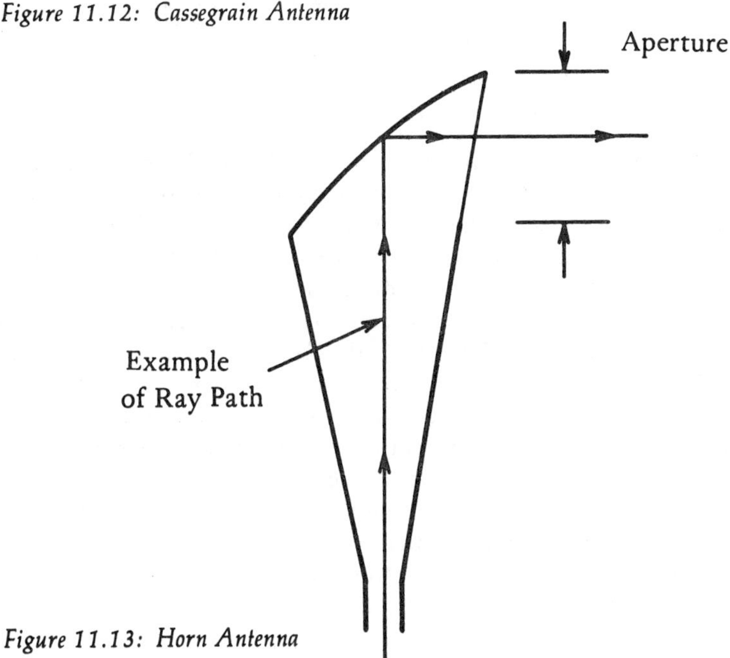

Figure 11.13: Horn Antenna

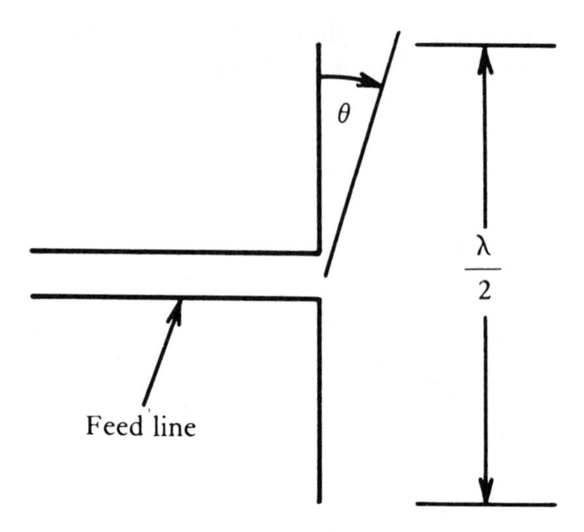

Figure 11.14: Half-wavelength Electrical Dipole Antenna

A basic antenna type for use at lower frequencies than microwave frequencies is the electric dipole antenna, commonly λ/2 in length and fed at the center as in Figure 11.14. The electric field intensity of a λ/2 dipole at a distance r in the far field is in the θ direction and given by

$$| E_\theta | = \frac{60 I_m}{r} \frac{\cos\left(\frac{\pi}{2}\cos\theta\right)}{\sin\theta} \tag{11.46}$$

where I_m is the maximum (input) current of the dipole. $|E_\theta|$ is seen to have its maximum value when $\theta = 90°$ or in the broadside direction. The dipole is an example of an antenna constructed from relatively thin wires or rods. It has a broad beamwidth in the θ direction and is omnidirectional (constant signal intensity) in the ϕ direction (referring to a spherical coordinate system with θ measured from the axis of the dipole). The directivity of a single λ/2 dipole is 1.64, but dipole antennas can be arranged in linear or two-dimensional arrays to provide narrower beamwidths. For use of HF frequencies for long-distance commercial communications, rhombic and log-periodic antenna have been commonly used. The rhombic antenna can be regarded as an array of four horizontal long-wire antennas arranged as in Figure 11.15. The presence of the terminating impedance Z_L causes the antenna to have maximum radiation in the forward direction (to the right in the Figure 11.15) at an angle above the horizontal, which is a function of the height-to-wavelength ratio of the antenna. Large rhombic antenna farms on the Atlantic and Pacific coasts of the U.S. formerly handled a large fraction of overseas traffic before the advent of satellites.

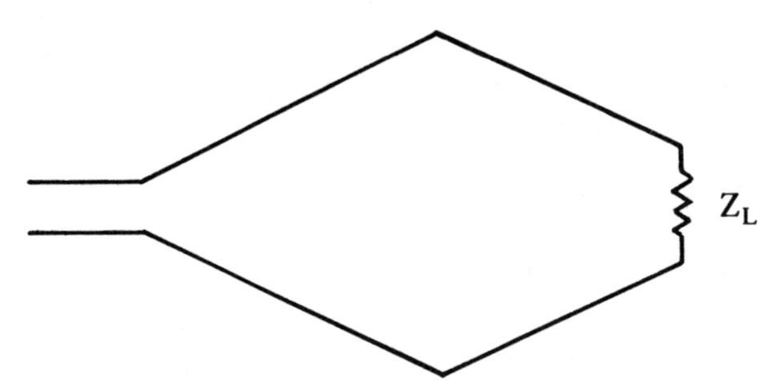

Figure 11.15: Rhombic Antenna

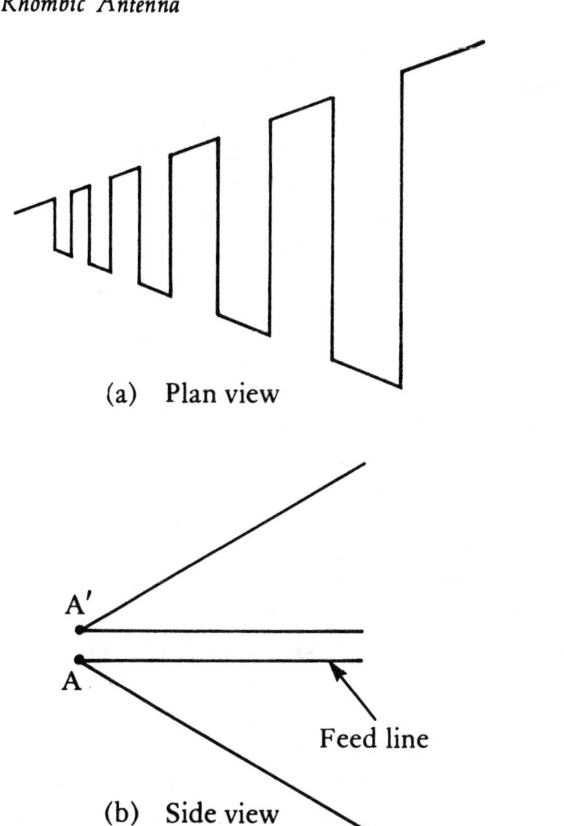

Figure 11.16: Log-periodic antenna. (a) Plan view of one element (b) Side view showing relative position of the two elements.

The log periodic antenna has the advantage of being able to operate over an especially wide frequency range. There are a variety of forms for log-periodic antennas, one form being shown in Figure 11.16. Such an antenna is fed at the apex (at the left between A and A′ in Figure 11.16), and a wave propagates from the feed point to the right until a resonant $\lambda/2$ condition is encountered. Reflection then takes place and radiation from the antenna is to the left.[2]

11.3 TROPOSPHERIC PROPAGATION

11.3.1 System Considerations

Electromagnetic waves that are utilized for terrestrial and satellite communications, and that propagate in or through the troposhere at frequencies that are too high to experience reflection in the ionosphere, follow paths that can be classified into line-of-sight, diffraction, and troposcatter categories. Line-of-sight paths are used extensively for terrestrial and earth-satellite communications. In some cases, such as a path over a mountain peak or other obstruction, communication is maintained with electromagnetic fields that reach the opposite side of the obstruction by a process of diffraction (scattering or reradiation) from near the top or edges of the obstruction. Troposcatter paths depend upon scattering from inhomogeneities in the atmosphere itself.

To a first approximation, radio waves travel in straight lines and one might think that little consideration need be given to the effect of the troposphere on communication systems. The atmosphere in a nonuniform or inhomogenious medium, however, and its properties vary with altitude, location, and time with the result that radio waves tend to travel in curved paths, the exact path or paths changing with time. Also, higher-frequency radio waves experience attenuation and scattering due to precipitation and gases. In the case of paths in the clear atmosphere, the parameter that affects propagation is the index of refraction of the air. Variations in index of refraction are small in magnitude, but nevertheless affect propagation significantly. Raindrops, however, constitute inhomogeneities that scatter electromagnetic radiation and that have an index of refraction that differs from that of the surrounding air by approximately a factor of 9.

For satisfactory operation of a telecommunication system, it is necessary to have a sufficient carrier-power-to-noise-power ratio, designated here by C/X with C representing carrier-signal power and X representing noise power. The carrier power C (or W_R) at the receiving location is given by

$$C = \frac{W_T G_T A_R}{4\pi d^2 L} \tag{11.47}$$

where W_T is transmitter power, G_T is the gain of the transmitting antenna, A_R is the effective area of the receiving antenna, and d is the path length. L is a loss factor, defined to be greater than unity, which may be used to account for any and all losses, including attenuation in the propagation medium. Attenuation in the appropriate waveguides or transmission lines to the transmitting and/or receiving antennas may be included in L if W_T and/or C are taken to be the output power of the transmitting tube and the receiver input power, respectively, rather than the input power to the transmitting antenna and the output power of the receiving antenna. In this treatment, however, powers will normally be defined as applying to the antenna terminals. Using Equation (11.45) to relate antenna gain and effective area, Equation (11.47) can be converted to the form

$$L_t = \frac{W_T}{C} = \left(\frac{4\pi d}{\lambda}\right)^2 \left(\frac{\lambda^2}{4\pi A_T}\right) \left(\frac{\lambda^2}{4\pi A_R}\right) L = \frac{L_{FS}\, L}{G_T\, G_R} \tag{11.48}$$

Expressing the relation in decibel values, it becomes

$$(L_t)_{dB} = 10 \log (W_T/C) = (L_{FS})_{dB} - (G_T)_{dB} - (G_R)_{dB} + L_{dB} \tag{11.49}$$

Also

$$C_{dBW} = (W_T)_{dBW} - (L_{FS})_{dB} + (G_T)_{dB} + (G_R)_{dB} - L_{dB} \tag{11.50}$$

where the notation dBW refers to the power level in dB relative to a watt. The transmitted and received powers may also be expressed in dBm (dB relative to a milliwatt).

For terrestrial line-of-sight systems, the noise power X may be given in terms of the standard reference temperature T_o (290 K) and bandwidth B in Hz by

$$X = k\, T_o\, BF \tag{11.51}$$

where k is Boltzmann's constant $(1.38) \times 10^{-23}$ J/K) and F is the receiver noise figure. The noise power X may also be expressed as being equal to the noise power density X_o times bandwidth B. That is

$$X = X_o\, B \tag{11.52}$$

The quantity C/X is given in decibel values by

$$\begin{aligned}(C/X)_{dB} = {} & (W_T)_{dBW} + (G_T)_{dB} - (L_{FS})_{dB} \\ & + (G_R)_{dB} - L_{dB} - k_{dBW} - (T_o)_{dB} - B_{dB} - F_{dB}\end{aligned} \tag{11.53}$$

For k, use has been made of k in units of J/K (joules/kelvins) times 1 K times 1 Hz, so that units of dBW are obtained and T_0 and B are treated as nondimensional. The expression for noise power, Equation (11.51) is based on the assumption that the noise temperature of the field of view of the receiving antenna is the ambient temperature,taken as the reference temperature of 290 K. This assumption is reasonably accurate though perhaps a somewhat conservative assumption, for microwave paths between two terrestrial stations. Further discussion of systems can be found in chapter 10, and in references by Flock,[11] Freeman,[12] and Panter.[13] In the above K stands for temperature in Kelvins.

11.3.2 Index of Refraction of the Troposphere

The index of refraction or refractivity n of a particular type of electromagnetic wave in a given medium is by definition the ratio of the velocity c about 3×10^8 m/s, to the phase velocity v_p. Thus

$$n = c/v_p \tag{11.54}$$

In a lossless medium, n is related to K, the relative dielectric constant, by

$$n = \sqrt{K} \tag{11.55}$$

Thus any effects upon propagation due to index of refraction can be interpreted in terms of relative dielectric constant and vice versa. (The symbols ϵ_r and ϵ' are sometimes used for relative dielectric constant.) The index of refraction of the troposphere is only slightly greater than unity, and for a measure of refractivity it has become standard pratice to use N units such that the refractivity in N units is given in terms of n by

$$N = (n - 1) \times 10^6 \tag{11.56}$$

If $n = 1.000300$, for example, the refractivity in N units is 300. The index of refraction of the troposphere is a function of pressure, temperature, and water vapor content and is given by

$$N = \frac{77.6p}{T} + \frac{3.73 \times 10^5 e}{T^2} \tag{11.57}$$

where p is atmospheric pressure in millibars (mb), T is temperature in kelvins, and e is partial pressure of water vapor in mb. Note the major effect of e in determining N. Sea level pressure p is 1013 mb or 1.013×10^5 N/m^2. The water vapor pressure e can be determined by taking the product of e_s, the saturation water vapor pressure, and the relative humidity.

Propagation in the troposphere is influenced by the vertical index of refraction profile. The profile can be determined at a particular location and time by radiosonde data or by use of an airborne microwave refractometer. The latter utilizes a resonant microwave cavity, having holes to admit the ambient air. The resonant frequency of the cavity is determined in part by the index of refraction of the air filling the cavity. The refractivity tends to decrease with altitude, and in the absence of specific detailed information it is helpful to have available a model describing the typical decrease of refractivity with height. The CRPL exponential radio refractivity atmosphere[14] having the form

$$N = N_s e^{-h/H} \tag{11.58}$$

has been found to correspond closely to average atmospheric conditions. N_s is surface refractivity, H is a constant, and h is height above the surface. The average value of N_s for the U.S. is said to be 313, and a value of H of 7 km is appropriate for the U.S.[14] Average surface refractivities vary, however, from as low as about 240 for dry, mountainous or high-altitude areas to about 400 for humid tropical areas.

11.3.3 Ray Paths in the Troposphere

Tropospheric propagation can be analyzed by considering the paths taken by rays launched with initial angles β from the horizontal. When the index of refraction varies with altitude the ray paths tend to be curved rather than straight. By definition, curvature C is equal to $1/\rho$ where ρ is the radius of curvature. That is

$$C = \frac{1}{\rho} \tag{11.59}$$

(Curvature has no relation to signal power which was also designated by the symbol C.) If curvature is constant for a distance d along a ray path, the total change in direction in this distance is Cd. It can be shown (11,14) that when the index of refraction varies with altitude at a constant rate dn/dh the curvature of a ray is given by

$$C = -\frac{dn}{dh}\cos\beta \tag{11.60}$$

where h is height. For a nearly horizontal path, $\cos\beta \cong 1$ and $C \cong -dn/dh$.

For a path over the surface of the earth, the difference between the curvature of the earth's surface and the ray curvature is important. Neglecting surface topography, the curvature of the earth's surface is $1/r_0$, where r_0 = 6370 km, the radius of the earth. The difference in curvatures is

$$\frac{1}{r_0} - C = \frac{1}{r_0} + \frac{dn}{dh} \tag{11.61}$$

The same relative curvature can be maintained if, instead of using the actual earth radius and actual ray curvature, one uses an effective earth radius kr_0 and a ray of zero curvature as illustrated by

$$\frac{1}{r_0} + \frac{dn}{dh} = \frac{1}{kr_0} + 0 \tag{11.62}$$

A typical value of dn/dh is –40 N units/km, corresponding to a k value of 4/3. If it is assumed that this k value is applicable, path profiles can be drawn using 4/3 times the true earth radius. The advantage of this procedure is that ray paths can then be drawn as straight lines, corresponding to the effective value of 0 for dn/dh on the right side of Equation (11.62). This procedure has been widely used.

A limitation of this approach, however, is that dn/dh and k may take on a range of values such as those of Table 11.1.

Table 11.1
***dN/dh* and *k* Values**

dN/dh, N/km	j·
220	5/12
157	1/2
78	2/3
0	1
- 40	4/3
- 52	3/2
–100	2.75
–157	∞
–200	–3.65

It is impractical to draw separate charts for each k value, but another procedure that also retains the correct relative curvature between the earth and the ray paths is useful. This approach involves making the earth flat and drawing a different ray path for each k value of interest. The basis for flat-earth plots is as follows. As shown in the Appendix, a ray that is launched horizontally in a uniform atmosphere and propagates for a distance l reaches a height above the spherical earth's surface given by $h = l^2/2r_0$, where r_0 is the radius of the earth. When refractivity varies with height corresponding to a particular k value, the relation becomes $h = l^2/2kr_0$. Or if l is expressed in kilometers and h in meters, we have

$$h = \frac{l^2}{12.75\,k} \tag{11.63}$$

Consider now the flat-earth plot of Figure 11.17. The curve is constructed by setting $h' = h_{max} - h$ where h_{max} and h are calculated by use of Equation (11.63). For calculating h_{max}, let l in Equation (11.63) be $(d_1 + d_2)/2 = d/2$, where d is the total path length and d_1 and d_2 are the distances from the two ends to any point along the path. For calculating h, let l be $d_2 - (d_1 + d_2)/2$ or $d_1 - (d_1 + d_2)/2$.

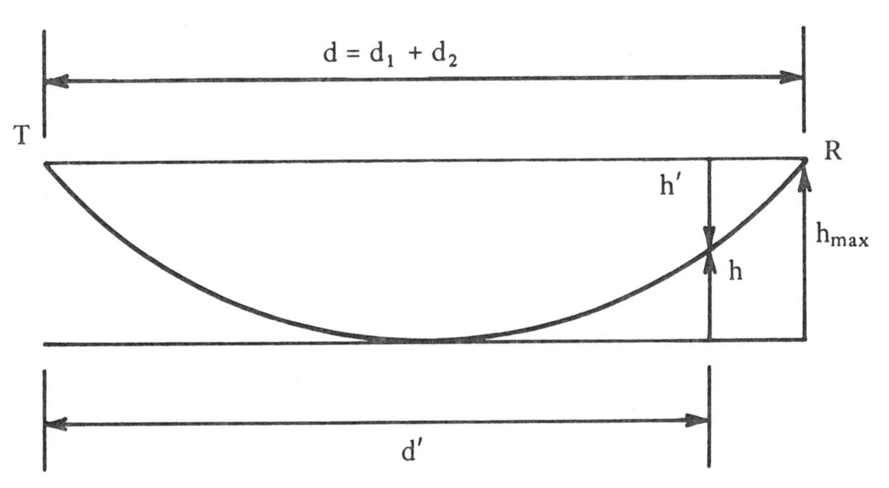

Figure 11.17: Flat-earth Plot

This procedure amounts to letting l be the distance from the center of the path to the point for which the calculation is made. If the above procedure is followed, algebraic manipulation provides the expression

$$h' = \frac{d_1\,d_2}{12.75\,k} \tag{11.64}$$

As discussed more fully in a well-known reference by GTE-Lenkurt, [15] curves can be constructed for any k value of interest, and the path profile can then be drawn to the same scale to determine path clearance (the spacing between the ray path, or electromagnetic-wave path, and any topographic feature or other obstacle along the path). The GTE-Lenkurt reference includes a set of such curves for general use, but it is easy to construct such curves as needed by utilizing Equation (11.64).

Clearance and Reflections

Obviously, it is necessary that a line-of-sight microwave path not be blocked by obstacles. But if a ray path just barely misses an obstacle, such as a mountain top or building, for the expected values of k, will the system operate satisfactorily? The answer to this question lies in an application of Huygens principle, which says that every elementary area of an electromagnetic wavefront acts as a source of spherical waves. Thus the antenna of a telecommunications systems tends to receive radiation from the full extent of the wavefront and not merely the radiation that follows the most direct path. This subject can be dealt with in terms of Fresnel zones.

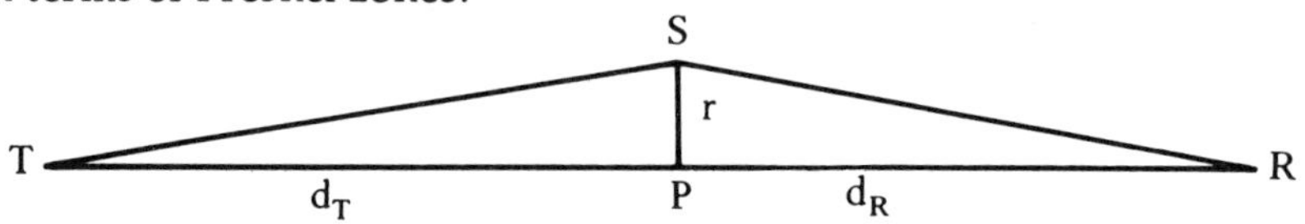

Figure 11.18: Geometry for Calculation of Fresnel Zone Radii

Consider Figure 11.18 which shows two ray paths between a transmitter and a receiver. $TPR = d_T + d_R$ is the direct path and $TSR = r_T + r_R$, idealized as being composed of the diagonals of two right triangles, is greater than TPR by half a wavelength. That is

$$r_T + r_R = d_T + d_R + \lambda/2 \tag{11.65}$$

The distance F_1 actually extends in all directions in the plane perpendicular to TPR and is defined as the first Fresnel-zone radius. Taking into account that $F_1 \ll d_T$ and $F_1 \ll d_R$, the following expression for F_1 can be obtained.

$$F_1 = 17.3 \sqrt{\frac{d_T\, d_R}{f_{GHz}\, d}} \tag{11.66}$$

with distances in km and F_1 in m.

The significance of F_1 is that all of the radiation passing within a distance F_1 of the direct path and reaching the receiver contributes constructively to the received signal power. Consider also a second path for which $r_T + r_R = d_T + d_R + \lambda$, with F_2 being the corresponding distance from the direct path. The radiation passing between F_1 and F_2 will interfere destructively with the radiation passing between the direct path and F_1. The region within F_1 is known as the first Fresnel zone, that between F_1 and F_2 is known as the second Fresnel zone, etc. Radiation from adjacent zones interferes destructively.

There is no precise value for the clearance that is needed on a line-of-sight path but the GTE-Lenkurt reference[15] quotes a value of $0.6\ F_1 + 3$m at $k = 1.0$ for light-route or medium reliability systems and at least $0.3F_1$ at $k = 2/3$ and $1.0F_1$ at $k = 4/3$ for heavy-route or highest reliability systems.

In addition to consideration of path clearance, possible reflections of rays from the earth's surface need to be taken into account. Reflected rays may add constructively or interfere destructively with the direct ray. Assuming perfect reflection and a reversal of phase of 180° upon reflection as is appropriate for a perfectly smooth, perfectly conducting earth, assuming equal power for direct and reflected paths, and neglecting the earth's curvature, it develops that the total electric field intensity, E, at a receiving location is given by

$$E = 2E_o \sin\left(\frac{2\pi h_T\, h_R}{\lambda d}\right) \tag{11.67}$$

where E_0 is the field intensity due to one path alone, h_T is the height of the transmitting antenna above the point of reflection, h_R is the corresponding height for the receiving antenna, d is path length, and λ is wavelength. Insofar as this equation is applicable, it indicates that E varies from 0 to 2 E_0, depending upon the values of h_T and h_R. The equation also indicates that if h_T is fixed, h_R can be chosen to maximize E. The limitations of the equation, however, should be kept in mind. The surface may not be perfectly smooth or highly conducting, in which case the amplitude of the reflected wave will be less than that of the direct wave. Also there may or may not be any reflection, depending on the k value. Finally when the curvature of the earth is taken into account, both the location of the point of reflection [15] and the phase of the reflected wave will vary with the k value.

Because of the possibility of destructive interference and variation of the phase and amplitude of the reflected wave, reflections should be avoided or minimized if possible by arranging for obstructions in the way of potentially reflecting paths or arranging for reflection to take place from rough, poorly reflecting surfaces rather than from smooth, highly reflecting ones.

Fading

In section 11.3.1 a procedure for calculating the signal level on a terrestrial path was outlined. It is necessary to allow for fading of the received signal, the usual allowance in the range of 35 to 45 dB. Fading may be caused by reflections from the earth's surface, as discussed in the previous paragraphs, or it may arise from strictly atmospheric effects. A common type of fading involves multipath propagation. Energy reaches the receiving antenna by two or more atmospheric paths, and the signals on the two paths alternately add constructively and interfere destructively. This type of fading tends to take place on nearly horizontal paths through or near temperature inversion layers, especially when the air below the inversion has a high moisture content as along the southern California coast. In that location fading tends to be intense when temperature inversions are present, but the signal is extremely steady when cyclonic storms eliminate the temperature inversion. Fading of this type can be minimized by avoiding horizontal paths and can be combatted if necessary by the use of space or frequency diversity. Space diversity involves using two receiving antennas, one spaced vertically above the other. Frequency diversity involves operation on two different frequencies, usually with the same receiving antenna.

A severe form of fading that may not be amenable to the remedy of space or frequency diversity has been referred to as blackout fading.[16] This type of fading involves very large negative gradients of dN/dh which cause the beam to bend downwards sharply so that radiation misses the receiving antenna nearly completely or falls outside the main beam of the receiving antenna. A rather shallow surface layer of extremely high water vapor content, which decreases rapidly with height, is the usual cause of blackout fading. Possible means of avoiding blackout fading include avoiding the immediate area where it tends to occur, utilizing shorter paths than usual, and increasing antenna heights.

Paths between earth stations and satellites pass through the lower atmosphere at rather large angles from the horizontal and are not subject to fading to the same extent as terrestrial paths. A fading allowance of only about 4 dB or less has been considered to be sufficient. It has developed unexpectedly, however, that even the 6 and 4 GHz signals commonly used for satellite operations are subject to scintillation of ionospheric origin in equatorial and auroral latitudes.[17]

11.3.4 Attenuation Due to Precipitation and Atmospheric Gases

Prominent radar backscatter echoes may be received from precipitation at lower frequencies, including the S (10 cm) and L (23 cm) bands and attenuation due to precipitation may become serious for frequencies of about 8 GHz (3.75 cm) and higher. A water vapor absorption line causing attenuation up to about 2 dB/km is centered on 22.235 GHz and was responsible for the fact that the performance of K band (1.25 cm) radars developed during World War II did not live up to expectations. A strong oxygen absorption line centered on 60 GHz (0.5 cm) causes attenuation over 10 dB/km at sea level, and a second oxygen peak near 120 GHz

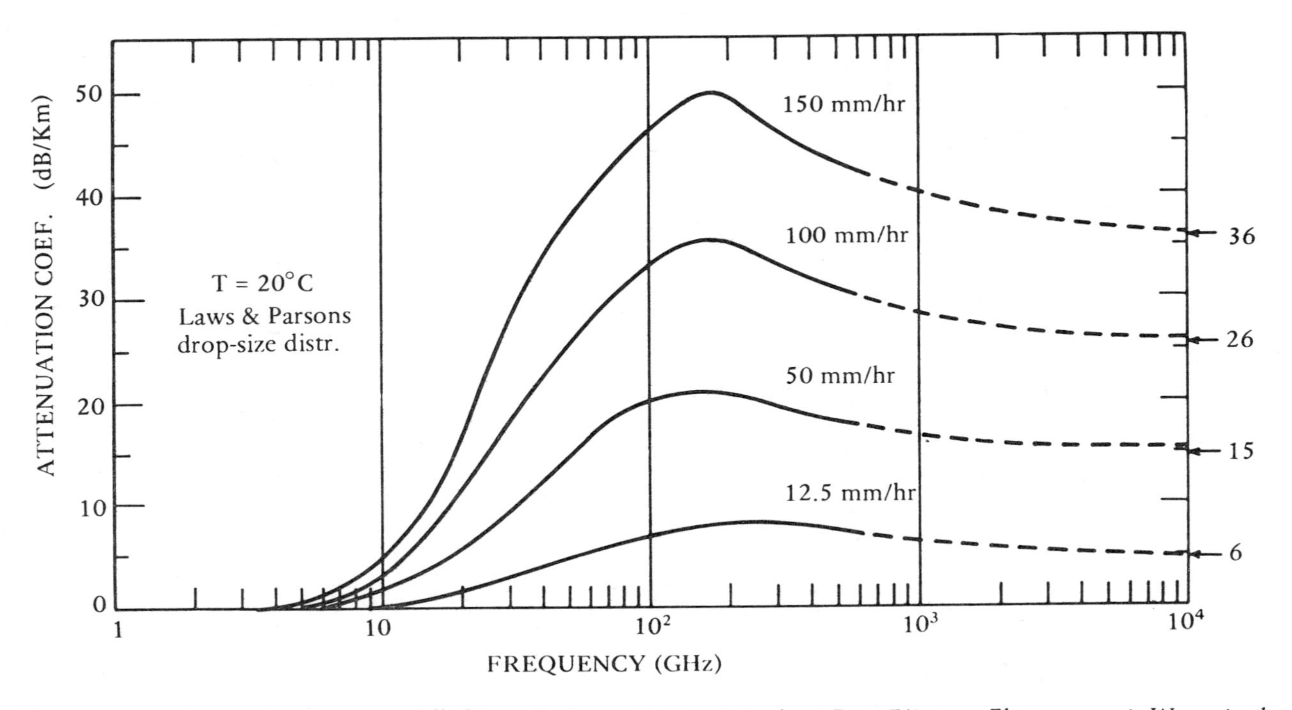

Figure 11.19: Attenuation due to rainfall. (From Zufferey, C. H., A Study of Rain Effects on Electromagnetic Waves in the 1-600 GHz Range.[21]*)*

causes attenuation near 2 dB/km. Attenuation due to water vapor reaches about this same value near 150 GHz.

Attenuation caused by precipitation increases up to frequencies near 100 GHz and is a factor that must be taken into account in the application of millimeter waves. Much interest is presently directed to the use of frequencies above 10 GHz for satellite communication. A program to investigate the use of 30 GHz and 20 GHz for satellite uplink and downlink operation, respectively, is being undertaken by NASA. Paths between the earth and satellites have the advantage that they tend to pass steeply through regions of rainfall, rather than extending through rainfall for a long, nearly horizontal distance. It was asserted at an early stage by Feldman(18) and others that because of normal delays in placing phone calls, especially when rapidity of service is sacrified to obtain a lower cost, occasional outages on millimeter paths should not be allowed to preclude their widespread use for voice communication.

The calculation of attenuation caused by given rates of rainfall is complicated but tractable.(19, 20) Figure 11.19 shows the result of calculations by Zufferey.(21) The results shown in Figure 11.19 are based upon the use of the Laws and Parsons drop-size distribution.(22) A problem in the past has been the lack of sufficiently extensive statistical data concerning rainfall,(23) but considerable progress in remedying this condition has been made in recent years (see section 11.5).

11.3.5 Troposcatter

Microwave telecommunication systems are an outgrowth of research on microwave radar systems during World War II. The first commercial microwave service was put into operation in 1945. In the early 1950s it was discovered that weak but reliable signals were propagated beyond the horizon at microwave and lower frequencies. Controversy existed for some time concerning whether these troposcatter signals were reflected from layers or scattered from atmospheric turbulence. The question has not been resolved completely; the present trend is to emphasize scattering from turbulence but to recognize that the turbulence may be limited in spatial extent.

Signal intensity can be calculated as for a line-of-sight link except that an additional loss is involved. This loss can be expressed as a function of the "angular distance" θ which in the case of a smooth spherical earth is simply d/r where d is the length of the path, and $r = kr_0$, with r_0 the radius of the earth and k the k factor of Table 11.1. The value of θ is modified as needed to take account of surface topography. Several methods for estimating the loss factor have been utilized and are described by Panter

(13) and Freeman (12). One approach developed by ITT is to use an empirical curve which presents troposcatter loss at 900 MHz as a function of angular distance.[13]

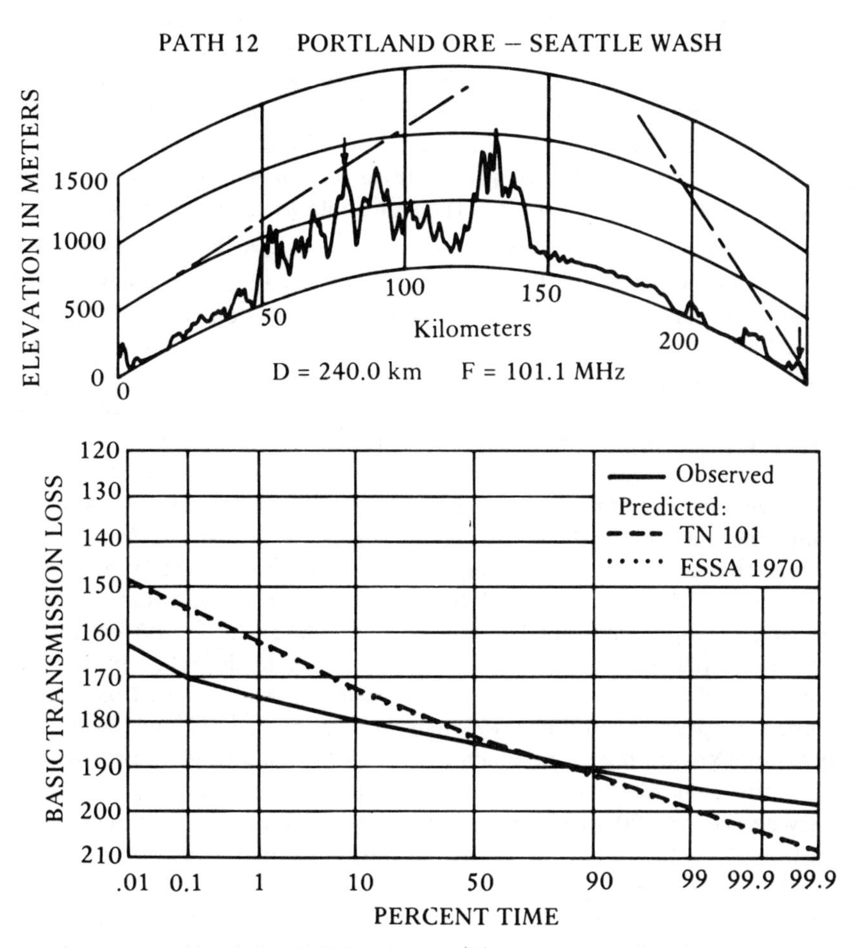

Figure 11.20: Troposcatter Path Performance[24]

Data concerning propagation on radio paths, whether line-of-sight or troposcatter, are best presented in statistical form as in Figure 11.20.[24] The solid curve of this figure shows the observed percentage of time that the transmission loss is at a certain level or less. The basic transmission loss is, for example, 193 dB or less for 99% of the time, about 173 dB or less for 1% of the time, etc.

11.4 IONOSPHERIC PROPAGATION

11.4.1 Introduction

Before the advent of satellites, HF transmissions, propagated via ionospheric reflection, supplied a large fraction of long-distance communications. Satellites now handle an increasingly large proportion of long-distance communications, but HF and lower frequency systems that utilize or are affected by the ionosphere still fulfill important needs. Submarine cable previously shared transoceanic service in some cases with HF systems and now plays the same role with respect to satellites.

The earth's ionosphere,[25] extending from perhaps 50 km as a lower limit to 1000 km as an upper limit, is a weakly ionized gas or plasma. This plasma is formed largely by ultraviolet and X-ray emissions from the sun. Different portions of the solar spectrum interact with different constituents of the atmosphere to form several different layers which have been designated as the D, E, and F layers, the latter often separated in the daytime into F_1 and F_2 layers. Ionization of the previously neutral molecules and atoms of the ionosphere results in free electrons and positive ions. We are concerned primarily with the free electrons of the ionosphere as they interact the most readily with the electromagnetic waves that are used for telecommunications. The lowest layer, the D layer, extends from about 50 to 90 km. It has a maximum electron density of approximately $10^3/cm^3$ between about 75 and 80 km in the daytime. This density is the lowest of the ionospheric layers, but a high electron collision frequency causes high attenuation of AM broadcast signals propagating in the D region in the daytime. The D layer essentially disappears at night. The E layer extends from about 90 km to 140 km with the peak electron concentration of $1 - 1.5 \times 10^5/cm^3$ occurring between about 100 and 110 km. The peak ionospheric electron concentration of up to $2 \times 10^6/cm^3$ occurs in the F_2 region in the 200 to 400 km height range. The daytime F_1 region occurs at lower altitudes with lower densities. Electron densities of the F_2 layer remain rather high, perhaps up to $4 \times 10^5/cm^3$, at night. Long-range HF propagation commonly involves reflection from the F_2 layer, but electromagnetic waves are also reflected from the E and F_1 layers as well. The D region forms part of the earth-ionosphere waveguide that influences the propagation of VLF (3 to 30 kHz) waves.

11.4.2 Characteristic Waves

Propagation in the ionosphere is influenced by the earth's magnetic field, as well as by the free electrons of the ionosphere. Wave propagation parallel to the magnetic field is different from propagation perpendicular to the magnetic field. In considering propagation in the ionosphere, the concept of characteristic waves is important and to discuss characteristic waves it is necessary to discuss wave polarization. It was mentioned above in section 11.1.1 that if the electric field intensity **E** is vertical the wave is vertically polarized, whereas if **E** is horizontal the wave is horizontally polarized. In both cases, and whenever the **E** vector is in a fixed direction, the wave polarization is linear. In this section we will not be much concerned with whether the polarization is vertical or horizontal, but rather, with whether it is linear or otherwise.

What kinds of polarization occur other than linear? Waves can have circular or elliptical polarizations. What is meant by the term circular polarization? This term refers to a wave having an electric field intensity vector that has a fixed length but rotates with angular velocity ω where $\omega = 2\pi f$ and f is the frequency. There are two possible directions of rotation, the right circular direction and the left circular direction. Right circular rotation is in the direction of the fingers of the right hand if they are pictured as encircling the thumb when the thumb points in the direction of wave propagation. Left circular rotation is in the opposite direction. Corresponding to these two directions, we speak of electromagnetic waves that have right circular (rc) and left circular (lc) polarizations.

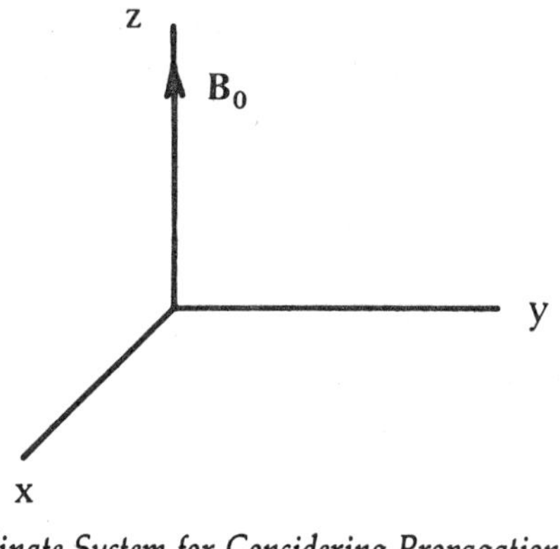

Figure 11.21: Coordinate System for Considering Propagation in Plasma

We now define what the term characteristic wave means. A characteristic wave is one that retains its same original polarization as it propagates. rc and lc waves have this characteristic when propagating parallel to the magnetic field in a plasma. An rc wave remains an rc wave as it propagates and likewise for an lc wave. What happens if a linearly polarized wave is launched parallel to the magnetic field in a plasma? It remains linear, assuming attenuation can be neglected, but the orientation of the **E** vector changes (rotates) as the wave propagates, and this change in orientation constitutes a change in polarization. That is, the concept of polarization involves not only the distinction between linear and circular but the direction of the **E** vector in the case of linear polarization. The characteristic waves for propagation parallel to the magnetic field in a plasma are rc and lc waves. A linearly polarized wave is not a characteristic wave and experiences rotation, known as Faraday rotation, as it propagates. The characteristic waves for propagation perpendicular to the magnetic field are two in number also but are linearly polarized. They retain the same linear polarization as they propagate.

To consider ionospheric propagation further, we utilize Figure 11.21, which shows a rectangular coordinate system with the earth's magnetic field $\mathbf{B}_0$ directed along the z axis. Propagation will be taken to be in the x–z plane. The simplest type of wave is a linearly polarized wave propagating in the x direction, perpendicular to the magnetic field, and having its electric field intensity vector in the z direction. The velocity v imparted to the free electrons of the plasma by the electric field is thus in the z direction; consequently the electrons are unaffected by $\mathbf{B}_0$ as the force on electrons due to $\mathbf{B}_0$ is proportional to $\mathbf{v} \times \mathbf{B}_0$ (Equation 11.2) the vector product of which is zero when $\mathbf{v}$ and $\mathbf{B}_0$ are in the same direction. Thus the analysis proceeds as if there were no magnetic field. The force $\mathbf{f}$ on the electrons is due to the electric field alone and is equated to mass times acceleration, i.e. $\mathbf{f} = q\mathbf{E} = m\mathbf{a}$; but $\mathbf{a} = d\mathbf{v}/dt$ and as $\mathbf{E}$ varies sinusoidally $d\mathbf{v}/dt = j\omega\mathbf{v}$ where $\mathbf{v}$ is now understood to be a complex quantity. Thus $mj\omega\mathbf{v} = q\mathbf{E}$ and

$$\mathbf{v} = \frac{q\mathbf{E}}{mj\omega} = \frac{-jq\mathbf{E}}{m\omega} \qquad (11.68)$$

The electrons in motion with velocity $\mathbf{v}$ constitute an electric current of density $\mathbf{J}$ having units of amperes/m², where $\mathbf{J} = Nq\mathbf{v}$, with N being the electron density. Therefore we obtain

$$\mathbf{J} = \frac{-jNq^2\mathbf{E}}{m\omega} \qquad (11.69)$$

This current density is 180° out of phase with the vacuum "displacement current density" $j\omega\epsilon_0\mathbf{E}$, whereas polarization current in a dielectric is in phase with vacuum displacement current. The various current densities appear in the right-hand side of Maxwell's $\nabla \times \mathbf{H}$ equation which can be written as

$$\nabla \times \mathbf{H} = \mathbf{J}_t \qquad (11.70)$$

where $\mathbf{J}_t$ represents the total current density (the sum of any and all current densities). The relative dielectric constant K is obtained by setting the right-hand side equal to $j\omega\epsilon_0 K \mathbf{E}$. The resulting expression for K is

$$K_{ord} = 1 - \omega_p^2/\omega^2 \tag{11.71}$$

where $\omega_p^2 = Nq^2/m\epsilon_0$ and the subscript "ord" indicates the "ordinary" wave, which is one of the linearly polarized characterisctic waves that propagate perpendicular to the magnetic field in a plasma. The same value of K applies if there is no magnetic field or if the frequency ω is so high that ω_B (section 11.1.1) is negligible by comparison. The second linearly polarized characteristic wave for propagation perpendicular to the magnetic field has a y component of electric field intensity and thus is affected by the magnetic field. The relative dielectric constant for this wave is given by

$$K_{ex} = K_l K_r / K_\perp \tag{11.72}$$

where the subscript "ex" stands for "extraordinary", the name given to this wave. Expressions for K_l and K_r are given in Equations (11.73) and (11.74) and $K_\perp = (K_l + K_r)/2$.

It was stated previously that the characteristic waves for propagation parallel to the magnetic field are left and right circularly polarized waves. The K values for these waves will be designated as K_l and K_r. They are given by

$$K_l = 1 - \frac{\omega_p^2}{\omega(\omega + \omega_B)} \tag{11.73}$$

and

$$K_r = 1 - \frac{\omega_p^2}{\omega(\omega - \omega_B)} \tag{11.74}$$

Expressions for the relative dielectric constants for the two characteristic waves that propagate perpendicular to the magnetic field and for the two characteristic waves that propagate parallel to the magnetic field have now been presented. Indices of refraction n can be obtained from the dielectric constant values by use of $n = \sqrt{K}$ (Equation 11.55). Phase velocities can be obtained by using $v_p = c/n$ (Equation 11.54), and wavelength can be obtained by noting that $\lambda = \lambda_o/n$ where λ_o is the free space value. (As $\lambda f = v_p$ and f does not depend on the propagation medium, λ must vary with n in the same way as v_p.) Also as β, the phase constant, equals $2\pi/\lambda, \beta = \beta_o n$ (Equation 11.11). Thus knowledge of the K values leads to the needed information about the characteristic waves.

As the lc and rc waves propagating parallel to the magnetic field have different K values, they also have different β values. It is the difference in the β values that is responsible for Faraday rotation, as a linearly polarized wave consists of and can be decomposed into lc and rc components. This concept is suggested in Figure 11.22 which shows instantaneous positions of lc and rc waves at a particular instant of time. The directions of rotation are shown by the small auxiliary arrows. As the lc and rc waves rotate their projections on the x axis cancel, but their projections on the y axis add to give a vector that varies sinusoidally in amplitude. In the usual vector diagram the **E** vector of a linearly polarized wave is shown as a vector of fixed length, but the instantaneous amplitude varies sinusoidally with time.

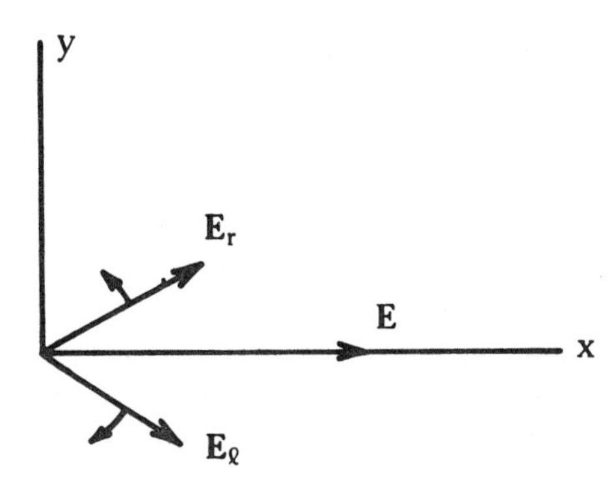

Figure 11.22: Vector diagram showing instantaneous positions of **E**, $\mathbf{E}_l$, *and* $\mathbf{E}_r$. *The linearly polarized electric field intensity* **E** *consists of lc and rc components,* $\mathbf{E}_l$ *and* $\mathbf{E}_r$.

If a linearly polarized wave propagates for a distance z parallel to the magnetic field in a region with constant electron density and magnetic field intensity, the rotation of **E** is given by

$$\phi = (\beta_l - \beta_r)\, z/2 \qquad \text{rad} \tag{11.75}$$

A wave that is launched originally as a vertically polarized wave can become horizontally polarized (or have any angle in between) and vice versa. Thus, the occurence of Faraday rotation is of considerable practical importance. It affects both waves which are used for communication by ionospheric reflection and waves which have a sufficiently high frequency to pass through the ionosphere without reflection. Faraday rotation is inversely proportional to frequency squared, however, for the higher frequencies, and frequencies in the microwave range that are utilized for satellite systems experience very little rotation.

In the discussion to this point we have referred to propagation as being either parallel or perpendicular to the magnetic field. What about propagation at an angle θ with respect to the field? Fortunately it works out that if

$$4\left(1-\frac{\omega_p^2}{\omega^2}\right)\omega_B^2\cos^2\theta \gg \omega_B^4\sin^4\theta \qquad (11.76)$$

the quasi-longitudinal (QL) approximation applies, and the characteristic waves are circularly polarized as for strictly parallel propagation. The expressions for K_l and K_r can be utilized when the QL approximation applies if ω_B in the expressions is replaced by $\omega_B \cos\theta$. If the inequality is the reverse of that shown in Equation (11.76) the quasi-transverse (QT) approximation applies and the characteristic waves are linearly polarized.

11.4.3 Reflection of Electromagnetic Waves from the Ionosphere

In this section, we give emphasis to the ordinary wave, having a relative dielectric constant given by Equation (11.71). One reason for doing so is that near the reflection location, ω_p is close to ω and the QT approximation applies, rather than the QL approximation that is specified by Equation (11.76). Thus near the reflection location, the ordinary wave is one of the two characteristic waves. This wave has a K_{ord} value less than unity. Note that there is nothing to prevent K_{ord} from decreasing to zero or becoming negative. In this respect, ω_p in a plasma plays somewhat the same role as f_c for a waveguide. Only waves having angular frequencies ω greater than ω_p can propagate in a plasma as K becomes a negative and the index of refraction becomes imaginary when ω is less than ω_p. In a waveguide, only waves having frequencies f which are greater than f_c can propagate. In both cases, when the operating frequency is above a critical value, wave propagation occurs with electric intensity varying as

$$\mathbf{E} = \mathbf{E}_o e^{-j\beta z} \qquad (11.11)$$

When the frequency is less than the critical frequency, however, E varies as

$$\mathbf{E} = \mathbf{E}_o e^{-\alpha z} \qquad (11.77)$$

β in Equation (11.11) is a phase constant having units of radian/m. α in Equation (11.77) is an attenuation coefficient having units of Nepers/m. Equation (11.77) describes what is referred to as an evanescent "wave" or field. Attenuation in the evanescent mode does not involve the conversion of electromagnetic energy to thermal energy as in a resistive medium but takes account of the fact that reflection is occuring. The reflection process does not occur at a discrete surface as in the case of reflection from a perfectly conducting surface, but it involves some penetration into the evanescent region as indicated by Equation (11.77).

Consider an ordinary wave that propagates vertically upward from the surface of the earth into the ionosphere. As the wave travels upwards through the lower part of the ionosphere the electron density N and the plasma frequency ω_p increase with height. Depending on the frequency utilized, it is possible that the condition $\omega_p = \omega$ corresponding to $K_{ord} = 0$, may be realized. If so, the wave will be reflected at the level where $\omega_p = \omega$. (As stated in the previous paragraph reflection really does not all take place at a precise level, but that is of little consequence for present considerations; we will refer to reflection as taking place when $\omega_p = \omega$.) If the condition $\omega_p = \omega$ is never reached as a wave propagates upwards (if $\omega > \omega_p$ along the entire length of the ionospheric path), the wave will not be reflected but will continue on beyond the ionosphere.

A sounding system known as an ionosonde, actually a special-purpose radar system, is used to obtain information about electron density profiles in the ionosphere on the basis of the height at which a wave of a given frequency is reflected. The typical ionosonde sweeps over about a 0.5 to 25 MHz frequency range in 10 to 15 s, transmitting pulses at a repetition rate of about 100 pps during the time interval of the sweep. The echo pulse received from the ionosphere forms a spot on a cathode-ray tube at a distance above a reference level proportional to the delay time of the echo, and the delay time is proportional to the virtual height h' of the ionospheric reflection layer. h' differs from true height h because the velocity of propagation in the ionosphere is different from c, but procedures are available for recovering true height from virtual height. (The Environmental Data Service of the National Geophysical and Solar-Terrestrial Data Center in Boulder, Co. will perform this function as a service.) Another technique that can be used to obtain information about electron density in the ionosphere, including the region above the peak the F_2 layer for which the ionosonde provides no information, is the incoherent scatter technique.[26] In this case a frequency too high for reflection of the type we have discussed is used, but a weak backscatter signal proportional to electron density is received from any height for which the system has sufficient sensitivity.

In carrying out communication from one location to another, one does not utilize waves that are vertically incident upon the ionosphere but waves that are obliquely incident. In this case a wave of higher frequency can be reflected from the ionosphere than for vertical incidence. In particular, reflection at a frequency of f can occur where the plasma frequency is f_p when f and f_p are related by

$$f = f_p \sec \phi_o \tag{11.78}$$

where ϕ_0 is the original launch angle of the wave measured from the vertical. If f_p corresponds to the highest electron density of the ionosphere, f is the maximum usable frequency.

The prediction of the performance of HF systems and the choice of operating frequencies has long been of interest to thè Institute of Telecommunication Science (ITS) and the National Bureau of Standards (NBS) which had responsibility for such matters before the formation of ITS.[27,28] Oblique sounders, similar to ionosondes but operating with a transmitter at one location and receiver at a distant location, are highly useful tools for obtaining both long-term and real-time data on propagation conditions. Oblique sounder terminals can be located at essentially the same positions as the terminals of a particular HF path and used to select suitable operating frequencies for the path.

Communication by ionospheric reflection at HF frequencies was formerly a principal means of long distance communication, and it is still an important and useful method. Communication by satellite, however, has taken over a large share of long-distance communication, and ionospheric propagation is not so widely used as it once was. Disadvantages of communication using HF frequencies and ionospheric reflection include: the variability of the ionosphere and possible disruptions of service due to ionospheric storms and disturbances, the low bandwidth available, and the susceptibility to mutual interference between various users. Advantages are low cost and the ability to communicate between a number of rather widely scattered locations. An example of an application that favors HF techniques is that of obtaining information from a number of widely scattered remote sensors such as low-cost data buoys which can be dropped at sea. The location of the buoys can be determined by HF radar and data can be transmitted from the buoys at HF frequencies. HF radar is also of interest for other over-the-horizon applications such as air-traffic control over the oceans and remote monitoring of sea-state.

Propagation by ionospheric reflection sometimes involves more than one hop so that the signals are alternately reflected by the ionosphere and the ground. Attenuation is experienced at each reflection. Actually the reflection process in the ionosphere is really a refraction process, and attenuation is encountered along the length of the path in the ionosphere. Data concerning the reflection and other loss factors, the recommended fade margins for HF systems, and noise in the HF band, together with sample calculations are given by Davies, chapters 5 and 7.[26] The propagation of ionospherically reflected waves is a large and interesting subject which is treated more fully by Budden,[29] Davies,[27] Kelso,[30] and Ratcliffe.[31]

11.5 PROPAGATION EFFECTS ON SATELLITE AND DEEP-SPACE TELECOMMUNICATIONS

For satellite communications the expression for the carrier-power-to-noise-power ratio, designated here by C/X, has the same general form as that of Equation (11.53) but is commonly written as

$$(C/X)_{dB} = (EIRP)_{dBW} - (L_{FS})_{dB} - L_{dB} + (G_R/T_{sys})_{dB} - k_{dBW} - B_{dB} \tag{11.79}$$

where W_T and G_T have been combined into EIRP (Effective Isotropic Radiated Power). The noise power X in this case is given by

$$X = k\,T_{sys}\,B \tag{11.80}$$

where T_{sys} is the system noise temperature. The quantities C, X, and T_{sys} are defined at the antenna terminals, as suggested for the case of T_{sys} in Figure 11.23. The system noise temperature T_{sys} is given in terms of the other receiving system parameters by

$$T_{sys} = T_A + (l_a - 1)T_o + l_a T_R \tag{11.81}$$

where T_A is the antenna noise temperature and T_R is the receiver noise temperature. The quantities T_A and T_R are measures of the noise introduced by the receiving antenna and the receiver, respectively. The quantity $l_a = 1/g_a$ where g_a is the "gain" of the transmission line between the antenna and the receiver. The quantity g_a is less than unity, and l_a is consequently greater than unity. T_o is the standard reference temperature of 290 K. The earth's atmosphere affects both the loss factor L and the system noise temperature T_{sys} of Equation (11.79). In the case of T_{sys}, the atmospheric effect appears in the value of T_A. In addition the atmosphere may cause excess time delay, Doppler frequency variations, variations in the direction of arrival, elevation angle error, and dispersion. This last term refers to the variation of time delay with frequency and has the potential to limit the data rate of high-capacity digital systems. Excess time delay degrades the precision of range measurements, which are required for deep-space missions and satellite positioning systems such as the Global Positioning System.[32] Attenuation due to rain increases with frequency up to about 100 GHz and has values in dB/km as a function of rain rate as shown by Figure 11.19. The attenuation caused by the atmospheric gases, shown in Figure 11.24, is characterized by a peak in absorption at about 22.2 GHz due to water vapor and peak absorption near 60 GHz and 118 GHz due to oxygen.[33]

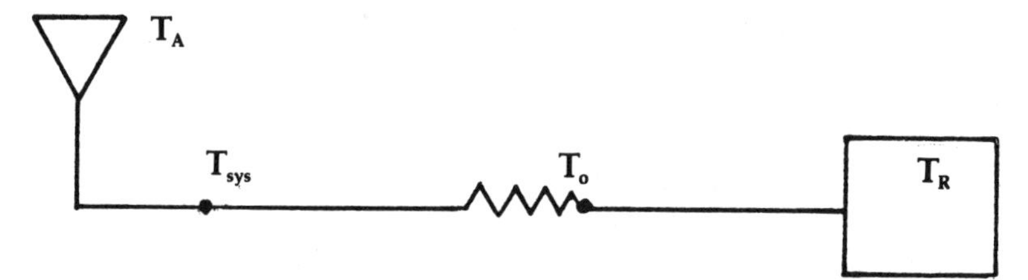

Figure 11.23: Noise Temperatures of Receiving System

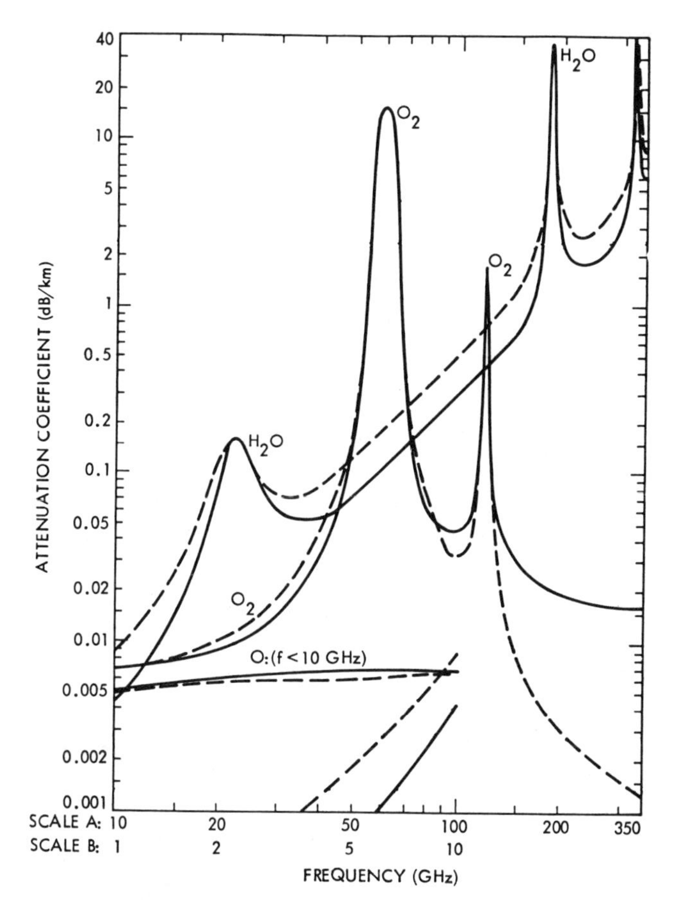

Figure 11.24: Attenuation due to atmospheric gases. Comparison of JPL values (dashed curves) with those of CCIR Report 719 for a pressure of 1 atmosphere, a temperature of 20°C, and a water vapor content of 7.5 g/m³.[33]

For treating attenuation due to rain in system design, a statistical approach is needed, and considerable attention has been devoted to modeling the occurence of rain over the earth in terms of the rain rates R in mm/h that are exceeded for p percent of the time. The percentage p is commonly taken to be a small quantity such as 0.01 or 0.1, but in other cases larger values such as 20 may be of interest. Where detailed rain data are available and have been analyzed for a particular location they should be used but otherwise a model such as the 1980 Global Model[34] or the 1982 CCIR Model[35] can be employed. The 1982 CCIR Model is illustrated by Figure 11.25, and by Table 11.2.

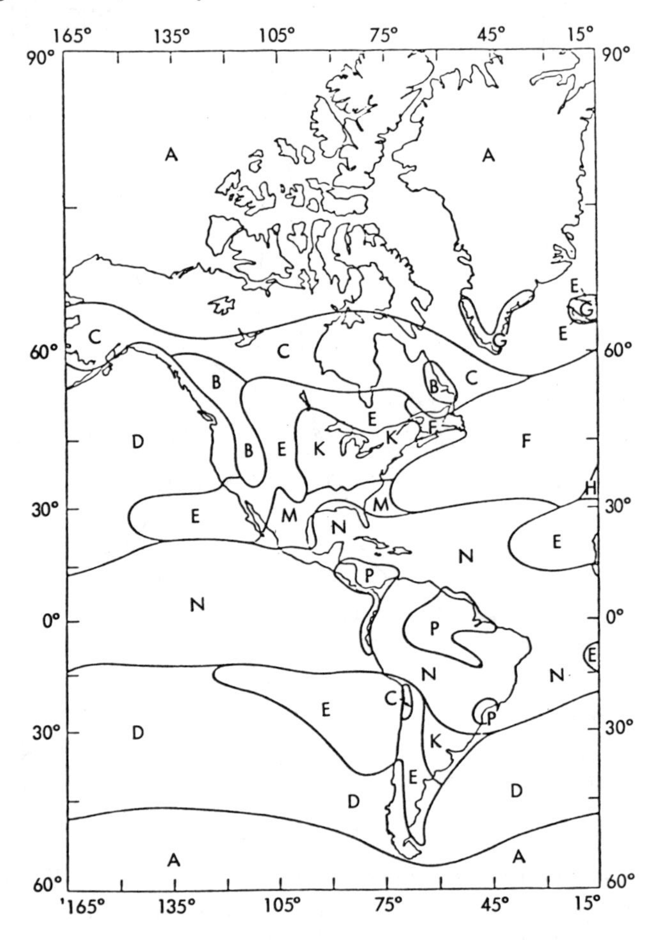

Figure 11.25: Rain-rate regions of the western hemisphere, according to the 1982 CCIR Model.[35]

TABLE 11.2

Rain rates exceeded for p = 0.01 for the regions of the 1982 CCIR Model[35]

Region of Fig. 11.25	A	B	C	D	E	F	G	H	J	K	L	M	N	P
Rain rate (mm/h)	8	12	15	19	22	28	30	32	35	42	60	63	95	145

The 1980 Global Model, utilizing extensive experimental data obtained in the United States, is suitable for application to the United States. Canada should use its own model,[36] and we recommend the 1982 CCIR Model for the rest of the world other than the United States and Canada.

Once the appropriate rain rate R has been determined from one of the models mentioned or otherwise, the power attenuation constant can be determined by use of the empirical expression

$$\alpha_p = a R^b \tag{11.82}$$

where a and b are functions of frequency. This relation has been analyzed by Olsen, Rogers, and Hodge[37] who have also provided values of a and b. Separate values of the coefficients for horizontally and vertically polarized waves are included in CCIR Report 721[38] and some values from this report are reproduced in Table 11.3. Total attenuation A in dB may be determined by

$$A = \alpha_p L r_p \tag{11.83}$$

where L is path length through rain and r_p is a path reduction factor to take account of the fact that intense rain tends to be restricted in extent.[39] For elevation angles above about 10°, the path length L equals $H/\sin\theta$

TABLE 11.3

Coefficients a and b of Equation (11.82) from CCIR Report 721[38]

The subscripts H and V refer to horizontal and vertical polarization, respectively.

Frequency (GHz)	6	10	20	30
a_H	0.00175	0.0101	0.0751	0.187
a_V	0.00155	0.00887	0.0691	0.167
b_H	1.308	1.276	1.099	1.021
b_V	1.265	1.264	1.065	1.000

where H is the height of the 0°C isotherm and θ is the elevation angle of the path. The height H is a function of percentage of time and latitude. Figure 11.26 shows values of H from CCIR Report 563.[35]

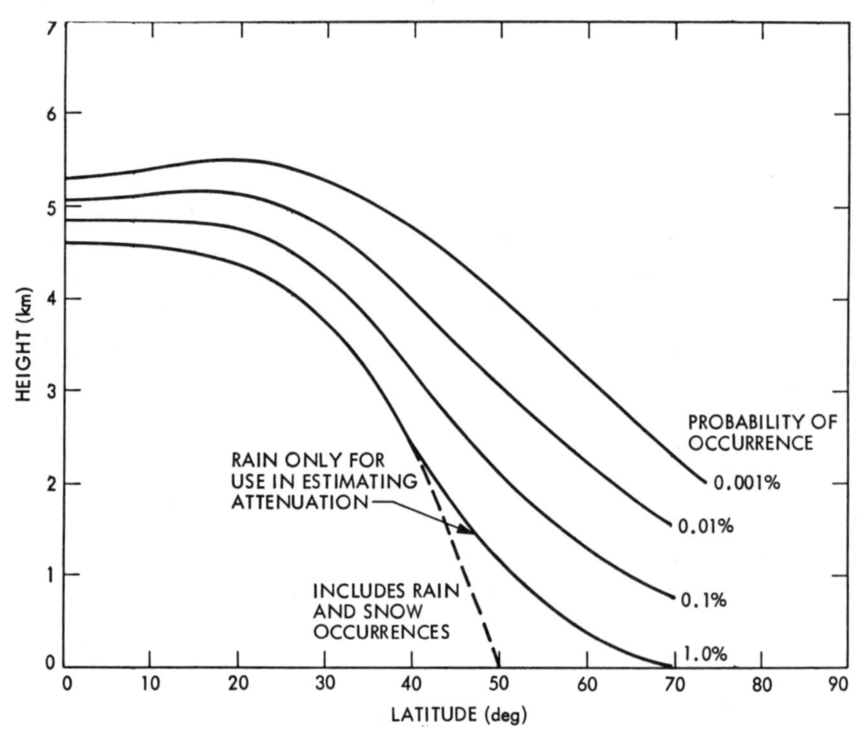

Figure 11.26: Heights of 0°C isotherm from CCIR Report 563.[35]

The factor r_p is given in the 1982 CCIR Model by

$$r_p = \frac{90}{90 + C_p D} \tag{11.84}$$

where $D = L\cos\theta$ and is the horizontal extent of the path. C_p has the value of 4 for $p = 0.01$. Procedures for estimating attenuation A as a function of location and percentage p are under continuing review. It appears that the heights shown in Figure 11.25 may be excessive for latitudes below 40°. For further details see CCIR Report 564.[39]

The attenuator of Figure 11.23 introduces noise accounted for by the second term of Equation (11.81). Likewise an attenuating region of the atmosphere introduces noise and makes a contribution to T_A of Equation (11.81). This contribution T_B is given by

$$T_B = T_i(l - e^{-\tau}) \tag{11.85}$$

where T_i is the intrinsic temperature of the medium and is commonly taken as 280 K, or lower for propagation through precipitation. The quantity τ is called optical depth and represents the integral of the attenuation constant along the path in the atmosphere. The factor $e^{-\tau}$ corresponds to g_a of the attenuator of Figure 11.23. Clouds make important contributions to attenuation and radio noise for higher frequencies above 10 GHz.[41]

The larger raindrops are not spherical in form as assumed in earlier analyses but have a greater extent in the horizontal direction than in the vertical. As a consequence, horizontally polarized waves experience slightly greater attenuation than vertically polarized waves (Table 11.3). Another consideration is that raindrops tend to be canted with their long axes oriented over a range of angles with respect to the horizontal. An important result of the departure from sphericity and the canting is that some of the energy in a wave propagating through a region where precipitation is occurring is converted from the original polarization (e.g. horizontal or left circular) to the corresponding orthogonal polarization (vertical or right circular). This process, referred to as depolarization and as resulting in the generation of cross polarized components, limits the ability to use two orthogonal polarizations simultaneously over the same path at the same frequency.[42, 43]

More detailed information on propagation effects on satellite communications above 10 GHz can be found in NASA Reference Publication 1082.[44] Propagation effects at frequencies below 10 GHz have been treated in a similar way by Flock.[45] Hall[46] has provided a good general reference on tropospheric propagation, and Miya[47] includes consideration of propagation effects in his treatment of satellite systems engineering.

REFERENCES

1. Johnk, C.T.A., *Engineering Electromagnetic Fields and Waves*, New York: Wiley, 1975.

2. Jordan, E.C. and K.G. Balmain, *Electromagnetic Waves and Radiating Systems*, 2nd ed. Englewood Cliffs, NJ: Prentice-Hall, 1968.

3. Ramo, S., J. R. Whinnery and T. Van Duzer, *Fields and Waves in Communication Electronics*, New York: Wiley, 1965.

4. Joint Technical Advisory Committee, *Radio Spectrum Utilization*. New York: IEEE, 1964.

5. Joint Technical Advisory Committee, *Spectrum Engineering — the Key to Progress*. New York: IEEE, 1968.
6. Hinchman, W.R., "Use and Management of the Electrospace; a New Concept of the Radio Resource", IEEE, Int. Conference on Communications, *69C29-COM*, pp. 13-1 to 13-5. New York: IEEE, 1969.
7. Joint Technical Advisory Committee, *Radio Spectrum Utilization in Space*. New York: IEEE, 1970.
8. Miller, S.E., E.A.J. Marcatili and T. Li, "Research Toward Optical Fiber Transmission Systems", *Proc. IEEE*, vol. 61, pp. 1703-1751, Dec., 1973.
9. Blake, L.V., *Antennas*. New York: Wiley, 1966.
10. Kraus, J.D., *Antennas*. New York: McGraw-Hill, 1950.
11. Flock, W.L., *Electromagnetics and the Environment: Remote Sensing and Telecommunications*. Englewood Cliffs, NJ: Prentice-Hall, 1979.
12. Freeman, R.L., *Telecommunication Transmission Handbook*, 2nd Edition, New York: Wiley, 1981.
13. Panter, P.F., *Communication Systems Design*. New York: McGraw-Hill, 1972.
14. Bean, B.R., and E.J. Dutton, *Radio Meteorology*. Washington, D.C.: Supt. of Documents, U.S. Government Printing Office, 1966.
15. GTE Lenkurt, *Engineering Considerations for Microwave Communication Systems*. San Carlos, CA: GTE Lenkurt, Inc., 1972.
16. Laine, R.V., "Blackout Fading in Line-of-sight Microwave Links", presented at PIEA-PESA-PEPA Conference, April 22, 1975, Dallas, TX, pp. 1-15. San Carlos, CA: GTE Lenkurt, Inc., 1975.
17. Taur, R.R., "Ionospheric Scintillation at 4 and 6 GHz", *COMSAT Technical Review*, vol. 3, pp. 145-163, 1973.
18. Feldman, N. and S.J. Dudzinsky, "A New Approach to Millimeter-Wave Communications," *R-1936-RC*, The Rand Corporation, Santa Monica, CA. April, 1977.
19. Kerker, M., *The Scattering of Light and Other Electromagnetic Radiation*. New York: Academic Press, 1969.
20. Kerr, D.E., ed., *Propagation of Short Radio Waves*. New York: McGraw-Hill, 1951.

21. Zufferey, C.H., "A Study of Rain Effects on Electromagnetic Waves in the 1-600 GHz Range." M.S. Thesis, Department of Electrical Engineering, University of Colorado, Boulder, CO, 1972.

22. Laws, J.O. and D.A. Parsons, "The Relation of Drop Size to Intensity", *Transactions American Geophysical Union*, pp. 452-460, 1943.

23. Crane, R.K., "Prediction of the Effects of Rain on Satellite Communication Systems", *Proc. IEEE*, vol. 65, pp. 456-474, March 1977.

24. Longley, A.G., R.K. Reasoner and V.L. Fuller, "Measured and Predicted Long-Term Distributions of Tropospheric Transmission Loss," *OT/TRER 16*, ITS, Boulder, CO., Washington, D.C.: Supt. of Documents, U.S. Government Printing Office, July 1971.

25. Rishbeth, H. and O.K. Garriott, *Introduction to Ionospheric Physics*. New York: Academic Press, 1969.

26. Evans, J.V. "Theory and Practice of Ionospheric Study by Thomson Scatter Radar", *Proc. IEEE*, vol. 57, pp. 496-530, April 1969.

27. Davies, K., *Ionospheric Radio Propagation*. Washington, D.C.: Supt. of Documents, U.S. Government Printing Office, 1965.

28. Haydon, G.W., M. Leftin and R. Rosich, "Predicting the Performance of High Frequency Sky-wave Telecommunication Systems," *OT Report 76-102*. Boulder, CO: Institute for Telecommunication Sciences, Sept., 1975.

29. Budden, K.G., *Radio Waves in the Ionosphere*. Cambridge: Cambridge University Press, 1961.

30. Kelso, J.M., *Radio Ray Propagation in the Ionosphere*. New York: McGraw-Hill, 1964.

31. Ratcliffe, J.A., *An Introduction to the Ionosphere and Magnetosphere*. Cambridge: Cambridge University Press, 1972.

32. Flock, W.L., S.D. Slobin, and E.K. Smith, "Propagation Effects on Radio Range and Noise in earth-Space Telecommunication," *Radio Science*, vol. 17, pp. 1411-1424, Nov.-Dec. 1982.

33. Smith, E., "Centimeter and Millimeter Wave Attenuation and Brightness Temperature due to Atmospheric Oxygen and Water Vapor," *Radio Science*, vol. 17, pp. 1455-1464, Nov.-Dec. 1982.

34. Crane, R.K., "Prediction of Attenuation by Rain," *IEEE Trans. Comm.*, vol. COM-28, pp. 1717-1733, Sept. 1980.

35. International Radio Consultative Committee, "Radiometeorological Data," Report 563-2 in Recommendations and Reports of the CCIR, 1982, Vol. V, *Propagation in Non-Ionized Media*, pp. 96-123, 1982.

36. Segal, B., "A New Procedure for the Determination and Classification of Rainfall Rate Climatic Zones," *Ann. Telecomm.*, vol. 35, pp. 411-417, 1980.

37. Olsen, R.L., D.V. Rogers, and D.B. Hodge, "The aR^b Relation in the Calculation of Rain Attenuation," *IEEE Trans. Antennas and Propagation*, vol. AP-26, pp. 318-329, March 1978.

38. International Radio Consultative Committee, "Attenuation by Hydrometeors, in Particular Precipitation, and other Atmospheric Particles," Report 721-1 in Recommendations and Reports of the CCIR, 1982, vol. V, *Propagation* in *Non-Ionized Media*, pp. 167-182, 1982.

39. International Radio Consultative Committee, "Propagation Data Required for Space Telecommunication Systems," Report 564-2 in Recommendations and Reports of the CCIR, 1982, Vol. V, *Propagation in Non-ionized Media*, pp. 331-373, 1982.

40. International Radio Consultative Committee, "Radio Emission from Natural Sources above about 50 MHz," Report 720-1 in Recommendations and Reports of the CCIR, 1982, Vol. V, *Propagation in Non-ionized Media*, pp. 151-166, 1982.

41. Slobin, S.D., "Microwave Noise Temperature and Attenuation of Clouds: Statistics of These Effects at Various Sites in the United States, Alaska, and Hawaii," *Radio Science*, Vol. 17, pp. 1443-1454, Nov-Dec. 1982.

42. Cox, D.C., "Depolarization of Radio Waves by Atmospheric Hydrometeors in Earth-Space Paths: A Review," *Radio Science*, Vol. 16, pp. 781-812, Sept.-Oct. 1981.

43. International Radio Consultative Committee, "Cross-polarization due to the Atmosphere," Report 722-1 in Recommendations and Reports of the CCIR, 1982, Vol. V, *Propagation in Non-ionized Media*, pp. 185-193, 1982.

44. Ippolito, L.J., R. Kaul, and R. Wallace, *Propagation Effects Handbook for Satellite System Design*, NASA Reference Publication 1082, 3rd ed., Washington, D.C.: NASA Headquarters, 1983.

45. Flock. W.L., *Propagation Effects on Satellite Systems at Frequencies Below 10 GHz*, NASA Reference Publication 1108. Washington, D.C.: NASA Headquarters, 1983.

46. Hall, M.P.M., *Effects of the Troposphere on Radio Communication*. Stevenage, United Kingdom: Peter Peregrinus, 1979.

47. Miya, K., *Satellite Communications Technology*. KDD Bldg. 3-2, Nishi-Shinjuku 2-chome, Shinjuku-ku, Tokyo 60, Japan, 1981.

chapter 12

SATELLITE COMMUNICATIONS

H. A. Haddad
Ball Aerospace Systems Division
Boulder, Colorado

12.0 INTRODUCTION

Communication by satellite is changing our perception of world events. Sports and world news are transmitted daily across the globe through satellites. The news and television media, as a normal course of their operation, are expected to provide daily coverage of national and international events which in some cases requires live coverage. Without satellites, live reporting of world events would be next to impossible. In the short period of time since the introduction of the first commercial satellite, Early Bird, in 1965, communication via satellites has evolved rapidly to become the dominant global and regional medium of communications.

Also, it is of particular importance to mention the large growth in high quality international telephone service. This growth trend is expected to continue at least until the end of this century.

In recent years, the interest has grown in the application of satellite communications to teleconferencing, local area network and direct broadcast services (DBS). These services have great impact on present day telephone and TV services that are provided by the common carriers and the local TV networks. In a society where face-to-face business meetings are occasionally held in widely dispersed locations, teleconferencing should help increase the occurrence of these meetings as well as improve their effectiveness at a reduced cost. Because of the wider bandwidths and the broadly separated area required for teleconferencing, satellites are the ideal medium for interconnections. Similarly satellites can provide the interconnection that is necessary in data or information transfer from one local area network to another. Recently, interest is also growing in DBS for providing direct home TV services with minimum complexity to the home receiver. This system will have far reaching political and legal ramifications as television networks access the home directly without going through the local stations.

Satellites can be thought of as active repeaters in the sky. They can receive and amplify thousands of telephone channels and several video channels simultaneously for transmission to different ground users. These satellites are more than simple microwave repeaters in that transmitted satellite signals can cover a large geographical area with multiple users sharing a single satellite. Satellite beams are shaped to cover the needs of a given geographical area with more effective spectrum usage. This progress was made possible by the rapid improvement in launching vehicles, satellite construction, and components. This medium of signal transmission has become highly reliable.

Figure 12.1 shows the trends in satellite communication costs per voice channel up to the early 1990's.[1] Although at present, the major contributor to the total annual cost per voice channel is the cost for local networks, it is expected that through the continuous progress in space technology, the space and ground segment costs will continue to decline, adding to the reduction in total voice circuit cost. The recent successes in launching satellites from the space shuttle to geostationary orbit will lower even further the cost of the space segment.

What follows provides some brief insights into the general model of satellite communications with some descriptions of space and ground systems. A detailed review will be made of the communication link, followed by current applications of satellite technology and its future outlook.

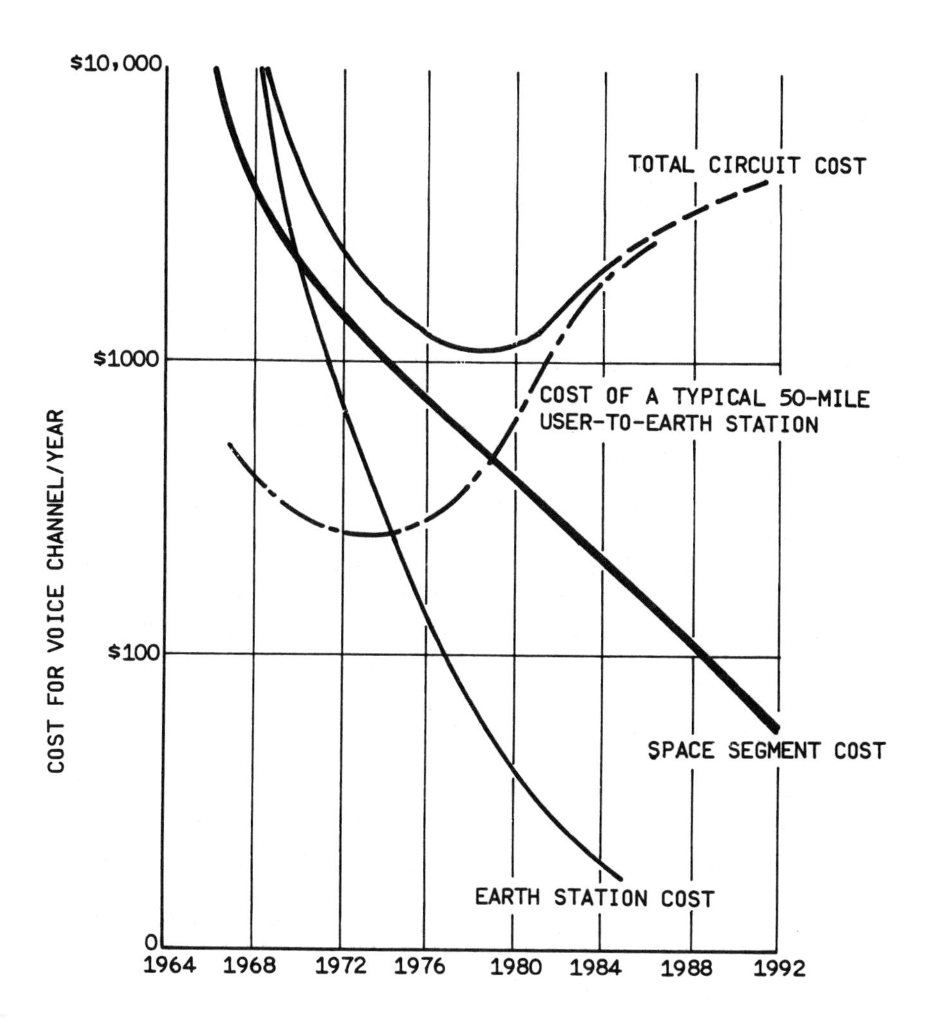

Fig. 12.1.: Approximate Cost Trends for Satellite, Earth Station and Terrestrial Interconnect Voice Circuits[3]

12.1 AN OVERVIEW OF SATELLITE COMMUNICATIONS SYSTEMS

A general model for a satellite communications system is shown in Figure 12.2. As in line-of-sight microwave repeaters, the satellite receives the up-link signal in a given frequency band, amplifies it, and then changes the frequency for re-transmission on the down-link. The equipment that performs this operation is called a transponder. Presently, most commercial satellites use the frequencies of the microwave terrestrial networks. While the distances between radio or microwave repeaters are on the order of 30 to 60 km, the distance of a communication satellite in the geostationary orbit is approximately 35,800 km. Since the received signal power is reduced by the inverse square of the distance to the satellite, space losses are high. In addition, the signal encounters a delay of approximately 250 milliseconds for a one way path from the earth to the satellite and back to earth.

The advantages of a satellite link outweigh the negative effects of path losses and signal delay. The satellite can provide a large geographical area coverage. One satellite placed in the geostationaroy orbit can cover over one-third of the earth surface. In addition, a satellite network offers the flexibility of interconnecting a large number of users with differing data requirements over a wide geographical location. Also, satellite systems provide broadcasting capability — that is a single earth station can transmit through a satellite to many receivers. At the same time, multiple users can transmit to a single or multiple user of the system. This mode of operation makes satellite systems a very powerful medium of communication especially in a multi-point distributed network.

As was mentioned earlier, most domestic and international communication satellites are placed in a geostationary circular orbit. In this orbit, the satellites are positioned in the equatorial plane at an altitude of 35,800 km from the earth's surface. These satellites look stationary to an earth observer. The advantage of this orbit is that no tracking or handover operation is necessary for maintaining continuous communication link. Three satellites would be adequate for full global communications.

The major components of the satellite communication system are the satellite subsystem, the receiving/transmitting earth station and the associated terrestrial links, and the telemetry, tracking and command (TT&C) controller. User's traffic, such as voice, data, or television may be connected to an earth station with direct satellite access or connected by a terrestrial network.

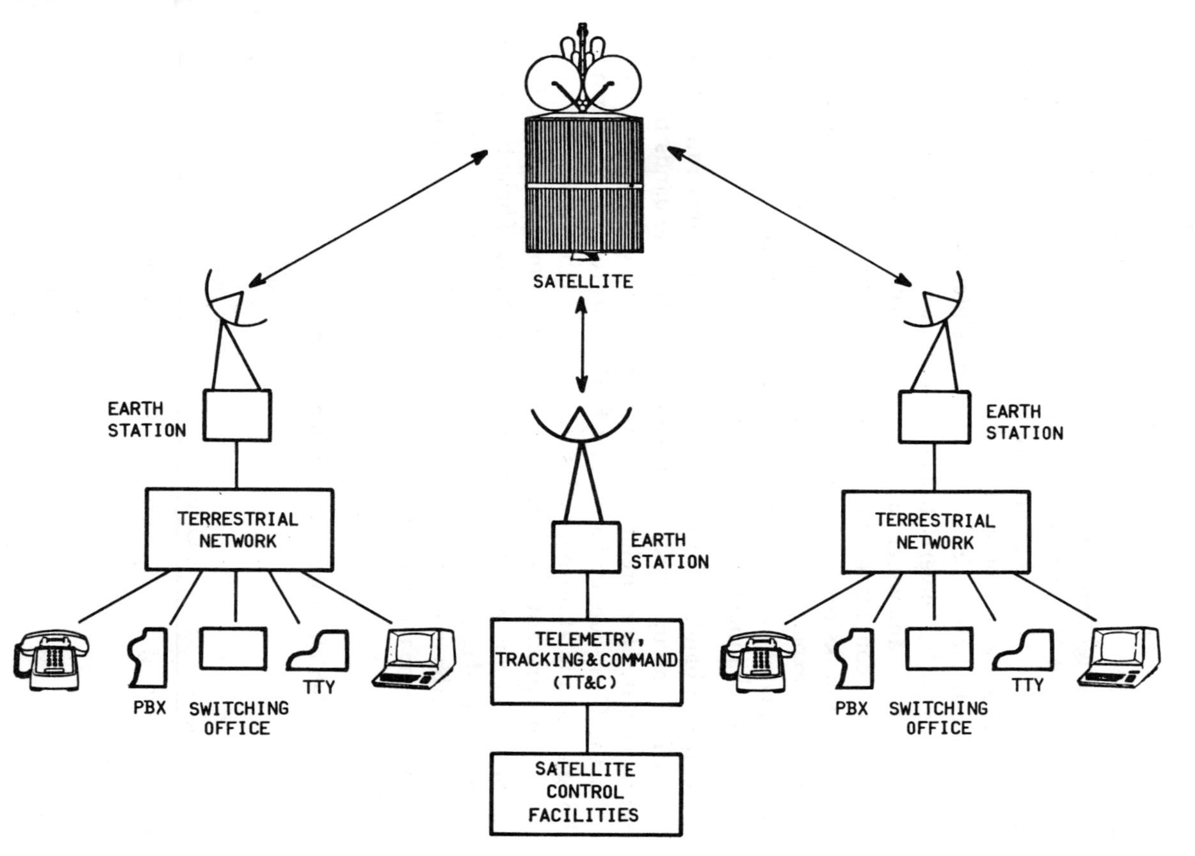

Fig. 12.2.: A Satellite Communications System Model

Communication satellite design starts with a trade-off study between space segments requirements and the earth station conditions. Once the technical objectives of the two segments are defined, another trade-off study is made between satellite communication requirements and the constraints of weight and envelope conditions of launch vehicles. Table 12.1 lists some of the vehicles used for launching satellites into geosynchronous orbit. It is of interest to mention the space shuttle which can deliver to a lower orbit a payload of 29,500 kg with a length of 18 meters and a diameter of 5 meters. Four satellites with Delta size accommodation can be delivered to a lower orbit using the space shuttle. In fact, the large size and weight accommodation will add the flexibility of designing larger communication satellites with higher power capability, expanded fuel tanks for longer lifetime, and elaborate communication subsystem. It is expected that the cost per pound will drop continuously for delivering satellites to geostationary orbit and that can translate into lower cost per channel using a satellite link.

The frequency bands used for satellite communications range from lower VHF to the upper Ka band. Table 12.2 lists the bands that are currently allocated for satellite systems. The technology is highly developed in the C frequency bands and fairly mature in the Ku-bands. Although rain and atmospheric losses are very high at the higher frequency bands, presently an interest is developing in the 20 to 40 GHz band for military as well as commercial applications. These bands can provide over 1 GHz of bandwidth with higher data rate capability. The wider bandwidths in some applications can provide a jam resistant system with low probability of intercept

TABLE 12.1

Vehicles for Launch into Synchronous Orbit

Vehicle	Origin	Payload Into Synchronous (*kg*)	Envelope Diameter (*m*)
Delta (2914 or 3914)	US	340 to 460	2.2
Atlas/Centaur	US	870	2.9
Titan III C	US	1470 to 1940	3.1
Space Shuttle	US	29,500*	5
Ariane	Europe	900	3.0
N Rocket	Japan	130	1.4
Soyuz SL-4	USSR	7,500*	-
Proton SL-9	USSR	18,000*	-

*Low earth parking orbit

TABLE 12.2

Frequency Bands Commonly Used in Satellite Communications

Designation	Range in MHz		Type of Service
Military UHF	240.0-328.6	Up & Down	Mobile-Satellite (LES -5,6,8 and 9; MARISAT, FLEETSAT)
L	1535.0-1543.5	Down	Maritime Mobile Satellite
	1636.5-1645.0	Up	(MARISAT)
	1542.5-1558.5	Down	Aeronautical Mobile-
	1644.0-1660.0	Up	Satellite (AEROSAT)
C	3700-4200	Down	Fixed-Satellite (INTELSAT series, (DOMSATs)
	5925-6425	Up	
X or Military SHF	7250-7750	Down	Fixed-Satellite (DSCS, NATO, GALS)
	7900-8400	Up	
Ku	10,950-11,200 & 11,450-11,700	Down	Fixed-Satellite (INTELSAT V, planned DOMSATs)
	14,000-14,500	Up	
Ka	17,700-21,200	Down	Fixed-Satellite (Japanese CS)
	27,500-31,000	Up	

12.1.1 Basic Satellite System

As shown in Figure 12.3 the communication satellite consists of the communication subsystem, the equipment for power generation and conditioning, thrust and stabilization, thermal control, and telemetry, tracking and command (TT&C). The communications subsystem consists of the antennas and the transponders. Early in the program development, trade-offs are made between the communications requirements and the conditions for acceptable dc power levels. This trade-off is made in order to meet the necessary operation of the satellite as well as maintain the proper thrust and stabilization conditions for accurate satellite positioning in its proper orbit. Generally, the selection is based on two types of stabilizations systems; one is spin stabilized and the other is body stabilized. Spin stabilized satellites are maintained in space by spinning the satellite at the rate of 30 to 100 revolutions per minute with periodic corrections using the thrusters. The main type of spin stabilized satellite used for communication purposes is the dual-spin satellite as shown in Figure 12.4a. This satellite uses an outer spinning drum with a despun section for maintaining constant earth pointing direction. The dual-spin satellite has simplicity in construction and stabilization. However, it suffers from the limitation in delivering low dc power levels, which is due to the limited effective area on a drum that faces the sun. Presently, the dual-spin satellite can deliver up to 1 kW of power for a seven year lifetime. To achieve higher power levels, body stabilized satellites are presently used. These types of satellites are attitude controlled about one or more axes using momentum wheel devices. Figure 12.4b shows a body-stabilized satellite. Except for satellite eclipse periods the body-stabilized satellite has a large continuous sun-tracking solar array that delivers more power per unit area than a solar array mounted on a cylinder. Although the trends are toward higher power levels, presently both types of stabilization are used in satellite construction.

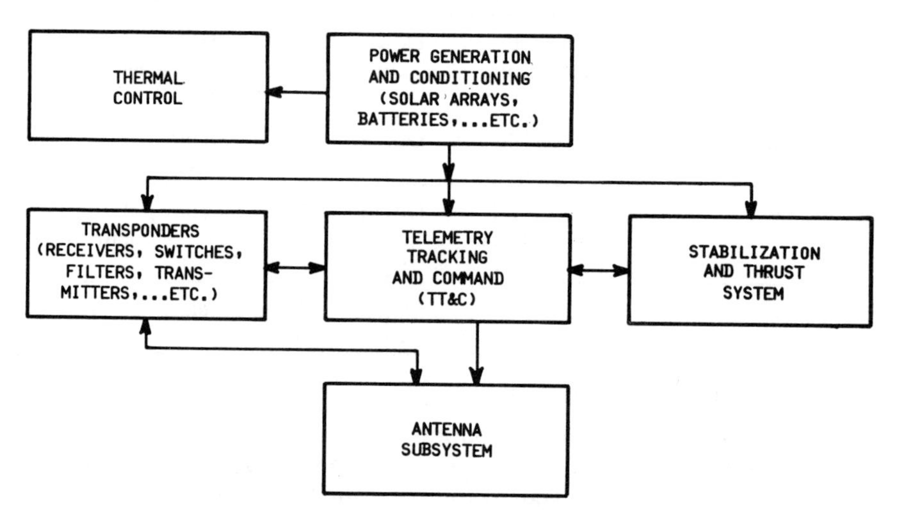

Fig. 12.3: Maior Functional Elements of a Satellite System

Because of drifts due to the gravitational effect of sun-earth-moon interactions, a satellite must use a propulsion system for continuous position corrections. Hydrazine thrusters are used for this purpose because of their low weight, high density, and high thrust level.

Photo voltaic solar cells are the main source for dc power generation. They are fixed to the outer body of a satellite or fixed to a solar panel which is continuously oriented toward the sun for maximum power level. During two 45 day periods, once in the fall and once in the spring where the earth shadows the satellite, a satellite will be in the dark for up to 70 minutes. To maintain operation of the satellite during this period, nickelcadmium or nickel hydrogen batteries are used.

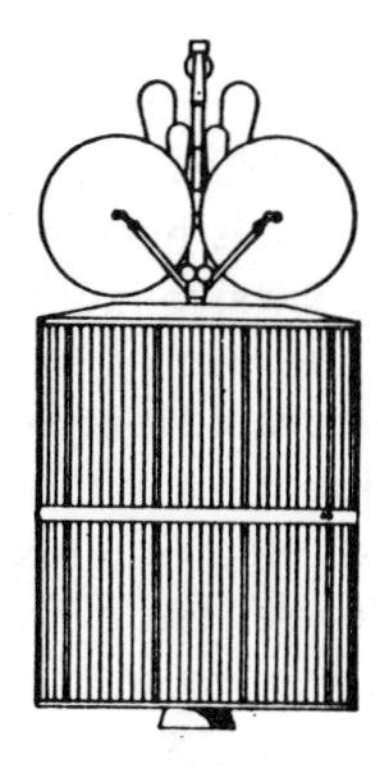

Fig. 12.4a: Spin-Stabilized Satellite (INTELSAT IV)

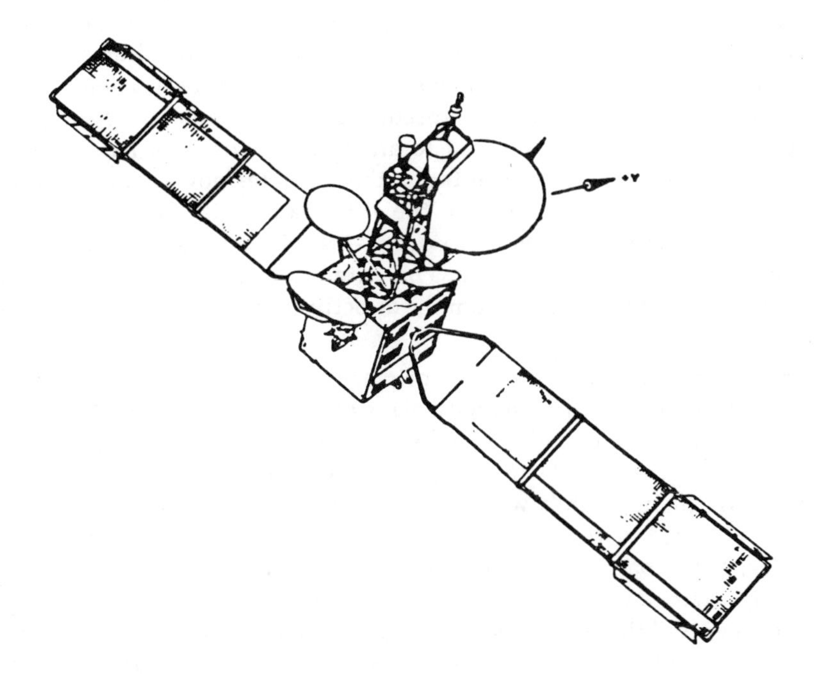

Fig. 12.4b: Body-Stabilized Satellite (INTELSAT V)

In a space environment, the temperature change can have wide variations. To protect equipment from temporary inoperation or damages, some form of thermal control is necessary. There are two methods of thermal control, one is the passive method and the other is active. The passive method involves the use of multi-layer thermal blankets on equipment to protect it from rapid temperature variations. Also, surface finishes are used to confine heat within the payload area. The active method involves the use of electrical heaters or heat pipes in order to maintain certain heating levels. The active heaters usually are used as a supplement to the passive heating method.

The antennas and transponder systems are the major components of the communication subsystem. A simplified system configuration of the antenna and transponder subsystem is shown in Figure 12.5. There are different types of antennas used in satellite communication systems. The majority are horns or reflectors. However, in some applications helics, dipoles, phased array, and lenses are also used. Antennas are the front interface to the communication system. Generally, antennas with directivity are selected to provide proper earth coverage. The SBS satellite has a shaped beam that covers the continental United States (CONUS) and the INTELSAT system has a shaped beam that has hemispherical/zonal coverage. A directive antenna is one that radiates its energy in a preferred direction compared to an isotropic antenna. An isotropic antenna is an ideal lossless antenna that radiates power equally in all directions. The directive gain of the antenna over isotropic in a given direction is referred to as "gain over isotropic" and it is given in dBi, where the i refers to isotropic. The antenna generally has losses and these should be subtracted from the directivity in order to have the total antenna gain.

A reflector antenna consists of a primary reflecting surface and a feed system that are placed at the focal plane of the main reflector. The main reflector can be a cylindrical, spherical, or parabolic surface. Surface tolerance of these reflectors requires good control, especially at the higher frequency bands where gain reduction can become noticeably large. The feed system can be single or multiple horns, open-end waveguides, cavity backed dipoles, phase arrays, or lenses. Beam shaping is usually attained through the proper selection of the field distributions in the elements of the feed network. The primary beam steering mechanism in reflector antennas is mechanical; however, in some applications where rapid limited scanning is required, electronic beam-steering may be used. Parabolic reflectors were used on many satellites such as ATS-6, TDRSS/WESTAR and INTELSAT V satellites.

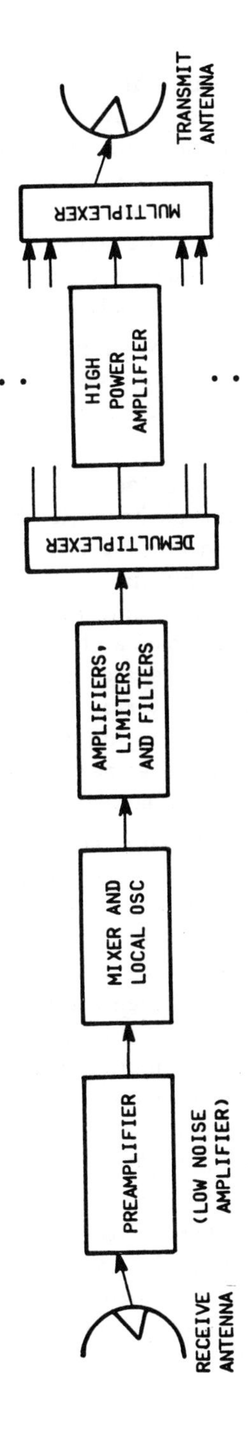

Fig. 12.5: Simplified Satellite Transponder Subsystem

Phased arrays are another form of antenna radiators. A phased array basically consists of radiating elements, phase shifters, power dividers, and a feeding network. The radiating elements can be waveguides, dipoles, microstrip, or any other type of radiating elements. The phase shifters are placed at each element level, and they are used to create the required phase front for steering the beam. The phase shift can be in discrete steps (digital) or continuous. In general, digital phase shifters are used with phased array for their compatibility with computer logics.

Because of higher losses and larger weight, phased arrays do not have wide applications as do reflector antennas to satellite systems. A small scale phased array may be used as a feed system to a reflector antenna for limited scanning applications.

Lens antennas are another type that focuses electromagnetic energy in a given direction. They offer multiple beam flexibility with a capability to null out undesirable interference in a given direction. Three types of antennas can be identified for this application. They are the dielectric lens, the metallic waveguide lens, and TEM transmission lens. The dielectric lens can be used in broadband applications. However, it is heavy and therefore has limited space applications. Like waveguide lenses, TEM transmission lenses operate on the different propagation delay for transmission lines with varying lengths. By the proper selection of the phases in each line, a desired beam can be generated in a given direction. These types of antennas have wide military applications.

Figure 12.5 shows the transponder subsystem which basically consists of the receiver, frequency converters, and the transmitter. The low noise amplifier (LNA) stage, which can be considered as part of the receiving system, performs the amplification of RF frequencies with a very low added noise to the system. These devices are characterized by their equivalent noise figure, which is defined as the ratio of the signal to noise at the input to the signal to noise at the output of a device. The noise figure can vary depending on frequency and device type. FET (Field Effect Transistor) devices have equivalent noise figures of 1.5 dB (1.41) or more at 4GHz, and they go up to 3.0 dB at 11 GHz. In addition to the FET type amplifier, tunnel diode amplifiers (TDA's) and parametric amplifiers are also used in this application.

The down conversion process, for a C-band system is simply the lowering of the up-link frequency of 6 GHz to the down-link frequency of 4 GHz with a mixer and a local oscillator operating a 2 GHz. The pre-amplificaion, mixing and post amplification and filtering are designed to operate over a wide bandwidth (500 MHz or larger for a C-band system).

A de-multiplexing process is performed prior to the amplification of each transponder channel through a high power traveling wave tube (TWT). Presently, TWT's are the commonly used amplification device. However, that is changing as the technology for solid state amplifiers improves. One of the disadvantages of TWT's is their nonlinear behavior with increasing input power. As we shall discuss in later sections, this non-linearity behavior increases the intermodulation interference between various channels for a multiple user system.

The TWT is basically a device that performs amplification through the transfer of power from an electron beam which travels along the center of a tube, to the RF field that travels in the helix that surrounds the electron beam. These devices can provide 50 to 65 dB of gain with a power added efficiency of 25% or better (power added efficiency is defined as the ratio of the power delivered to the rf field to the input dc power required to operate the device).

Finally, the telemetry, tracking and command (TT&C) subsystem is the key functional unit that continuously monitors the operation of the satellite and reports back the status of all essential equipment. It accepts ground controller command for routine position keeping, switches between redundant systems, and performs any number of other corrective measures. The telemetry and command channels have a very low data rate compared to the main transponder system. It is designed to be highly reliable.

12.1.2 Earth Station System

Figure 12.6 shows a block diagram of an earth station system with both transmit and receive channels. The earth stations can vary in size and complexity. In the early days of satellite communications, the satellites were small and delivered low rf power, and therefore, the earth stations were large and complex in order to attain the necessary condition for good performance. Typical antenna dish diameters were 30 meters or larger, which required a complex supporting structure and steering mechanism. Nowadays, communication by satellite is made using a variety of antenna sizes with different data requirements. The antenna sizes can be as large as the INTELSAT 33-meter diameter dish that handles high data rates with different traffic conditions and as small as a one-meter antenna that can be mounted on the rooftop of a house for television reception.

To avoid interference from terrestrial sources and from satellites that are in close proximity to the desired satellite (within 3° to 6°), a lower limit must be placed on the earth station antenna size, as well as strict

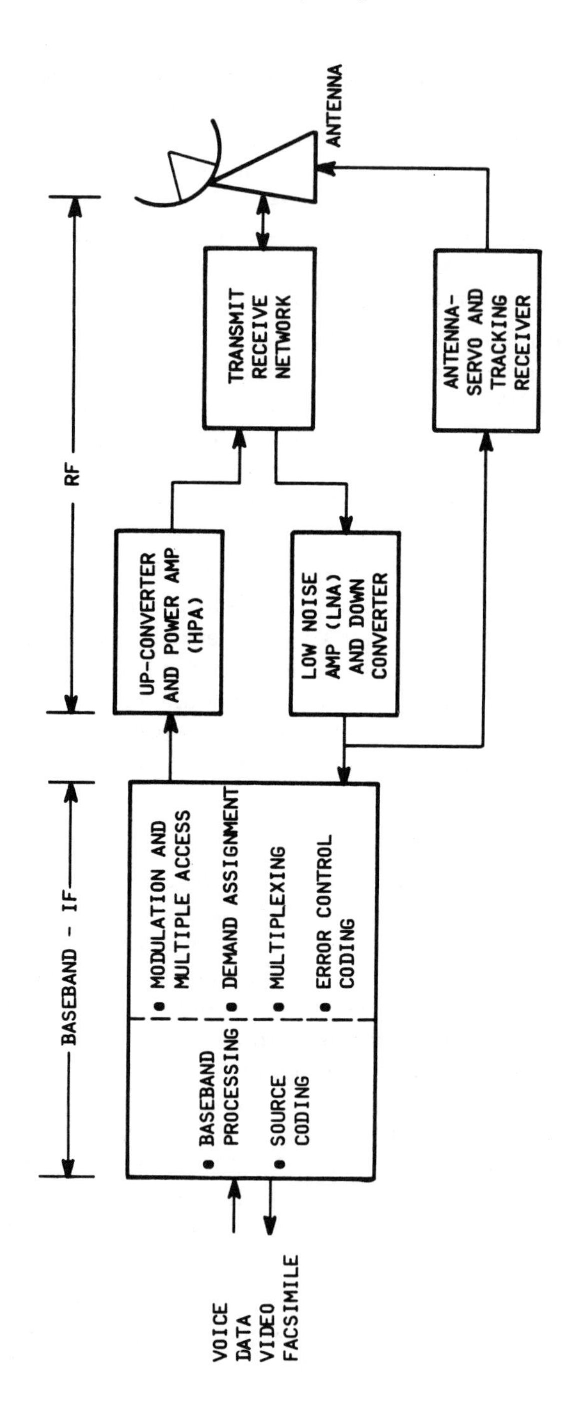

Fig. 12.6: Block Diagram of a Typical Transmit/Receive Earth Station with Tracking Capability

requirements on sidelobe level interference. Generally, the requirement for low sidelobes imposes design constraints on antenna construction which can translate into higher antenna cost.

Again, antennas are used to perform the front link to the communication system. The transmitter uses very high power TWT's or Klystrons. Typical transmitter power output ranges from a few hundred to a few thousand watts. As in the satellite system, the earth station system has low noise amplifiers in its receiving system. These are cooled or uncooled parametric, bipolar, or FET amplifiers.

Through a down conversion process in the receiving section, the RF signal is reduced to a lower frequency IF signal. The center of the IF signal is typically equal to 70 MHz for a 4/6 GHz system. The baseband processor then extracts the baseband information from this IF signal. The reverse process occurs when transmitting — that is baseband to IF then to RF transmission.

12.2 THE COMMUNICATION LINK

The previous section discussed the typical subsystem in a spacecraft and earth station communication link. It focused primarily on the general configuration of the satellite communications system rather than on a particular part of the system. This section specifically discusses the communication subsystem and the associated modulation and accessing schemes. It establishes the communication link requirement for some given sets of earth and satellite system parameters.

12.2.1 Link Budget Analysis

Figure 12.7 shows the basic communication model for the up-link system. The performance of this link can better be understood in terms of the carrier-to-noise density ratio (C/N_o) observed at the satellite, which is given by

$$\left(\frac{C}{N_o}\right)_u = (\mathrm{EIRP})_e \left(\frac{\lambda}{4\pi R}\right)_u \left(\frac{G}{T}\right)_s \frac{1}{kL_u} \tag{12.1}$$

where the subscripts *u, e* and *s* refer to *up*-link, *e*arth station and *s*atellite system, respectively. The other parameters are defined as follows:

C = received carrier power
N_o = noise density at the receiving end
$EIRP$ = effective isotropic radiated power
$EIRP = P_t G_t$
P_t = transmitted power of antenna input
G_t = transmitted antenna gain
λ = free space wavelength

R = distance from the transmitting to the receiving system
G = receive antenna gain
T = receive equivalent system noise temperature
L_u = atmospheric losses in the up-link
k = Boltzmann constant = 1.38 10^{-23} joule/°K

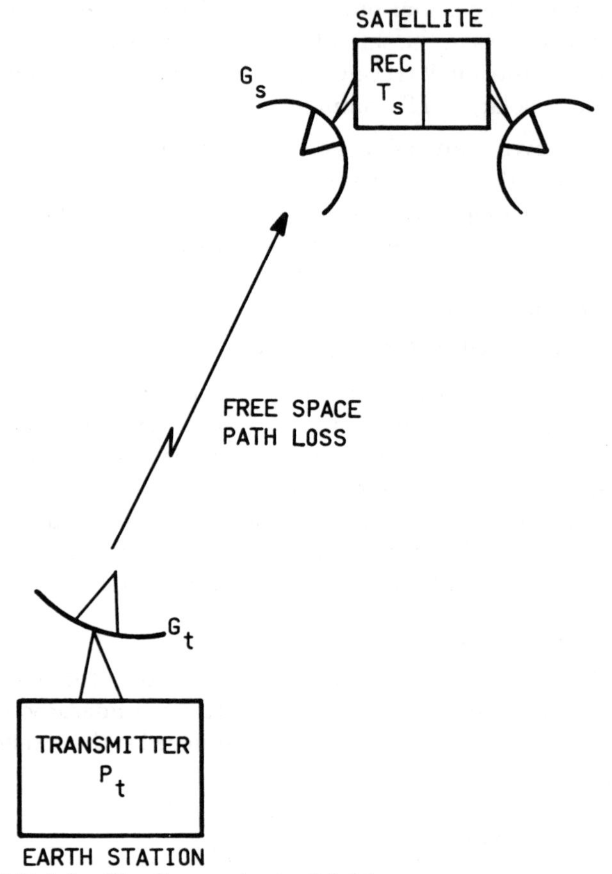

Fig. 12.7: Uplink Satellite-Communication Model

A similar equation may be written for the down-link system where the subscripts *u*, *e*, and *s* are replaced by *d*, *s*, and *e*, respectively. The second term in the right hand side of the above equation is usually referred to as the free space path loss and the third term as the figure of merit gain-to-noise temperature ratio of the receiving system.

A design trade-off is often made between the earth station EIRP and the satellite figure of merit G/T as well as the satellite EIRP and earth station G/T. This trade-off involves either placing more complexity into the earth station system with higher associated cost versus a more complex satellite system. Such a trade-off, however, is limited by the output power capability and the intermodulation noise generated by the satellite system. Earlier satellite systems development put heavy emphasis on the ground segment with modest complexity in the satellite system. With the recent successes of launching heavier and larger satellites into space, the trend is toward better space segment receiver and transmitter systems with simpler and less costly earth station systems.

When noises from all sources are assumed to be additive, the total carrier-to-noise density ratio of the receiving earth station can be written in terms of both the up-link and down-link C/N_o

$$\frac{1}{\left(\frac{C}{N_o}\right)_t} = \frac{1}{\left(\frac{C}{N_o}\right)_u} + \frac{1}{\left(\frac{C}{N_o}\right)_d} + \frac{1}{\left(\frac{C}{N_o}\right)_i} \qquad (12.2)$$

The third term refers to the equivalent intermodulation interference that is caused by the amplitude and phase non-linearity of the satellite Travelling Wave Tube Amplifier (TWTA). Typical input-output amplitude and phase characteristics of this amplifier are shown in Figure 12.8. A power backoff of 10 dB or more would place the system operation into a linear range having a low level of intermodulation products. However, this system would have a low power efficiency. In a multi-carrier system where a large number of users access the satellite simultaneously, there is an optimum point for efficient TWTA operation. The up-link carrier-to-noise density ratio is selected such that an acceptable level of intermodulation noise is introduced with a smaller TWTA power backoff. The intermodulation and cross-talk noise are very negligible when a single carrier accesses the satellite at any one moment.

Table 12.3 shows a typical link budget calculation for a down-link satellite system. Similar link budget estimates are usually made for the up-link system. The total carrier-to-noise density ratio is then estimated using equation 12.2. If the system is assumed to be limited by the down-link C/N_o, and depending on the type of modulation and access schemes used, this system may support over 800 voice channels with a transponder bandwidth of 36 MHz.

In an FM analog system, the demodulated signal-to-noise ratio (S/N) is related to the carrier-to-noise density ratio (C/N_o) discussed earlier, the carrier bandwidth, and the FM modulation index. Once the desired

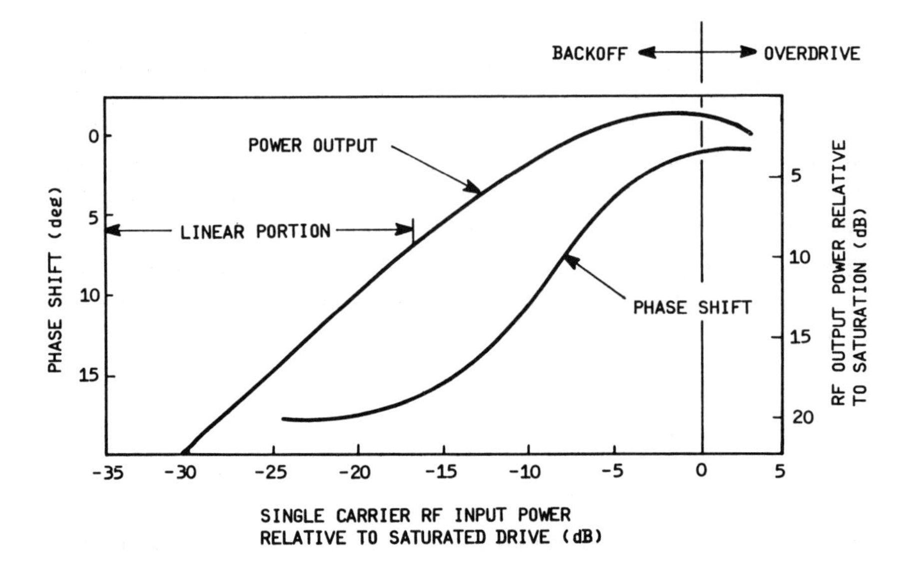

Fig. 12.8: Typical Power Output and Phase Characteristics of a TWT Amplifier

TABLE 12.3
Down-Link Budget for a 36 MHz Transponder Operating at 12 GHz

Spacecraft *EIRP*	48	dBW
Output Backoff & Losses	-3	dB
Free Space Path Losses	-205	dB
Atmospheric and Network Losses	-3	dB
Receiver Gain (2.5 m, 12 GHz)	47	dB
Receiver Noise Temperature (315°K)	-25	dBK
Earth Station G/T	22	dB/°K
Boltzmann Constant	228.6	dBW/°K/Hz
Down Link C/N_o*	87.6	dB-Hz

*This transponder system can provide 800 one-way telephone channel with each channel having a 45 kHz bandwidth.

signal-to-noise ratio S/N and the number of channels per carrier are established, a trade-off is made between C/N_o and the carrier bandwidth to determine the level of complexity in the system. In order to ensure an acceptable level of demodulation, the carrier power at the input of the receiver must be sufficient to overcome the noise in the demodulator. Typically, a minimum carrier-to-noise ratio of 13 dB is specified for an FM demodulator.

In a digital system, the channel data rate R is usually written in terms of the carrier-to-noise density ratio C/N_o and the energy per bit — per noise density ratio (E_b/N_o). This result is given as follows:

$$R = \frac{C/N_o}{E_b/N_o} \tag{12.3}$$

It is well known that the probability of bit error rate (BER) of a digital system is related to E_b/N_o. For example, in a coherent BPSK or QPSK modem, the required E_b/N_o to achieve BER of 10^{-5} is approximately 9.6 dB. In general, the functional relation between BER and E_b/N_o is a complex one, and usually a graphic representation of it is made for different modulation and coding schemes. The data rate R is therefore based on the type of modulation scheme used as well as on the carrier system performance — that is the value of C/N_o.

We have inserted in the link equation of 12.1, loss factor to account for atmospheric losses. These losses are path as well as weather dependent. At low elevation angles (i.e., close to the horizon), the path length through the atmosphere is very long; therefore, a higher signal attenuation is expected compared to higher elevation angles. Specific attenuation by atmospheric gases (uncondensed water vapor and oxygen) is shown in Figure 12.9.

The attenuation is also dependent on the yearly rain rate and density. The eastern part of the United States has higher rain rates than the western side. This requires the satellite link to have a higher system margin to account for the losses. In general, rain attenuation is not critical to system operation for frequencies below 10 GHz; however, it is very serious in the frequency range 20/40 GHz.

12.2.2 Modulation and Multiple Access

In communication systems, information is transmitted through modulation of a carrier signal. The carrier is a sinusoidal signal with amplitude, frequency, and phase as parameters for variations. If the variations in these parameters are continuous, the modulation is analog. If specific discrete variations are allowed, then the modulation is digital.

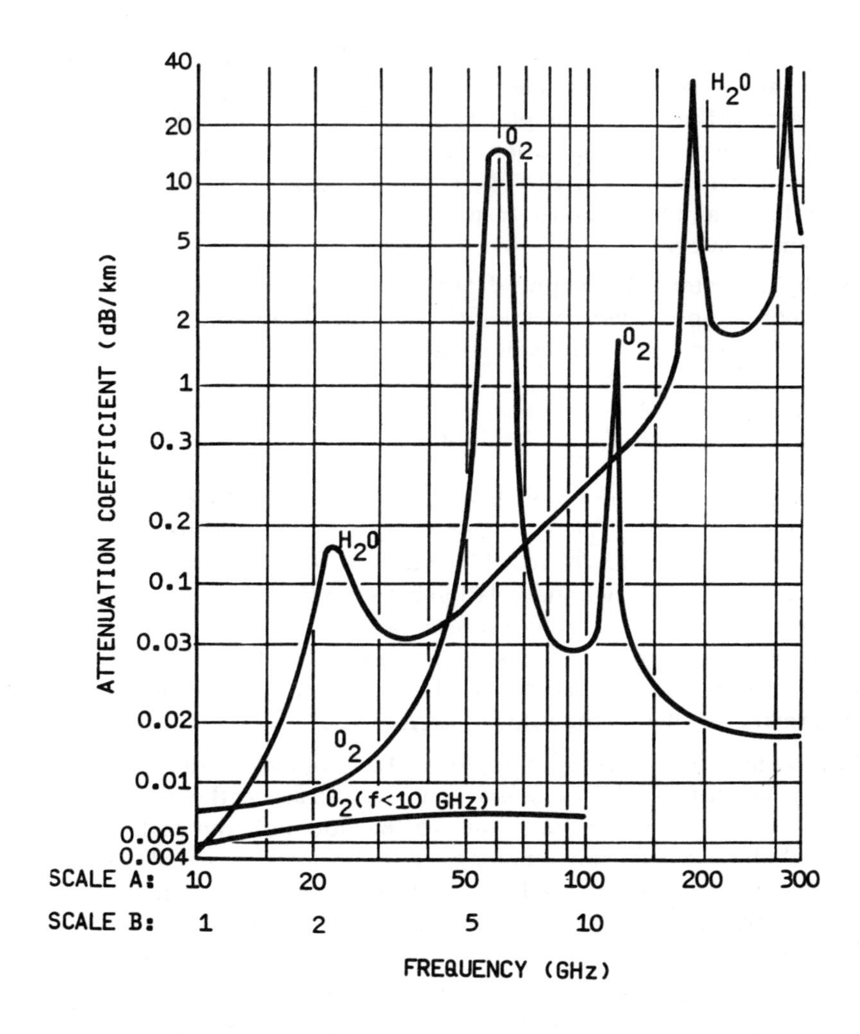

Fig. 12.9: Absorption by Atmospheric Gases

In satellite communications, the preferred types of modulation schemes are the FM and PM. This is due to the constant envelope of the FM, PM carrier compared to the time varying envelope of the AM carrier. A time varying envelope carrier would have a serious distortion through a nonlinear satellite channel. Both FM and PM are used in a digital satellite transmission while FM is generally used in an analog transmission.

Before a signal is modulated to a carrier, multiplexing methods are used to combine the various small channels. The two most commonly used methods of multiplexing are frequency-division multiplexing (FDM) and time-division multiplexing (TDM). Individual multiplexing techniques may be used with any modulation schemes such as FDM/FM, TDM/FM, etc. As will be shown later, a third method can be used in satellite systems with FM modulating a single carrier and with no multiplexing method. This method is called FM-SCPC where SCPC refers to single channel per carrrier. In a multiple access system where many earth station users access a single satellite, there are three primary methods of separating the communication link users:

i. Frequency-division multiple access (FDMA)
ii. Time-division multiple access (TDMA)
iii. Code-division multiple access (CDMA)

Each of the above access schemes is briefly described below.

Frequency-Division Multiple Access (FDMA).

This is the most widely used method. In this technique, each transmitting earth station uses a different carrier frequency to communicate to the satellite (i.e., to access the satellite). The isolation between channels of the various earth stations is achieved through frequency separation. The FDMA is popular because of its compatibility with existing terrestrial FDM/FM systems. The main disadvantage of this method is the creation of intermodulation and crosstalk interference. This interference noise is caused by the non-linearity of the satellite TWTA. In order to reduce the intermodulation noise level, the satellite amplifier must be backed off. This would cause a reduction in system efficiency. Generally, an optimized system would have a back off of 6 dB compared to a single carrier transponder.

Presently, there are two FDMA techniques in operation. In one technique, earth station frequency division multiplexes (FDM) several single side-band suppressed carriers into one baseband carrier that are FM modulated into an RF carrier and then transmitted through the satellite network using the FDMA method. This technique is called FDM/FM/FDMA. The other method is called Single-Channel-Per-Carrier (SCPC), where each telephone channel modulates a separate frequency carrier. This type of operation is referred to as SCPC-FDMA. It was developed for efficient usage of the satellite transponder as well as providing acceptable services to the smaller earth station users such as bush terminals.

There are two types of services using this technique. One uses the usual analog FM modulated signal on each carrier. The other uses digital modulation with Phase Shift Keying (PSK). The digital system is called SPADE system, which refers to SCPC, PCM, Multiple-Access, Demand-Assignment Equipment. The frequency allocations of a typical INTELSAT transponder system is shown in Figure 12.10. This system consists of 800 one-way telephone channels that are shared among all users of the satellite. The circuits are assigned on a demand basis which forms a temporary connection between any two earth stations using the satellite system. The SPADE system is therefore a demand assignment system for satellites.

Time Division Multiple Access (TDMA).

In TDMA, each earth station is assigned a time slot to access the satellite with all participant earth stations using the full-transponder frequency during their allocated time slots. At any one time, there is only one carrier that enters the satellite TWTA. In this way, the intermodulation interference is eliminated when the satellite is operated near saturation resulting in an increase in satellite capacity.

The increase in satellite capacity is achieved at the expense of having a more complex earth station system with additional equipment for time synchronization and frame control. Clearly, the TDMA system is more suited for terrestrial systems using digital transmission, although a hybrid system is possible. Satellite Switch-TDMA (SS-TDMA) will be implemented on future satellite systems. In such a system the satellite acts as a switch for directing the traffic from one zone or spot beam to a different intended zone beam, such as switching between the data leaving a spot beam to be routed into another spot beam. Figure 12.11 shows a conceptual design of satellite switched TDMA.

Code Division Multiple Access (CDMA).

In this method, the signal in each earth station is combined with its own pseudo-random code in such a way that they occupy the whole transponder bandwidth. Since all earth stations transmitting to the satellite have their carriers overlap each other, the intended earth station must cross-corrolate the received signal with the same pseudo-random code that is transmitted in order to recover the information content. In CDMA, precise frequency or time separation is not required as in the previous two cases. Since the signal is spread over the full bandwidth, the technique is called Spread Spectrum Multiple Access.

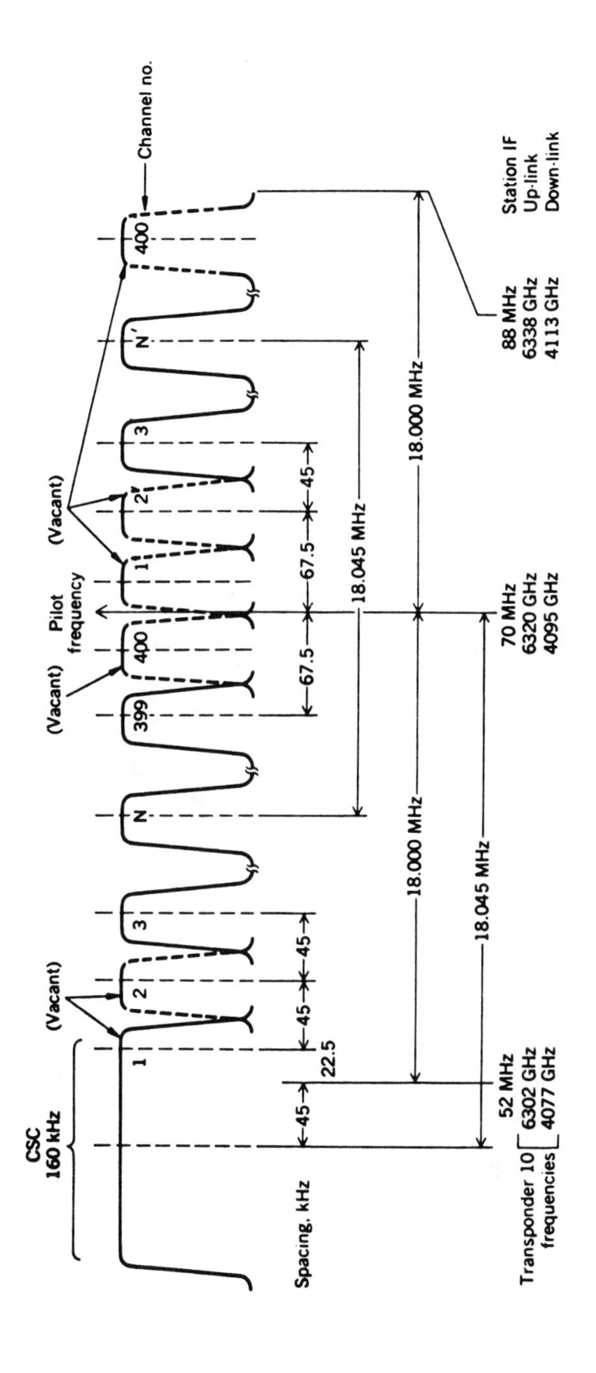

Fig. 12.10: The SPADE System Frequency-Allocation

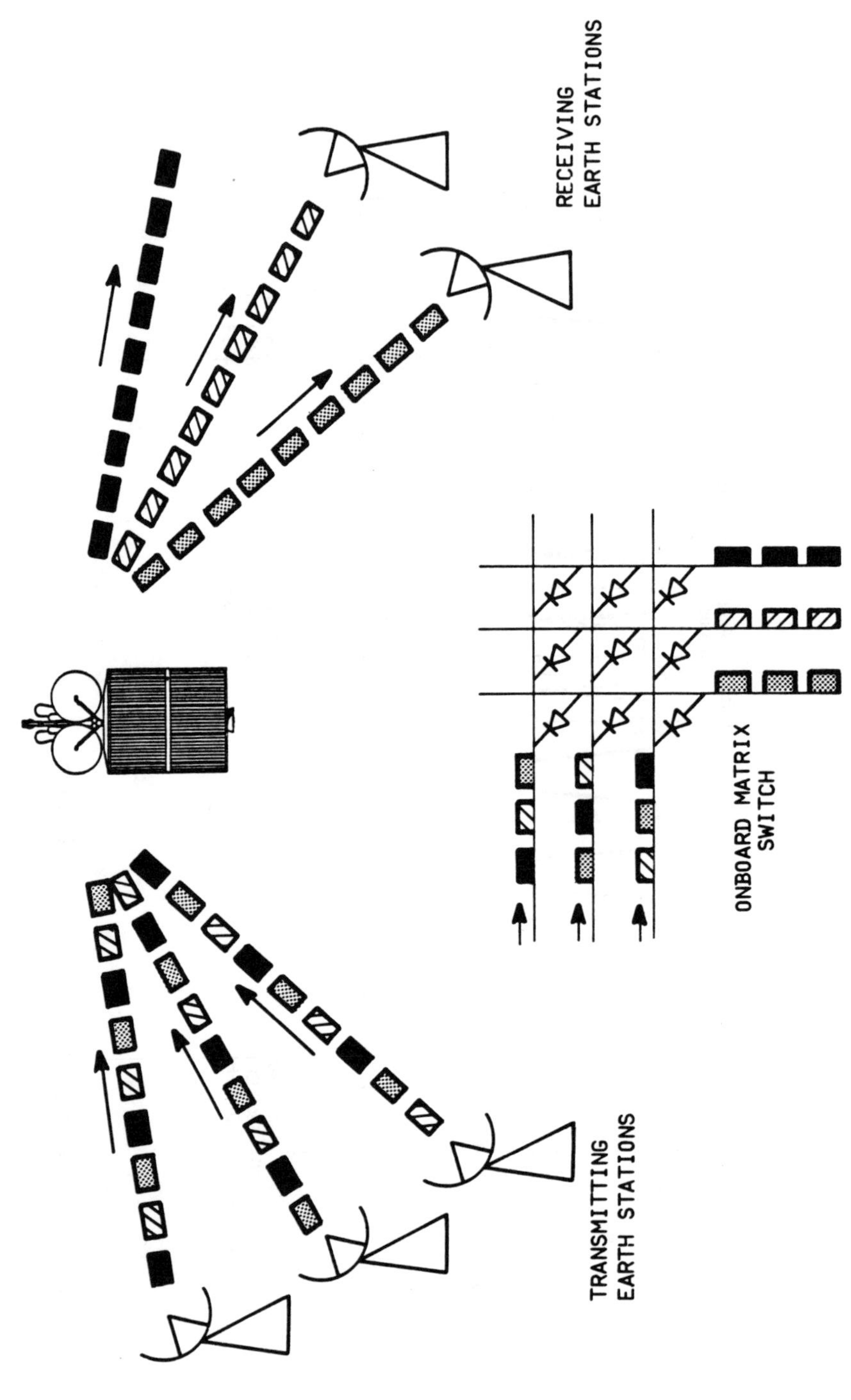

Fig. 12.11: Satellite-Switched Time-Division Multiple Access (SS-TDMA)

The CDMA has wide applications in military systems, although recently its popularity has been extended to small commercial earth station receivers with higher interference noise level. The use of spread-spectrum techniques increases the sytem resistance to intentional jamming and the use of encryptions add more security to the communication link.

Although the system provides jamming and interference resistant capability, its inefficient usage of the bandwidth makes it less practical in commercial applications. It is also a more complex system to implement than the simpler FDMA system. This system however will continue to be used in military communications. Figure 12.12 shows all three types of multiple access techniques.

12.2.3 Fixed and Dynamic Assignment

Channel assignment through a satellite network can be accomplished through a fixed or demand assignment. In the fixed or preassignment case, the channels are permanently assigned to the earth station users. In this case when the traffic is heavy, full utilization of the channel is accomplished. However, when the traffic is light, specifically during non-busy hours, lines which are idle are not available for other stations with heavy traffic. This results in poor utilization of the system.

Increased efficiency is achieved with dynamic assignment, where satellite channels are continuously pooled, and channels are seized on a demand basis. When the demand for a channel ends, it becomes available to others users. This allows efficient utilization of the satellite channels. The SPADE system is an example of a demand assignment system. In this system, each earth station puts its order through a common signaling channel shared by all participant earth stations and then seizes a vacant line on a first come first serve basis.

There are a variety of demand assignment schemes. Traditional ones are time assignment and polling techniques. A random access scheme which is a form of demand-assignment system for interactive data transmission is the ALOHA system. This scheme is highly suited for medium traffic with high peak to average ratio and many earth station users. There are three types of ALOHA Systems: pure random, slotted, and reservation. In the pure random, the system implementation is very simple. Each earth station transmits to a satellite at any time with no coordination with other earth station users. When the number of active users increases, the system throughput decreases because of the continuous collisions of packets at the satellite. The maximum channel utilization is 18.4%.

In the slotted ALOHA System, packets transmitted from an earth station start on a timing slot. This system requires a synchronized clock for the control of each starting slot. The maximum throughput of this system is twice that of the purely random ALOHA — that is 36.8%.

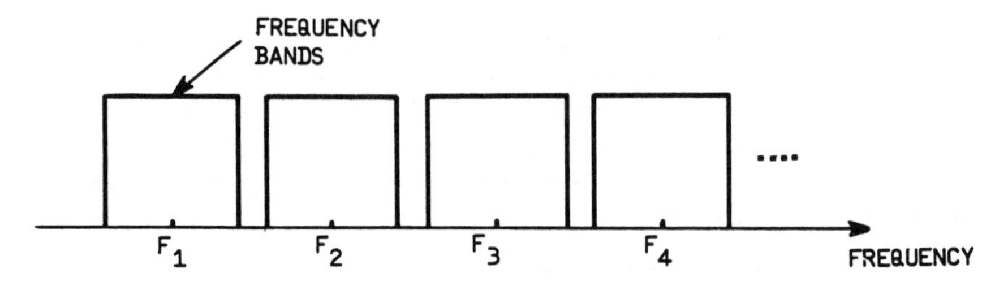

Fig. 12.12a: Frequency Division Multiple Access (FDMA)

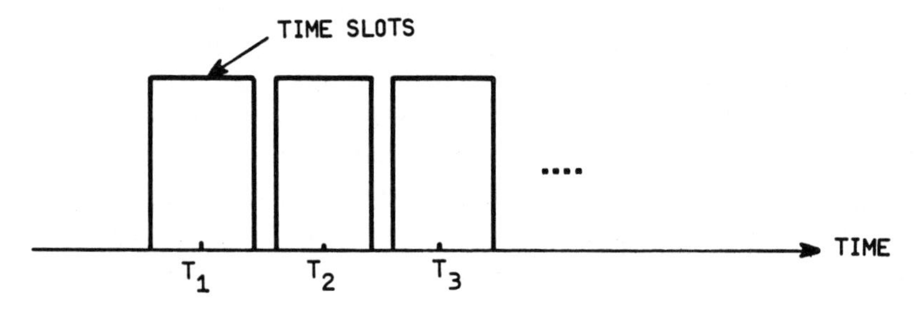

Fig. 12.12b: Time Division Multiple Access (TDMA)

Fig. 12.12c: Code Division Multiple Access (CDMA) or Spread-Spectrum

Both ALOHA Systems share the problem of increased transmission delays as the channel utilization reaches the maximum throughput. In order to keep the time delay low, the system throughput must be below approximately 15% for the purely random ALOHA and 30 % for the Slotted System.

If some form of reservation is made in part of the channel, a better channel efficiency is achieved. In this system however, the transmission delay is higher than in the other two ALOHA for low channel throughput (below 30%). If a full fixed reservation system is implemented in the system, the channel would act as if it is a TDMA with fixed slots dedicated to the participating earth station users. However, if a portion of the system is used in reservation and the remainder stays as in the conventional ALOHA, a higher utilization is expected. For example, if a system has 60% fixed reservation assignment with 100% full factor, and the remainder of the channel (i.e., 40%)is left to operate as slotted ALOHA with an efficiency of 30%, then the total system efficiency is 72%. Compare this with the regular slotted ALOHA with a peak efficiency of 36.8%. The reservation ALOHA offers good utilization of the system at the expnese of a more complex control mechanism.

In satellite systems ALOHA techniques are primarily used for computer communications (data transmission). It is less desirable in voice communication where real time interaction is required. A collided voice packet would require at least ½ second to be retransmitted. This time delay is intolerable for active conversation.

Other forms of packet transmission are the Round-Robin and the Priority-Oriented Demand-Assignment (PODA) protocols. In the Round-Robin protocol system each earth station, through a defined rotation, is allowed to transmit one packet in its allocated time slot. The PODA protocol is a demand assignment protocol which supports a variety of services such as variable message lengths, variable message destination, multiple delay classes, etc.

The PODA method can be implemented in TDMA, FDMA, or CDMA systems. SATNET is an example of a PODA system that was implemented on a single 64 Kb/s channel in the SPADE transponder using the TDMA channelization method. A frame structure of SATNET is shown in Figure 12.13. Each PODA frame is divided into an information subframe and a control subframe. The control subframe is used for channel reservations and assignments.

When a large number of low duty cycle earth stations or stations with unknown traffic requirements desire to access the satellite, a slotted

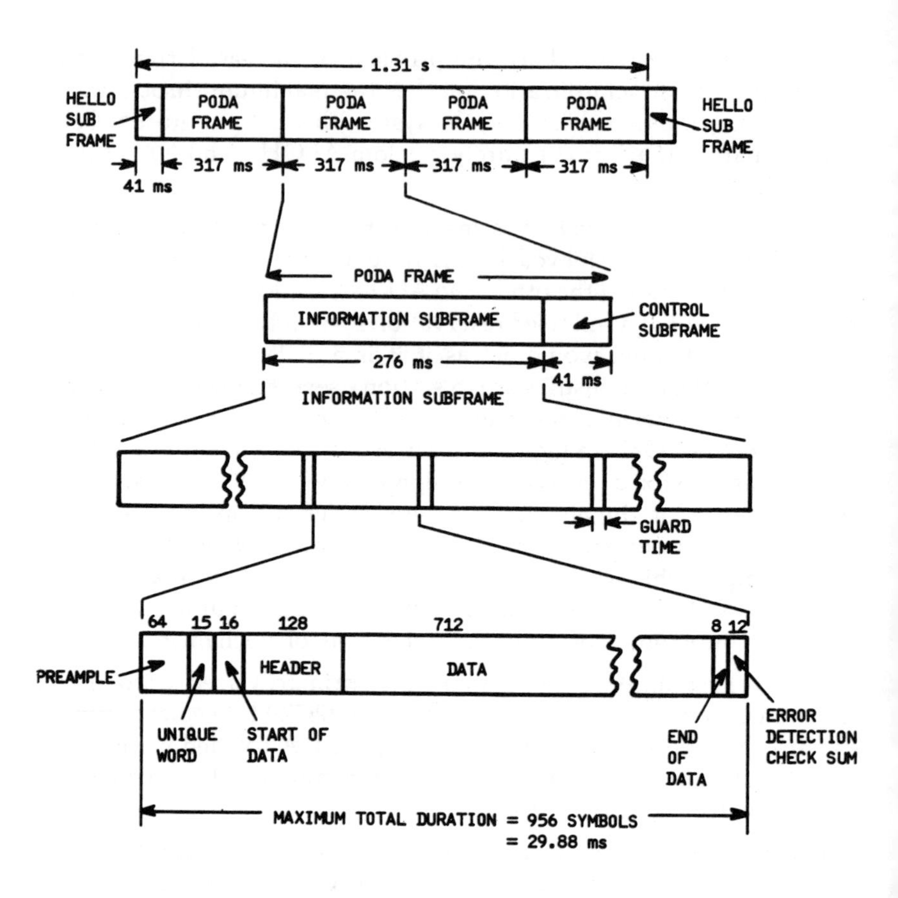

Fig. 12.13: The SATNET Frame and the PODA Subframe

ALOHA system is implemented in the control subframe. In this case the system is called Contention-PODA or CPODA. However, if the number of earth stations is small, fixed slot assignments are made in the control subframe and the system is called Fixed-PODA or FPODA. The advantages of this system are its flexibility and adaptability to varying demand requriements.

Table 12.4 summarizes the performance of the major methods of satellite techniques. The static or preassigned multiple access methods such as FDMA or TDMA have no data collisions and no major transmission overheads, but there are more empty slots when channel activities are low.

TABLE 12.4
Trade-off Performance Between the Various Multiple-access Techniques

Access Method	System	Collisions	Overhead	Empty Slots
Preassigned (Static)	FDMA TDMA	No	No	Yes
Random Access (Uncontrolled)	Pure ALOHA Slotted ALOHA	Yes	No	No
Dynamic Reservations	Polling Reservation	No	Yes	No

The random access methods such as pure random and the slotted ALOHA are simple methods to access the satellite with practically no overhead or empty time slots when large numbers of users are accessing the system. However, these random methods have a high collision rate which reduces channel efficiency. Finally, the dynamic reservation method, which has high control and transmission overhead, is a more complex method to implement, but the system is efficient with no wasted capacity to collision or empty slots.

12.2.4 Error Control Coding

In a digital satellite system one-bit error in 10^3 bits can be tolerated for voice communication. This level of error rate is unacceptable, however, for data transmission between two computers. Therefore, the design of satellite communication systems requires some form of error control mechanism.

In terrestrial links, the commonly used method of error control is the automatic repeat request (ARQ). Once an error is detected, the receiving system asks for retransmission from the sender. There are two types of ARQ systems: (1) stop-and-wait ARQ and (2) continuous ARQ. In the stop-and-wait ARQ, the sender waits for an acknowledgment from the receiver on the status of the transmitted block. If it is positive, then the sender transmits the next block; however, if it is negative, it repeats the transmission of the blocks with errors. This system is highly undesirable in a satellite system where a round trip delay of half a second is required for the reception of the acknowledgment. A more acceptable form of ARQ is the continuous ARQ. In this system, the sender sends the blocks and receives the acknowledgment continuously. Once a negative acknowledgment is received, the transmitter sends either the block with an error and all blocks that follow it, or sends only the block that has the error. In the case of retransmission of the block with the error and the block that follows, the system is called go-back N ARQ; the method which sends only the block with the error is called selective-repeat ARQ. The selective-repeat ARQ requires large buffers at the transmitting and receiving terminals as well as control for blocks reordering. The go-back N ARQ has its complexity at the transmitting terminal only where large buffer and control logics are required.

The other form of error control which is commonly used in satellite systems is the forward error control (FEC). In this system, extra bits of data are added to the blocks for error checking and correction. There are two types of FEC methods, the block codes and the convolutional codes. In block codes, the stream of data is broken into k symbols of information and n-k redundant symbols for error control where n is the total length of the code word. The resultant system is referred to (n, k) block code with a symbol rate of k/n.

There are many forms of block codes, and the most popular ones are the cyclic codes. These codes can simply be formed from cyclic shifting of bits of data in a block, thus creating a new code word. The two most popular cyclic codes are the BCH and the Golay codes. Their performances are shown in Figure 12.14, where for a probability of error rate of 10^{-5}, the E_b/No required in BCH code (rate 7/15) is 8.8 dB and for the Golay code (rate ½) is 7.7 dB. The gain advantage over no coding is 0.8 dB and 1.9 dB for the BCH and Golay codes, respectively. It should be noted that the BCH codes have better gain advantage for larger blocks of data. However, in overall system, performance coding offers a gain advantage over no coding. This gain advantage is achieved at the expense of a larger bandwidth requirement.

Convolutional codes, which are sometimes called tree codes, are another type of FEC coding. In convolutional coding, the encoder consists of a K-stage shift register and v module-2 adder. Each of the v module-2 adders are connected to some part of the shift register which determines the type of codes. The information symbols are inputed to the shift register one symbol at a time. The output of the v module-2 adders are sampled by a commutator to produce v output symbols. The coder rate is therefore $1/v$. Now, if k symbols are shifted in the K-stage shift register, then the code rate would be k/v. An example of a convolutional encoder with a 3-stage shift register and rate ½ is shown in Figure 12.15a. Its associated tree output is shown in Figure 12.15b. As an example, the encoder generates the sequence 11100001 for an input sequence of 1011.

Decoding of convolutional codes is the most complex operation because of the lengthy search operation required to determine the likely path the code has followed. The most widely used decoding techniques are the sequential and Viterbi. The performance of these decoders is shown in Figure 12.14. Both systems offer high performance and an efficient decoding mechanism with high gain advantage.

The rapid progress in solid state technology with the fast development of VLSI chips, makes it very attractive to have systems with complex but highly reliable codes. This is particularly useful in power limited systems, where bandwidth can be traded off for gain in E_b/N_o or, in other words, gain in signal to noise ratio.

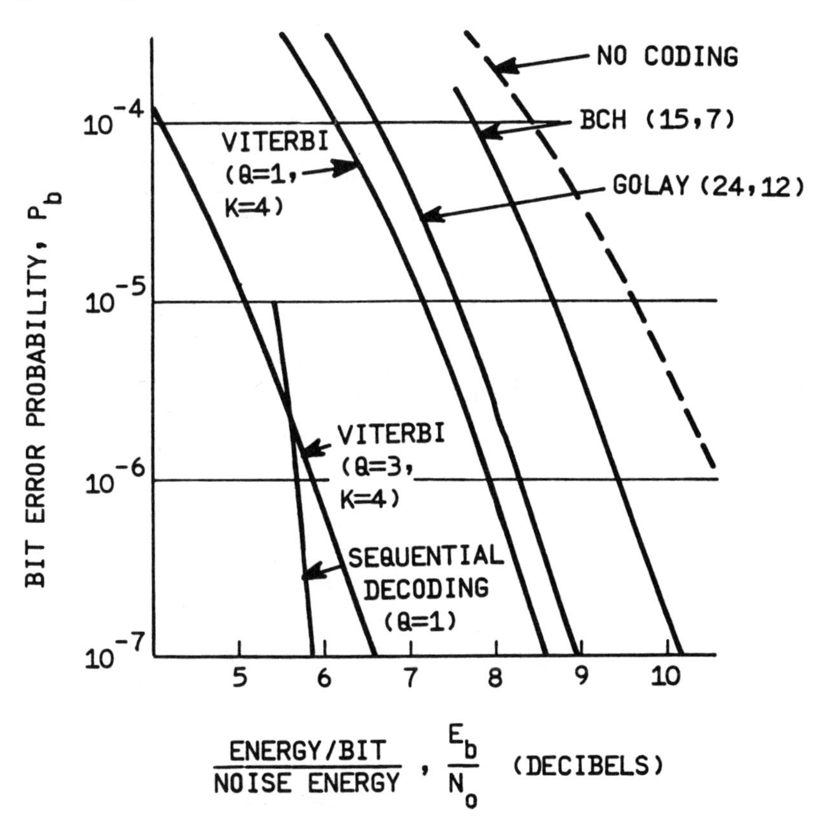

Fig. 12.14: The Performance of Various Error Correcting Codes

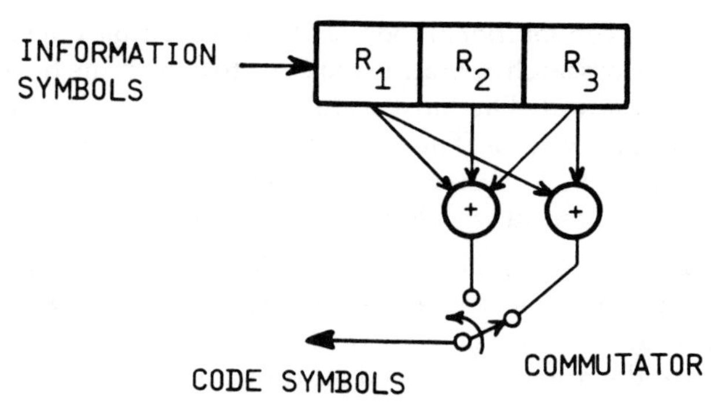

Fig. 12.15a: Convolutional Encoder with 3 Shift Registers and Rate — 1/2

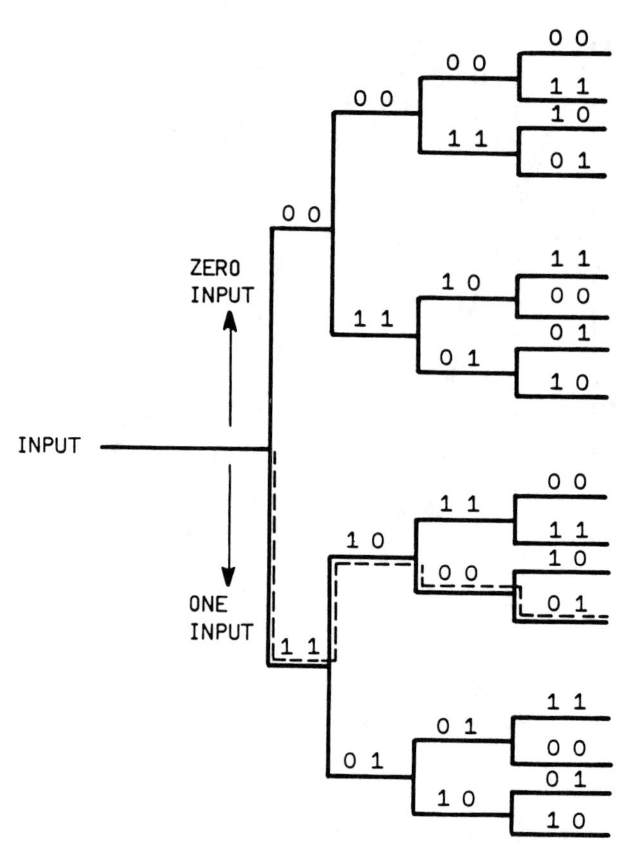

Fig. 12.15b: A Code Tree Diagram. An Input Sequence of 1011 Generates the Sequence 11100001

12.2.5 Digital Source Coding

The future trend in satellite communications is towards using digital systems. This is because digital systems are more resistant to random noise and interference and because of the availability of wider bandwidths in satellite channels. Voice or video transmission through a satellite channel is usually achieved using analog methods. However, the rapid progress in VLSI development makes it possible to digitize voice and transmit the voice data rather than the full analog voice signal. There are many forms of digital coding of analog, but the most widely used is PCM (pulse code modulation). In this method, the waveform is sampled at twice the highest frequency in the wave (the so-called Nyquist frequency) and for each sample 8 bit levels are used to sample the amplitude. In voice communication where the highest frequency is about 4000 Hz, the Nyquist sampling rate would be 8000 samples per second. Now if each sample has 8 bit levels, then the total required bits of information is 64000 bits/sec (i.e., 8bits/samples x 8000 samples/sec = 64000 bits/sec). This PCM digital coding method is widely used in present day terrestrial and satellite systems.

Two other methods which require lower numbers of bits per second to achieve the equivalent voice quality of PCM are Differential Pulse Coded Modulation (DPCM) and Delta Modulation (DM). DPCM takes advantage of the strong correlation between successive samples of the speech signal and transmits only the difference between present and previous samples. One form of the DPCM coding scheme is the ADPCM or adaptive DPCM. In the ADPCM the sample levels are adaptively selected rather than having equal fixed sampling levels as in the standard PCM. The adaptive DPCM has a toll quality with only 24000 bits/sec with no major complexity in the hardware logics.

The DM is another form of digital coding with wide application in satellite systems. The DM modulation scheme exploits further the signal correlation in speech by transmitting only an indication of the difference between samples —that is, if it is positive or negative. Therefore, what is required is a single bit level to determine the signal slope; however, a sampling rate higher than the Nyquist rate is required for better signal transmission. One form that has excellent signal performance is the Adaptive DM (ADM) with a rate of 32000 bits/sec or lower which achieves the toll quality of the 64000 bits/sec PCM. The ADM has very low hardware complexity.

With increased complexity, toll quality at 16000 bits per second is possible. Presently, with the rapid progress in integrated circuits, the hardware is not a prime concern compared to the savings in channel bandwidth and signal-to-noise advantage. The future holds great promise for digital telephony and digital video transmission.

12.3 APPLICATIONS

Presently there are many types of services in business and government that use satellite networks either directly or indirectly. Those selected in this section represent the major applications in this technology. Specific military or mobile satellite systems are not addressed here.

It is anticipated that satellite service will continue to have wider applications in long distance telephone services with potential growth into the video market, such as television transmission or video teleconferencing. Direct broadcast satellites will have a major impact on the home television market with a projected user population of over 30 million. It is also anticipated that in the late 1980's or 1990's an integrated digital network with a satellite as the long-haul link will be implemented. Such a system is conceptualized in Figure 12.16. The following section addresses the major applications of satellite communication technology.

12.3.1 Voice and Data Transmission

Earlier satellites have been used mainly for analog voice transmissions. The typical capacity of a 36 MHz transponder using FDM/FM/FDMA ranges from 400 to 800 one-way voice channels. The standard FDM multiplexing with FM modulation was used for its simplicity and compatibility to existing terrestrial networks. An outgrowth of this system was the SCPC network with its wider application to thin route terminals.

Presently, the trend is toward digital transmission using TDM/PSK/ TDMA systems. Such systems have the advantage of efficient usage of the satellite channel as well as accommodating the growing demand for data transmissions. This growth is spurred by the demand for fast computer transmission, facsimile, electronic mail, and video telconferencing.

Digital coding of speech such as ADM and ADPCM have provided excellent voice quality at a low data rate (16to 32 kb/s) with essentially no complexity in the coded design. Digital Speech Interpolation (DSI) such as TASI or SPEC improves the channel capacity by a factor of better than 2. These two methods are presently used for efficient utilization of the satellite channel for digital voice transmissions.

Reliable data transmission with a probability of error rate lower than 10^{-8} can currently be achieved using a convolutional encoder with a soft decision Viberbi decoder or hard decision sequential decoder. Block codes such as BCH or Golay are still used in satellite channels with higher carrier-to-noise density ratios. Generally, using error coding would increase channel resistance to noise, but at the expense of wider

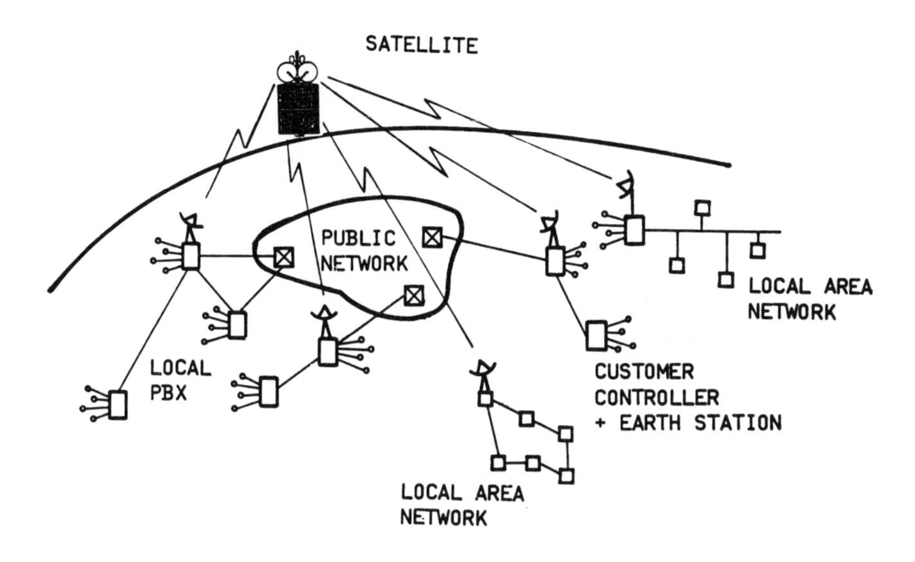

Fig. 12.16: An Integrated Digital Satellite Network[3]

bandwidth requirements. Error coding is used in many satellite systems, however, it is very beneficial to small power limited terminals.

With the current low cost of computer-based processors, an integrated digital satellite network with voice, data, and video transmissions can be expected at the turn of the century. Although voice and data flow requirements are different, some type of priority system may be implemented to accommodate both modes of operations. This can be accomplished at the expense of more sophisticated control logics which are presently at very low cost.

The INTELSAT V is a system which can support both analog and digital data transmission using FM and QPSK modulations and both FDMA and TDMA access schemes. The system operates in both the 4/6 and 11/14 GHz with fourfold reuse (beam separation and orthogonal polarization) in the 4/6 GHz and twofold reuse (beam separation only) in the 11/14 GHz. The system can support over 12,000 voice circuits plus two TV channels. An example of an all digital system is SBS (Satellite Business System). This system provides voice, data and image services. It utilizes a QPSK/TDMA transmission method with Demand Assignment to achieve efficient channel capacity. The system operates in the 12/14GHz bands using 5 or 7 meter earth station antennas. It has 10 transponders with each transponder supporting a data rate of 43 Mb/s.

12.3.2 Teleconferencing

With the higher cost of travel, video teleconferencing is emerging as an alternative method for direct face-to-face business meetings. The projected demand for video conferencing using satellite channels is expected to increase from 25 in 1984 to 40,000 channels in 1995. If each channel would occupy 7.2 MHz of bandwidth, then the increase is approximately 5 transponders to 8000 transponders by 1995. (One transponder is assumed to be 36 MHz of bandwidth). Such tremendous growth can only be met using more satellites per orbits, frequency re-use, and use of the higher frequency band — that is the Ku and the millimeter wave bands with the availability of wider bandwidths.

There are presently two types of video conferencing, the slow scan and full motion pictures. In the former system, no large bandwidth is required, so that regular telephone lines can be used for tansmitting the very low data rate signal. Nevertheless, in the general business or government applications, full motion teleconferencing is of prime interest for having effective business meetings. The full motion video/audio signal occupies approximately 7 MHz of bandwidth. This bandwidth may not be easily available using terrestrial lines to simultaneously access many other users. Therefore, satellite networks with their broadcasting capability and their wide band channels are the most suitable systems for video conferencing. It is expected that larger capacity satellites with multiple frequency bands and multiple-spot beams with SS/TDMA switching will be implemented to accommodate the anticipated growth in this area.

The main component of the system is the coder/decoder, which exploits the small change in the video signal between frame to frame (Interframe Coding) to reduce the high data rate required for video transmissions. A compression ratio of better than 10:1 over the standard PCM rate can be achieved using these codes. Typically, these codes operate at data rates as low as 6.3 Mb/s. The earth station terminal can be installed on the user's premises or in some cases, terrestrial lines with possible satellite interconnections are used to transmit the video signals.

12.3.3 Local Area Networks

Satellite systems can play an important role in connecting various types of local area networks. This allows efficient sharing of the resources in the network using the wide band, low cost, satellite services. One way of interconnecting local area networks is to use network gateways. These gateways can accept data from the local distribution networks, add the necessary heading, and transmit the data to a distant gateway using satellites as an interface. The receiving end in the distant gateway converts the information back to the signal local protocol for delivery to its proper destination. The use of gateways usually simplifies the long distance network architecture and minimizes the number of different devices and networks that interface to gateways. Figure 12.16 shows several local area networks tied together through gateways to the satellite networks.

Generally, local area networks use the inexpensive wideband cables for their interconnection with less concern for bit saving for internetwork protocol operation. However, in using satellite networks, the emphasis is on address bits saving for efficient channel utilization. The two systems can be made compatible by using the standard long-haul addressing system. However, additional bits may be added to the address once the data are in the local distribution network.

Satellite networks offer the necessary speed matching to the local networks, A satellite network has the wide bandwidth for achieving high data rates suitable for long distance communication between local networks. Earth station gateways with reasonably large buffers may be used to maintain speed matching between networks with different speed requirements.

The economic incentive for using satellite systems in local area networks is great. Terrestrial line connections are costly, and they do not offer the network flexibility found in satellite systems. A low cost dedicated earth station to a local area network will eliminate the need for costly carrier-provided lines. This earth station would have a direct access to other local area networks within the satellite footprint. As was discussed in previous sections, the cost for a satellite channel is continuously dropping with the implication for more satellite usage in long distance communication.

12.3.4 Broadcast Services

Satellites can provide a wide ground area coverage. This capability makes satellites different from line of sight communication in that they use the broadcast mode. Until recently satellites had low or intermediate power transmitters. The limitations were due to the difficulty of launching larger satellites with larger solar array for generating the necessary dc power as well as implementing larger antennas with high gains. Presently the technology for launching these types of satellites has matured. Higher power satellites reduce the burden on the design for complex ground receivers. Currently simple ground receivers with a 1 meter diameter antenna operating at 12 GHz can be designed to receive direct satellite signals with good quality television pictures.

In this application, the major design burden was placed on the satellite with minimum complexity at the earth station receivers. This type of satellite application is referred to as Direct Broadcast Satellites (DBS). These services are planned for the year 1985 and beyond.

The ground system has the small rooftop antenna and a simple receiver which converts the FM satellite signals to the standard AM commercial TV receiver. The receiver figure of merit G/T is expected to be *12* dB/°K with a satellite EIRP of 56 dBW. The receiver's noise figure is about 4 to 5 dB. The cost for these receivers as well as simplicity in installation will play a major role in attaining the necessary market.

In the international arena this medium of communications will have broader political and cultural implications, where such broadcast services if transmitted with no international regulation, can create problems in countries or regions of the world which are not interested in the free flow of information. In the United States, DBS will allow the large networks direct access to the home television receivers, thus bypassing local television stations. This will provide greater economic benefits to the major networks.

12.4 FUTURE OUTLOOK

In the commercial sector, the demand for high quality voice and data transmission is expected to grow for the next 20 years. This growth trend is also expected to continue in services using satellite transmissions. It is projected that there will be more demand for new types of services that use satellites such as video teleconferencing, electronic mail, or electronic fund transfer.

In order to meet this growning demand, the 1979 WARC allocated more bandwidth for satellite services at the higher frequency bands. Presently, the technology is moving toward using the 11/14 GHz and 20/40 GHz bands with a frequency reuse of the present allocation of 4/6 GHz band.

The INTELSAT had only two transponder systems with 240 telephone circuits. This compared to the large capacity of present day satellites such as INTELESAT V that can support up to 12000 voice circuits in addition to 2 TV channels. It is expected that larger satellites with higher power capability and switched multiple spot beams will continue to service the long distance communication users. Currently, the trend is toward implementing processing on board satellites to improve the overall communication system efficiency.

Intersatellite links using the millimeter wave band (60 GHz) or optical links will increase the effective utilization and connectivity of future satellite links. The implementation of such systems is expected in future satellites.

Satellite systems will have wider legal and political implications when used for direct television transmission to the home users. Direct television broadcasting using intermediate power satellites is already in service. It is expected that full direct-broadcasting services will be available to every home user by 1986. The expansion of this service will have a major impact on the services of the local networks.

Commercial mobile satellite communication is also becoming of prime interest in interconnecting fixed and mobile terminals in ALASKA and CONUS to each other and to the standard telephone network. The WARC 79 has already designated bands in the 800 MHz frequency band for mobile satellite communications. This system is expected to operate on a DAMA basis (i.e., Demand Assignment Multiple Access) similar to the INTELSAT SPADE system. This system can have wide application in the areas complementing the presently growing area of cellular radio.

In military systems, satellite services are increasing the pace of connection as well as improving the security between widely separated communication users. Anti-jam (AJ) with low probability of intercept (LPI) satellite system operating in the millimeter wave band will dramatically improve the communication system to a hostile environment.

The global implication in the usage of satellite systems is to increase the information exchange between nations as well as to understand one another better. The projected growth to over 400,000 two-way voice circuits by 1995 provided by the INTELSAT system is an indication in the world community's desire for a better communication medium.

REFERENCES

1. Van Trees, Harry L., Editor, *Satellite Communications,* IEEE Press, 1979.
2. Martin, James, *Satellite Communications Systems,* Prentice Hall, 1978.
3. Roy, Rosner D., *Distributed Telecommunications Network,* Lifetime Learning Publications, 1982.
4. Feher, Kamilo, *Digital Communications,* Prentice Hall, 1983.
5. Gould, R. G. and Y. F. Lum, Editors, *Communications Satellite Systems,* IEEE Press, 1976.
6. Gagliardi, Robert, *Satellite Communications,* Lifetime Learning Publications, 1984.
7. Tugal, Dogan and Osman Tugal, *Data Transmission,* McGraw-Hill, 1982.
8. Special Issue On "Packet Communication Networks," Proceedings of the IEEE, Vol. 66, No. 11, Nov. 1978.
9. Lin, Shu and Daniel J. Costello, Jr., *Error Control Coding,* Prentice-Hall, 1983.

APPENDIX

BASIC MATHEMATICAL AND ENGINEERING CONCEPTS

Leonard Lewin
Department of Electrical and Computer Engineering
University of Colorado
Boulder, Colorado

A.0. INTRODUCTION TO THE APPENDIX

It has been our experience that many students, not excluding those with degrees in technical subjects, often lack a necessary fluency in relatively elementary operations in mathematics and physics. It is not that these individuals haven't been exposed to the material at some time or other (although occasionally this may have been so) but that oftentimes it has been little used, or has been largely forgotten, or perhaps was never properly understood in the first place. This imposes an undesirable handicap on the individual, and in order to remedy the shortcoming, and also to provide essential material for revision, a course entitled "Introduction to Communication Systems Theory" was instituted. The material of this course is summarized in the nine sections of the following appendix. It should be made clear that the purpose is to introduce the use of mathematics as a working tool, and that the key feature of the course is to gain a working familiarity with the subject. It is no part of the intention to provide rigorous mathematical proofs; and in fact it is probably the case that for many students a premature involvement in rigorous theorem proving may have been off-putting, and the cause of much aversion to mathematics. This is a pity, because mathematics can be used very powerfully as a tool in many ways, and a competency in this area can prevent both misunderstandings and also an awareness of possible misuse. (The suspicion that many people have for statistics is but one example of this.) Here is probably not the place to discuss the general teaching of mathematics, but the remarks by one student that "for the first time I can see what the calculus is all about — the explanations are not all bogged down in a mass of 'epsilonics' and other things" may indicate that, for engineering students, at least, a more pragmatic approach to teaching mathematics could be potentially rewarding. In line with this, the "proofs" in the following sections should be viewed more as heuristic demonstrations of plausibility, and as indications as to where results come from, and what their relevance to the subject matter may be. It is not my wish in any way to disparage the rigorous mathematical treatment; in advanced work the subtleties are often needed, and failure to observe the finer points is often a cause of error. But I feel that mathematical rigor, for many students, is something to be built up later on a foundation that has maybe been laid in a different way. To some degree this appendix may be helpful to those students whose earlier involvement in this area may have been incomplete, or whose fluency in the subject may be deficient.

A.1. BASIC PHYSICAL RELATIONSHIPS

A.1.1 Units and Dimensions

The three basic mechanical units are of length, mass, and time. In the MKS system they are measured in meters (m), kilograms (kg) and seconds (s) respectively. Occasionally, particularly in magnetic units, one meets features derived from the earlier cgs (centimeter, gram, second) system, and this difference is largely responsible for powers of 10 that enter into some formulas.

Although these basic units are now defined in terms of atomic constants they are specified from semi-arbitrary physical standards.

Two other units needed are of temperature and electricity. The former is the degree Celsius (°C) and the latter the ampere (A).

These five basic units are all independent in character; none can be derived from the others. In contrast, many other units, e.g., velocity, can be expressed in terms of them. In particular, no independent magnetic unit is needed, since magnetic effects are expressible in terms of the others. Units so expressible are called *derived* units.

A.1.2 Multiples and Submultiples

The following multiples and subdivisions of units are the ones most commonly encountered. They are based on powers of a thousand, and are used in order to avoid small numbers, strings of zeroes or powers of ten in statements of results. When using a formula which is self-consistent in a certain set of units (practical units) the subdivision or multiple should ALWAYS be first converted into the relevent unit times a power of ten. Otherwise the formula, based on a self-consistent set, will yield incorrect results. (This stipulation does not apply to some practical engineering formulas where the dimensions may be *explicity* required to be in some other unit, e.g., MHz, miles, etc.). Needless to say, non-metric units must first be converted to metric values before MKS formulas are used.

When using a formula, the units of the quantity being derived should be stated *after* the formula or calculation. It is a source of confusion to allow the units themselves to be caught up in the calculations, and is also unnecessary since the formula, being self-consistent in character, ensures that the units of the calculated quantity are automatically looked after.

TABLE A.1 Multiples and Submultiples

Multiple			Submultiple		
k	kilo	10^3	*m*	milli	10^{-3}
M	mega	10^6	μ	micro	10^{-6}
G	giga	10^9	*n*	nano	10^{-9}
T	terra	10^{12}	*p*	pico	10^{-12}

Do not confuse M as designated here with M as used for mass (in a formula) or (as occasionally used) to denote miles. Similarly avoid confusing m for milli with m used to denote meter. (They occur together as mm, millimeter.) Submultiples are denoted by lower-case letters and multiples by capital letters (k for kilo is an historical exception).

Occasionally, other subdivisions are encountered, such as centi (c) = 10^{-2}, as in centimeter (cm); deci (d) = 10^{-1}, as in decibel (dB).

The unit of mass in the MKS system is the kilogram, and is treated *as if* it were a basic integral unit. (Historically the gram was the basic mass unit and the kilogram is therefore, really composite.)

A.1.3 Derived Mechanical Units

The important derived mechanical units are

(a) Frequency, denoted by f, measured in hertz, Hz. A hertz is one cycle per second, s^{-1}.

(b) Velocity, measured in meters per second, m/s or ms^{-1}.

(c) Acceleration, measured in meters per second per second, (m/s)/s or m/s^2 or ms^{-2}

(d) Force, measured in newtons, N. One newton is the force necessary to produce an acceleration of one meter per second per second in a mass of one kilogram; its units are kg ms^{-2}.

(e) Work or Energy, denoted by W, measured in joules, J. One joule of work is done if a force of one newton is applied, in the direction of the force, for a distance of one meter; its units are kg m^2 s^{-2}.

The cgs unit of energy is the erg. Since there are 10^2 centimeters in a meter and 10^3 grams in a kilogram, one joule = 10^7 ergs.

(f) Power, denoted by P, measured in watts, W. Power is the rate of doing work. One watt is the power when one joule of work is performed per second; its units are kg m^2 s^{-3}.

A.1.4 Derived Electrical Units

The speed of light in a vacuum is a mechanical measure which provides a link with electromagnetic phenomena. It has the value

$$c = 2.99792458 \times 10^8 \text{ m/s} \tag{A.1}$$

which is independent of the frequency (color) of the light, and is also independent of the state of motion of the observer. The value in (A.1) is usually approximated by the abbreviated value $c = 3 \cdot 10^8$ m/s, and is the source of the number $30 (= c \cdot 10^{-7})$ that occurs frequently in electromagnetic formulas. The main electrical units are defined in relation to mechanical units as follows:

(a) Electric current, denoted by I or i, measured in amperes, A. (This is the *basic* electrical unit). Two parallel wires carrying equal currents in the same direction feel a mutual force of attraction. If the wires are each one meter long and spaced one meter apart, and if the current is adjusted such that the attractive force is made equal to 2×10^{-7} newtons*, then the strength of the current in each wire is defined as being one ampere. Experimentally it is found that this corresponds to a flow of 6.23×10^{18} electrons per second.

(b) Electric charge, denoted by Q, measured in coulombs, C. If a current of one A flows for one second, the total charge carried is defined as one coulomb. Its units are As. Since a coulomb is equal to the charge on 6.23×10^{18} electrons, the charge on one electron, usually denoted by e, is equal to 1.60×10^{-19} C. It is conventionally taken as *negative*. Since a (positive) current is considered to be a flow of positive electric charge, this means that the direction of current flow is *opposite* to the actual direction of *electron flow*. Note that a positive current flow from, say, left to right, is equivalent to a negative current flow from right to left. Each is an *equally valid* representation. In any case, with alternating current, the current changes direction twice per

*The factor 2×10^{-7} comes from the need to provide continuity with prior definitions of the amp. It equals $\mu_0/2\pi$, where $\mu_0 = 4\pi \times 10^{-7}$ is defined in section A.1.5(c).

period, so there is no absolute sense in which one can specify the direction of current flow.

(c) Potential difference, denoted by V, (voltage, electromotive force) is measured in volts, V. It is the cause of electric current flow. If a potential difference causes current to flow, then work is being done at a certain rate. The volt is defined as that value of potential difference which results in work being done at the rate of 1 joule per second when the current strength is one amp. Hence, we can write

1 watt = 1 volt × 1 amp

This equation can be re-written dimensionally in the form

$$J/s = V \cdot C/s$$

or

$$J = [(V/m)C]m$$

(d) Electric fieldstrength, denoted by E, is measured in volts per meter, V/m. In the previous equation the left-hand side is work = force × distance (dimensionally), so it is apparent that (V/m)C is a force. The quantity V/m is called electric field-strength and is the force *per unit charge* caused by a voltage difference operating over a distance.

Coulomb's law says that the force in free space between two electric charges Q_1 and Q_2 is proportional to their product and inversely proportional to the square of the distance between them: Force $\propto Q_1 Q_2 / r^2 = (Q_2/r^2)Q_1$. Clearly, then, the expression Q_2/r^2 appears here as a force per unit charge, and can be thought of as due to the electric field produced by the charge Q_2 at distance r. The relation can be written with an equality sign provided a suitable constant is incorporated to take into account the units used. In the rationalized MKS system the equation is written

$$\text{Force} = \frac{Q_1 Q_2}{4\pi\epsilon_0 r^2} \tag{A.2}$$

where the force is in newtons, Q in coulombs and r in meters.

(e) Permittivity. The quantity ϵ_0 in (A.2) is called the permittivity of free space, and from the equation it is clear that its units are C^2/Nm^2, but this representation is not usually utilized. Rather, a new unit, the farad, F, is defined, and the units of ϵ_0 are F/m.

The definition of F is given later under section f) on capacitance. The numerical value of ϵ_0 is

$$\epsilon_o = \frac{1}{4\pi 9 \times 10^9} \text{ F/m} \tag{A.3}$$

$$= 8.854 \text{ pF/m}$$

In a more general medium, (A.2) is modified by the replacement of ϵ_0 by ϵ where ϵ is the permittivity of the medium. It is usually between about 2 to 100 times ϵ_0, and this multiplying factor is called the *relative permittivity* and is denoted by ϵ_r. It is dimensionless, since it is defined by

$$\epsilon_r = \epsilon/\epsilon_o \tag{A.4}$$

(f) Capacitance, denoted by C, is measured in farads, F. If a body is charged, its potential increases. If the potential becomes one volt when the charge is one coulomb, the capacitance is defined to be one farad. Capacitance is the charge per unit potential, and in practice one volt would produce only a minute charge on most bodies. Thus the farad is a *very* large unit, and most capacitances are measured in microfarads or picofarads. Very approximately a sphere 1 cm in radius and removed more than a few cm from neighboring bodies, has a capacitance of about 1 pF.

(g) Resistance, denoted by R, is measured in ohms, Ω. If a circuit requires a voltage of one volt to drive a current of one amp, it is said to have a resistance of one ohm. If the response of the circuit is linear (i.e., doubling the voltage doubles the current, etc.) we have Ohm's law:

$$V/I = R \text{ or } V = IR \text{ or } I = V/R \tag{A.5}$$

Since the power $P = VI$ we also get the corresponding relationships

$$P = VI = I^2R = V^2/R \tag{A.6}$$

for the power dissipated in a resistor.

(h) Resistivity, denoted by ρ, is measured in ohm meter, Ωm. A resistor made from a certain material has a resistance proportional to the length of the resistor and inversely proportional to the cross-section area. Hence, we can write

$$R = \rho \cdot \text{length/cross section area} \quad \text{(A.7)}$$

The proportionality constant ρ is called the *resistivity* and is a property of the material, not the size and shape of the resistor. (It will also vary with the temperature, however.)

(i) Conductivity, denoted by σ, is measured in mho/meter, $1/\Omega\text{m}$, where mho, the inverse of ohm is also called Siemens. It is defined as the inverse of the resistivity, so that $\sigma = 1/\rho$. For pure copper at room temperature its value is approximately $\sigma = 5.8 \times 10^7$.

(j) Electric Displacement, denoted by D, is measured in coulombs per square meter. It is equal to ϵE, and may be thought of as the density of electric flow lines emanating from an electric charge.

A.1.5 Derived Magnetic Units

It might at first sight appear that a separate basic unit for magnetic quantities is needed, and in the earlier development of magnetism this was implicit. It was, moreover, based on the cgs system of units, and two of these units, the oersted and the gauss, are still with us. They are not part of the MKS system, and since the self-consistent equations require all units to be in the MKS system, measurements in the cgs system must be converted. A further difficulty of the earlier development of magnetism is that it presumed the existence of an isolated magnetic pole analagous to the electric charge, for which Coulomb's law (A.2) applies. In practice, magnetic poles always occur in equal and opposite pairs, though the properties of one end of a long bar magnet approximate those of the hypothetical isolated pole.

We commence, therefore, with an analogue of (A.2):

$$\text{Force} = M_1 M_2 / r^2 \quad \text{(A.8)}$$

where M is a measure of a (hypothetical) magnetic pole. The formula can also be written force = $(M_1/r^2) \cdot M_2$ where (M_1/r^2) can be interpreted as the force per unit pole and is the analogue of the electric field which is the electrically caused force per unit charge. The quantity (M/r^2) is therefore, called the magnetic field and, if the units in (A.8) are based on the cgs system, the unit of magnetic field is called the oersted, Oe. As mentioned above, it is not part of the MKS system.

(a) Magnetic field, denoted by H. In the MKS system the unit (analagous to volt/meter for electric field) is the amp per meter. It is the field at a distance of one meter from a long straight wire carrying a current of a certain strength. It would be neat if this current strength were one amp, but in fact it has to be 2π amps. The factor 2π comes from the rationalization process used in setting up the MKS system. Basically it comes from the circumference of a unit circle, which is 2π meters, and this factor 2π (and also 4π, from the surface area of the unit sphere) has to appear in one place or another in the system. The conversion factor between the two magnetic fields is

$$1 \text{ A/m} = (4\pi/10^3) \text{ Oe}$$

(b) Magnetic induction, or magnetic flux density, denoted by B. It is analagous to the electric displacement D, and is a measure of the density of magnetic flux lines. In the cgs system it is called a gauss, and in the MKS system the unit is weber/(meter)^2 where the weber is the analogue of the coulomb, and is a measure of magnetic pole strength. Its magnitude is such that, placed a distance of 1 meter from a long wire carrying a current of 2π amps, its experiences a force of 1 newton. The conversion factor between gauss and weber/m^2 is

$$1 \text{ gauss} = 10^{-4} \text{ Wb/m}^2$$

(c) Permeability, denoted by μ. The relation between magnetic induction and magnetic field can be written (at least for simple materials) $B = \mu H$ where the proportionality quantity μ is either a) independent of the field and equal to, or very nearly equal to, the value μ_0 in free space, or b) is a highly non-linear function of the field and is typically many times the value of μ_0. Most substances fall into the first category, while many substances containing iron or certain other metals, fall into the second. The material of a permanent magnet would be a typical example of this group.

According to the above definition, the units for μ would be (Wb/m^2)/(A/m) but this representation is not usually utilized. Rather a new unit, the henry, H, is defined, and the units of μ are H/m. The definition of H is given later under section d) on inductance. The numerical value of μ_0 is

$$\begin{aligned} \mu_0 &= 4\pi 10^{-7} \text{ H/m} \\ &= 1.256 \ \mu\text{H/m} \end{aligned} \qquad \text{(A.9)}$$

(d) Inductance, denoted by L, is measured in henries, H. If a magnetic flux density penetrates (at right angles) a circuit of area S, then the total flux is BS. If this changes with time its rate of change produces a voltage around the circuit. This law was discovered by Faraday. A change of 1 weber/sec induces a voltage change of 1 volt.

Now the change of magnetic flux could be due either to the movement of a magnet near the circuit or a change of current in the circuit, since the current itself produces a magnetic field. The inductance is a property of the circuit which relates the voltage induced by a change of current in it. If the current changes at 1 A/s and induces a voltage of 1 V, then the inductance of the circuit is said to be 1 henry.

A.2 TRIGONOMETRIC FUNCTIONS

A.2.1 The Sine

The sine (abbreviated to sin) of an angle in a right-angled triangle is defined as the ratio of the *opposite* side to the *hypoteneuse*. It is a function of the angle only, and does not depend on the absolute size of the triangle. Thus, in Figure A.1 ABC is a triangle, with right-angle at B. If the angle at A is denoted by θ, then

$$\sin\theta = BC/AC \qquad \text{(A.10)}$$

As $\theta \rightarrow 0$, $BC \rightarrow 0$, so $\sin 0 = 0$

As $\theta \rightarrow 90°$, $BC \rightarrow AC$, so $\sin 90° = 1$

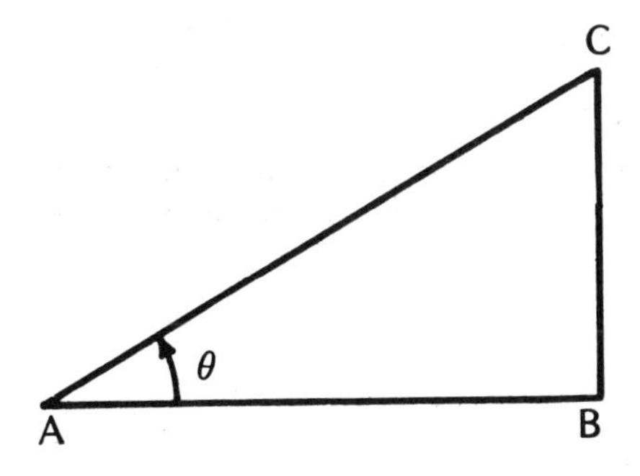

A.1: Triangle for Definition of the Sine: Acute Angle

This definition can be extended beyond 90° by thinking of AC as a rotating arm, pivoted about A, and CB as a "plumb line" dropped onto AB. Thus, for angles up to 180° we get Figure A.2 with $\sin\theta = BC/AC$ as before. Also, from the triangle, $BC/AC = \sin\theta'$ where $\theta + \theta' = 180°$.

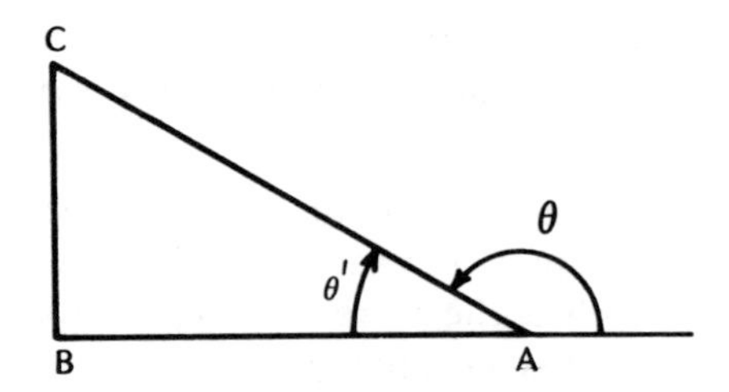

A.2: Triangle for Definition of the Sine: $90° < Angle < 180°$

Hence, $\sin \theta = \sin \theta' = \sin (180° - \theta)$. Accordingly $\sin \theta$ is symmetrical about $\theta = 90°$. Beyond $\theta = 180°$ the rotor arm is below *AB*, so the line *CB* has to extend in the opposite direction as before. Hence, $\sin \theta$ is negative for $180° < \theta < 360°$, as in Figure A.3. Otherwise the values repeat the 0 to 180° range. Thus $\sin (\theta) = -\sin (\theta - 180°)$. The graph of $\sin \theta$ repeats every 360°, and is as in Figure A.4. Notice that $\sin \theta$ always lies between the limits ± 1, and that $\sin (-\theta) = -\sin (\theta)$.

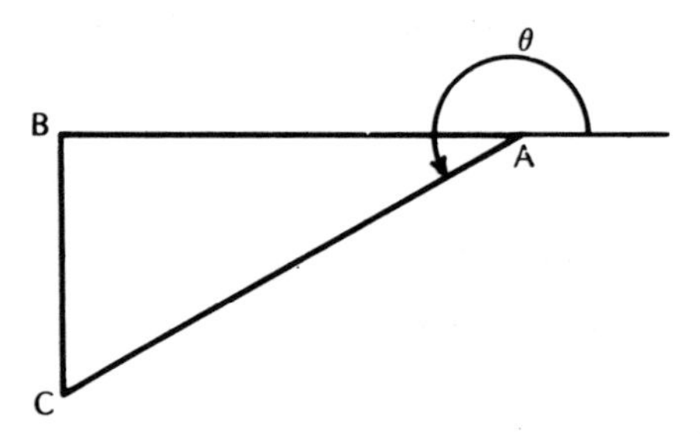

A.3: Triangle for Definition of the Sine: $180° < Angle < 270°$

A.2.2 The Cosine

The cosine (abbreviated to cos) of an angle in a right-angled triangle is defined as the ratio of the *adjacent* side to the *hypoteneuse*. In Figure A.1

$$\cos \theta = \frac{AB}{AC} \tag{A.11}$$

As $\theta \to 0$, $AB \to AC$, so $\cos 0 = 1$

As $\theta \to 90°$, $AB \to 0$, so $\cos 90° = 0$

The definition can be extended beyond 90° by noting that *AB* reverses sign the other side of the vertical, so that $\cos \theta$ is negative from 90° to 270°. The graph of $\cos \theta$ is as in Figure A.5. Note that this is like a graph of $\sin \theta$ shifted 90° to the left. It repeats every 360°.

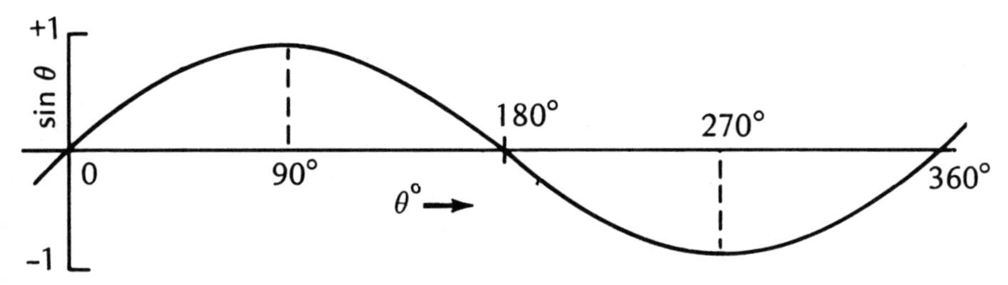

A.4: Graph of the Sine Function

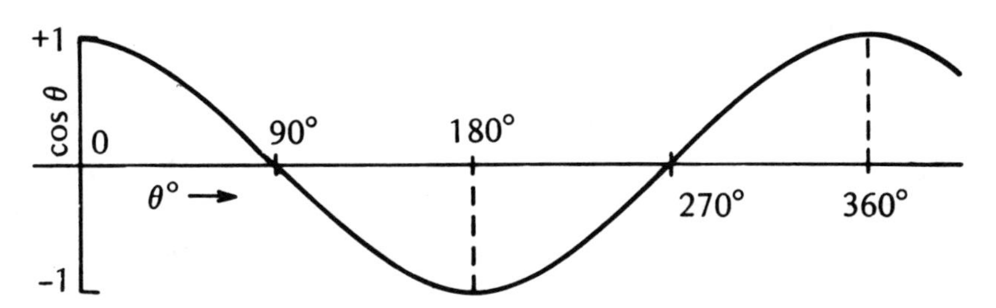

A.5: Graph of the Cosine Function

A.2.3 The Tangent

The tangent (abbreviated to tan) of an angle in a right-angled triangle is defined as the ratio of the *opposite* side to the *adjacent* side. In Figure A.1

$$\tan\theta = \frac{BC}{AB} \tag{A.12}$$

As $\theta \to 0$, $BC \to 0$, so $\tan 0 = 0$

As $\theta \to 90°$, $AB \to 0$, so $\tan 90° = BC/0 = \infty$ (infinity)

From 90° to 180°, BC is positive and AB is negative. Hence, the tangent is negative. From 180° to 270° both BC and AB are negative; their ratio is therefore, positive. From 270° to 360° BC is negative and AB is positive; the tangent is therefore, negative. The graph of $\tan\theta$ is as in Figure A.6. It repeats every 180°. The dotted line at 90° is known as an asymptote: $\tan\theta$ is $+\infty$ on one side and $-\infty$ on the other.

A.2.4 Important Relations Between These Functions

(a) $$\tan\theta = \frac{BC}{AB} = \frac{BC}{AC} \cdot \frac{AC}{AB} = \frac{BC/AC}{AB/AC} = \frac{\sin\theta}{\cos\theta} \tag{A.13}$$

(b) In Figure A.7 $\cos\theta = AB/AC$ and $\sin\theta' = AB/AC$. But these ratios are the same. Hence, $\sin\theta' = \cos\theta$. Now the three angles of a triangle add up to 180°. This gives $\theta + \theta' + 90° = 180°$, or

$\theta' = 90° - \theta$. Hence

$$\sin(90° - \theta) = \cos\theta, \text{ or} \tag{A.14}$$

$$\cos(90° - \theta') = \sin\theta' \tag{A.15}$$

(c) From Pythagoras's theorem we get

$$(AC)^2 = (AB)^2 + (BC)^2 \tag{A.16}$$

Dividing by $(AC)^2$ throughout gives

$$\left(\frac{AC}{AC}\right)^2 = \left(\frac{AB}{AC}\right)^2 + \left(\frac{BC}{AC}\right)^2, \text{ or}$$

$$1 = \cos^2\theta + \sin^2\theta \tag{A.17}$$

(d) If θ and ϕ are any two angles, the sin or cos of combinations of them can be expressed in terms of sines and cosines of the separate angles. There is an unlimited number of such relations. But the following are *particularly important*:

$$\sin(\theta \pm \phi) = \sin\theta\cos\phi \pm \cos\theta\sin\phi \tag{A.18}$$

$$\cos(\theta \pm \phi) = \cos\theta\cos\phi \mp \sin\theta\sin\phi \tag{A.19}$$

Note that the + sign for the cosine on the left of (A.19) goes with the – sign on the right, and vice versa.

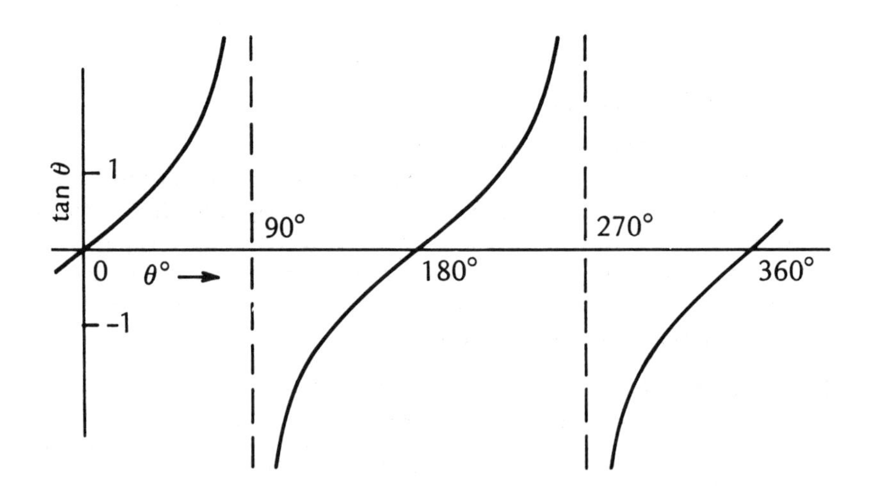

A.6: Graph of the Tangent Function

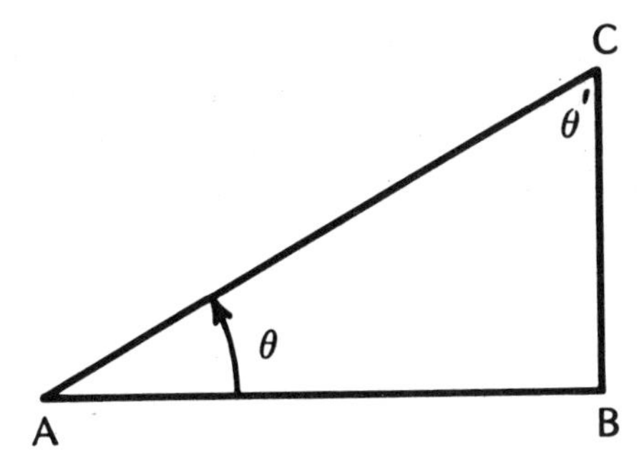

A.7: Triangle Illustrating Complementary Angles

Particular cases come from taking $\theta = \phi$, and using the + sign.

$$\sin(2\theta) = 2\sin\theta\cos\theta \tag{A.20}$$

$$\cos(2\theta) = \cos^2\theta - \sin^2\theta \tag{A.21}$$

Two variants of (A.21) come from using (A.17) in (A.21)

$$\cos(2\theta) = 2\cos^2\theta - 1 = 1 - 2\sin^2\theta \tag{A.22}$$

This last equation can be presented in a form which gives the *squares* of sines or cosines in terms of the cosine of the double angle,

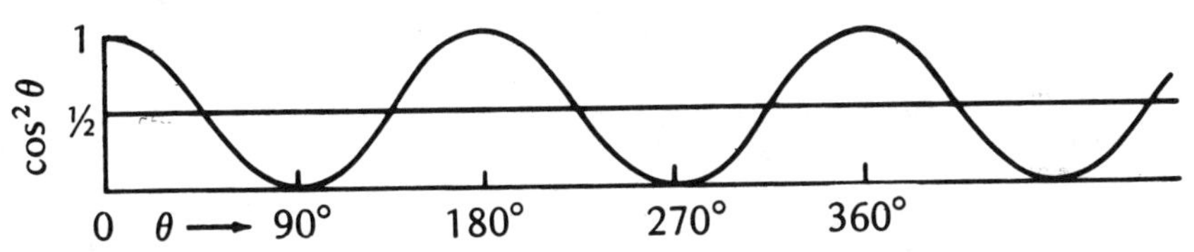

A.8: Graph Illustrating the Average Value of the Cosine Squared

$$\cos^2\theta = \frac{1 + \cos(2\theta)}{2} \quad ; \quad \sin^2\theta = \frac{1 - \cos(2\theta)}{2} \tag{A.23}$$

It is seen from Figure A.8 how $\cos^2\theta$ behaves like a half-amplitude $\cos(2\theta)$ curve, displaced in level by 1/2.

Since $\cos(2\theta)$ consists of equal sections that are alternately positive and negative, it averages out to zero. Hence, from the figure, we see that the *average value* of $\cos^2\theta$ is exactly 1/2.

A.2.5 Special Values

Certain simple values recur frequently. The sines and cosines of 0 and 90° have already been given. If, in Figure A.9, $AB = BC$, then the angles at

A and C are equal, from symmetry, and each will be 45° since the sum of all the angles has to be 180°. Let AB and BC each be of unit length. Then, from Pythagoras's theorem, $(AC)^2 = (AB)^2 + (BC)^2 = 1 + 1 = 2$. Hence, $AC = \sqrt{2}$. From the triangle, and considering the angle at A,

$$\sin 45^\circ = \frac{BC}{AC} = \frac{1}{\sqrt{2}} = 0.707\ldots$$

$$\cos 45^\circ = \frac{AB}{AC} = \frac{1}{\sqrt{2}}$$

$$\tan 45^\circ = \frac{BC}{AB} = 1 \qquad \text{(A.24)}$$

Consider now an *equilateral* triangle, of unit side as in Figure A.10. The angles are all equal, and hence, are 60° each. If we construct a perpendicular bisector from C to AB, cutting AB at D, then $AD = 1/2$ and the angle at C in the triangle ACD will be 30°. Moreover, $(AC)^2 = (AD)^2 + (DC)^2$ gives $1 = 1/4 + (DC)^2$. Hence, $(DC)^2 = 3/4$ and $DC = \sqrt{3/2}$. Hence

$$\sin 60^\circ = \frac{DC}{AC} = \frac{\sqrt{3}}{2} = 0.866\ldots = \cos 30^\circ$$

$$\sin 30^\circ = \frac{AD}{AC} = \frac{1}{2} = 0.5 = \sin 60^\circ$$

$$\tan 30^\circ = \frac{AD}{DC} = \frac{1}{2}\Big/(\sqrt{3/2}) = 1\Big/\sqrt{3} = 0.577\ldots$$

$$\tan 60^\circ = \frac{DC}{AC} = \frac{\sqrt{3}}{2}\Big/\frac{1}{2} = \sqrt{3} = 1.732\ldots \qquad \text{(A.25)}$$

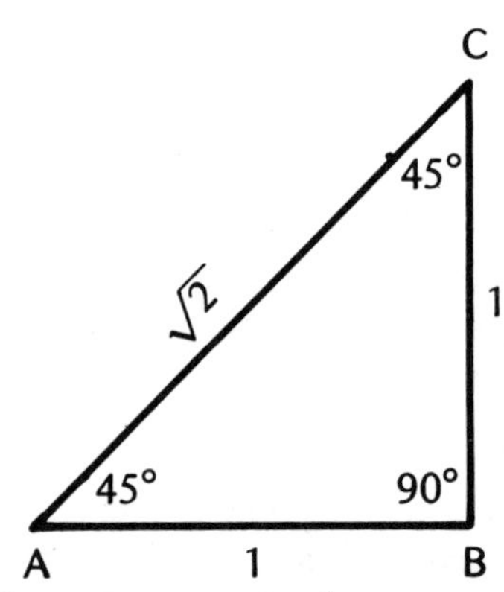

A.9: Triangle for Calculating Functions of 45°

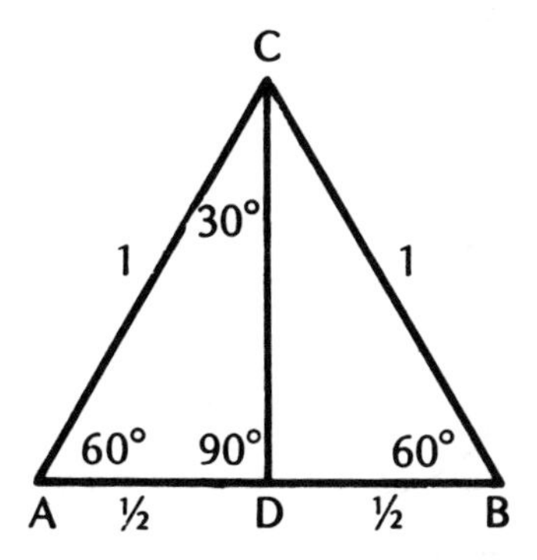

A.10: Triangle for Calculating Functions of 30° and 60°

A.2.6 Reciprocal Functions

These are the secant, cosecant and cotangent, abbreviated to sec, cosec, and cot respectively. They are defined by

$$\sec\theta = \frac{1}{\cos\theta}, \quad \operatorname{cosec}\theta = \frac{1}{\sin\theta}, \quad \cot\theta = \frac{1}{\tan\theta} \qquad \text{(A.26)}$$

Their graphs are shown in Figure A.11. Clearly the have the same periodicity as their corresponding functions have.

A.2.7 Combinations of Sinusoidal Functions

Provided they have the same periodicity, sines and cosines can be combined to form a composite function which is also sinusoidal. Consider the equation

$$A\cos\theta + B\sin\theta = C\cos(\theta + \phi) \qquad \text{(A.27)}$$

Is it possible to choose a value for C and for ϕ such that this is true for *all* values of θ? It is not immediately obvious that this is so. Using (A.19) on the right-hand side we get

$$A\cos\theta + B\sin\theta = (C\cos\phi)\cos\theta - (C\sin\phi)\sin\theta.$$

For this to be true for *all* values of θ we *must* have

$$A = C\cos\phi, \text{ and, } B = -C\sin\phi \qquad \text{(A.28)}$$

By squaring and adding we get

$$A^2 + B^2 = C^2\cos^2\phi + C^2\sin^2\phi = C^2(\cos^2\phi + \sin^2\phi) = C^2$$

by (A.17)

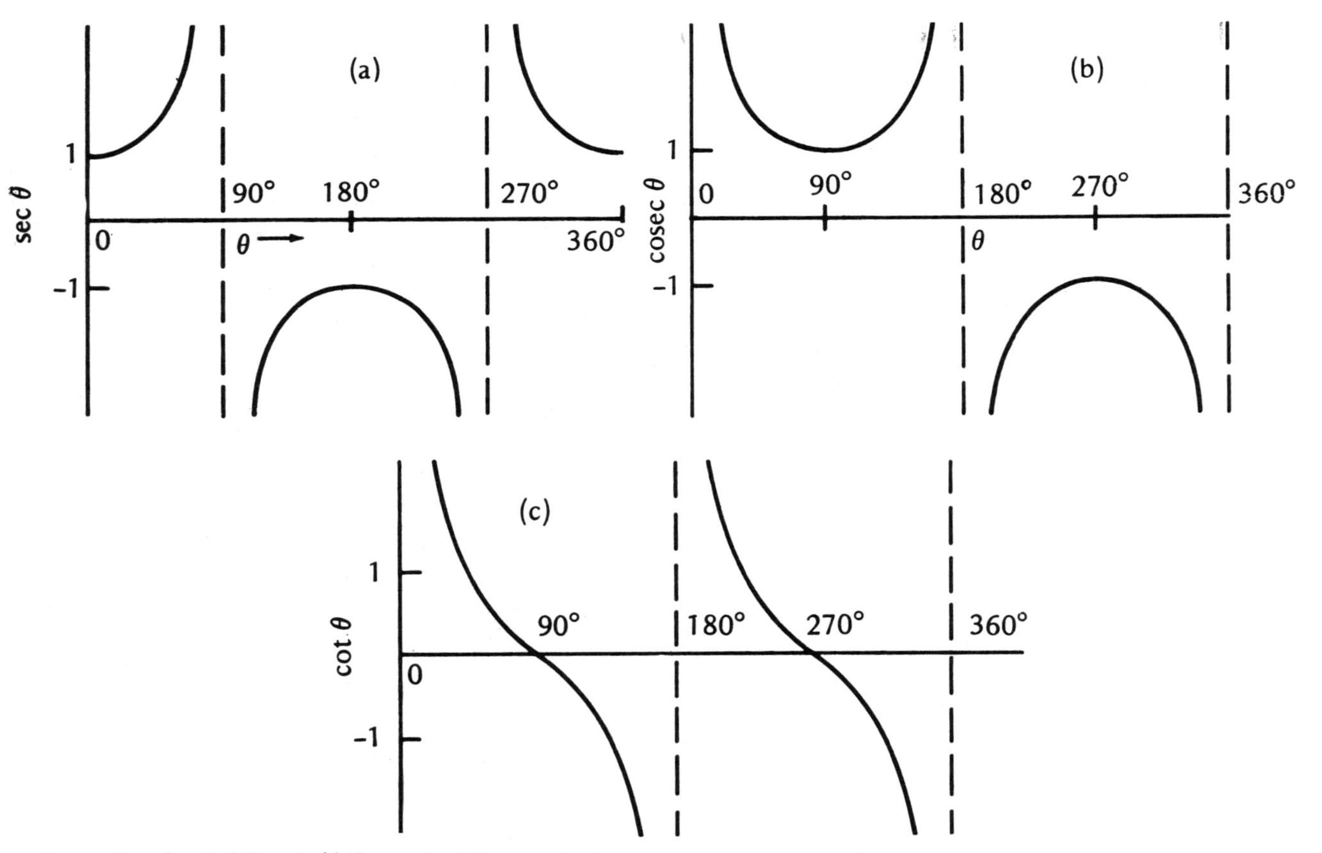

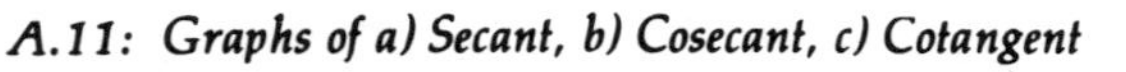
A.11: Graphs of a) Secant, b) Cosecant, c) Cotangent

Hence,

$$C = \sqrt{(A^2 + B^2)} \tag{A.29}$$

It is usual to take the positive square root in this equation. By dividing the two relations in (A.28) we get

$$\frac{B}{A} = \frac{-C \sin \phi}{C \cos \phi} = -\tan \phi \tag{A.30}$$

This equation will give *two* values of ϕ separated by 180°. But only *one of these* will satisfy (A.28). *It is always necessary* to return to (A.28) to *check* which is the correct value.

More generally, let

$$A \cos \theta + B \cos (\theta + \psi) = C \cos (\theta + \phi) \tag{A.31}$$

Applying (A.19) to both the B and C terms we get

$$\cos\theta\, (A + B \cos \psi) - \sin \theta\, (B \sin \psi) = \cos \theta\, (C \cos \phi) - \sin \theta\, (C \sin \phi)$$

Hence,

$$\begin{aligned} A + B \cos \psi &= C \cos \phi \\ B \sin \psi &= C \sin \phi \end{aligned} \tag{A.32}$$

Squaring and adding gives

$$C^2 = (A + B \cos \psi)^2 + (B \sin \psi)^2 = A^2 + 2AB \cos \psi + B^2 (\cos^2 \psi + \sin^2 \psi)$$

whence

$$C = \sqrt{A^2 + 2AB \cos \psi + B^2} \tag{A.33}$$

Equation (A.29) is the special case $\psi = -90°$, since $\cos (\theta - 90°) = \sin \theta$. Dividing the two relations of (A.32) gives, similarly to (A.30),

$$\tan \phi = \frac{B \sin \psi}{A + B \cos \psi} \tag{A.34}$$

As before, this equation gives *two* values of ϕ, separated by 180°, and they have to be checked with (A.32) to select the correct one.

A.2.8 Radians

Angular measure can be expressed in cycles, degrees or radians. One cycle = 360° = 2π radians. To see the relation between degrees and radians, consider the circle in Figure A.12, for which the ratio of the arc AB to the radius is $AB/OA = \theta$ where θ is *defined* as the angle in radians. For

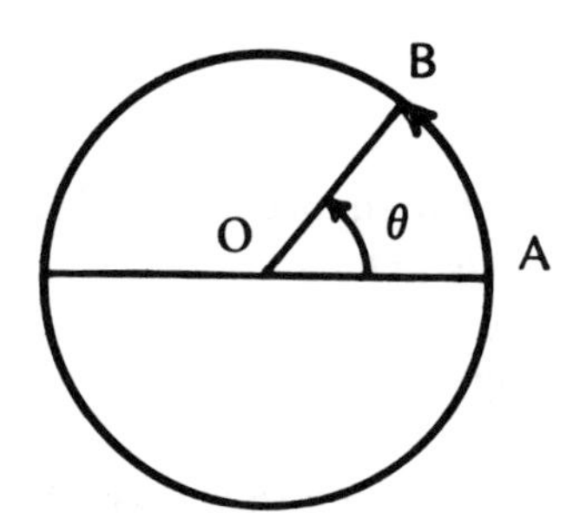

A.12: Circle for Definition of Radian Measure

a complete cycle the arc AB becomes the circumference of the circle, = 2π times the radius. Hence, 1 cycle = 360° = 2π, where π = 3.14159...

$$\pi \text{ radians} = 180° \text{ or } 1 \text{ radian} = 57.3° \ldots \quad \text{(A.35)}$$

A.2.9 Angles in Excess of 90°

Tabulated angles usually cover the range 0 to 90° only. From Figure A.13 we can get sines of angles greater than 90°. Thus $\sin\theta = \sin(180° - \theta) = -\sin(180° + \theta) = -\sin(360° - \theta)$, etc. Similarly, if we imagine a new axis erected at $\theta = 90°$, the curve from there is a cosine curve. Thus $\sin(90° + \theta) = \cos\theta$. Sines or cosines shifted by 180° or 360° become sines and cosines, with a possible sign change that needs to be ascertained. Sines or cosines shifted by 90° or 270° become, respectively, cosines or sines, again with a possible sign change. Note, in particular,

$$\sin(-\theta) = -\sin\theta \qquad \cos(-\theta) = \cos\theta \quad \text{(A.36)}$$

$$\sin(90° - \theta) = \cos\theta \qquad \cos(90° - \theta) = \sin\theta \quad \text{(A.37)}$$

The quadrants where the respective functions $\sin\theta$, $\cos\theta$ and $\tan\theta$ are *positive* are shown in Figure A.14. Note that the angles are conventionally measured from the horizontal axis in a *counter-clockwise* sense.

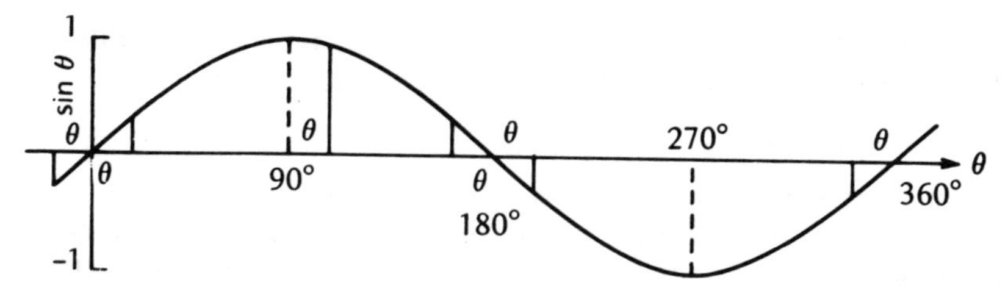

A.13 Graph for Sine Function of Angles in Excess of 90°

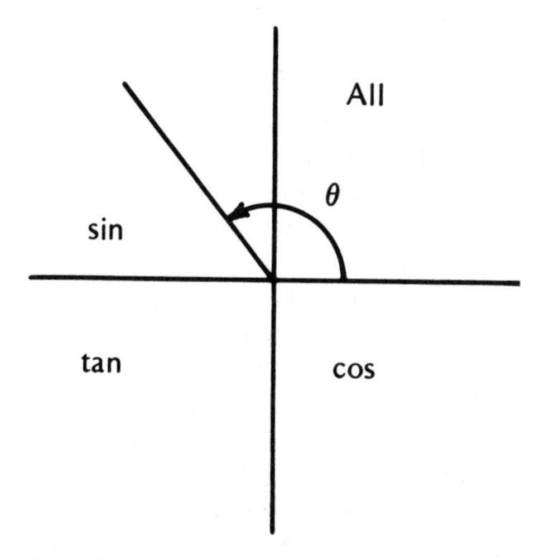

A.14 Quadrants where the Trigonometrical Functions are Positive

A.2.10 Approximations for Small Angles

It can be seen from Figure A.15 that the arc *AB* and the side *BC* of the triangle *OBC* are nearly equal when the angles is small. This leads to the approximation

$$\sin\theta \approx \theta, \theta \text{ small and measured in radians} \tag{A.38}$$

From (A.17) it can then be shown that

$$\cos\theta \approx 1 - \theta^2/2 \tag{A.39}$$

and from (A.13)

$$\tan\theta \approx \theta \tag{A.40}$$

It should be emphasized that these three equations require that the angle be measured in radians. If it is in degrees, θ must be replaced by $(\pi\theta/180)$ on the right hand sides.

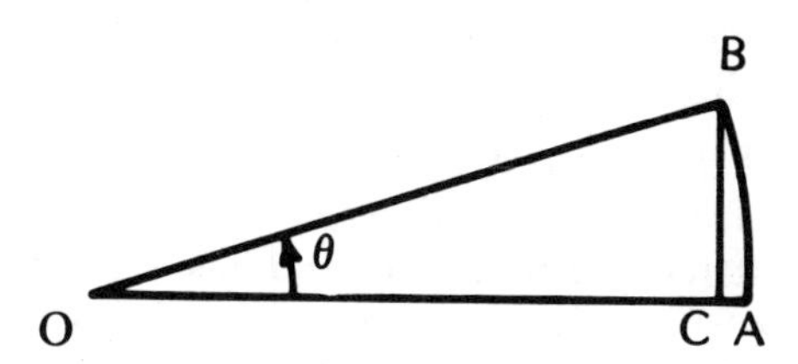

A.15: Approximate Equality of Arc to Chord for Small Angles

A.3 ELECTROMAGNETIC WAVE TRANSMISSION

A.3.1 Maxwell's Equations and Electromagnetic Waves

Under static conditions it is known that the magnetic field produced by a current flowing through a small area is proportional to that current. Maxwell surmised that, under time-varying conditions, the current should be supplemented by the time rate of change of the electric displacement through the area. This term is somewhat analagous to the term in Faraday's law which says that the voltage round a circuit is equal to the time rate of change of magnetic induction through the circuit. These two laws form the basis of Maxwell's equations which, when set up mathematically, lead to solutions which are interpreted physically as electromagnetic waves, a form of oscillatory energy in which the electric and magnetic fields are coupled together and propagate with a velocity given by $(\epsilon_0\mu_0)^{-1/2}$ in free space. This velocity turns out to be 3×10^8 m/s, the velocity of light, and leads to the conclusion that light is an example of an electromagnetic wave.

A.3.2 Plane Electromagnetic Waves

An idealized form of electromagnetic wave is the *plane wave*. A long distance away from a transmitter in free space the wave is a good approximation to a plane wave. It has the following important properties.

(a) It propagates in a given direction with the velocity of light, denoted by c. In vacuum c has the value 2.9979×10^8 m/s. It is usually approximated by 3×10^8 m/s, and this is good enough for most purposes except for very accurate conversion of wavelength to frequency. It is very slightly slower in air, depending on, among other things, temperature, pressure and moisture content.

(b) In a plane transverse to the direction of propagation the value of electric field E is everywhere the same, as is also the value of the magnetic field H.

(c) E is perpendicular to H, and they are related such that a right handed rotation from E to H points in the direction of propagation, as indicated in Figure A.16.

(d) The ratio E/H has the dimensions (V/m)/(A/m) = V/A or ohms. In free space it has the value $(\mu_0/\epsilon_0)^{1/2} \approx 377$ ohms. Here, ϵ_0 is the permittivity of free space and has the approximate value $1/(4\pi \cdot 9 \times 10^9)$ F/M and μ_0 is the permeability of free space and has the value $4\pi \times 10^{-7}$ H/m. Hence, the square root of the ratio can also be expressed as 120π, an expression frequently encountered.

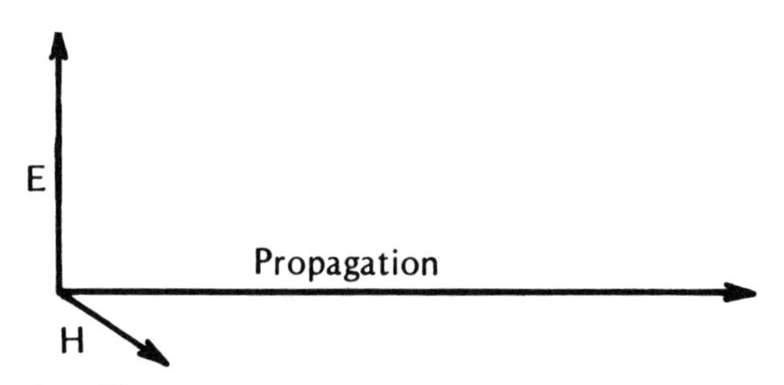

A.16 Direction of Propagation

(e) We shall not consider magnetic materials for which the permeability differs from μ_0 since they barely affect practical transmission. But for materials whose relative permittivity is ϵ_r we have $E/H = 377 / \sqrt{\epsilon_r}$, and wave velocity $v = 3.10^8 / \sqrt{\epsilon_r}$. For various soils ϵ_r varies from about 4 to 20. For seawater it is about 80, though dropping at microwave frequencies.

A.3.3 Propagation

A typical plot of the instantaneous field in a plane wave is given in Figure A.17. The peak-to-peak separation *AB* is called the wave-length, and is denoted by λ. At a later instant the wave has moved to the right, as shown by the dotted curve. The time it takes to move one wavelength is the *period* of the wave, and is denoted by T. The frequency, f, is the number of cycles per second (hertz) so that $1/f$ is the time for one cycle, and therefore, equals T. Hence

$$T = 1/f \qquad \text{or} \qquad fT = 1 \tag{A.41}$$

Since the wave moves a distance λ in a time T, the velocity is

$$\begin{aligned} v &= \lambda/T \\ &= \lambda f \qquad \text{from (A.41)} \end{aligned} \tag{A.42}$$

A.3.4 Sine Waves

The angular frequency ω is the number of radians per second. Since a complete cycle represents 2π radians we get

$$\omega = 2\pi f \qquad \text{radians/sec} \tag{A.43}$$

At a fixed position the wave amplitude can be written

$$E = E_o \cos(\omega t + \phi) \tag{A.44}$$

where E_o is the maximum value of the field

t is the time (seconds)
ϕ is the initial phase of the wave.

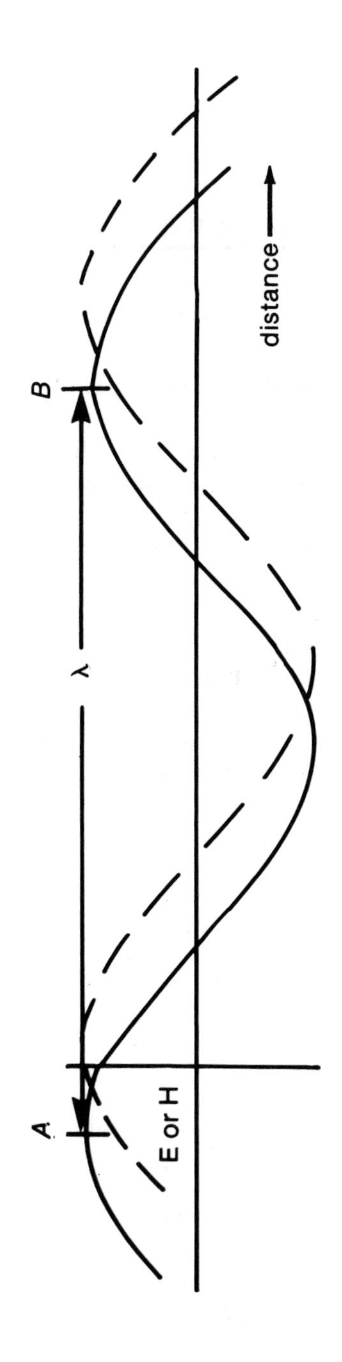

A.17 Propagation of a Radio Wave.

The total angle is usually expressed in radians but for the convenience of use of tables it can be converted to degrees by using 2π radians = 360°.

At a position distant D from the reference implied in (A.44) the wave is delayed by the time it takes to propogate the distance D, i.e., D/v. Hence, the wave at that position has t replaced by $t - D/v$. The wave representation is therefore,

$$E = E_o \cos [\omega (t - D/v) + \phi] \tag{A.45}$$

From (A.42) and (A.43) we get $\omega/v = 2\pi f/\lambda f = 2\pi/\lambda$. Hence, (A.45) can also be written,

$$E = E_o \cos [\omega t + \phi - 2\pi D/\lambda] \tag{A.46}$$

The following formula combining cosines of different angles should be noted:

$$\cos A + \cos B = 2 \cos \left(\frac{A + B}{2} \right) \cos \left(\frac{A - B}{2} \right) \tag{A.47}$$

Thus if two equal waves arrive at a point having transversed distances D_1 and D_2, the composite wave is represented by

$$\begin{aligned} &E_o \cos [\omega t + \phi - 2\pi D_1/\lambda] + E_o \cos [\omega t + \phi - 2\pi D_2/\lambda] \\ &= 2E_o \cos [\omega t + \phi - 2\pi (D_1 + D_2)/2\lambda] \cos [\pi (D_1 - D_2)/\lambda] \end{aligned} \tag{A.48}$$

The final factor, $\cos[\pi(D_1 - D_2)/\lambda]$, would be unity if the distance difference were an integral multiple of the wavelength, leading to a total signal of double amplitude. But if the difference were an odd multiple of the half-wavelength, the cosine would be zero. This result exemplifies the phenomenon of *wave interference*. Since the effect depends on λ, it is frequency-dependent, and the interference is known as frequency-selective fading.

A.3.5 Power Flow

The Poynting vector gives the power flow per unit area. In a plane wave, if the area S is perpendicular to the direction of propagation, the power flow is

$$P = EHS \quad \text{watts (W)} \tag{A.49}$$

In free space this gives the value

$$P = \frac{E^2 S}{377} \tag{A.50}$$

Hence, the power is proportional to the square of the electric field-strength. In these equations, E or H vary sinusoidally with time, so that E^2, say varies between its maximum value and zero. The time average value is half the maximum, so that if E is given by, say, (A.45) then the *average* power flow is seen, as in figure A.8, to be given by

$$P_{av} = \frac{1}{2} \frac{E_o^2}{377} \cdot S \tag{A.51}$$

A.4 INDICES, LOGARITHMS AND THE EXPONENTIAL FUNCTION

A.4.1 The Law of Addition of Indices

Let us consider the following sequence which both defines the index or exponent for positive integers, and extends the definition to zero and negative integral values.

$$\begin{aligned} 1000 &= 10^3 \\ 100 &= 10^2 \\ 10 &= 10^1 \\ 1 &= 10^0 \\ 0.1 &= 1/10 = 10^{-1} \\ 0.01 &= 1/100 = 10^{-2} \end{aligned}$$

and so on.

In the relation $10 \times 10 \times 10 \ldots\ldots \times 10 = 10^n$ (in which n tens are multiplied together), 10 is called the *base* and n the *exponent*, or *index*. Each time we divide by ten we reduce the exponent by unity. The relations $10 = 10^1$ and $1 = 10^0$ can be thought of as self-consistent definitions of 10^1 and 10^0. Similarly the relation

$$1/10^n = 10^{-n} \tag{A.52}$$

forms a self-consistent continuation of the meaning of the index to negative integers.

From the relation $(10 \times 10 \times 10 \ldots\ldots \times 10)_{n\ terms} \times (10 \times 10 \ldots\ldots \times 10)_{m\ terms} = (10 \times 10 \times 10 \ldots\ldots \times 10)_{(m+n)\ terms}$ we get the important equation

$$10^n \times 10^m = 10^{n+m} \tag{A.53}$$

This is the law of addition of indices, and it clearly holds for any base b

$$b^n \times b^m = b^{n+m} \tag{A.54}$$

A.4.2 Extension to Fractional Indices

Clearly (A.54) holds for any number of terms; e.g., $b^n \times b^m \times b^p = b^{n+m+p}$, etc. To extend the definition to fractional powers, suppose $y = b^{p/q}$ where p and q are integers. Then $y^q = (b^{p/q} \times b^{p/q} \ldots \times b^{p/q})_{q\ \text{terms}} = b^p$. Hence, $y^q = b^p$ or y = the qth root of b^p.

$$b^{p/q} = \sqrt[q]{b^p} \tag{A.55}$$

In particular $b^{1/2} = \sqrt{b}$, an important special case. Note also

$$(b^m)^n = b^{\overbrace{(m+m+\ldots+m)}_{n \text{ terms}}} = b^{mn} \tag{A.56}$$

A.4.3 The Exponential Function

Figure A.18 shows a plot of the function $y = b^x$ where the base, b, is any number greater than unity. (If b were less than unity the graph would have been the mirror image about the y-axis.)

The curve is seen to go through the point ($x = 0, y = 1$) irrespective of the value of b. For large positive x, y gets very large. Conversely, for large negative x, y gets very small.

(Strictly speaking, the graph is only plotted for rational values of x. The magnitude of y for irrational values of x is found as the limit for closely bounding rational values.)

A.4.4 The Logarithmic Function

If $y = b^x$ the value of x which solves this equation (given y and b) is called the logarithm of y to base b, and is written

$$x = \log_b y \tag{A.57}$$

If $y = b^x$ and $z = b^w$ then $yz = b^x b^w = b^{x+w}$, by the law of indices. Hence, $x + w = \log_b(yz)$ or, since $x = \log_b y$ and $w = \log_b z$

$$\log_b(yz) = \log_b y + \log_b z. \tag{A.58}$$

The law of *addition of logarithms* states that the logarithm of a product of factors is the sum of the logarithms of the factors. Clearly, (A.58) can be extended to several terms. Note also,

$$\begin{aligned} \log(1) &= 0 \\ \log(1/y) &= -\log y \\ \log(y^n) &= n \log y \end{aligned} \tag{A.59}$$

These results are true independent of the base, which need not be shown.

The graph of the logarithm is obtained from the graph of $y = b^x$, and is shown in Figure A.19 from which we note that $\log 1 = 0$, $\log(0) = -\infty$, $\log(\infty) = \infty$, and that the logarithm increases *very* slowly to infinity. Thus $N = 10^{30}$ is an *enormous* number, but $\log_{10} N = 30$, a very moderate value only.

In almost all mathematical and numerical work, only three bases are used

(a) Base 10 for numerical work
(b) Base $e = 2.71828$....for mathematical formulas
(c) Base 2 for (binary) information theory.

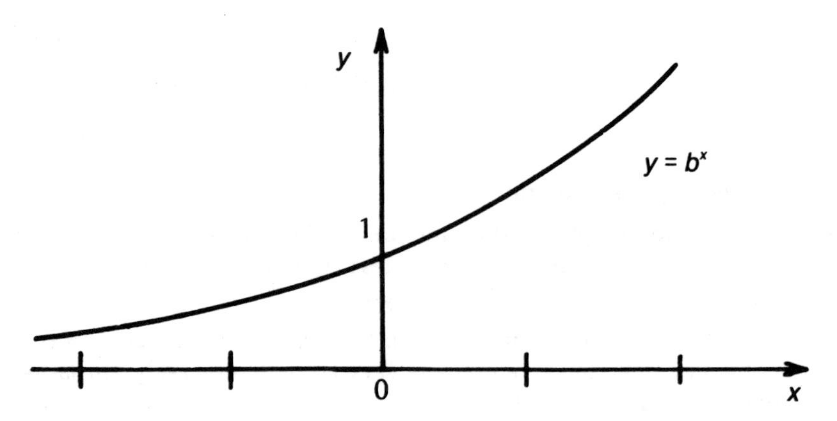

A.18: The Curve $y = b^x$

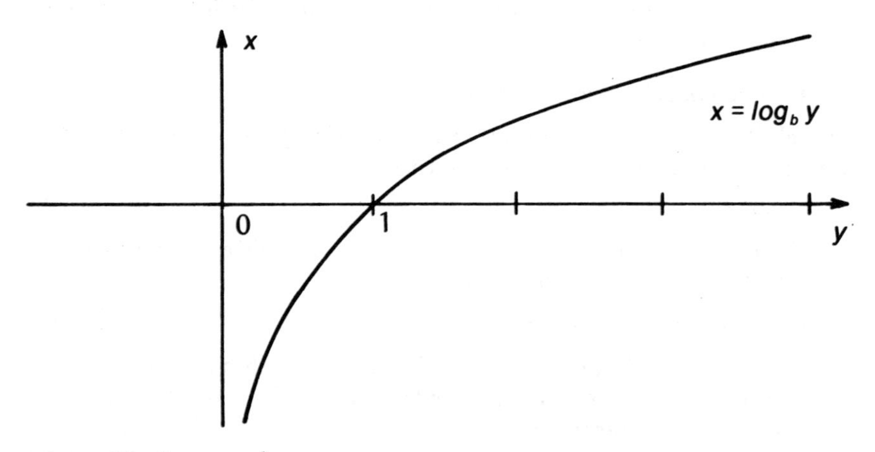

A.19: The Curve $x = \log_b y$

Logarithms to base e(the significance of e is considered in detail later in section A.4.5) are called *natural logarithms* and the notation $\log_e$ is sometimes replaced by ln.

It is possible to transform logarithms from one base to another in the following way:

Let $y = b^x$ (so that $x = \log_b y$) and take logarithms of both sides to base a.

$$\log_a y = \log_a b^x = x \log_a b \text{ (by(A.59))} = \log_b y \log_a b.$$

Hence,

$$\log_a y = \log_b y \log_a b \quad \text{(A.60)}$$

This relation enables logarithms to base b to be calculated when tables to base a are given. In particular, by taking $y = a$ we get, since $\log_a a = 1$, $\log_b(a) = 1/(\log_a(b))$.

Note that the *positions* of a, b and y in (A.60) are the same as in the formula $y/a = y/b \cdot b/a$. This can be a helpful reminder to (A.60).

A.4.5 The Law of Continuous Growth

The expotential function to base e is best approached from a consideration of the growth of compound interest. Capital X_0, invested for one year at a rate r becomes at the end of the first year,

$$X_1 = X_0 + rX_0 = X_0(1 + r) \tag{A.61}$$

At the end of the second year, X_1 becomes $X_2 = X_1 + rX_1 = X_0(1 + r)^2$. And similarly, at the end of m years, X_0 has grown to $X_m = X_0(1 + r)^m$. This is known as the law of compound interest.

Take now the formula (A.61) for the capital at the end of one year, and suppose that instead of rate r per twelve months, the interest is compounded at rate $r/2$ per six months. At the end of one year we get, instead of (A.61), the value $X_0(1 + r/2)^2$. Similarly, if the interest is compounded n times per year at a rate r/n, the capital at the end of the year is $X_1 = X_0(1 + r/n)^n$. For *continuous* growth the interval for adding on the earned interest becomes indefinitely small, and n approaches infinity. Thus, for continuous growth,

$$X_1 = \lim_{N \to \infty} X_0(1 + r/n)^n \tag{A.62}$$

It can be shown that this limit does exist, and that it is finite. If we write $n = rN$ in (A.62) we get

$$X_1 = \lim_{N \to \infty} X_0(1 + 1/N)^{rN} = \lim_{N \to \infty} X_0[(1 + 1/N)^N]^r \tag{A.63}$$

by (A.56).

The quantity $\lim_{N \to \infty}(1 + 1/N)^N$ is denoted by e, so that the expression in (A.63) is $X_0 e^r$. Since $(1 + 1/n)^n$ with $n = 1$ is equal to 2, clearly $e > 2$. With $n = 2$ we get $(1 + 1/2)^2 = 2.25$, and with $n = 3$ we get $(1 + 1/3)^3 = 2.37$, and so on. The expression increases as n increases, but settles down to the value 2.718....

We can use the *binominal theorem* to evaluate e. For all values of x and N we have

$$(1 + x)^N = 1 + Nx + \frac{N(N-1)}{1.2}x^2 + \frac{N(N-1)(N-2)}{1.2.3}x^3 + \ldots \tag{A.64}$$

With $x = 1/N$ this gives

$$(1 + 1/N)^N = 1 + 1 + \frac{(1 - 1/N)}{1.2} + \frac{(1 - 1/N)(1 - 2/N)}{1.2.3} + \ldots$$

Taking N very large, so that $1/N$, $2/N$ etc., are negligible, we get

$$e = 1 + 1 + 1/1.2 + 1/1.2.3 + \ldots \quad \text{(A.65)}$$

from which e can be calculated.

The product of the first n natural numbers, which occurs here in the denominators of the various terms, is known as *factorial n*, and is written $n!$ Thus, $2! = 1 \cdot 2 = 2$, $3! = 1 \cdot 2 \cdot 3 = 6$, $4! = 1 \cdot 2 \cdot 3 \cdot 4 = 24$, etc. Clearly, for $n > 1$.

$$n! = n \cdot (n - 1)! \quad \text{(A.66)}$$

Taking $n = 1$ in this result would give $0! = 1$ as a *self-consistent definition* of $0!$, thus extending the range of (A.66) to $n = 1$.

From (A.67) and (A.63), $\lim_{n \to \infty} (1 + r/n)^n = e^r$, and by using (A.64) with $x = r/N$

we see that

$$e^r = 1 + r + r^2/2! + r^3/3! + \ldots \quad \text{(A.67)}$$

In particular, for small values of r we have the approximation

$$e^r \approx 1 + r \quad \text{(A.68)}$$

Thus the graph of $y = e^r$, near the origin, looks like a straight line through the point $y = 1$, and with unit slope. It is shown in Figure A.20. The graph of $y = e^{-r}$ is the mirror image of the graph of $y = e^r$ reflected in the line $r = 0$.

The expression $y = e^{-r}$ at $r = 1$ drops to a value $e^{-1} = 0.36788\ldots$ and is a commonly used level at which to assess the decay of an exponential process. Because e enters "naturally" in all proportionate growth and decay calculations, it occurs in most mathematical formulas for these phenomena. As already mentioned, logarithms to base e are called natural logarithms, and are denoted by $\log_e$ or ln.

A.5 DECIBELS AND NEPERS

A.5.1 The Decibel

Power ratios, for example between a transmitter power and a receiver power, may involve very large numbers. By taking the logarithm a more managable number is obtained. Moreover, if the power reduction is due to a number of separable factors, the logarithms of each part can be

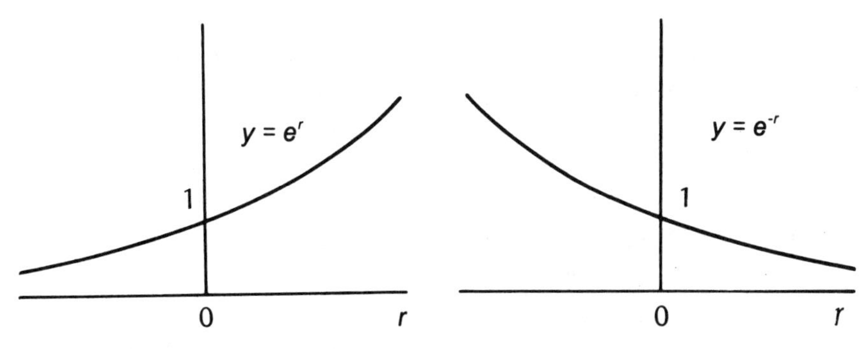

A.20: The Curves $y = e^r$ *and* $y = e^{-r}$

added to get the final figure. This may make it easier to visualize the relative significance of the various contributions. In these calculations the logarithms are usually taken to base ten.

If P_1 and P_2 are two powers, then $\log_{10}(P_1/P_2)$ is a measure of their ratio in a unit known as a bel. The conventionally used unit is a tenth of a bel, or decibel, symbolized by dB. Thus

$$10 \log_{10} (P_1/P_2) = \text{measure of power ratio in decibels} \tag{A.69}$$

As an example, a path loss of 120 dB corresponds to a power ratio given by $10 \log (P_1/P_2) = -120$, or $P_1 = P_2 \times 10^{-12}$.

Since power is proportional to fieldstrength (or voltage or current) *squared*, the decibel measure for these quantities is obtained by taking *twenty* times the logarithm. Thus a path loss of 120 dB corresponds to a fieldstrength ratio given by

$$20 \log (E_1/E_2) = -120, \text{or } E_1 = E_2 \times 10^{-6}.$$

Decibels refer to power *ratios*. Sometimes the *reference level* may be stated explicitly, sometimes it may be implied. The symbol dBW means decibels relative to a power level of 1 watt. The symbol dBmW, often contracted to dBm, means decibels relative to one mW. Decibels of sound are relative to the threshold power level of hearing for the normal ear.

A.5.2 The Neper

Many physical quantities grow or attenuate exponentially. Thus, the decay of the electric field through a lossy material would be given by

$$E_x = E_o e^{-\alpha x} \tag{A.70}$$

where E_x = fieldstrength at depth x(x in meters), E_o = fieldstrength at the surface (x = 0), and α = attenuation coefficient.

The units for α are neper per meter. At one meter depth $E_x/E_0 = e^{-1}$. The decibel measure of this is $20 \log_{10} e^{-1} = -8.686$. This attenuation is also, by definition, one neper. Thus 1 neper = 8.686 decibels.

The depth at which the field has attenuated by one neper is known as the *skin depth*, sometimes denoted by δ. Thus $\alpha\delta = 1$ or $\delta = 1/\alpha$: the skin depth is the inverse of the attenuation constant (in nepers per meter).

A.6 GEOMETRICAL FEATURES

A.6.1 Horizon Distance

Figure A.21 depicts a position at height h above the surface of the earth, radius R, with distance D to the horizon. By applying Pythagoras's theorem we get

$$(h + R)^2 = R^2 + D^2 \text{ or } D^2 = 2h(R + h) \quad \text{(A.71)}$$

Usually, h is completely negligible compared to R, and (A.71) is approximated by

$$D = (2Rh)^{1/2} \quad \text{A.72)}$$

This relation can also be solved for h, giving the height necessary for a given horizon distance

$$h = D^2/2R \quad \text{(A.73)}$$

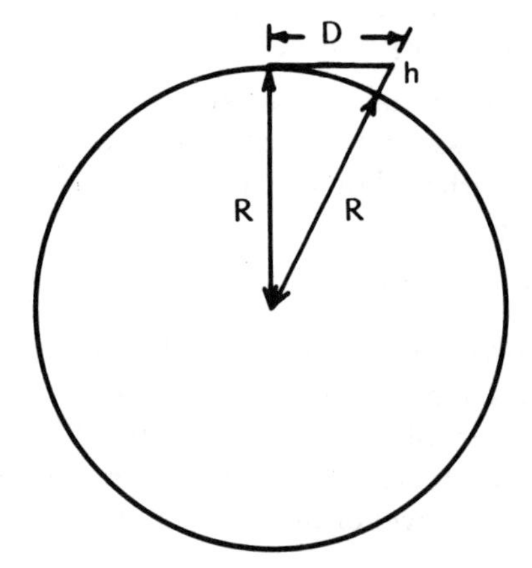

A.21: Horizon Distance

In these formulas, all distances should be in the same units, such as meters. However, in engineering applications it is often the case that D is given in miles and h in feet. The value of R is 3960 miles, a number which conveniently happens to be $3/4 \times 5280$; and 5280 is the number of feet in a mile. Hence, (A.72) can be re-written

$$D = (3h/2)^{1/2} \quad \text{(A.74)}$$

with D in the miles and h in feet.

When two observation heights are given, as in Figure A.22, the separate line-of-sight distances D_1 and D_2 must each be calculated from the heights h_1 and h_2, using (A.72). The maximum separation distance D is then the sum of the two grazing incidence distances:

$$D = D_1 + D_2 = (2Rh_1)^{1/2} + (2Rh_2)^{1/2} \qquad \text{(A.75)}$$

If a clearance d, as shown in Figure A.23, is needed, it must be subtracted from *each* height before doing the calculation indicated in (A.75)

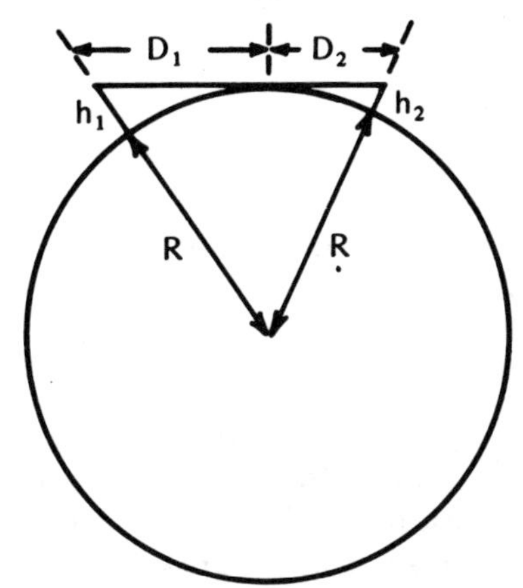

A.22: Grazing Range

A.6.2 Atmospheric Refraction of Radio Waves

The preceding results hold for calculations of the *optical* horizon. The radio horizon for line-of-sight links is somewhat different due to atmospheric bending of the rays. This effect is more pronounced for radio waves due to the relatively higher dielectric constant of water vapor at the radio frequencies. It is the gradient of the refractive index of the air, due mainly to variations with height of the water vapor content, that causes the radio waves to be refracted and to follow curved rather than straight lines. It happens that the effect can be allowed for by using a *fictitious* or *effective* earth radius, R_{eff}, in the preceding formulas, where R_{eff} differs from the actual earth radius R by a curvature factor K such that one can write

$$R_{eff} = KR \qquad \text{(A.76)}$$

The value of K depends on atmospheric conditions, but its *average* value is about 4/3. Hence, the radio horizon, *under average conditions* is $(4/3)^{1/2}$, or about 15%, greater than the optical horizon. However, K changes greatly, depending on conditions, and allowance for its range must be made in any practical calculations of clearance or tower height.

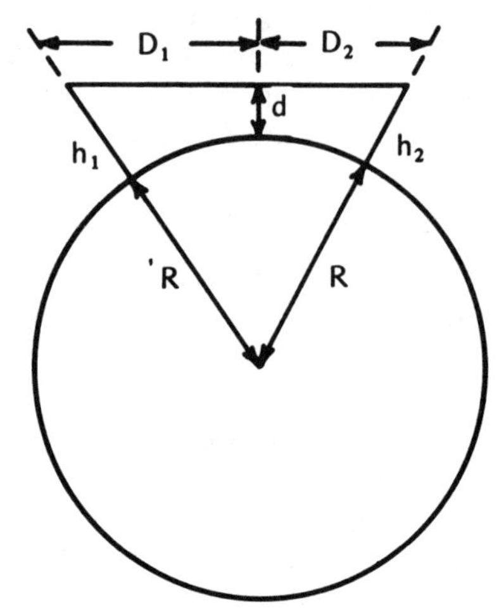

A.23: Clearance Range

A.6.3 Triangles

By inspection of the triangle in Figure A.24 and comparison with the surrounding rectangle it is seen that the area is half that of the rectangle; i.e., the area of a triangle is half its base times its height. From the figure, $h/a = \sin B$ where B is the angle at the apex B. Hence

$$\text{Area} = (1/2)\, ac \sin B \tag{A.77}$$

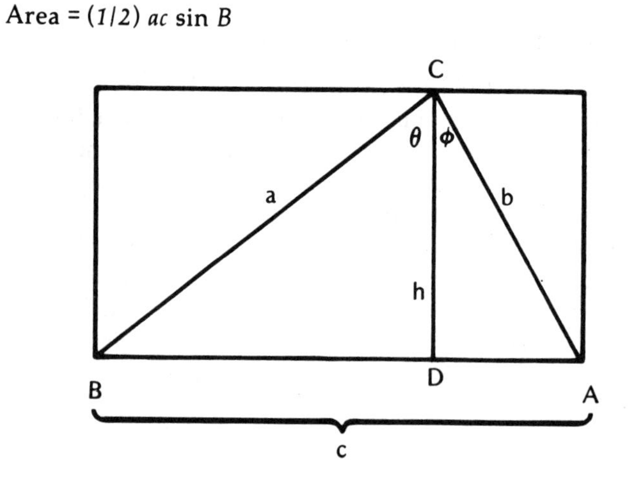

A.24: Triangle and Circumscribing Rectangle

By symmetry the area also equals (1/2) bc sin A or (1/2) ab sin C. Equating these results gives, on re-arrangement

$$\frac{a}{\sin A} = \frac{b}{\sin B} = \frac{c}{\sin C} \tag{A.78}$$

This equation can be used for calculating the remaining sides and angles if two sides and the included angle, *or* two angles and the adjacent side are known. Note that *always* $A + B + C = 180°$ for a triangle. The largest angle is opposite the largest side. Equation (A.78) gives the sine of the angle, leading to *two* possible vlaues between 0 and 180°. The correct value has to be selected from a consideration of the above information. The above area formula
can be used to prove the addition formula for the sine function, as displayed in (A.18). In Figure A.24 $C = \theta + \phi$ and the area formula can be written

$$\text{Area} = (1/2)\, ab \sin (\theta + \phi)$$

It also equals area BCD + area CDA = (1/2) ah sin θ + (1/2) bh sin θ. But $h = a \cos \theta$ and it also equals $b \cos \phi$. Hence, area BCD + area CDA = (1/2) ab cos ϕ sin θ + (1/2) ab cos θ sin ϕ. Equating the two results and cancelling (1/2) ab gives the addition formula for the sine function:

$$\sin (\theta + \phi) = \sin \theta \cos \phi + \cos \theta \sin \phi$$

An extension of Pythagoras's theorem to non-right angle triangles is as follows. From triangle BDC, since $BD = a \cos B$,

$$a^2 = h^2 + (a \cos B)^2 \tag{A.79}$$

Similarly, from triangle CDA, since $DA = c - BD = c - a \cos B$

$$b^2 = h^2 + (c - a \cos B)^2 \tag{A.80}$$

Subtracting (A.80) from (A.79) gives

$$a^2 - b^2 = (a \cos B)^2 - (c - a \cos B)^2 = 2\, ac \cos B - c^2$$

Hence,

$$b^2 = a^2 + c^2 - 2\, ac \cos B \tag{A.81}$$

This equation, and the two similar ones obtained by cyclic interchange of the letters, also enables the angles of a triangle to be calculated when all three sides are given. Since $\cos B = 0$ when $B = 90°$, (A.81) reduces to the usual Pythagorean form for a right-angled triangle.

A.6.4 Areas and Volumes

A few useful results will be quoted here. Most of them require the calculus for their proof, and a further discussion will be given in section A.9.

Area

(a) Circle, radius R. Area = πR^2

(b) Circular sector, radius R, angle of sector α (*in radians*), Area = $\alpha R^2/2$

(c) Sphere, radius R. Area of surface = $4\pi R^2$

(d) Cylinder, radius R, height h. The lateral surface area can be calculated by "unrolling" the surface, and gives Area = $2\pi Rh$

(e) Cone, base radius R, *slant* height h. Again the lateral surface area can be calculated by unrolling the surface, and gives Area = πRh

Volumes

(a) Sphere, radius R. Volume = $4\pi R^3/3$

(b) Cylinder, radius R, height h. Volume = $\pi R^2 h$

(c) Cone, base radius R, *vertical* height h. Volume = $\pi R^2 h/3$

A.7 COMPLEX NUMBERS

A.7.1 The Quantity j

Since $2 \times 2 = 4$ and $(-2) \times (-2) = 4$ also, the square root of 4 can be either plus or minus two. The square root of *minus* four can therefore, be neither of these, and involves a new quantity denoted by j (or sometimes i) which is equal to the square root of minus one;

$$j^2 = -1 \text{ or } j = \sqrt{-1} = (-1)^{1/2} \tag{A.82}$$

In terms of it we can write $\sqrt{-4} = \sqrt{(-1 \times 4)} = \sqrt{(-1)} \times \sqrt{4} = \pm j2$.

The *real* numbers can be plotted along a single axis, such as the x-axis. *Imaginary* numbers can be plotted along a perpendicular axis, such as the y-axis, as shown in Figure A.25.

To see that this representation is *self-consistent*, we note that multiplying 1 by j rotates the position of the point P, representing 1, to the point Q,

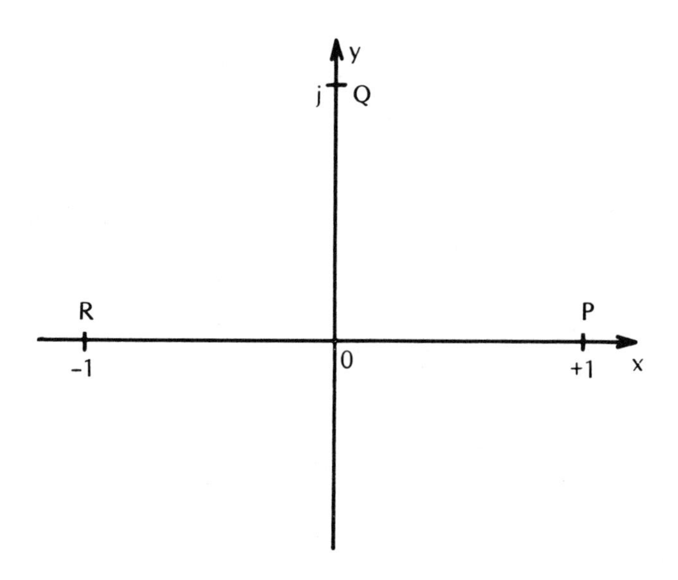

A.25: The Argand Diagram

representing j, a counter-clockwise rotation of 90°. A further multiplication by j, or a further counter-clockwise rotation by 90° takes P to the point R represented by –1. Thus $j \times j \times 1 = -1$ or $j^2 = -1$.

A.7.2 Complex Numbers

The representation of imaginary numbers in this way is a permissible, self-consistent visual aid, and is called the *Argand diagram*. Points off the axes are called *complex numbers*. Thus both real and imaginary numbers are special cases of complex numbers. A complex number is of the form $(a + jb)$, and is represented by the point $x = a, y = b$. Thus $(a + jb)$ and (a, b) are equivalent representations, but, in the former, the two real numbers a and b are *combined together* to form a *single* complex number. Four examples are shown in Figure A.26.

If z is the complex number $a + jb$, then we write

$$\begin{aligned} \text{Re}\, z &= a \\ \text{Im}\, z &= b \ (\text{not } jb) \end{aligned} \tag{A.83}$$

where Re and Im stand, respectively, for "the real part of" and "the imaginary part of." When adding complex numbers, the real parts are added and the imaginary parts are added; $(2 + j4) + (1 - j2) = (2 + 1) + j(4 - 2) = 3 + j2$. When multiplying, the expressions are multiplied as if they were ordinary numbers, and use is made of $j^2 = -1$. Thus,

$$(5 + j7) \cdot (2 + j3) = 5 \times 2 + j7 \times 2 + 5 \times j3 + j^2\, 7 \times 3$$
$$= 10 + j14 + j15 - 21$$
$$= -11 + j29$$

Division is more complicated, and is achieved by multiplying both numerator and denominator by the *complex conjugate* of the denominator; i.e., the expression in the denominator *with the sign of j changed*. Thus,

$$\frac{50 + j10}{3 - j4} = \frac{(50 + j10)(3 + j4)}{(3 - j4)(3 + j4)}$$

$$= \frac{50 \times 3 + j10 \times 3 + j50 \times 4 + j^2\, 10 \times 4}{3 \times 3 + j3 \times 4 - j4 \times 3 - j^2\, 4 \times 4}$$

$$= \frac{(150 - 40) + j(30 + 200)}{9 + 16} = \frac{110 + j230}{25} = 4.4 + j9.2$$

Note that, since $-j^2 = +1$ the *denominator* will *always* consist of a *sum* of two squares.

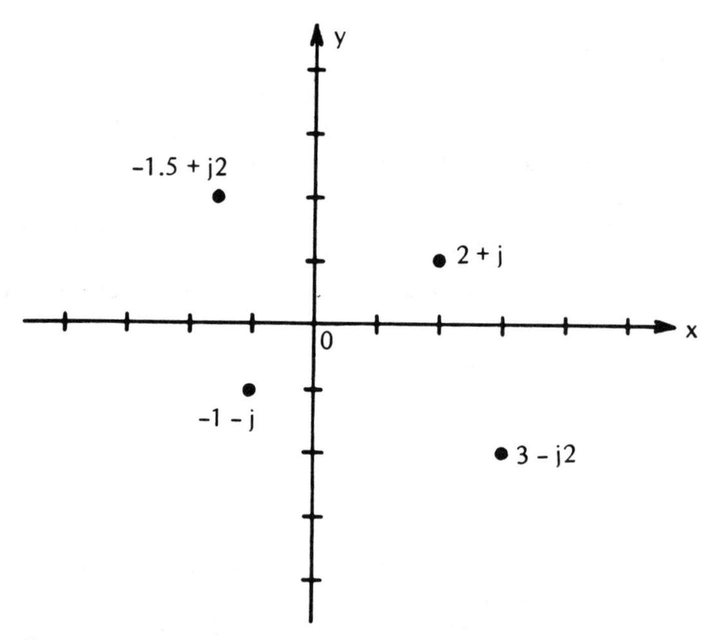

A.26: Illustration of Complex Numbers on the Argand Diagram

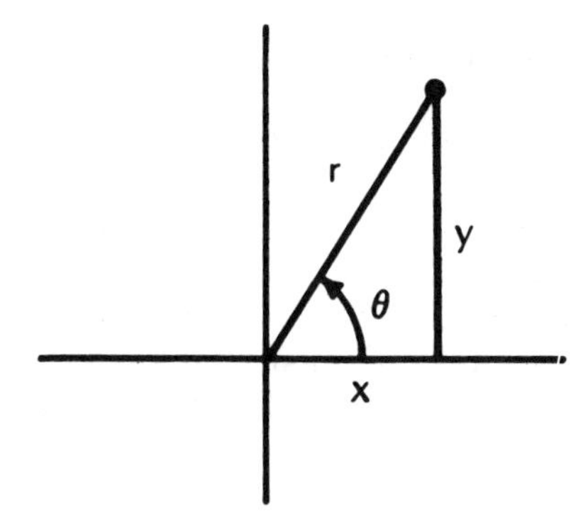

A.27: Polar Representation

A.7.3 Polar Representation

A point can also be represented by the distance r (a positive quantity) from the origin, and the angle θ of a counter-clockwise rotation. Thus we can write $x + jy$, or (x, y) in the form (r, θ). *This* form does *not* combine r and θ into a single quantity. This will be done later. Meanwhile, we see from Figure A.27 that

$$x = r \cos \theta$$
$$y = r \sin \theta \qquad \text{(A.84)}$$

This equation gives x and y in terms of r and θ. Correspondingly,

$$r = + \sqrt{(x^2 + y^2)}$$
$$\theta = \tan^{-1} (y/x) \qquad \text{(A.85)}$$

Since the tangent function is periodic with period 180°, the value of θ obtained from (A.85) *has to be checked for the correct quadrant*.

A.7.4 The Complex Exponential

The combination $f(\theta) = \cos \theta + j \sin \theta$ is *very* important in subsequent developments. By direct multiplication, we see that

$$\begin{aligned} f(\theta) \times f(\phi) &= (\cos \theta + j \sin \theta)(\cos \phi + j \sin \phi) \\ &= (\cos\theta \cos\phi - \sin\theta \sin\phi) + j(\sin\theta \cos\phi + \cos\theta \sin\phi) \\ &= \cos(\theta + \phi) + j \sin(\theta + \phi) \end{aligned}$$

(on using the formulas for cosine and sine of a sum of two angles)

$= f(\theta + \phi)$, from the definition of f

Since $e^{K\theta} \times e^{K\phi} = e^{K(\theta+\phi)}$, where K is any constant we see that $f(\theta)$ and $e^{K\theta}$ both satisfy the same functional form, namely, the law of indices. This makes it plausible that there is a value of K for which $\cos\theta + j\sin\theta = e^{K\theta}$. To see what value this is we note that, for small θ, $\cos\theta \approx 1$ and $\sin\theta \approx \theta$, while $e^{K\theta} \approx 1 + K\theta$. Hence, $\cos\theta + j\sin\theta \approx 1 + j\theta \approx 1 + K\theta$, giving $K = j$. Hence

$$\cos\theta + j\sin\theta = e^{j\theta} \tag{A.86}$$

(This is not, of course, a rigorous mathematical proof.)

We are now in a position to combine (r, θ) into a single complex form. From (A.84)

$$x + jy = r\cos\theta + jr\sin\theta = r(\cos\theta + j\sin\theta) = re^{j\theta} \tag{A.87}$$

Thus, if two complex numbers can be represented in polar form, they can be multiplied by multiplying their amplitudes and adding the angles.

$$(r_1 e^{j\theta_1}) \times (r_2 e^{j\theta_2}) = r_1 r_2 e^{j(\theta_1 + \theta_2)} \tag{A.88}$$

A.7.5 Square Roots

Equation (A.88) can be used for finding the square roots of complex numbers. Thus, if

$$(R e^{j\theta})^{1/2} = r e,^{j\phi} \text{ then } (r e^{j\phi})^2 = R e^{j\theta}$$

Using (A.88) we get $r^2 e^{j2\phi} = R e^{j\theta}$ so that $r^2 = R$ and $2\phi = \theta$. Hence

$$(R e^{j\theta})^{1/2} = R^{1/2} e^{j\theta/2} \tag{A.89}$$

Thus one takes the square root of the amplitude and halves the angle. Since the angle could also have been increased by 360° without altering the original quantity, (A.89) could also be interpreted as involving $e^{j(\theta+360°)/2} = e^{j\theta/2} e^{j180°}$. But $e^{j180°} = \cos 180° + j\sin 180° = -1$. Hence, another square root is $-R^{1/2} e^{j\theta/2}$. This gives the familiar $\pm$ for square roots. In a physical situation, the physical requirements may determine which root to use. They are both equally valid mathematically.

A.7.6 The Complex Propagation Constant

Since $\cos\theta + j\sin\theta = e^{j\theta}$ we can put

$$\cos\theta = \text{Re}\,(e^{j\theta}) \tag{A.90}$$

This important relation enables us to replace trigonometric functions by exponentials. Thus,

$$E_o \cos(\omega t + \phi - 2\pi x/\lambda) = E_o\,\text{Re}\,(e^{j(\omega t + \phi - 2\pi x/\lambda)})$$

$$= \text{Re}\,\{E_o e^{j(\omega t + \phi)} \times e^{-j2\pi x/\lambda}\}$$

In this way the time and distance-dependent terms can be *separated*.

If a wave is attenuated with attenuation constant α, then if E_x is the value of E at depth x we can write, since $E_0 e^{-\alpha x}$ is real,

$$E_x = E_o e^{-\alpha x} \cos(\omega t + \phi - 2\pi x/\lambda) \tag{A.91}$$

$$= \text{Re}\,\{E_o e^{j(\omega t + \phi)} \times e^{-(\alpha + j2\pi/\lambda)x}\} \tag{A.92}$$

It is seen that the attenuation and distance-dependent phase terms have been *combined* and also separated from the time-varying term. The quantity $e^{-(\alpha + j2\pi/\lambda)x}$ contains important valuable information on the behavior of the wave. The *interpretation* is via (A.91) and (A.92), but it is convenient to work with this term isolated.

The quantity $\alpha + j2\pi/\lambda$ is sometimes denoted by γ and is called the *complex propagation constant*. The *real part* gives the *attenuation*, and the *imaginary part* gives the *velocity of the wave*. Since

$\cos(\omega t - 2\pi x/\lambda) = \cos[\omega(t - x/v)]$ we get $\omega/v = 2\pi/\lambda$ or

$$v = \omega/(2\pi/\lambda) = \omega/(\text{Im}\,\gamma) \tag{A.93}$$

which, together with

$$\alpha = \text{Re}\,(\gamma) \tag{A.94}$$

give important wave properties in terms of γ.

The quantity $2\pi/\lambda$ is sometimes denoted by β or by k, so that we can write

$$\gamma = \alpha + j\beta = \alpha + jk \tag{A.95}$$

Equation (A.93) can thus also be written $v = \omega/k$.

A.7.7 The Complex Permittivity

Currents in a material involve two aspects, in general. *Conduction currents* involve flow of electrons, and are associated with the conductivity σ of the material.

Displacement currents involve *charges* and are associated with the dielectric constant ϵ of the material. Suppose a block of material sandwiched between two parallel plates of area A and separation D. If C is the capacitance of the plates then $C = \epsilon A/D$, and the charge for a voltage V is $Q = VC = \epsilon V\ A/D$. The displacement current is the rate of change of charge. If V is sinusoidal, and given by $V_0 \cos \omega t$, its rate of change (see later) is $-\omega V \times \sin \omega t$, so that the displacement current is $(V_0\, A/D)\, [-\omega\epsilon \sin \omega t]$.

The conduction current is $V/R = V/(A/\sigma D) = (V_0\, A/D)\, \lfloor \sigma \cos \omega t \rfloor$. Hence, the *total current* involves $\sigma \cos \omega t - \omega\epsilon \sin \omega t = Re[e^{j\omega t}\,(\sigma + j\omega\epsilon)]$. Thus, using complex number notation, the effect of conductivity can be taken into account by augmenting the term $j\omega\epsilon$ by σ, or alternatively, by augmenting ϵ by $\sigma/j\omega$. The quantity

$$\epsilon_c = \epsilon + \sigma/j\omega \tag{A.96}$$

is called the *complex permittivity*. It can also be written

$$\epsilon_c = \epsilon_0\,(\epsilon_r + \sigma/j\omega\epsilon_0) = \epsilon_0\,(\epsilon_r - j\,60\lambda_0\,\sigma) \tag{A.97}$$

where λ_0 is the free-space wavelength and the factor 60(ohms) comes as

$1/\omega\lambda_0\epsilon_0 = 1/[2\pi f\lambda_0/\,(4\pi\cdot 9\cdot 10^9)] = 1/[2\pi\cdot 3\cdot 10^8\,/(4\pi\cdot 9\cdot 10^9)] = 1/[1/60] = 60.$

In free space the complex propagation constant is simply $j2\pi/\lambda_0$, but in a medium whose properties are described by ϵ_c, the value of γ is $j(2\pi/\lambda_0)\,(\epsilon_c/\epsilon_0)^{1/2}$.

Hence

$$\alpha + jk = j\,(2\pi/\lambda_0)\,(\epsilon_c/\epsilon_0)^{1/2} \tag{A.98}$$

so that

$$k = \mathrm{Re}\,[\,(2\pi/\lambda_0)\,(\epsilon_c/\epsilon_0)^{1/2}\,] \tag{A.99}$$

$$\alpha = -\mathrm{Im}\,[\,(2\pi/\lambda_0)\,(\epsilon_c/\epsilon_0)^{1/2}\,] \tag{A.100}$$

The last two equations are extremely important for determining velocity and attenuation in a material.

A.7.8 Radio Propagation in Water

As an example of the application of the above results, let us consider the evaluation of the velocity and attenuation of a radio wave in fresh water at a frequency of 300 kHz. The electrical constants are $\epsilon_r = 80$ and $\sigma =$ 1m mho/m. Hence,

$$\begin{aligned}
\epsilon_c/\epsilon_o &= 80 - j60\,(3 \cdot 10^8/3 \cdot 10^5) \cdot 10^{-3} \\
&= 80 - j60 \\
&= (80^2 + 60^2)^{1/2}\, e^{-j\tan^{-1}(60/80)} \\
&= 100\, e^{-j36.8^\circ} \\
(\epsilon_c/\epsilon_o)^{1/2} &\; 100^{1/2}\, e^{-j(36.8/2)^\circ} \\
&= 10\, e^{-j18.4^\circ} = 10\,[\cos 18.4^\circ - j \sin 18.4^\circ] \\
&= 9.49 - j3.16
\end{aligned}$$

This gives

$$\gamma = (6.28/1000)\,(9.49 - j3.16) = 0.06 - j0.02$$

Hence, from (A.99) and (A.100),

$$k = 0.06 \text{ radian/m}$$

$$\alpha = 0.02 \text{ neper/m} = 0.17 \text{ dB/m}$$

The velocity of the wave is

$$v = \omega/k = (6.28 \cdot 3 \cdot 10^5)/0.06 = 3.16\; 10^7 \text{ m/s}$$

This reduction in velocity relative to free space is largely due to the dielectric constant. However, for appreciably higher conductivities or lower frequencies, the term $60\,\lambda_o\,\sigma$ becomes dominating over ϵ_r, and it is the former that then mainly determines the properties. Since, for sea water, $\epsilon_r = 81$ and $\sigma = 4$ mho/m, the two terms are equal when $81 = 160 \times 4 \times \lambda_o$ or $\lambda_o = 34$ cm. Thus, for frequencies less than 1 GHz it is the conductivity that mainly determines both the attenuation *and* the velocity in the sea water. The usual expression for refractive index $n = \epsilon_r^{1/2}$ has to be modified to $n = Re(\epsilon_c/\epsilon_0)^{1/2}$ and, in the case of sea water below about 1 GHz, the dielectric constant plays little part. Such a material is said to possess metallic properties, since, in a metal, it is *only* the conductivity which determines the response.

A.8 DIFFERENTIAL AND INTEGRAL CALCULUS

A.8.1 Differentiation

The process of differentiation is concerned with finding the slope or gradient of a curve at a point, and is defined as the tangent of the angle of slope. It is obtained by approximating an arc of the curve by a chord. As the chord and arc become smaller and smaller they approach equality; in the limit of a vanishingly small arc, the process yields the slope of the curve.

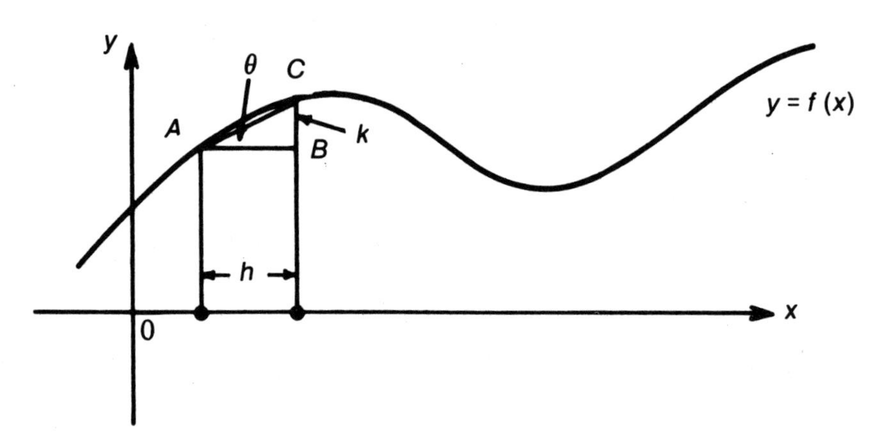

A.28: Illustration of Differentiation

Figure A.28 shows a triangle ABC in which AC is the chord joining the points A and C on the curve. The side AB is denoted by h and BC by k. The tangent of the angle BAC is therefore, k/h, and we need to find the limiting value of this as k and h approach zero. If the equation of the curve is $y = f(x)$ then at the point A we have simple $y = f(x)$ where x and y are the coordinates of A. The coordinates of C are $x + h$ and $y + k$, and these are related by $y + k = f(x + h)$ since C is also on the curve. Hence, from the triangle ABC we get

$$\tan\theta = \frac{BC}{AB} = \frac{k}{h} = \frac{f(x+h) - f(x)}{h} \tag{A.101}$$

This is true for any size triangle ABC. If we now take the limit for vanishingly small h, θ becomes the angle of slope of the curve at A and we get

$$\tan \theta = \lim_{h \to 0} \frac{f(x+h) - f(x)}{h} \quad \text{(A.102)}$$

The limiting form of h is usually written dx meaning "a differential increment in x." The d is not to be treated as a multiplier and cannot be separated from the x; dx is an entity. Similarly k is written as dy. Hence, (A.101) and (A.102) can be written

$$\tan \theta = \frac{dy}{dx} = \lim_{dx \to 0} \frac{f(x+dx) - f(x)}{dx} \quad \text{(A.103)}$$

A few important examples follow.

A.8.2 Examples of Differentiation

By following the process indicated in (A.103) the following results are deduced form first principles:

(a) $y = Cx^n$, where C and n are constants
Then

$$y + du = C\,(x + dx)^n$$
$$= C\,(x^n + nx^{n-1}\,dx + \ldots.)$$

where the binomial theorem has been used to expand $(x + dx)^n$, and higher order terms in (dx) have not been explicity shown. The next step is to subtract $y = Cx^n$ from both sides, divide by dx, and take the limit as $dx \to 0$. A term such as $(dx)^2$ will thus be reduced to dx and will vanish in the limit, as will still higher-order terms. Only the term initially proportional to dx will survive the process. Hence, we get $dy = C(x^n + nx^{n-1}\,dx + \ldots.) - Cx^n$

$$\frac{dy}{dx} = Cnx^{n-1} + \text{terms vanaishing with dx}$$

On taking the limit as $dx \to 0$ this gives

$$\frac{dy}{dx} = Cnx^{n-1} \quad \text{(A.104)}$$

This method can be applied to a sum of terms, and leads to the formulation that differentiation of a sum equals the sum of the individual differentiated terms.

Note that in the special case $n = 0$, $y = C$, a constant and (A.104) then gives $dy/dx = 0$.

(b)

$$
\begin{aligned}
y &= Ce^{ax};\ \text{C and } a \text{ are constants.} \\
y + dy &= Ce^{a(x+dx)} = Ce^{ax}e^{adx} \text{ from (A.54)} \\
dy &= Ce^{ax}e^{adx} - Ce^{ax} \\
&= Ce^{ax}(e^{adx} - 1) \\
&= Ce^{ax}(adx + \ldots) \text{ from (A.67)}
\end{aligned}
$$

Hence

$$\frac{dy}{dx} = aCe^{ax} \text{ as } dx \to 0 \tag{A.105}$$

(c)

$$
\begin{aligned}
y &= \sin(\omega x + \theta);\ A,\ \omega \text{ and } \theta \text{ are constants.} \\
y + dy &= A\sin(\omega x + \theta + \omega dx) \\
&= A\sin(\omega x + \theta)\cos(\omega dx) + A\cos(\omega x + \theta)\sin(\omega dx) \text{ from (A.18)} \\
dy &= A\sin(\omega x + \theta)[\cos(\omega dx) - 1] + A\cos(\omega x + \theta)\sin(\omega dx)
\end{aligned}
$$

Dividing by dx and using the results from section A.2.10 that $\lim_{\phi \to 0} (1 - \cos\phi)/\phi \to 0$ and $\lim \sin(\phi)/\phi = 1$ we get

$$\frac{dy}{dx} = \lim_{dx \to 0} A\cos(\omega x + \theta)\frac{\sin(\omega dx)}{\omega dx} \cdot \omega = A\omega\cos(\omega x + \theta) \tag{A.106}$$

In a similar way, or by increasing θ in (A.106) by $\pi/2$ it can be shown that

(d) $y = A\cos(\omega x + \theta)$

leads to

$$\frac{dy}{dx} = -A\omega\sin(\omega x + \theta) \tag{A.107}$$

(e) $y = \log_e x$

Then $x = e^y$, and by taking $C = a = 1$ in (A.105) and interchanging the roles of x and y, we get

$$\frac{dx}{dy} = e^y$$

$$\frac{dy}{dx} = \frac{1}{x} \tag{A.108}$$

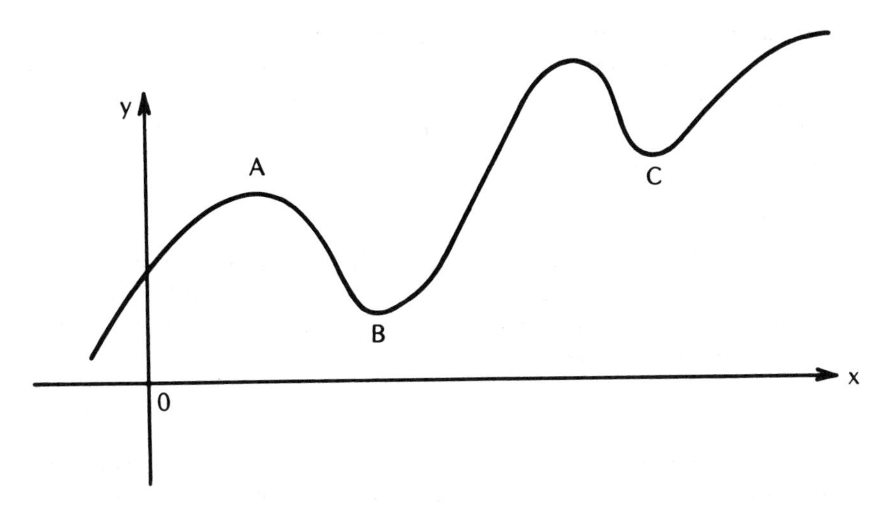

A.29: Maxima and Minima

A.8.3 Maxima and Minima

Figure A.29 shows a curve with a maximum at A and a minimum at B. Both positions are known as *turning points*, and at a turning point the gradient is zero. Hence, the equation $dy/dx = 0$ *locates* the values of x for which turning points occur. To find the value of y corresponding to a value of x, i.e., to actually determine the height of the maximum or minimum, the value of x has to be inserted in the equation $y = f(x)$ for the curve. It is usually clear by inspection whether a particular turning point is a maximum but in curves with several turning points it is possible to have a minimum which is actually higher than a maximum. This would be so for the position C, as compared to A, in Figure A.29.

As an example, take the curve $y = \cos x$ for which $dy/dx = -\sin x$. Since $\sin x = 0$ when $x = n\pi$ where n is any integer, including 0, we see that the cosine function has turning points at these positions. The value of $\cos x$ at one of these positions is $\cos(n\pi) = (-1)^n$. Hence, the turning points are alternately positive and negative, and are numerically of unit magnitude.

A.8.4 Integration

Perhaps the most suitable approach to defining integration is in the determination of the area under a curve. In Figure A.30 the area is denoted by S up to a particular value of x. If x is increased to $x + dx$, the area increases by an amount dS where dS is represented by the (approximate) rectangle of height y and width dx. As the limit $dx \rightarrow 0$ is taken the approximation (due to the curved upper edge of the rectangle) becomes exact, and we get $dS = ydx$ or

$$dS/dx = y \tag{A.109}$$

Now S can be thought of as a summation of many strips, each of the form ydx, or $S = \Sigma ydx$ where the summation covers the area of interest. In the limit as $dx \rightarrow 0$ the summation sign Σ is replaced by the integral sign $\int$ and the formula for S becomes

$$S = \int ydx \tag{A.110}$$

This result should be compared with (A.109), and it shows that if we know the expression which, when differentiated, gives the function y, then this expression represents the area S. Thus integration may be thought of as the inverse process to differentiation, and the results of section A.8.2 may be used to set up a table of results for integration. Note that since the differentiation of a constant gives zero, an arbitrary constant should *always* be added when an integration is performed. Its value can be determined from physical considerations in an actual problem. Sometimes, but not always, its value is zero.

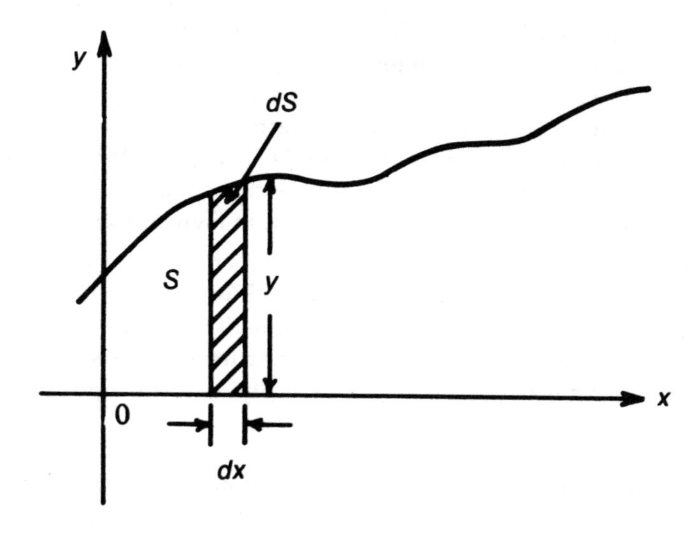

A.30 Area Under a Curve

Table A.II: Examples of Integration

y	Cx^n	$\frac{C}{x}$	Ae^{ax}	$A \sin(\omega x + \theta)$	$A \cos(\omega x + \theta)$
$\int y dx$	$\frac{Cx^{n+1}}{n+1}$	$C \log_e x$	$\frac{Ae^{ax}}{a}$	$-\frac{A}{\omega} \cos(\omega x + \theta)$	$\frac{A}{\omega} \sin(\omega x + \theta)$

In the above table the constant of integration has not been shown explicity. Note that the integration of x^{-1} is a special case that does not come under the first column but gives $\log_e x$ as in column 2.

A.8.5 Integration Limits

The expression in (A.110) is known as an *indefinite* integral, since the limits for determining the area S are not specified. If the lower limit is L_1 and the upper L_2, then (A.110) would be written

$$S = \int_{L_1}^{L_2} y dx \qquad \text{(A.111)}$$

and gives the area between the values L_1 and L_2 for x. If the function which, when differentiated, gives y, is denoted by $F(x)$ then (A.111) gives

$$S = F(L_2) - F(L_1) \qquad \text{(A.112)}$$

Note that a) the value at the lower limit L_1 is *subtracted* from the value at the upper limit L_2; and b) it is not necessary to show explicity the integration constant because it cancels out when the two functions are subtracted. What has really happened is that $F(L_1)$ is doing duty as the integration constant. This may be seen as follows: If we return to (A.110) we could write $S = F(x) + C$, and hence, $S(L_2) = F(L_2) + C$. What value is needed for C? Clearly $S = 0$ when $X = L_1$ so $S(L_1 1) = 0 = F(L_1) + C$, or $C = -F(L_1)$. Hence, $S(L_2) = F(L_2) - F(L_1)$, in agreement with (A.112). As an example, the area under the cubic curve $y = 2 + 4x^3$ between $x = 1$ and $x = 2$, shown in Figure A.31, is obtained from

$$S = \int_1^2 (2 + 4x^3)\, dx$$
$$= (2x + x^4)\Big|_1^2 \text{ from table A.1}$$
$$= (4 + 16) - (2 + 1) = 17 \text{ units of area}$$

A.8.6 Volumes of Revolution

In Figure A.32 the differential of volume, dV, is a disc of radius y and thickness dx. Hence, its value is $\pi y^2 dx$, and we get

$$V = \int \pi y^2\, dx \qquad \text{(A.113)}$$

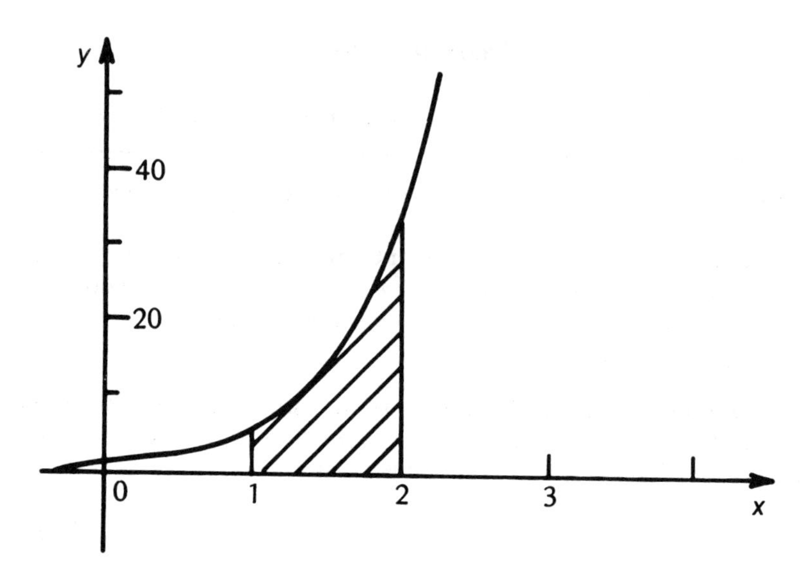

A.31: Area Under the Curve $y = 2 + 4x^3$ between $x = 1$ and 2

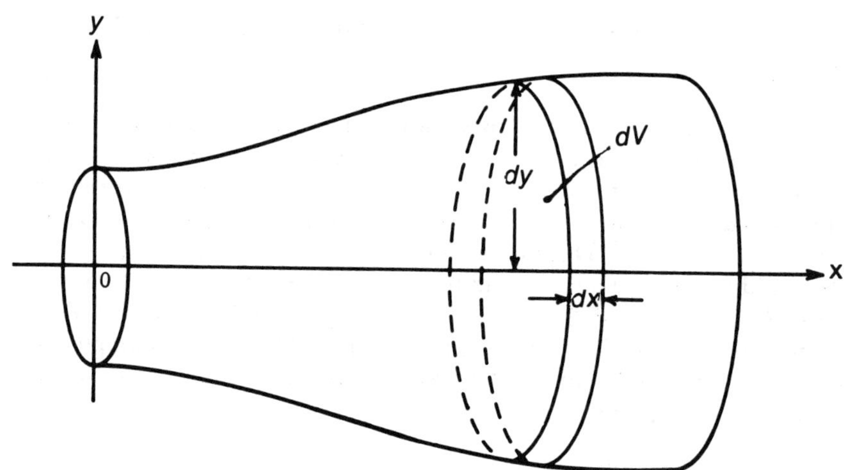

A.32: Volumes of Revolution

(a) Volume of a cone. The equation for the cone radius, shown in Figure A.33, is $y = x \tan \alpha$, where α is the half-angle at the cone apex. Hence

$$V = \int_0^L \pi x^2 \tan^2 \alpha \cdot dx = (\pi/3)\, x^3 \tan^2 \alpha \Big|_0^L = (\pi/3)\, L^3 \tan^2 \alpha$$

But $L \tan \alpha$ is the radius at the cone base, whose area is accordingly $A = \pi(L \tan \alpha)^2$. Hence

$$V = (1/3)\,(\text{Area of base}) \times (\text{perpendicular height}) \qquad (A.114)$$

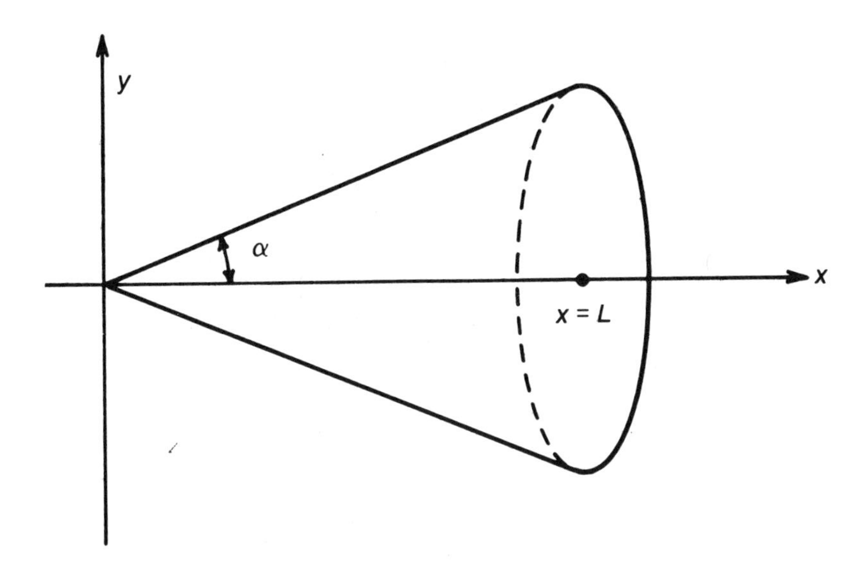

A.33: Cone of Semi-angle α

(b) Volume of a sphere. The equation of the circle which when rotated determines the spherical shape, is $y^2 = R^2 - x^2$ where R is the radius of the circle or sphere. Hence, as indicated in Figure A.34,

$$V = \int_{-R}^{R} \pi (R^2 - x^2)\, dx$$

$$= \pi (R^2 x - x^3/3) \Big|_{-R}^{R}$$

$$= \pi (R^3 - R^3/3) - \pi(-R^3 + R^3/3)$$

$$= 4\pi R^3/3 \qquad \text{(A.115)}$$

We can also get the volume of the sphere from its surface, since, if S is the surface, the volume dV of a spherical shell, as shown in Figure A.35, is given by $dV = SdR$. If we know S as a function of R, we can get V by integrating. Alternatively, knowing V we can calculate S by evaluating dV/dR. Thus, using (A.115) gives

$$S = dV/dR = 4\pi R^2 \qquad \text{(A.116)}$$

A.9 PROBABILITY

A.9.1 Permutations and Combinations

If there are n objects and we wish to select r of them, the *order of selection being important*, we can choose the first object in n ways and, for each of

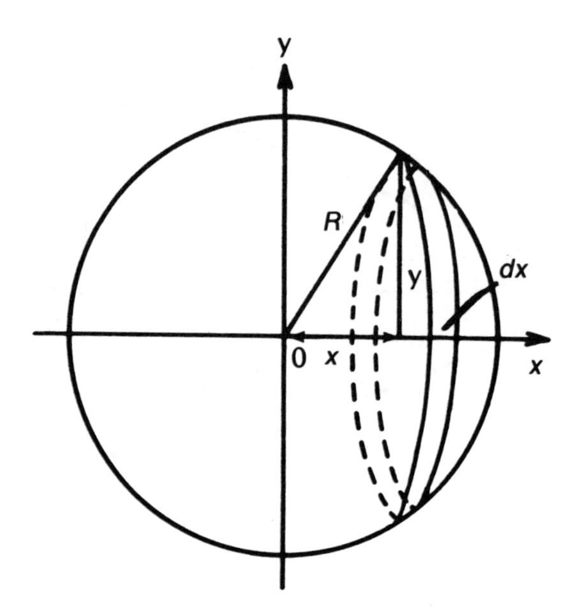

A.34: Calculation of Spherical Volume

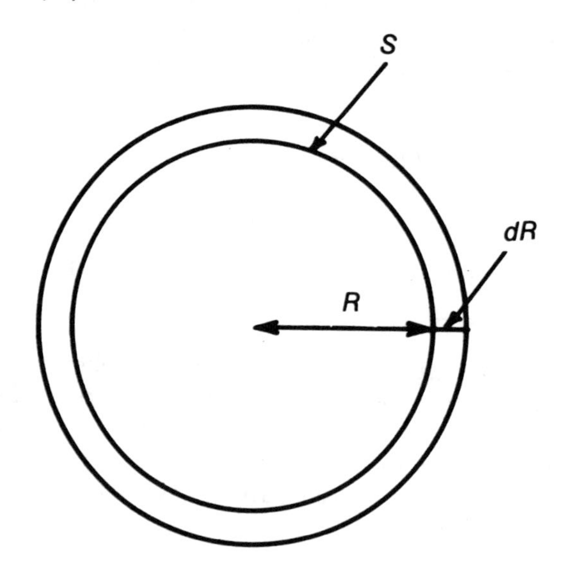

A.35: Calculation of Spherical Surface

these n ways, we can select the second object in $(n - 1)$ ways, and so on. The total number of ways is $n(n - 1)(n - 2)....(n - r + 1)$. This number is denoted by ^nP_r, the number of permutations of n objects r at a time.

$$^nP_r = n(n - 1).....(n - r + 1) = \frac{n!}{(n - r)!} \tag{A.117}$$

These r objects can be rearranged among themselves in $r!$ ways, since there is a choice of r for the first position, $(r - 1)$ for the second, etc. If we are interested in the number of ways of selecting r objects from n objects, when the *order of selection is of no importance,* then it is clear that (A.117) overestimates this number by a factor $r!$ Denoting by ^nC_r the number of ways of selecting r out of n objects, irrespective of order, then $^nC_r(r!) = {}^nP_r$, whence

$$^nC_r = \frac{n!}{r!(n - n!)} = \frac{n(n - 1)....(n - r + 1)}{r!} \tag{A.118}$$

Note that $^nC_n = 1$, since all n objects can be selected in only one way; i.e., by taking them all. This also follows from the formula, since $0! = 1$. From (A.118), by replacing r by $n - r$, we get the relation

$$^nC_r = {}^nC_{n-r} \tag{A.119}$$

which expresses a symmetry property of ^nC_r. For reasons that will shortly be apparent, ^nC_r is called a binomial coefficient. Alternative notations are C_r^n and $\binom{n}{r}$, the latter being the most frequent form encountered in mathematical work.

As an example of the application of these results, we shall now prove the binomial theorem for integer powers:

$$(x + y)^n = x^n + {}^nC_1 x^{n-1} y + {}^nC_2 x^{n-2} y^2 + ... + y^n \tag{A.120}$$

We can write $(x + y)^n = (x + y)(x + y)(x + y).....(x + y)$ (n factors). To get the coefficient of $x^{n-r}y^r$, we have to select $(n - r)$ symbols x and r symbols y when multiplying out the factors. The number of ways of selecting the x's is $^nC_{n-r}$, and the remaining factors automatically give the required number of y's. Hence, the required coefficient is $^nC_{n-r}$. From (A.119), $^nC_{n-r} = {}^nC_r$. Hence, we have the *binomial theorem* (A.120). As a check, $x^{n-1}y$ comes by selecting one y from any of the n factors, and x's from all of the others. There are n ways of doing this, agreeing with (A.120) since $^nC_1 = n$.

A.9.2 Examples of Permutation Calculations

Not all calculations are simple applications of (A.117) and (A.118). It is often necessary to break a problem up into parts to which the above formulas individually apply, and then combine the results appropriately. This can best be seen from a number of examples.

(a) There are 8 boys and 5 girls. How many ways can a group of 6 children be selected when

i) there is no restriction on the makeup of the group;
ii) there must be 4 boys and 2 girls;
iii) there must be at least 4 boys?

i) There are 13 children altogether, and the number of ways of selecting any 6 (the *order of selection* being of no consequence) is

$$^{13}C_6 = \frac{13.12.11.10.9.8}{6.5.4.3.2.1} = 1716.$$

ii) The number of ways of selecting 4 boys is $^8C_4$ and the number of ways of selecting 2 girls is $^5C_2$. These sets are *independent* so that the total number of ways is their product (for *each* of the $^8C_4$ ways of selecting the boys there are $^5C_2$ ways of selecting the girls — hence, the numbers are *multiplied*, not added)

$$^8C_4 \times {}^5C_2 = \frac{8.7.6.5}{4.3.2.1} \times \frac{5.4}{2.1} = 700.$$

iii) This can be done by selecting 4 boys and 2 girls, in $^8C_4 \times {}^5C_2$ ways; *or* 5 boys and 1 girl, in $^8C_5 \times {}^5C_1$ ways; *or* 6 boys, in $^8C_6$ ways. These are *mutually exclusive*, so that the total number of ways is their *sum*.

$$^8C_4 \times {}^5C_2 + {}^8C_5 \times {}^5C_1 + {}^8C_6 = 700 + 280 + 28 = 1008$$

Note carefully, that when the sets of numbers are *independent* the total number of ways is the *product* of the sets, but when they are *mutually exclusive* the total number is their sum.

(b) How many ways can 4 hands of 13 cards be dealt when i) there are no specifications on the deal; ii) one player (at least) should have all cards of one suit; iii) a specified player should have all the hearts?

i) The number of ways of dealing 4 hands in a particular order, with the order of the cards in the hands also specified is clearly 52! But each hand can be re-arranged in 13! ways, and the hands re-arranged amongst themselves in 4! ways. Hence, the total number of possible deals is $52!/(13!)^4\ 4!$ We take case iii) before ii). Since a specified player has all the hearts, the remaining 39 cards can be dealt to the remaining players (as in case i) in $39!/(13!)^3\ 3!$ ways.

ii) There are more ways of dealing ii) than iii) because any of the 4 suits could have been chosen, increasing the number of ways by 4; and *any* of the players could have been chosen for the special hand, again increasing the number of ways by 4. Hence, the total number of ways is $16 \cdot (39!)/(13!)^3\ 3!$.

A.9.3 Probability

If there are *n* independent ways favorable to the outcome of an event, and if there are *N* ways in which things could occur, and if all of these ways are equi-probable, then the probability of the event is n/N.

Example 1. What is the probability that at least one player will have a hand all one suit in a game of bridge? The total number of deals is as in i) above, and the number favorable deals is as in ii). Since these deals are all independent and equi-probable, the required probability is

$$\frac{16\,(39!)}{(13!)^3\ 3!} \div \frac{52!}{(13!)^4\ 4!} = \frac{64\,(13!)\,(39!)}{52!}$$

after some cancellation of factors (many more factors would also cancel). This number works out to about 1 chance in 10^{10}.

Example 2. What is the probability of throwing 10 or more with a throw of two dice? The total number of ways in which the dice can appear is $6 \times 6 = 36$. To get 10 or more we need

$12 = (6+6)$ 1 way only
$11 = (6+5)$ *or* $(5+6)$ 2 ways
$10 = (6+4)$ *or* $(5+5)$ *or* $(4+6)$ $= 3$ ways
10 or 11 or 12; $1+2+3$ ways $= 6$ ways
$p\,(10 \text{ or more}) = 6/36 = 1/6$

The probability of getting *exactly* 10 is $p(10) = 3/36 = 1/12$
The probability of *not* getting a ten is $(36-3)/36 = 11/12$
The probability of an event happening is denoted by p.
The probability of an event *not* happening is denoted by q.
Since the event must either happen or not happen, and these are *mutually exclusive possibilities*, one of which must happen, we get $p + q = 1$ (certainty).

Example 3. What is the probability of getting 3 heads *followed by* 3 tails in 6 spins of a penny? The probability that the first spin is heads is 1/2. The same for the second spin. These are *independent events*. Hence, the probabilty of the first two being heads is $1/2 \times 1/2 = 1/4$. Similarly for each spin, giving $(1/2)^6$ as the total probability. Note that the same result would have been obtained if the order of heads and tails had been different, so long as it were *specified.*

Example 4. What is the probability of 3 heads and 3 tails in 6 spins of a penny? This differs from the previous example in that the order is not specified and therefore, is of no importance. The three heads could have been selected in ${}^6C_3 = 20$ different ways, all mutually exclusive. Therefore we need $(1/2)^6 + \ldots (1/2)^6$, with 20 terms, i.e., $20 \times (1/2)^6 = 5/16$. (Having got 3 heads the rest *have to be tails*, and therefore, does not affect the calculation further.)

Example 5. What is the probability of at least 3 heads in 6 spins of a penny? The *mutually exclusive* possibilities are 3, 4, 5 or 6 heads. Their respective probability is therefore, the *sum* of these

$$p(\geq 3) = (1/2)^6 [{}^6C_3 + {}^6C_4 + {}^6C_5 + {}^6C_6] = \frac{20 + 15 + 6 + 1}{64} = \frac{21}{32}$$

In some cases it is easier to work with the non-occurence of an event, and to deduce its probability from $p = 1 - q$.

Example 6. What is the probability of getting at least one 6 in a throw of 4 dice? The probability of *not* getting a six in one throw is 5/6. Since the throws are independent, the probability of *no* sixes in 4 throws is $(5/6)^4$. Hence, the probability of *at least one six* (the mutually exclusive *and* exhaustive alternative to *no* sixes) is $1 - (5/6)^4 = 671/1296$.

A.9.4 Venn Diagrams

Probabilities can be represented as a ratio of two areas, a favorable area to a total area. The shape of the area is irrelevant. Favorable areas are shaded in the following. Figure A.36 shows eight examples of the representation of different types of events. One could imagine in the first example that the rectangle represents a dart board onto which darts are thrown in a uniform and random manner. Some of them will land in the shaded area A. The probability of this happening is proportional to the area A, and is equal to the area A to the total area. Examples of the use of this method of representation follow.

A.9.5 Joint Probability

We use $P(A \text{ or } B)$ to mean "the probability of event *A or B*," and $P(A, B)$ to mean "the probability of both *A and B*." From the third example in Figure A.36

$$P(A \text{ or } B) = \frac{(\text{area } A \text{ only}) + (\text{area } B \text{ only}) - \text{overlap area}}{S = \text{total area of rectangle}}$$

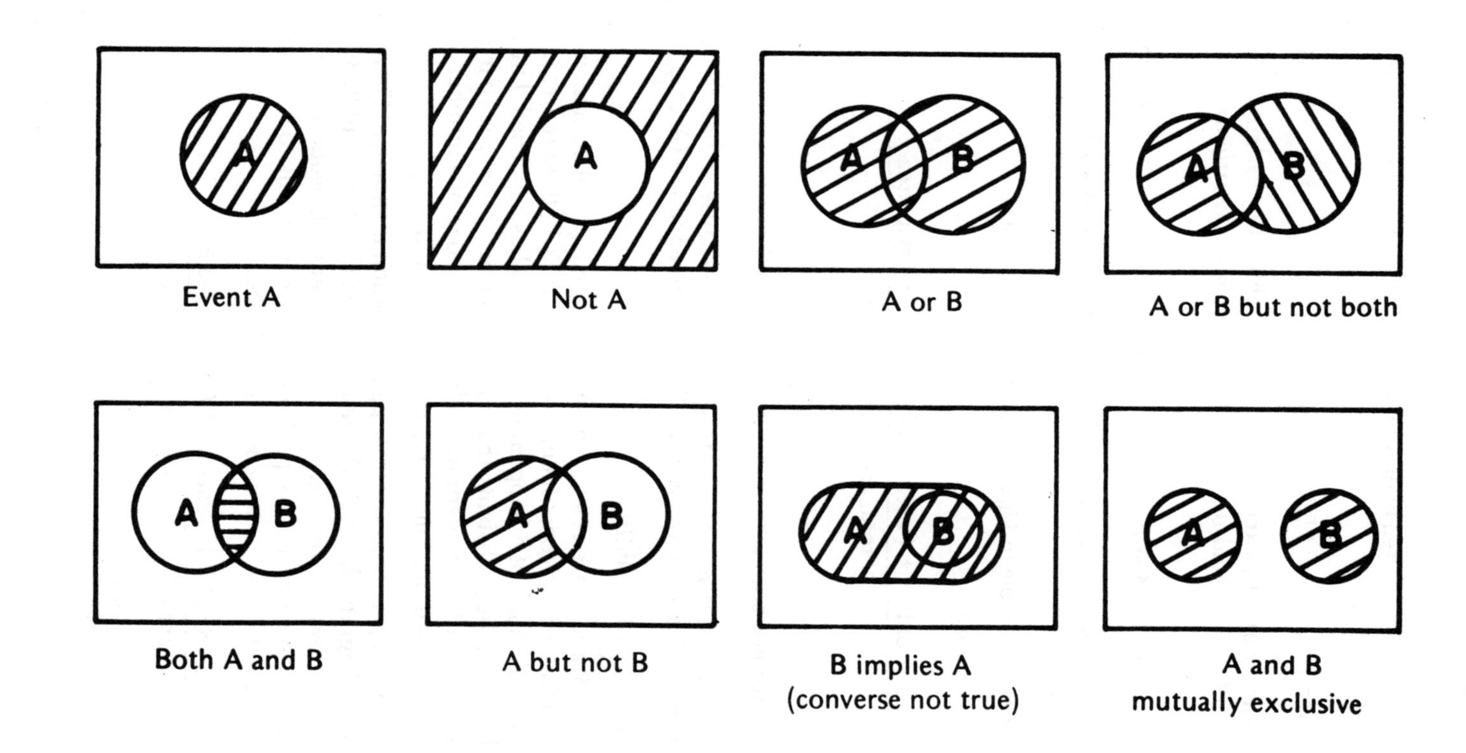

A.36: Venn Diagrams

The overlap area is subtracted in order *not to count it twice* in calculating the areas of A and B. Hence

$$P(A \text{ or } B) = P(A) + P(B) - P(A,B) \tag{A.121}$$

If the events are mutually exclusive or incompatible, they cannot occur together and $P(A, B)$ would be zero. Hence, for *mutually exclusive events* (only)

$$P(A,B) = 0 \text{ and } P(A \text{ or } B) = P(A) + P(B) \tag{A.122}$$

Note that this is the particular case of (A.121) when the events are mutually exclusive.

From the fourth example in Figure A.36 we deduce that

$P(A$ or B but not both)

$$= \frac{(\text{area } A \text{ only}) + (\text{area } B \text{ only}) - 2 \times \text{overlap area}}{S = \text{total area of rectangle}}$$

Hence,

$$P(A \text{ or } B \text{ but not both}) = P(A) + P(B) - 2P(A,B) \tag{A.123}$$

Note that the overlap is now subtracted twice because it has to be excluded from both area A and area B. Compare to equation (A.121).

A.9.6 Conditional Probability

If two events A and B are related in some way, e.g., total equipment failure, and failure of one component, we can write $P(A \mid B)$ to represent the *conditional* probability of A happening, *given* that B has happened. Thus one can ask what is the probability that a given transistor has failed if it is known that a piece of equipment has failed (for any one of a number of reasons including, but not limited to, a transistor failure).

The Venn diagram pertinent to this type of situation is shown in Figure A.37. The first diagram shows $P(A)$ when nothing in known about B. But if B is *known to have occurred*, the area favorable to event A is reduced to the overlap area of B. Also the field of possible events has shrunk to the area of B only (since B is *known* to have occurred). Hence

$$P(A \mid B) = \frac{\text{area }(A,B)}{\text{area } B} = \frac{\text{area }(A,B)}{\text{area } S} \div \frac{\text{area } B}{\text{area } S}$$

$$= \frac{P(A,B)}{P(B)}$$

This yields the theorem of *joint probability*:

$$P(A,B) = P(A|B)P(B) \quad \text{(A.124)}$$

A similar argument gives

$$P(B,A) = P(B|A)P(A) \quad \text{(A.125)}$$

But $P(A, B) = P(B, A)$ since both are the probability of A *and* B. Hence

$$P(A|B)P(B) = P(B|A)P(A) \quad \text{(A.126)}$$

This equation connects $P(A \mid B)$ with $P(B \mid A)$. Note that these probabilities are not the same thing. For example, if A stands for the event of a premature failure of a tube, and B for the event that its filament is defective, then $P(A \mid B)$ is the probability of a premature failure given that the filament is defective, whereas $P(B \mid A)$ is the probability that the filament is defective, given that the tube has prematurely failed. The first probability would be of interest for investigating the effects of a certain defect, the second for investigating the causes of a certain failure.

If A and B are *independent*, then A does not depend on B, so $P(A \mid B)$ is simply $P(A)$. Hence, (A.124) becomes

$$P(A,B) = P(A)P(B) \quad \text{(A.127)}$$

This is the formula for *multiplication* of probabilities for *independent* events.

If the event B can take place only in conjunction with n mutually exclusive events $A_1, A_2 \ldots A_n$ with conditional probabilities $P(B \mid A_r)$ for the rth event, then

$$P(B) = \sum_1^n P(B,A_r)$$

$$= \sum_1^n P(B|A_r)P(A_r) \quad \text{(A.128)}$$

This is called the theorem of *total probability*.

In many applications we know the value of $P(B \mid A)$ and need to get $P(A \mid B)$. This comes from (A.126), since

$$P(A|B) = \frac{P(B|A)P(A)}{P(B)} \quad \text{(A.129)}$$

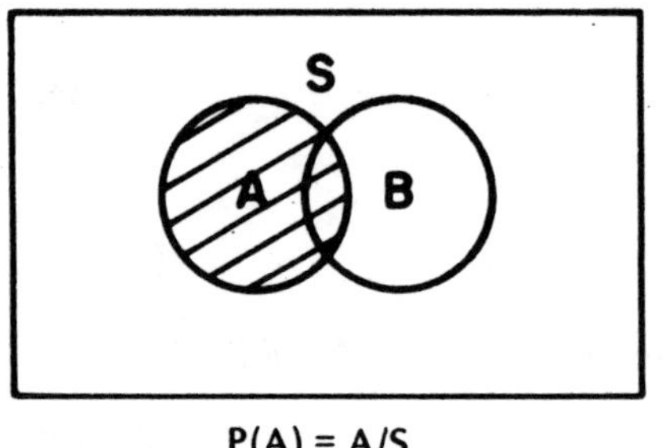

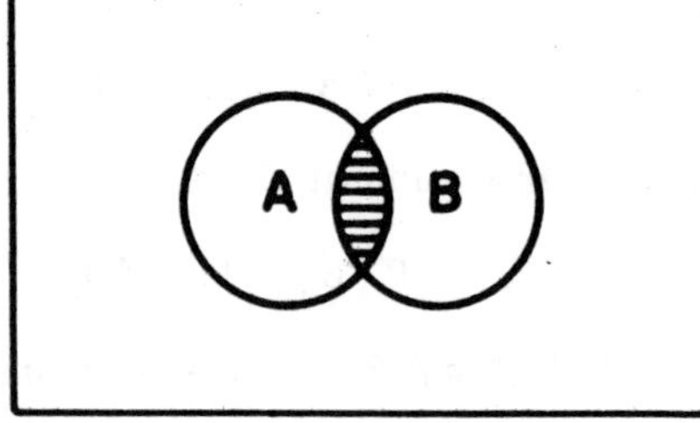

A.37: Venn Diagram Illustrating Conditional Probability

Using (A.128) to eliminate $P(B)$ we get

$$P(A \mid B) = \frac{P(B \mid A)P(A)}{\sum_{1}^{n} P(B \mid A_r)P(A_r)} \tag{A.130}$$

This extremely important theorem is known as Bayes' theorem.

A.9.7 Continuous Variables and Probability Density

A continuous random variable X may fall in the range $x < X < x + dx$ with a probability proportional to dx. The proportionality factor depends on x, and is known as a *probability density*. It is defined by

$$P(x < X < x + dx) = p(x)\,dx \tag{A.131}$$

Since X must take some value between $-\infty$ and $+\infty$ (in special cases within a more restricted range), we have

$$\int_{-\infty}^{\infty} p(x)\,dx = 1 \tag{A.132}$$

The *distribution function* $F(x)$ is defined by

$$F(x) = P(X < x) = \int_{-\infty}^{x} P(x)\,dx \tag{A.133}$$

The probability that X will fall in the finite range $a < x < b$ is accordingly

$$P(a < x < b) = \int_{a}^{b} p(x)\,dx = F(b) - F(a) \tag{A.134}$$

The joint probability density for two continuous random variables x and y is defined by

$$P(x < X < x + dx,\ y < Y < y + dy) = p(x,y)\,dx\,dy \tag{A.135}$$

In (A.130) the events may be associated with continuous variables X and Y rather than discrete sets A_r. In this case the summations are replaced by integrations, the range covering all permissible values of the variables. In the most general case this would be from $-\infty$ to $+\infty$, though in special cases the range would be restricted for physical reasons. The form taken by (A.130) for the continuous case is

$$P(x \mid y) = \frac{P(y \mid x) P(x)}{\int_{-\infty}^{\infty} P(y \mid x) P(x)\, dx} \tag{A.136}$$

A.9.8 Mean Value and Variance

The average value, or mean value, of a function $f(X)$ of a discrete random variable X is defined by

$$< f(X) > = \sum_{1}^{n} f(X_r) P(X_r) \tag{A.137}$$

where X takes the values X_r, $r = 1$ to n, with probabilities $P(X_r)$. If X is a continuous random variable the sum is replaced by an integral to give

$$< f(X) > = \int_{-\infty}^{\infty} f(x) p(x)\, dx \tag{A.138}$$

In particular, the average value $<X>$ is given by

$$< X > = \int_{-\infty}^{\infty} x p(x)\, dx \tag{A.139}$$

As before, physical restriction may reduce the integration range. For example, if x can take only positive values, the lower limit is zero instead of $-\infty$.

The *variance* $D(X)$ is defined as the difference between the average value of the square of X and the square of its average value:

$$D(X) = < X^2 > - (< X >)^2 \tag{A.140}$$

If X represents a noise signal, its average value is zero and the variance reduces to $<X^2>$, proportional to the power in the noise.

A.9.9 Normal Distribution

A "normal" (or Gaussian) probability density is defined by

$$p(x) = \frac{1}{\sqrt{2\pi\sigma}} e^{-(x - x_0)^2/2\sigma^2} \tag{A.141}$$

where x_0 is the average value of the random variable x, and σ^2 is the variance, as computed via (A.140). This probability distribution is the outcome of the contribution of a large number of small random variations. The shape of the curve is shown in figure A.38. It has a maximum at $x = x_0$, and a width between the $p = 1/e$ points of $\Delta x = 2\sqrt{2}\,\sigma$. The smaller the variance the sharper the curve, centered around the average value x_0. σ is called the standard deviation.

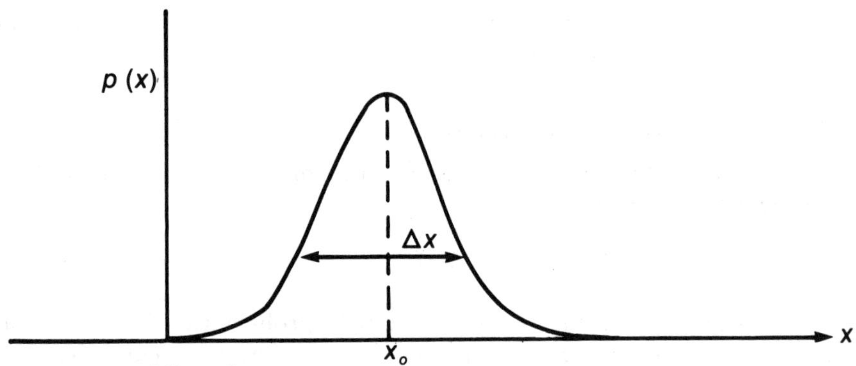

A.38: Normal Distribution

The probability that x lies between x_1 and x_2 is

$$p\,(x_1 < x < x_2) = \frac{1}{\sqrt{2\pi\sigma}} \int_{x_1}^{x_2} e^{-(x - x_0)^2/2\sigma^2}\, dx \tag{A.142}$$

The integral defines a new function, called the error function, defined by

$$\mathrm{Erf}\,(u) = \frac{2}{\sqrt{\pi}} \int_0^u e^{-v^2}\, dv \tag{A.143}$$

In terms of it, (A.142) can be written

$$p\,(x < x < x_2) = \frac{1}{2}\,\mathrm{Erf}\left(\frac{x_2 - x_0}{\sqrt{2}\,\sigma}\right) - \frac{1}{2}\,\mathrm{Erf}\left(\frac{x_1 - x_0}{\sqrt{2}\,\sigma}\right) \tag{A.144}$$

A.9.10 Reliability

Suppose that the probability that a component, having lasted a time t, has the probability $\lambda\, dt$ of failing in the interval dt. The quantity λ is a measure of the *quality* of the component and may itself vary with the component's age. But we shall here assume it is constant. This is approximately so in many cases. It exactly represents the situation in radioactive decay, and in many other examples where random processes are responsible for failure.

What is the probability that the component will reach at least an age t before failing?
Let

A = component will reach at least age t
B = component will fail in the interval t to $t + dt$
$\overline{B}$ = component will *not* fail in this interval

Then $P(A, \overline{B})$ is the probability that the component will reach age t *and* will survive the next interval dt. If $P(t)$ is the probability that the component survives at least to age t, then $P(A) = P(t)$ and $P(A, \overline{B}) = P(t + dt)$. Also $P(\overline{B}) = \lambda dt$ and $P(\overline{B}) = 1 - \lambda dt$. Hence

$$P(t + dt) = P(t)(1 + \lambda dt) \tag{A.145}$$

But from the definition of differentiation we have

$$\frac{P(t + dt) - P(t)}{dt} = \frac{dP}{dt}$$

$$P(t + dt) = P(t) + \frac{dP}{dt}\, dt \tag{A.146}$$

Combining this with (A.145) gives $P(t) + (dP/dt)\, dt = P(t) - P(t)\lambda dt$, so that

$$\frac{dP}{dt} - - \lambda P \tag{A.147}$$

This is a differential equation for P, and is of a form satisfied by the exponential function. Hence

$$P = Ke^{-\lambda t} \tag{A.148}$$

where K is a constant. Now when $t = 0$, (A.148) gives $P(t = 0) = K$. But $P(t = 0) = 1$ since there is certainty that the component will survive zero time. Hence, $K = 1$ and (A.148) reduces to

$$P(t) = e^{-\lambda t} \tag{A.149}$$

Many physical decay processes, e.g., radioactive decay, mortality statistics, etc., follow this relation either exactly (λ independent of age) or approximately. λ is a measure of *quality* in a component; small λ means long life or higher quality.

The *average life expectance* T_{av} is given by $\int_0^\infty te^{-\lambda t}\,dt / \int_0^\infty e^{-\lambda t}\,dt = 1/\lambda$. It is different from the *half-life* $T_{1/2}$, which is the time to produce a probability of ½ of failure. $T_{1/2}$ is accordingly determined by $1/2 = e^{-\lambda T_{1/2}}$, or $\lambda T_{1/2} = 0.6931$. Since the average life is $T_{av} = 1/\lambda$ the relation between the two is

$$T_{1/2} = 0.6931\, T_{av} \qquad \text{(A.150)}$$

T_{av} is also called the *mean time between failures*, MTBF.